Special Reconnaissance
and
Advanced Small
Unit Patrolling

Special Reconnaissance and Advanced Small Unit Patrolling

Tactics, Techniques and Procedures for Special Operations Forces

Edward Wolcoff

Pen & Sword
MILITARY

First published in Great Britain in 2021 and reprinted in this format in 2026 by
Pen & Sword Military
An imprint of
Pen & Sword Books Ltd
Yorkshire – Philadelphia

ISBN 978 1 39902 026 8

A CIP catalogue record for this book is available from the British Library.

Typeset by Mac Style
Printed and bound in the UK by CPI Group (UK) Ltd, Croydon, CR0 4YY.

The Publisher's authorised representative in the EU for product safety is
Authorised Rep Compliance Ltd., Ground Floor, 71 Lower Baggot Street, Dublin
D02 P593, Ireland | www.arccompliance.com

For a complete list of Pen & Sword titles please contact

PEN & SWORD BOOKS LIMITED
47 Church Street, Barnsley, South Yorkshire, S70 2AS, England
E-mail: enquiries@pen-and-sword.co.uk
Website: www.pen-and-sword.co.uk
or
PEN AND SWORD BOOKS
1950 Lawrence Rd, Havertown, PA 19083, USA
E-mail: uspen-and-sword@casematepublishers.com
Website: www.penandswordbooks.com

Contents

Dedication   x
Acknowledgements   xi

Introduction   xii
  The 'Learn and Forget' Cycle.   xii
  Purposes and Sources   xv
  Conventions Used in this Book   xvii
  How to Use This Book   xvii

Chapter 1: **Overview**   1
  Relevant MAC-V SOG Context and Terms   1
  Special Reconnaissance (SR) Leadership Considerations   9
    Effects of Senior Leadership Failures and Mistakes on SR Operations and Notional
      Remedies   9
    SR Paradoxes   11
    Implications of Evolving Strategic Doctrine and Selected Technologies on
      SR Operations   12
    Effects of Fatigue   15
    Countering Fatigue TTPs   16
    Team Leadership TTPs   17
      Leader Training TTPs   17
      Team Leader TTPs   17
      Conditioning Team Members to SR (TTPs)   22
      Study of Special Operations Missions (TTPs)   23
  Impacts and Utilization of SR in Future Conflicts   24
    General/Limited War   24
    Counter-Insurgency (COIN), Irregular Warfare, Operations Other Than War (OOTW)   25

Chapter 2: **Pre-Mission Activities**   26
  General Pre-Mission TTPs   26
  Security TTPs   26
  Intelligence Preparation TTPs   27
  Mission, Enemy, Terrain and Weather, Troops and Support Available, Time Available
    and Civil Considerations (METT-TC) TTPs   30
    Mission Analysis (METT-TC) TTPs   31
    Enemy Analysis (METT-TC) TTPs   35
    Terrain and Weather Analysis (METT-TC) TTPs   40
      Terrain-Based TTPs   40
      Weather-Based TTPs   45
    Troops and Support Available Analysis (METT-TC) TTPs   47
      Troops Available Analysis TTPs   47
    Unconventional (Guerrilla/Partisan) Warfare Mission Environment TTPs   49

Counter-Insurgency (COIN) Mission Environment TTPs   50
Indigenous Troops TTPs   50
Snipers   52
Support Available Analysis TTPs   52
Mission Preparation Phase   52
Aviation Support TTPs   53
Other Support TTPs   54
Time and Time Available Analysis (METT-TC) TTPs   54
  Time Analysis TTPs   54
  Time Available Analysis TTPs   55
  Sample SOG Timetable   55
Civil Considerations Analysis (METT-TC) TTPs   57
Pre-Mission Tactical and Technical Training TTPs   57
  Team Battle Drill and Combat Skill Training TTPs   59
Checklist TTPs   64
Aerial Visual Reconnaissance (VR) TTPs   64

Chapter 3: Employment/Execution   67
Pre-Launch TTPs   67
  Launch Site TTPs   68
  Bright Light (BLT)/Reaction Force (RF) TTPs   70
Infiltration/Exfiltration TTPs   72
  Extracted SOE (and OSS) Fixed-Wing Insertion/Extraction/Resupply Aircraft
  TTPs   76
Movement/Maneuver TTPs   82
  Land Navigation TTPs   82
    Reading Sand Dunes TTPs   84
    Reading Snow TTPs   84
    GPS TTPs   85
    Compass TTPs   85
  Local Weather Forecasting   86
  Stealth TTPs   88
    Stealth Formations TTPs   92
    Camouflage TTPs   92
  Movement Technique TTPs   94
    Mounted Movement TTPs   101
    Tactical Dismounted Formations TTPs   102
    Movement in Steppe, Savannah, Grassland, Tundra TTPs.   104
    Night Movement TTPs   105
    Rally Points (RP) TTPs   107
    At the Halt TTPs   108
  Night Defensive Perimeter (NDP) TTPs   108
Special Reconnaissance (SR) and Advanced Patrolling TTPs   113
  General SR/Patrolling TTPs   113
  Human Waste TTPs   123
  Operations During Night/Periods of Limited Visibility TTPs.   124
  Weapons Employment TTPs   125
    Weapons General TTPs   125
    Fire Flame and Incendiary Weapons TTPs   126
  Decoy or Deception Techniques TTPs   129
  Collection TTPs   131

Electrical Power Operational Indicators TTPs   137
Bomb Damage Assessment (BDA) Mission TTPs   137
Fixed Site Surveillance TTPs   138
Tracking and Reading Sign TTPs   138
Counter-Tracking TTPs   148
Nuclear, Biological and Chemical Reconnaissance TTPs   157
Operating Environment-Specific TTPs   158
Jungle/Rainforest Operations TTPs   159
Desert Operations TTPs   160
Mountain Operations TTPs   161
Cold Regions/Winter Warfare Operations TTPs   162
Operations in Swamp/Marsh TTPs   173
SR Operations Conducted in Built-Up Areas   173
Subterranean Operations   175
Offensive Operations TTPs   176
Counter-Reconnaissance Operations TTPs   177
SR in the Counter-Insurgency (COIN)/Counter-Guerilla (CG)/Counter-Terrorism
(CT) Role   180
German Anti-Guerrilla Operations in the Balkans (1941-1944)   187
Counter-Guerilla Operations During the US Civil War   187
The Emergence of Modern Era US Counter Guerilla Doctrine   188
United States Special Forces Mobile Guerrilla Force (MGF) Operations in
South Vietnam   189
MGF Operations TTPs   189
General Ambush TTPs   192
Deliberate Near Ambush TTPs   195
Hasty Near/Far Ambush   195
Point Near/Far Ambush TTPs   198
Ambush Site Selection TTPs.   199
Ambush Formations TTPs   199
Point Ambush Formation TTPs   199
Linear Ambush Formation TTPs   200
'L' Ambush Formation TTPs   200
'V' Ambush Formation TTPs   201
'Z' Ambush Formation TTPs   203
'Y' Ambush Formation TTPs   203
Area Ambush Formation TTPs   205
Advanced Ambush TTPs   205
POW Ambush TTPs   208
Deliberate vs Hasty POW Ambushes TTPs.   209
POW (Ambush/Raid) Snatch TTPs   213
Far Ambush TTPs   215
Opportunistic Ambush of Disabled Enemy Equipment (to include armor) TTPs   217
Ambush Considerations for Priority Targets TTPs.   218
Raid TTPs   228
Raid General TTPs   228
Rail Network Raids TTPs   233
Defensive Operations TTPs   236
Attacks from Enemy Air/Fire Support   236
SR Team in the Defense TTPs   238
Counter-Ambush TTPs   242

Breaking Out From Encirclement TTPs 242
What to do if Captured 245
Supporting Fires TTPs 247
Aviation Support TTPs 248
General Aviation Support TTPs 248
Forward Air Controller (FAC)/UAV and other Air Support TTPs 249

Chapter 4: Sustainment 254
Logistics 254
Team Resupply 254
Water TTPs 255
Food and Ration Discipline TTPs 255
General Food/Ration TTPs 255
Mealtime TTPs 256
Maintenance TTPs 257
General Maintenance TTPs 257
Primary and Secondary Individual Weapons Maintenance TTPs 257
Crew Served Weapons Maintenance TTPs 257
Mobility Equipment Sustainability/Maintenance TTPs 258
Communication Equipment Sustainability/Maintenance TTPs 258
Transportation TTPs 258
Supply and Equipment TTPs 260
Footwear and Foot Care TTPs 260
Common Individual Equipment and Supplies TTPs (US Personal Wear/Carry) 262
Carry on Person 262
Carry on Load Bearing Equipment (LBE) 266
Hanson Rig Donning Process 269
Carry in/on Rucksack/Haversack 271
Mission Support Site (MSS)/Cache TTPs 276
Weapons TTPs 282
Instinctive Shooting TTPs 288
Foreign Weapons TTPs 294
Other Weapons, Attachments and Accessories TTPs 294
Silencers/Suppressors TTPs 294
Hand Grenade TTPs 295
Rifle Grenade TTPs 301
Missile, Rocket Propelled Grenade (RPG) and Rocket TTPs 302
Mines, Booby-Traps and Explosive Devices TTPs 303
Claymore Mine Specific TTPs 311
Mortar TTPs 318
Personnel TTPs 322
US Personnel Selection 322
General Personnel TTPs 322
Mental and Character Attributes of SR Personnel 323
Physical Attributes of SR Personnel 323
Indigenous Team Members TTPs 323
Presumptive Indigenous Training Cycle. 324
Carrying Wounded While Being Pursued/Tracked TTPs 324
Graves Registration TTPs 327
Enemy POW TTPs 327
Medical TTPs 328

Medical Cross-Training TTPs   330
Medical Preventative TTPs   331
Immersion Foot (Trench Foot) TTPs   334
Miscellaneous TTPs   334
Survival TTPs   334
General Survival TTPs   334
Emergency Stealth Fire TTPs   336

Chapter 5: Command and Signal   338
Command and Control (C2) TTPs   338
Signal TTPs   339
General Signal TTPs   339
Cryptography TTPs   340
Team Internal Communications Capabilities, Modes and Devices TTPs   341
Interpreter TTPs   341
Passwords, Signs/Countersigns TTPs   341
Hand and Arm Signal TTPs   342
Audible Voice and Signal Device TTPs   343
Visual Signaling Device TTPs   344
Covert Signs TTPs   344
Intra-Squad (Team) Communicators/Radios TPPs   344
Photographic and Video Device TTPs   345
Team External Communications Capabilities, Modes and Devices TTPs   345
IT Devices TTPs   347
Pre-Launch Communications TTPs   347
Communications Support TTPs   348
Mission Execution Communication TTPs   348
Countering Enemy Signal Counter-Measures TTPs   350
Communication Equipment TTPs   351
Wiretapping TTPs   352
Orders, Reports and Communications Formats TTPs   353
Other Team and FOB Administration TTPs   353

Chapter 6: Post Mission Activities TTPs   354
Debriefing/After Action Report (AAR) TTPs   354
Joint FOB/Base Occupancy   354
Post-Mission Sustainment/Equipment Maintenance TTPs   355
FOB Security and Defense TTPs   355
General Security TTPs   355
Weapons and Munitions for FOB Perimeter Defense   356
More FOB Tips   357

Appendices
A: Glossary/Abbreviations   358
B: Notional Tactical Training and Range Complex   365
C: Local Weather Indicators   367
D: SR Team Orders, Communication and Report Formats (Samples)   373

Bibliography, Sources and Further Reading   378

Dedication

This book is dedicated to the magnificent Special Forces soldiers assigned to the Military Assistance Command, Vietnam – Studies and Observation Group (MACV-SOG), with special regard and respect to those who volunteered for the Reconnaissance Companies of the SOG Forward Operating Bases (FOBs). In particular, Lessons-Learned and accounts contained in this book are largely drawn from operations into southeastern Laos and Northeastern Cambodia that were conducted by SOG's Command and Control Central (CCC) based at FOB2. During an approximate two-year period, CCC Strategic Reconnaissance (SR) personnel earned a stunning five[1] Medals of Honor (MOH) and numerous awards of the Distinguished Service Cross. Recon Team Members from CCC's sister unit (CCN) were awarded two MOHs and several DSCs as well.

Several true accounts contained in this book were derived from operations conducted by RT California, under the Team leadership of then Sergeant First Class Joe Walker. In the estimation of this Author, SFC Walker was the most accomplished and respected RT Leader in all of SOG (and among other 'Special' Operations during the Vietnam era). Upon his retirement from the military, Joe continued his service to his nation in the Intelligence Community, where his operational performance might again be considered of legendary stature – but for the enduring secrecy attached to his missions.

Figure 1. SFC Bob Howard with his favorite weapon (modified M-14 with duplex rounds).

The book is also dedicated to Command Sergeant Major (CSM) Norman A. Doney, who is referred to in the 'Acknowledgments' page that follows, due to his Vietnam-era Lessons-Learned contribution contained within the pages of this book.

Lastly, the book is dedicated to the heroic Colonel Robert L. Howard (MOH recipient), who was a CCC Recon Team Leader, later the Recon Company First Sergeant (1stSgt) and subsequently the Recon Company Commander. His example inspired all of us to feats of courage.

---

1. FOB-2 Recon Company MOH Recipients: 1LT George K. Sisler, SP/4 John Kedenburg, SFC Fred Zabitosky, SFC Robert L. Howard, SSG Frankline D. Miller. Col Bob Howard was a MACV-SOG Recon Team Leader, Recon Company 1stSgt and then Recon Company Commander. He was awarded his MOH for an action where he was a 'strap-hanger' on a Hatchet Force operation.

Acknowledgements

My first acknowledgement is to CSM Norman A. Doney, Distinguished Member of the Special Forces Regiment, former First Sergeant, Recon Company, FOB2/CCC, Military Assistance Command, Vietnam – Studies and Observation Group (MACV-SOG), (now deceased) who was a superb mentor to SOG CCC Team Leaders.

In two earlier assignments, Doney served in Project Delta (also known as B-52, 5th Special Forces Group), to include Delta's Recon Section, conducting in-country SR operations similar to those conducted by SOG's cross-border Teams.

Prior to taking the position of First Sergeant of Recon Company at CCC, Doney led his own Recon Team on several SOG operations. His experiences during those operations, and during his previous Delta experience, convinced him to establish periodic, mandatory Lessons-Learned skull sessions for CCC US Team Leaders once he became the Recon Company First Sergeant. Without these 'chalk talks', Lessons-Learned

Figure 2. CSM Norman A. Doney. Source: Projectdelta.net

information would not have been broadly cultivated/shared across the approximately 20+ Teams at CCC. CCC Reconnaissance Team operations were improved, as Team Leaders thought through the Lessons-Learned and applied those that made sense to them – and doubtless, Team lives were spared as a result. During this B-52 assignment, he authored a 24-page Lessons-Learned document on reconnaissance topics that he later used and expanded to mentor CCC Team Leaders. Content from Doney's Lessons-Learned document has been incorporated in this book.

My next acknowledgement is to John E. Padgett, PA-C Emeritus, PhD, Major, US Special Forces (Ret), Emeritus Founding Professor, Touro University, Nevada, and Vice President, Refugee Relief International, Inc., who formerly served as the Non-Commissioned Officer In Charge (NCOIC) of SOG's FOB2/CCC dispensary in Kontum. John reviewed and provided additional input to the medical TTPs found in this book, TTPs drawn from his experiences as a Senior SOG medic and from his many years as a prominent education professional in the field of training and qualification of Physicians' Assistants/Emergency Medicine Technicians.

Further acknowledgement goes to Neil Thorne, who provided several photos and many of the drawings/tactical renderings found within this book. Note: Neil has vigorously pursued resurrection of missing Vietnam-era valor awards for Special Forces personnel. He has tirelessly prepared award packages, tracked down witnesses and lobbied Army agencies and Congressional staffs, resulting in a recent award of the Medal of Honor to Gary Michael Rose (CCC Exploitation Force)

> 'Wisdom too often never comes, and so one ought
> not to reject it merely because it comes late.'
> – US Supreme Court Justice Felix Frankfurter

Introduction

'In hemmed-in situations, you must resort to stratagem.'

Sun Tzu, The Art of War[1]

The 'Learn and Forget' Cycle

'All our ignorance brings us closer to death.'

T.S. Eliot

The US Military Services go to some lengths to capture tactical combat knowledge gained in previous wars and military operations. But this knowledge is often insufficiently embedded in a variety of training and doctrinal publications that are frequently only general in nature; and which seldom articulates detailed techniques and tradecraft. Some, more specific, tactical content may be found in a more narrow spectrum of training and doctrinal publications (e.g. Special Forces-specific manuals/handbooks, Ranger Handbook, etc.), but this content does not embrace the spectrum of Special Reconnaissance wisdom or convey this knowledge in sufficient granularity.

The Author has consulted topics contained in US Army Field Manual 31-20-5, Special Reconnaissance Tactics, Techniques, and Procedures for Special Forces dated 7 March 1993, the reader may compare the content of this book to that doctrinal publication. In general, the Author recommends reading FM 31-20-5, as it contains some useful information on Special Reconnaissance Tactics, Techniques, and Procedures (TTPs); however, some of the content is dated and yet other content is flawed, incomplete, mistaken and lacking in adequate detail.[2] The Author has occasionally identified points of dispute with the FM in several paragraphs throughout this book, but has decided not to make this book a critique of the existing FM. Instead, the Author presents TTPs that are much more abundant, detailed and comprehensive than those expressed in the FM; let the reader determine the relative merit. The reader will note that one chief area of dispute between the Author and the FM is that the FM largely portrays Special Reconnaissance (SR) operations as mere observation/collection, reporting of information and target acquisition. The Author, however, promotes a much more aggressive, multifaceted approach to SR, based partly on cross-border SR operations conducted by Military Assistance Command, Vietnam, Studies and Observations Group (MACV-SOG) during the Vietnam conflict, and by other successful deep-penetration operations conducted in other conflicts. Effectively, SR is not merely reconnaissance, but much more. This alternative approach is driven by rationale made evident throughout this book.

---

1. J.H. Huang, *Sun Tzu: The New Translation*, William Morrow and Company, Inc., New York, 1993.
2. It is important to note that the publication date of FM 31-20-5 (March 1993) predates the declassification of information and subsequent historical publications pertaining to MAC-V SOG. Had the authors and editors of FM 31-20-5 access to this information, the content of the FM might have been far different.

'As a role, SR is distinct from commando operations, but both are often carried out by the same units… Like other special forces, SR units may also carry out direct action (DA) and unconventional warfare (UW), including guerrilla operations…. Special forces units that perform SR are usually polyvalent, so SR missions may be intelligence gathering in support of another function, such as counter-insurgency, foreign internal defense (FID), guerrilla/unconventional warfare (UW), or direct action (DA) … Other missions may deal with locating targets and planning, guiding, and evaluating attacks against them … Every SR mission will collect intelligence, even incidentally.'[3]

Where this book and the FM are in accord, the Author may refer or defer to the FM and generally may not reprint FM TTPs or content with which he concurs. The Author does critique most content in the FM: where such criticism is warranted, that is associated with SR organizational or bureaucratic processes in a summary manner, with few details; for instance, staff activities articulated in the FM are generally much too time-consuming and unresponsive in a realistic operational setting and the typical Special Operations (SpecOps) Operating Tempo (OPTEMPO) environment, when lives and tactical opportunities are at stake. The historical bureaucratic drift in the US military toward unresponsive cycle-time has been the curse of timely and actionable intelligence for far too long, and it represents just one example of how we have tended to repeatedly, and inexcusably ignore Lessons-Learned – at a cost of lives and operational effectiveness.

The Author acknowledges that various members/units of the Special Operations and Intel communities, both past and present, may have had a spectrum of SR experiences in a variety of operational environments and that these experiences have undoubtedly yielded valuable Lessons-Learned, which may well have then been incorporated within unit Standard Operating Procedures (SOPs) and possibly into a TTP knowledge base. This book does not necessarily challenge the wisdom contained in those Lessons-Learned documents and SOPs, but it does offer a very substantial body of relevant knowledge to complement what knowledge may be contained in existing unit SOPs and TTPs – and much of this complementary knowledge has been gained through years of intensive Special/Strategic Reconnaissance operations executed in mature combat theaters, within austere environments, and against well-trained, highly motivated, well-led, well-equipped and even tactically and technically sophisticated enemy forces.

Publications that illuminate arcane recon techniques and tradecraft are much less commonly available than FMs. It takes some dedication and time to unearth relevant experiences and lessons that may be found in histories of past conflicts; the Author has attempted to cite a few examples of such experiences and lessons that are repeated through history. One would think that, given the enduring nature of these recurring experiences through time, the wisdom derived should consequently be considered as virtually foundational and even 'immutable' and therefore taught as core material. Unfortunately, enduring Lessons-Learned, techniques and tradecraft protocols are not widely studied by, taught to, or practiced by American servicemen, including US SpecOps elites. With successive conflicts, small unit leaders who are to be deployed to combat operations must often 're-invent the wheel' on TTPs and tradecraft, and consequent Lessons-Learned knowledge is once again collected and then again consigned to the ash-bin of historical studies after the conflict is ended. This 'Learn and Forget' cycle comes at a huge cost of combat effectiveness … and in the blood of our servicemen. Unfortunately, the study of tactical combat Lessons-Learned often then becomes the domain of academics, historians and other 'wonks' among us. Only occasionally will military officers possessing

---

3. Special Reconnaissance, https://en.m.wikipedia.org/wiki/Special_reconnaissance, n.d.

such an historical frame of reference ascend to positions of authority where they attempt to disseminate this valuable knowledge to where it will impact current and future operations. Other nations may enjoy significant success in applying Lessons-Learned; a prime example may be found in the continuum of conflicts in Malaysia.

> 'There is evidence that the success of security forces in various conflicts in Malaysia resulted from shared experiences and Lessons-Learned. J. Paul de B. Taillon, a professor of war studies at the Royal Military College of Canada, is convinced that due to the frequency of British involvement in irregular warfare operations, they were able to acquire and maintain a high level of combat skill among all ranks.[4] As a result, successful tactics and techniques evolved from earlier conflicts and grew in subsequent conflicts as well. In his study of the wars in Malaya and Vietnam, [John A.] Nagl explains that the superior performance of the British army in learning and implementing successful COIN in Malaya was due to its capabilities as a learning institution and its organizational culture.[5] Most of their tactics and techniques were continued and could be observed in later conflicts such as the Confrontation and the 1968–1989 insurgency.[6] Charters and Tugwell also write that armies do best in irregular warfare when they learn from experience, adapt their existing force structure and doctrine to the particular demands of a conflict, emphasize small-unit operations, and allowing initiative at the lowest levels.'[7]

It is worth noting that potential adversaries study US SpecOps doctrine and operations in some detail, with the intention of emulating some of the US capability and internalizing our Lessons-Learned.

> 'Russian special-operations forces typically serve high-intensity operational deployments of a few months, a rotation schedule that is modeled on the US military's elite special-operations teams. The Russians have closely studied the American experience as part of a multibillion-dollar military modernization project that began earlier in the decade…. From the helmets to the kit, they look almost identical…. Russia is using [the Syrian conflict] as an opportunity to test and refine doctrine for these special-operation forces, … [The] deployment to Syria is also a way for Russian special-operations forces to gain valuable combat experience…. [The] forces in Syria are likely comprised of three groups, including the special forces unit of Russia's military intelligence.'[8]

To further illustrate the point: commissioned in 1974 as an infantry officer, General David Petraeus served in a number of assignments that emphasized small unit, and subsequently, counter-insurgency tactics. He later served as Commanding General of Fort Leavenworth, Kansas, and the US Army Combined Arms Center (CAC), and therefore had oversight of the Army's Center for the Collection and Dissemination of Lessons-Learned. General Petraeus became well read in counter-insurgency theory and studied the theorists and the successful practitioners of both insurgency and counter-insurgency (including Sir Robert Grainger Ker Thompson) at a time when such doctrine was virtually ignored by the institutional Army.

---

4. J. Paul de B. Taillon, *The Evolution of Special Forces in Counter-Terrorism: The British* and *American Experiences* (Westport, CT: Praeger Publishers, 2001), p. 8.
5. John A. Nagl, *Learning to Eat Soup with a Knife: Counterinsurgency Lessons from Malaya and Vietnam* (Chicago: The University of Chicago Press, 2002), pp. 103–107.
6. Nagl, *Soup with a Knife*, pp. 103–107.
7. David A. Charters and Maurice Tugwell, eds., *Armies in Low-Intensity Conflict: A* Comparative *Analysis*, (McLean, VA: Pergamon-Brassey's International Defense Publishers, 1989), 252–253.
8. Thomas Grove, 'Russian Special Forces Seen as Key to Aleppo Victory: Low-profile ground deployments show importance of battle to Kremlin', *Wall Street Journal*, New York, 16 December, 2016.

During his time at CAC (2005–07), Petraeus (then a Lieutenant General) and Marine Lieutenant General James N. Mattis jointly became proponents of Field Manual 3–24, Counterinsurgency; Petraeus was then able to implement FM 3-24 doctrine in subsequent field assignments, culminating with his promotion to General and ultimately his assignment as Commanding General of Central Command (CENTCOM).The convergence of General Petraeus' unconventional focus, his series of relevant and accommodating assignments and his ascendancy through key promotions, came at an historical moment of consequence in Iraq with a payoff in successful counter-insurgency operations.

But the General Petraeus experience serves as an exception … not the rule.

Purposes and Sources

The primary purposes of this book are (1) to enhance the prospects of SpecOps mission success and to elevate Team and individual Operator lethality, by educating SpecOps personnel in esoteric close combat SR Lessons-Learned and advanced patrolling TTPs and (2) to save the lives of SpecOps personnel by providing Lessons-Learned and advanced TTPs/tradecraft, including survival in the most demanding combat environments. These purposes cannot be better expressed than in the following quote:

> 'The problem with being too busy to read is that you learn by experience (or by your men's experience, i.e. the hard way). By reading, you learn through others experiences, generally a better way to do business, especially in our line of work where the consequences of incompetence are so final for young men.
>
> 'Thanks to my reading, I have never been caught flat-footed by any situation, never at a loss for how any problem has been addressed (successfully or unsuccessfully) before. It doesn't give me all the answers, but it lights what is often a dark path ahead.
>
> '… Ultimately, a real understanding of history means that we face nothing new under the sun. For all the '4th Generation of War' intellectuals running around today saying that the nature of war has fundamentally changed, the tactics are wholly new, etc, I must respectfully say … 'Not Really.' Alex the Great would not be in the least bit perplexed by the enemy that we face right now in Iraq, and our leaders going into this fight do their troops a disservice by not studying (studying, vice just reading) the men who have gone before us. We have been fighting on this planet for 5,000 years and we should take advantage of their experience. 'Winging it' and filling body bags as we sort out what works reminds us of the moral dictates and the cost of incompetence in our profession.'
>
> Correspondence from General James N. Mattis, USMC
> to a colleague on 20 November 2003.

While the nature of war and tactics remains basically unchanged, the lethality of tactical and strategic weapons has changed dramatically. The United States and its allies are facing too many threats from too many actors. These are not trivial threats, but grave; and they have serious implications for SR. For instance, a serious cyber-attack or use of Weapons of Mass Destruction (WMD) is likely – it is not a matter of if, but when such an attack occurs. Such attacks may come from rogue states, non-state actors or terrorists, rather than a major power. As the stakes are so high in such an environment, the pressure on the Intelligence Community and SpecOps will grow immeasurably. US and allied SpecOps may have to operate routinely without technical superiority; and MUST then rely heavily on TTPs and tradecraft – better learned now than when circumstances are far more difficult and urgent.

One of the anticipated criticisms of this book, is its reflection back on history – and subsequently an accusation of orientation toward fighting the next war as we have fought wars of the past. General Mattis addressed this issue admirably in his comments above, but add these realities:

- It is a fact that there are fighters/soldiers, especially in primitive, austere environments where US SpecOps are expected to operate, who can overmatch US SpecOps Teams (man-for-man) in field/tradecraft, despite US technical superiority. To think otherwise is dangerously arrogant and the very definition of fatal hubris.
- Lethal, adaptive and dual-use technology is proliferating worldwide. The technology gap, once firmly held by the US and its western allies, is closing rapidly. So the technology advantage that once favored US SpecOps forces is eroding. This portends that enemy combatants that US SpecOps might face may have near parity technologically and either equivalency or a decisive overmatch in TTPs/fieldcraft.
- In deep penetration operations, many US technical advantages may largely be irrelevant. For instance, consider a situation where the US does not possess air superiority or lacks air assets or a robust support structure, or where C3I capabilities are limited or surpassed by those of an enemy. What is operationally left to SR Teams, but TTPs?
- Further, as WMD and advanced technology proliferates world-wide among smaller nations, conflicts will increasingly be waged in shadow wars that will not provoke a WMD response from a belligerent. SpecOps will be the tool of choice in such environments. It will not be enough to merely expand the ranks of SpecOps organizations; SpecOps personnel must become more lethal, more skilled, and more effective.

Do not misconstrue General Mattis' comments above, e.g. 'that we face nothing new under the sun'. The evolution of battlefield technology, rather than 'the nature of war', now more than ever, has had a substantial and fundamental impact on lethality, making an emphasis on TTPs so much more vital. How some of this technology may affect the conduct of SR operations will be explored in pages to follow.

This book is designed to educate the SpecOps soldier in advanced patrolling and Special Reconnaissance tactics, techniques and procedures, and in specialized tradecraft related thereto, with particular emphasis on what was formerly known as Strategic Reconnaissance (now Special Reconnaissance) Lessons-Learned. By absorbing this material through study and application in training and operations, the Author hopes that SR and other SpecOps personnel will develop operational intuition (or wisdom) that will greatly speed tactical decision-making and increase operational effectiveness while sparing the lives of Special Operators.

> 'Intuition is nothing more than a person's sense about a situation, influenced by experience and knowledge…. Some of the more significant studies regarding intuition have been developed by Gary Klein, who developed the idea of recognition-primed decision-making (RPD). RPD describes how people with expertise intuitively identify a pattern in a situation and quickly determine a course of responses without any analysis or comparing different courses of action … best done in the types of situations that are time constrained, high-stakes, uncertain and constantly changing.'[9]

In other words, to minimize instances where Team Member hesitation may often result in death.

Some portion of the enclosed material is gleaned from both current and vintage military FMs and other official military doctrinal and training publications, to establish a baseline of relevant knowledge and to provide a refresher on advanced TTPs taught to US combat forces;

---

9. Patrick van Horne and Jason Riley, *Left of Bang: How the Marine Corps' Combat Hunter Program Can Save Your Life*, 2014, Digital Version, no pagination.

but much of the remaining material is a compilation of relevant historical Lessons-Learned, and advanced TTPs and tradecraft which are not commonly available or cannot be found compiled in any other single source. This material also includes a dose of experience gained from years of intensive wartime SR operations of Project Delta (B-52) and SOG (and its antecedents: Projects Omega and Sigma). As noted previously, some of the information presented herein may depart from what is considered published doctrine or commonly accepted practice, and where the Author, based on his own experience, believes that the official military doctrine or practice is flawed. Some of this advanced information is produced from hard-earned Lessons-Learned derived from formerly highly classified cross-border SR operations conducted by SOG. Where appropriate, the Author will crosswalk Lessons-Learned, advanced tactics, techniques and tradecraft to other historical experiences to demonstrate their enduring nature and consistent value through time.

SpecOps SR personnel can learn as much, or more, from After Action Reviews (AARs)/ operational accounts of failed or poorly executed missions than from successful ones. The Author has peppered this book with some such accounts. Some particularly stark examples (that are not discussed in this book) may be found in commercially available books (and derivative films) that recount SOF operations gone very wrong. These books/films were particularly aggravating to the Author, as they were accounts of egregious and prolific TTP errors throughout all phases of operational planning, preparation and execution – serving as a litany of what NOT to do. That SpecOps personnel could be guilty of so many profound errors inspired the Author to write this book.

Conventions Used in this Book

The reader will note that I use unconventional capitalization throughout this book. This is to deliberately draw purposeful distinctions for the reader. For instance, when I capitalize 'Team', I am referring to an SR Team rather than to the fire team of an infantry squad or to a Special Forces 'A' Team (also referred to as a Special Forces Operational Detachment (SFOD)); when I capitalize position titles such as 'Team Leader' (T/L), this is to differentiate between the leader of an SR Team from the leader of a regular 'A' Team or fire team.

Additionally, acronyms, abbreviations, key terms and other points of emphasis may be capitalized, underlined or italicized for reasons that should be self-evident.

This book, as is usual in many military references, uses acronyms extensively. The convention is to spell out the first-time use of a term before the stand-alone use of its acronym when it is used later in the text. I use this convention, but I may occasionally spell out the term again in subsequent text for emphasis or for the sake of continuity or clarity. A glossary of terms and acronyms is provided.

Further, I use special text, indentations and other devices for illustrative examples, historical references and quoted materials to emphasize key points.

How to Use This Book

This book contains a great deal of information; arguably way too much information to absorb by simply reading (or even studying) the text. So, it is best to consider the most effective manner of using this information.

As a matter of context, SpecOps personnel may belong to a specific organization (e.g. Special Forces Group) that has been allocated a particular regional orientation. In a garrison setting, this would facilitate organizational concentration of languages peculiar to the region; it would also facilitate area studies and promote regional familiarity for assigned personnel. In practice, these intentions are spoiled by the necessities of major operational commitments, tour rotations, personnel reassignments, OPTEMPO and other factors. Subsequently, SpecOps personnel are often committed to regions and areas of operations outside their language specialization and area familiarity. It is therefore important for the SpecOps reader to read this book in its entirety, rather than 'cherry-pick' information that pertains to a specific region or environment. It is important to note, that many of the TTPs contained in this book are 'transferable' across the spectrum of regions and operational environments.

In garrison, while assigned to an organization with a regional orientation, the TTPs contained in this book can be used to formulate and formalize unit SOPs and training plans and to 'train as you will fight'.

By studying the entire content, the Special Operator may be able to recall TTPs/Tradecraft at a critical operational moment that may lead to operational success, while mitigating tactical risk. As many SR operations will involve unanticipated close combat engagements, Team Leaders must not hesitate, they must be decisive; this book will better inform their decisions.

So, reading the book in its entirety is appropriate and worthwhile; but retention of the knowledge contained in the TTPs can only be attained through application. Unit leaders MUST make a concerted effort, a commitment, to incorporate and emphasize relevant TTPs in field training and in operational planning, preparation and execution. This book offers some advice on how this may be done. A wise SR Team Leader (T/L) would build Training and Evaluation Plans around this content.

Last, if a Team is operationally deployed, and is equipped with tactical tablets, a digital (and secure) version of this book can be carried on deployments and used as a resource. The tactical tablet can also contain a database of enemy tactical equipment, survival information, common military terms in the local/enemy language or other study information that Team Members can resort to during operational commitments and during down-time.

What immediately follows is Chapter 1 – Overview. From my perspective, despite its mostly general nature, it is indispensable to the remainder of the book. Do not pass it by! The bulk of the TTPs and tradecraft are found in the remaining chapters and in the appendices.

This book contains no index. The Table of Contents should be sufficient to find the information sought. The content aligns with paragraphs of a standard field order.

Chapter 1

Overview

Relevant MAC-V SOG Context and Terms

MAC-V SOG conducted covert cross-border operations during the Vietnam conflict from 1964 to 1972. Prior to that period, US cross-border operations during the Vietnam conflict were conducted largely by the Central Intelligence Agency and its South Vietnamese counterpart; however, President John F. Kennedy, frustrated and dissatisfied with the Agency's lackluster performance, ordered the mission to be reallocated and executed by the Pentagon – and specifically US Army Special Forces. SOG was formed as a US Joint-Service (and US-South Vietnamese coalition) covert operation; SOG operations were conducted or supported by US servicemen of all four military Services and local national counterparts; but the largest contingent of US military personnel, by far, were drawn from Special Operations Forces (SOF), especially Army Special Forces (SF), and the indigenous commandos that were trained and mostly led by SOG SF personnel.

During its existence, SOG 'was the largest and most complex covert operation initiated by the United States since the days of the OSS.'[1] At its organizational peak, 'SOG's unconventional warfare forces were the size of an Army division and combined joint and multinational forces',[2] including Operational Control/Tactical Control of direct support attachments/forces and elements allocated by South Vietnam; many of these personnel operated from three Forward Operating Bases (FOBs): Command-And-Control North, Command-And-Control Central and Command-And-Control South in the late 1960s through the early 1970s. SOG launched its operations from South Vietnam (and other friendly countries in the region) into North Vietnam, Laos and Cambodia (the occasional SOG SR operations conducted within South Vietnam were often considered 'training missions', until cross-border operations ceased in 1971). SOG's four major mission areas included: inserting and running Covert Agent Teams; conducting Psychological Warfare; conducting Covert Maritime Operations, and executing SR and associated ground combat Exploitation Force operations against the North Vietnamese Army operating along the Ho Chi Minh Trail. And from 1966 until its deactivation, SOG also ran the Joint Personnel and Recovery Center, responsible for recovering downed airmen and allied prisoners from enemy territory. 'The 12,000 miles of trails, footpaths, and roads that made up the Ho Chi Minh Trail played a critical role in supplying communist forces operating in South Vietnam.'[3, 4] It is from SOG's SR mission area experience that much of this book is grounded.

---

1. Richard H. Shultz, Jr., *The Secret War Against Hanoi: The Untold Story of Spies, Saboteurs, and Covert Warriors in North Vietnam* (Harper Collins, New York, 2000).
2. John K. Singlaub, and Malcolm McConnell, *Hazardous Duty: An American Soldier in the Twentieth Century*, (Summit Books, New York, 1991), p. 292.
3. William Rosenau, *Special Operations Forces and Elusive Enemy Ground Targets : Lessons from Vietnam and the Persian Gulf War*, (RAND, Santa Monica, CA, 2001).
4. Note: Estimates of the trail's length vary. The one used here is from John Prados, *The Blood Road: The Ho Chi Minh Trail and the Vietnam War*, (John Wiley & Sons, Inc., New York, 1999), p. 374. The trail and its immediate surroundings covered an area of some 1,700 square miles.

Strategic Reconnaissance or Special Reconnaissance (terms used interchangeably in this book) is associated with the primary/core competencies allocated to current-day US Army Special Forces. These competencies include: the 'kinetic' mission sets of Unconventional Warfare (UW); Foreign Internal Defense (FID), including Counter Insurgency (COIN); Direct Action (DA); Counter-Terrorism (CT) and Combating Weapons of Mass Distraction (CWMD) which are all supported by SR and are often dependent on SR as a prerequisite to their conduct. It is essential to understand that SR units may also be expected to execute or integrate with Unconventional Warfare (UW), Direct Action (DA) and other Special Forces tasks in conjunction with SR mission assignments. A SR Team may often be the only capability in-place that is available to take out fleeting, opportunistic or high-priority targets, especially if friendly forces cannot provide immediate air support or lack air superiority over operational real estate. Non-kinetic mission sets include Psychological Operations (PSYOPS), Information Operations (IO), and Civil Affairs (CA); SR Teams may perform some PSYOPS and IO tasks coincident with its core missions. It is important to realize that SpecOps commitments and OPTEMPO in a theater of operations, or in operations conducted on an even broader scale, will substantially overtax limited SpecOps resources; so parsing mission competencies to specialized or specific Teams (as prescribed by FM 31-20-5) in such conditions is simply unrealistic and operationally ill-advised.

The term 'Strategic Reconnaissance' has been replaced in the US military lexicon by the term 'Special Reconnaissance'. Strategic/Special Reconnaissance may be defined as reconnaissance that is conducted to obtain information on the enemy, terrain, weather and other key elements of information for strategic-level planning and operational-level purposes. SR missions may be undertaken to gather new intelligence, and to confirm, verify or repudiate intelligence that was previously collected.

As of the writing of this book, the accepted definition of SR is: 'Reconnaissance and surveillance actions conducted as a special operation in hostile, denied, or politically sensitive environments to collect or verify information of strategic or operational significance, employing military capabilities not normally found in conventional forces.' (DoD Dictionary of Military and Associated Terms.)

As contrasted to SR units, Long Range Reconnaissance Patrol (LRRP) units, now known as Long Range Surveillance (LRS) units, operate beyond the main line of troops at Division and Brigade levels in their assigned areas of interest, and forward of battalion-level reconnaissance elements and cavalry scouts.

SR, however, is conducted by small units of highly trained Special Operations personnel, who generally operate far behind enemy lines at *strategic depth* – tens to hundreds of kilometers deeper than LRS missions. Beyond the depth of penetration, and the integrated relationships to other assigned Special Operations missions, and the exceptional skills and expertise required, the SR mission is further differentiated from the LRS mission by: political considerations attendant to the penetration and conduct of operations within foreign-friendly and hostile/ belligerent sovereign states, and the inherent capability of Special Operations to operate in the presence of sophisticated threat environments. These SR missions are frequently conducted under conditions of deniability, especially where the area of operations includes neutral or third-party states or prior to declared hostilities. Doctrinally, Special Forces SR Teams are conducted by 12-man 'A' detachment formations or in 6-man split 'A' detachments. However, this doctrinal organization for the conduct of SR missions is rarely optimal or even prudent, as explained later in this book.

As compared to the current-day SR portfolio, specific mission tasks within SOG's Strategic Reconnaissance/Exploitation Force mission portfolio included: Point and Area Reconnaissance; Road and Trail Watch (surveillance); use of Wiretaps, Mines, Sabotage Materials and Devices and Electronic Sensors; Target Acquisition; Rescue of Downed Aircrews and Brightlight Operations (rescue/recovery of Teams, Team Members and personnel of integrated supporting units); Ambush, Raid, Road Block Operations; Bomb Damage Assessments (BDAs); Prisoner of War (POW) Snatch Operations, insertion of Psychological Warfare materials and several other high-risk tasks. Furthermore, SOG Teams integrated American Special Forces and indigenous commando personnel. The benefits and challenges to an integrated Team are explored in subsequent text.

The 'nature and the size of the terrain, combined with adversary countermeasures, made it extremely difficult for the ground teams to achieve their tactical and operational objectives…, enemy forces operated in vast areas of difficult and unforgiving terrain. Lacking a thorough awareness of where the targets were likely to be, U.S.… ground reconnaissance teams were forced to patrol huge amounts of territory searching for well-hidden targets.'[5] Because enemy targets were so difficult to approach and often so fleeting in nature, SOG SR Teams normally cycled back and forth from reconnaissance to DA/combat patrol mode on any given mission, and attacked enemy targets opportunistically in meeting engagements, in ambushes, and with Close Air Support whenever enemy targets presented, at the Team Leader's discretion. Most SR Teams were heavily armed, acknowledging the realities and nature of the SOG operating environment, the fleeting nature of targets and the overwhelming likelihood of detection and subsequent necessity of close combat without fire support. Subsequently, some SOG SR Teams, depending on the operating environment and the temperament and Concept of Operations (CONOPS) of the Team Leader (henceforth referred to as the T/L) were geared for hunter-killer operations, while performing other mission priorities. Additionally, once intelligence analyses produced proximate locations of enemy base areas, SOG Teams were assigned repeated SR missions against those base areas, which were occupied by very large troop concentrations. Operations against base areas infested with high concentrations of enemy troops, who were typically expecting the SOG Teams, resulted in high SR casualty rates. All SR Teams were almost always assigned complementary, concurrent missions (beyond reconnaissance or surveillance) including insertion/distribution of PSYOPS materials and Sabotage Devices and conduct of opportunistic POW snatch operations; in fact, capture of enemy personnel generally superseded all other mission taskings except rescue/recovery missions. Other missions simultaneously assigned to SR Teams, on a routine basis, included: insertion of Wiretaps and Electronic Sensors/Beacons and conducting of Bomb Damage Assessments (BDAs). FM 31-20-5 indicates that SR Teams should be assigned BDA missions 'only by exception', relying instead on satellite/aerial photography, etc.; however, the Author proposes that immediate post-strike exploitation may yield opportunities to capture disoriented/wounded enemy personnel and quantities of intelligence materials seldom obtainable by other means.

> 'The US military and many of its allies consider DA one of the basic special operations missions. Some units specialize in it, such as the 75th Ranger Regiment, and other units, such as US Army Special Forces, have DA capabilities but focus more on other operations. Unconventional warfare, special reconnaissance and direct action roles have merged throughout the decades and are typically performed primarily by the same units. For instance, while US special operations forces were originally created for unconventional warfare (UW) missions and gradually added other capabilities, the US Navy SEALs, and the UK Special Air Service (SAS) and Special Boat Service (SBS) continue to perform a primary DA role with special reconnaissance (SR) as original missions. The SEALs, SAS, and SBS added additional capabilities over time, responding to the needs of modern conflict. Russia's Spetsnaz combines DA and SR units.…

5. Rosenau, *Special Operations Forces.*

There is a line between Special Reconnaissance units that never directly attack a target with their own weapons, instead directing air and missile strikes onto a target, and Direct Action, where the soldiers will physically attack the target with their own resources, and possibly with other support. Some special operations forces have doctrine that allowed them to attack targets of opportunity; Soviet Spetsnaz, while on SR during a war, were expected to attack any tactical nuclear delivery systems, such as surface-to-surface missiles, that they encountered.'[6]

Given SpecOps resource constraints, OPTEMPO and the spectrum of missions assigned to SpecOps units within an Area of Operations (AO), Teams conducting SR missions must also be trained and prepared to multi-task and execute other missions simultaneously or on an alternating basis, similar to the manner in which SOG SR Teams operated – as opposed to the mandate of mission specialization prescribed in FM 31-20-5. This is particularly relevant to deep penetration, long duration operations; given limited SpecOps and especially SR-trained assets, and the limitations and risks associated with long-range air insertions and extractions, it makes no sense to deploy single-purpose teams versus flexible, multi-mission teams.

'Around 75 men had been recruited for Blue Light, which was now organized into three assault teams which were still structured as 12-man ODAs with one exception ... the final team was a plussed up 24-man element ... which also had an intelligence collection mission.'[7]

SOG US casualty rates were substantial. 'SOG's all-volunteer Special Forces elements suffered casualties not comparable with those of any other US units of the Vietnam War.'[8] The high casualties were not so much caused by inadequate operational preparation or execution by the Teams, but were often attributable to genuine failures of senior civilian and military leadership. A brief accounting of these blunders is found later in the book. Since the Author strongly advocates SOG SR Lessons-Learned and TTPs, an accounting is offered to the reader to explain why some of SOG's operations were so costly and less effective than they might otherwise have been, and what remedies are available to avert similar consequences. Especially so, in that these very same failure modes were repeated during the Obama administration, further illustrating the point of a US proclivity to ignore the lessons of history.

'SOG missions were so sensitive that the White House retained mission approval authority and maintained tight oversight of SOG activities. The operations were both highly classified and compartmented ... and extremely hazardous; by 1969, the casualty (Killed, Wounded or Missing in Action) rate for United States Special Forces reconnaissance operations in Laos was 50 per cent per mission – overall, MAC-V SOG recon casualties exceeded 100 per cent, the highest sustained American loss rate since the Civil War. In 1968, every MAC-V SOG recon man [on average] was wounded at least once, and about half were killed. But despite such high losses, MAC-V SOG boasted the highest 'kill ratio' in US military history, topping out at 150-to-1 in 1969.'[9]

This kill data infers at least two things: (1) SOG Teams routinely and simultaneously operated in both Reconnaissance and Direct Action roles and (2) they operated in an environment that was so hostile, so densely occupied with enemy forces, that they routinely had to employ all

6. Wikipedia, Direct Action (military),https://en.wikipedia.org/wiki/Direct_action_(military). 4 August 2017.
7. Jack Murphy, 'Blue Light: America's First Counter-Terrorism Unit', (*The Drop*, Special Forces Association, Fayetteville, N.C., Summer 2016), p. 40.
8. John L. Plaster, *SOG: A Photo History of the Secret Wars*, (Boulder, CO, Paladin Press, 2000), p. 466.
9. Plaster, *SOG: The Secret Wars of America's Commandos in Vietnam* (Simon & Schuster, New York, 1997).

their resources and skills (TTPs and tradecraft) to survive and prevail against numerically superior forces.

It is sometimes debated that the classic UW mission, as practiced, for instance, in the Second World War, may rarely be employed again due to US policy issues, apart from the possibility of a *general* war. Notwithstanding this view, the Author believes that there are UW employment opportunities that exist today in regional conflict scenarios; US adversaries would agree with this notion wholeheartedly. The question emerges: how much political will is required for the US to support a full-blown UW campaign in *limited* regional conflicts? This is an important and relevant SR issue, as guerilla/partisan bases may serve as staging areas/Launch Sites or support for SR Team deep-penetration missions. More likely, in today's political climate, are deep-penetration SR operations, launched directly from allied/friendly nation territory, into insurgent sanctuaries and into unfriendly/hostile states.

Along the Ho Chi Minh Trail, Reconnaissance Teams and Exploitation Forces were facing approximately 50,000 (but much higher according to several sources) rear area operations forces, including dedicated security units; thousands of anti-aircraft weapons, combat support and service support units and North Vietnamese Special Operations Forces. And many of these forces were concentrated in the vicinity of enemy base areas, which were naturally the particular focus of SOG SR mission activity. As many of these North Vietnamese troops could otherwise have manned additional combat formations in South Vietnam; SOG was accordingly highly successful as an Economy-of-Force operation. Further, SOG SR Teams and Exploitation Forces frequently faced battle-hardened front-line North Vietnamese Army (NVA) regimental and battalion combat formations moving along the Trail or occupying these same base areas that were also serving as sanctuary locations in Laos and Cambodia.

> 'At any given time, approximately 100,000 people were employed along the trail as drivers, mechanics, engineers, and porters and in ground security and anti-aircraft units.[10] Anti-aircraft artillery appeared in 1965,[11] and by 1970, the entire trail was protected by anti-aircraft guns, some equipped with radar.[12] The PAVN's employment of 'hunter-killer' teams and tribal scouts also protected the trail against enemy incursions.
>
> 'By 1971, the North Vietnamese Army devoted almost 4 divisions' worth of troops and 10,000 air defense weapons to protect the Ho Chi Minh Trail against no more than 50 US led SOG personnel at any one time…. SOG's investment of less than a company-sized US force tied down the equivalent of four plus divisions in Laos and Cambodia, an economy of force unparalleled in US history, perhaps without precedent in world military history '[13]

By this measure, SOG's SR Teams of typically three US SF personnel leading five to seven indigenous commandos was also a stunning force-multiplier success.

SR Teams operating in Laos or Cambodia were operating almost entirely outside the range of friendly long-range (e.g. 175mm) artillery fire support. The only available supporting fires came from Close Air Support (CAS) provided by Army, USAF (United States Air Force), Navy or Marine rotary and fixed-wing aviation platforms. These air assets were successfully

---

10. Gregory T. Banner, 'The War for the Ho Chi Minh Trail,' Master's Thesis, US Army Command and General Staff College, 1993, p. 12 (DTIC, AD-A272 827).

11. BDM Corporation, 'A Study of Strategic Lessons-Learned in Vietnam, Vol. 1, The Enemy', (McLean, VA, 1979). p. 5-19.

12. Brigadier General Soutchay Vongsavanh, RLG [Royal Laotian Government], 'Military Operations and Activities in the Laotian Panhandle', (*Indochina Monographs*, US Army Center of Military History, Washington, DC, 1981), p. 17.

13. Plaster, *SOG: A Photo History*, p. 466.

employed by SOG ground elements to inflict substantial casualties on the NVA and to interdict logistics, transport and materiel along the Ho Chi Minh Trail. As a result, the North Vietnam placed a high priority on counter-reconnaissance operations; they employed espionage operatives against SOG Headquarters and its operational organizations; and North Vietnam was supported by Chinese and Soviet 'advisors', who deployed and supported sophisticated capabilities (e.g. Radio Direction Finding (RDF), signal intercept, anti-aircraft systems, etc.); they also employed novel tactics and techniques (discussed later in this book) to counter SOG operations that posed such a grave threat to their war efforts.

> 'The nature of the Ho Chi Minh Trail environment, and the North Vietnamese efforts to defend their logistical lifeline, combined with the need to maintain strict secrecy, helped to make [SOG] OP 35's cross-border operations among the most demanding, stressful, and dangerous of the Vietnam War. The jungle that shrouded the trail was a formidable obstacle for the … teams. Forward movement was often extremely difficult and sometimes impossible. … [Teams] often were forced to crawl on their hands and knees to get through the tangled vines that choked much of the trail's environs.'[14]
>
> 'As noted by a former Laotian military commander, the trail passed through some of Southeast Asia's most inhospitable terrain: The trail runs through tropical, dense forests….The jungles along these trails are almost impenetrable primeval forests; the mountains are steep and rocky.[15]
>
> 'Adding to the challenge was the need to maintain absolute silence, since PAVN 'Route Protection Battalions' and 'Rear Security Units' constantly patrolled the trail looking for American and South Vietnamese interlopers.'[16]
>
> 'Hanoi devoted tremendous human intelligence resources to penetrating MACVSOG operations. Communist agents served as drivers at MACVSOG headquarters, and as bartenders and waitresses at MACVSOG compounds, where they were able to gather useful and highly sensitive information about personnel, operations, and tactics.'[17]

In the context of what SOG SR Teams endured in its operational environment, it is obvious that advanced patrolling skills taught in Special Forces qualification training, the Ranger program, etc., are mere starting points for the TTPs/tradecraft required of SR personnel. In order to prevail in such a lethal environment, SOG Reconnaissance Teams (RTs) elevated reconnaissance and patrolling tactics and techniques to new levels of tradecraft. This book contains many TTPs and Lessons-Learned drawn from the SOG experience.

US-led SOG RTs were typically comprised of three US Special Forces soldiers and five Indigenous Troops; however, some Teams ran 'heavy' and some ran 'light'. Team composition and equipment was almost entirely tailored by the T/L to mission needs.

Mission duration was generally programmed for seven days. T/Ls were most often Non-Commissioned Officers who were almost always selected on the basis of recon experience (merit); consequently, a T/L, perhaps at the rank of Sergeant or Staff Sergeant, would lead a Team with Senior NCOs, or even an officer, as subordinate Team Members. The skills of the indigenous commandos were indispensable to the RTs and to SOG operations; furthermore, US SpecOps personnel resources were simply not available in the numbers required.

During SOG's existence, TTPs and tradecraft were embodied in curricula and taught at an in-country training location and delivered to personnel who were newly assigned to SOG's

---

14. Harve Saal, SOG, *MACV Studies and Observations Group: Behind Enemy Lines*, (Edwards Brothers; First Editions edition 1990), pp. 256-259.
15. Brigadier General Soutchay Vongsavanh, 'Military Operations and Activities', p. 4.
16. Plaster, *SOG: The Secret Wars*, p. 86.
17. Richard H. Shultz, Jr., *The Secret War*, pp. 244-246.

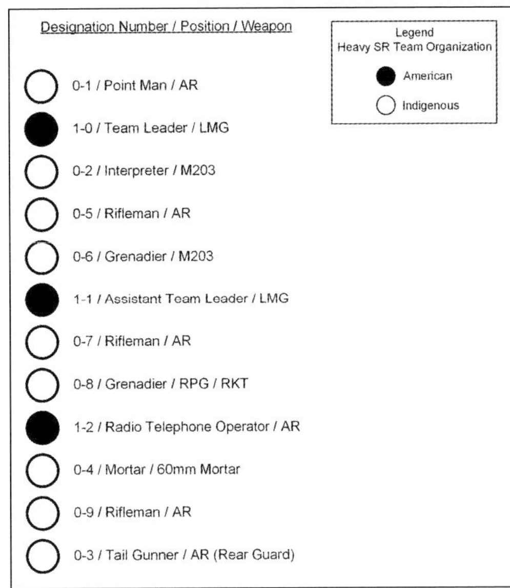

Figure 3. Typical "Heavy" Team Composition.

Forward Operating Bases (FOBs) and their Reconnaissance and Exploitation Force units. Later, similar, if not identical, curricula was institutionalized at the Strategic Reconnaissance Course located at the Special Forces training activity at Camp Mackall, North Carolina, near Fort Bragg. Once SOG was deactivated, specialized Strategic Reconnaissance training was conducted exclusively at Camp Mackall. Initially, instructor cadres at Camp Mackall were SOG veterans, so a small portion of the legacy of SOG Lessons-Learned was conveyed to SR classes and preserved to some degree. But this legacy quickly eroded as SOG-veteran SR Course training cadre personnel moved on to other assignments or retired from military service and were replaced by less experienced cadre. And while some retired SR cadre and veterans were subsequently hired as contractors to assist in other Special Forces training activities (e.g. Robin Sage Culmination Exercises), the Strategic Reconnaissance Course was ultimately terminated and the SR curricula nearly vanished, except for some short duration classes (to include some basics/fundamentals in the Special Forces Qualification Course) or at unit level. Other reconnaissance and advanced patrolling courses were taught by SF personnel during the Vietnam War, most notably,

> '… the US Army's 5th Special Forces Group held an advanced course in the art of patrolling for potential Army and Marine team leaders at their Recondo School in Nha Trang, Vietnam, for the purpose of locating enemy guerrilla and main force North Vietnamese Army units, as well as artillery spotting, intelligence gathering, forward air control, and bomb damage assessment.'[18]
>
> 'Other evolutions of the Recondo School proliferated through to the 80s before establishment at Ft. Benning.'

Since the deactivation of the SF Strategic Reconnaissance Course, reconnaissance and advanced patrolling TTPs, as of this writing, are only taught at Service Sniper schools and the United States Army Reconnaissance Course (ARC), which is currently taught by a conventional Army unit at Ft. Benning, GA and which is designed to teach recon fundamentals and related matters such as tactical intelligence collection, 'surveillance, target acquisition, battle damage assessment, communications, planning, foreign vehicle identification, and other skills.'[19] Until such time as the SpecOps community reestablishes a SR course, ARC and Ranger School may have the only reconnaissance and 'advanced patrolling' content available to SpecOps personnel.

In 1995, Richard Shultz, author of *The Secret War Against Hanoi: The Untold Story of Spies, Saboteurs, and Covert Warriors in North Vietnam*, 'received access to formerly classified data from the Commander of the US Army Special Operations Command, Lieutenant General Terry Scott, who "realized that SOG's Lessons-Learned were being kept secret even from those charged with conducting similar operations today".'[20]

---

18. Robert C Ankony, *Lurps: A Ranger's Diary of Tet, Khe Sanh, A Shau, and Quang Tri*, revised ed., (Rowman & Littlefield Publishing Group, Lanham, MD 2009)
19. Army Reconnaissance Course, https://www.benning.army.mil/Armor/316thCav/ARC/
20. Shultz, *The Secret War*, p. x.

There were a few exceptions to the unfortunate learn-and-forget cycle; where some Lessons-Learned/SOG TTPs endured. Notably,

'… special operations aviation units routinely conduct selected SOG-type operations in the GWOT … the aviation close air support assault, and extraction techniques developed by SOG have not changed much…. SOG operators also pioneered the high altitude low opening (HALO) airborne insertion techniques still in use by Special Operations Forces today. Further contributions involved weapons handling and break contact battle drills still taught today to reconnaissance personnel at various schools. In addition, the criticality of having forward air controllers with ground experience supporting troops is still as true now as it was then. Further study of SOG tactics, techniques, and procedures proves useful to operators faced with similar mission profiles.'[21]

As of this writing, US high-level SR missions are conducted by detachments/teams of the 75th Ranger Regiment Regimental Reconnaissance Company (RRC), a Special Mission Unit under the control of an element of Special Operations Command – when deployed as part of a SpecOps Task Force. The unit's primary tasks include Active Reconnaissance, Surveillance and Direct Action. Notionally, detachments/teams of the RRC have inherited the SR mantle from SOG, as they carry out many (but not all) of the same types of missions and mission tasks that SOG SR Teams conducted during the Vietnam conflict. While it is certain that the RRC Teams are fully expert in advanced-combat patrolling techniques and have developed their own mission-focused TTPs based on mission-experience Lessons-Learned, and possibly information gleaned from the Vietnam-era Recondo School, it is doubtful that the RRC has acquired full knowledge of the SR TTPs used by SOG SR Teams, as the Vietnam-era One-Zero School and Strategic Reconnaissance Course lesson plans and training materials no longer exist. Hopefully this book will supply the RRC Teams with that knowledge, making them even more capable than they already are.

While distinct American SR capabilities and disciplines were generally eroding, it is ironic to note that the British Armed Forces established its Special Reconnaissance Regiment (SRR) in 2005, perhaps in reaction to the SAS experience in Iraq, under the command of the United Kingdom Special Forces. SRR sister units include the Special Air Service (SAS), the Special Boat Service (SBS) and the Special Forces Support Group (SFSG). The SRR conducts covert surveillance and Special Reconnaissance.

'Like their MACVSOG predecessors, SAS personnel did more than find targets and call in air strikes. They were multipurpose forces, capable of taking direct action, conducting BDA on targets previously hit by coalition aircraft, and capturing Iraqi prisoners. Teams destroyed fiber-optic links that carried targeting data for the Scud missile crews, and used plastic explosives to blow up microwave relay towers and communications bunkers. Frustrated with the relatively long delays involved in calling in air strikes, SAS troopers also attacked Iraqi vehicles and other targets directly, usually at night. Using thermal imagers, the teams employed shoulder-fired Milan missiles to engage Iraqi mobile TELs. As the Iraqis began moving Scud-related equipment in 10- to 20-vehicle convoys as a defensive measure, SAS teams mounted ambushes using bar mines and bulk explosives.'[22]

The UK has had a distinguished history in SpecOps and has been a pioneer in SpecOps (and SR) doctrine.

---

21. Danny M. Kelly, Major (USA), Master's Thesis: 'The Misuse of the Studies and Observation Group as a National Asset in Vietnam, (Command and General Staff College, Ft. Leavenworth, KS. 2005), p. 68.
22. Rosenau, *Special Operations Forces*, p. 39-40. https://www.rand.org/pubs/monograph_reports/MR1408.html.

The British, during the Malaysian Emergency of 1948–1960, the Indonesian Confrontation of 1963–1966, and subsequently, the Malaysian military, and during the Malaysian Insurgency of 1969–1989, created and/or deployed an array of special operations organizations (e.g. Ferret Forces (Malay and Ghurka), Malayan Scouts (later known as the 22nd SAS after 1958), the Senoi Praaq and Sarawak Rangers (including SAS, Malay aboriginal personnel and Borneo headhunters), and Police Commandos), against the Communist Terrorists (CTs). The SAS played a dominant role in forming, training and frequently leading these units. Over the course of time, some Malaysian SpecOps units were deactivated, to be succeeded by other SpecOps organizations. These organizational iterations served as field laboratories in a combat continuum, allowing the SAS and the Malaysians to develop counter-terrorism; and unconventional warfare doctrine – each organization benefiting from the Lessons-Learned from predecessor units.

The rebirth of the SAS in Malaya was the catalyst that enabled the British SAS to gain a permanent position in the UK forces' order of battle.[23]

Key lessons of the recurring Malaysia SpecOps/SAS experiences (four conflicts during 1941–1989 spanning 42 years) included:

- Operational experience facilitates integration of Lessons-Learned. The SAS emphasized the Use-Learn-Adapt-Train cycle.
- Integration of Intelligence with Operations.
- Strong Leadership.
- Realism in Training.
- Integration of Aborigines.
- Integration of Reconnaissance with Direct Action
- Learn from the Enemy.

Special Reconnaissance (SR) Leadership Considerations:

Effects of Senior Leadership Failures and Mistakes on SR Operations and Notional Remedies:
The Johnson Administration, and its most senior political and military leadership, was responsible for world-class blunders that were to have grave consequences for the conduct of the Vietnam War. These problems would cascade down to SOG and to all of its operations, to include its SR mission, resulting in high SR casualties and extraordinary obstacles to mission execution that limited SOG's contribution to the war effort.[24] These blunders included:

- Presidential timidity in political and operational decision-making.
- Failure to establish a strategy and supporting objectives for the war effort.
- Micro-management of operations down to the tactical level. In the case of SOG, mission approval, to a large degree, resided with the President. The mission/target approval process hand-carried the target list through the SecDef, to the Secretary of State, to the Executive Office of the President, where the proposed operations were 'individually weighed, approved and often modified'. The process was so agonizing and time-consuming that if 'there ever was an opportunity, quite frankly, it passed. It was too late, but they [SOG elements] were still compelled to execute.'[25]

23. Shamsul Afkar bin Abd Rahman, *History of Special Operations Forces in Malaysia*, (Calhoun: The Naval Postgraduate School Institutional Archive, Monterey, CA, June 2013), p. 38 [quoting Alastair Mackenzie, *Special Force*, (I.B. Tauris & Co Ltd: London, 2011), p. 49].
24. Derivative/extract: Danny M. Kelly, Major (USA), Master's Thesis: 'The Misuse of the Studies and Observation Group as a National Asset in Vietnam', (Command and General Staff College, Ft. Leavenworth, KS, 2005).
25. Derivative/extract: Major Danny M. Kelly (USA) tape recorded interview of John L. Plaster, (Command and General Staff College, Ft. Leavenworth, KS, 7 January 2005).

- Emphasis on tactics rather than strategy and the enemy Centers of Gravity. Success was often gauged by 'body count'.
- State Department interference in military operations. In the case of SOG, effectively ambassadorial veto power over SOG operations.
- Substantial limitations exercised over the operational and tactical levels of the conflict.
- Intelligence and political failure to understand the enemy; an underestimation of his will/commitment and an overestimation of our own capabilities (Sun Tzu '101').
- Institutional bias against unconventional approaches to warfare.
- Denial of appropriate resources to accomplish assigned missions and to defeat the enemy.

These failures affected some SOG staff divisions more than others, and some divisions were therefore more operationally successful. Of 'all SOG activities, this division [OP-35 (Ground Operations, e.g. SR)] experienced the greatest degree of tactical success against enemy targets that mattered to the North Vietnamese war effort.'[26]

The reader should note that failures listed above were repeated during the Obama Administration during the 'Global War On Terrorism' (GWOT) in all theaters of the worldwide conflict. It is entirely possible that these very same failures will be replicated in future conflicts, as long as Senior Leaders ignore history/Lessons-Learned. So what can SpecOps leaders do to insulate themselves from such catastrophic incoherence and incompetence, while executing high risk operations and while preserving the lives of their men? Some thoughts:

- Senior SpecOps officers should fight tooth and nail for maximum operational flexibility, minimal restraints and freedom from micromanagement.
- Exploit whatever operational flexibility may be found in the 'Commander's Intent' within planning and operational guidance from higher headquarters. Use the least restrictive interpretation of that guidance, while being at least nominally compliant with it. Then allow SR Teams to execute within whatever cover the Commander's Intent interpretation might imply and without tactical micromanagement.
- Use classification and compartmentalization to strictly limit access to SpecOps plans, operations, reports, communications, etc. using the strictest of need-to-know restraints. Ensure these controls are embedded in program and/or operational Security Classification Guidance (SCG). Ensure these controls are extended down to OPCON/TACON and supporting units. Ensure that these controls are iron-bound when dealing with supporting units from outside the SpecOps community, especially where certain other US agencies (e.g. DoS), or where multinational participation is involved.
- Use combined operations collaboration with intelligence agencies 'cautiously' (to eliminate the potential for leaks) to further restrict access to plans, operations, reports, communications, etc., and to thwart interference.
- Wherever operational compromise has occurred or is likely (or even possible), use deception in planning, operations, reports and communications.
- Implement and/or reinforce a rigorous Lessons-Learned program. Use best judgment to ensure widest dissemination consistent with need-to-know.

Subsequently, 'unconventional warfare unit commanders should … study SOG to understand and hopefully avert the strategic and operational blunders that might be repeated by senior policy makers unfamiliar with unconventional warfare operations.'[27]

---

26. Major Danny M. Kelly (USA) tape recorded interview of General (Ret) John K. Sinlaub, (Command and General Staff College, Ft. Leavenworth, KS, 4 March 2005).
27. Kelly, Major (USA), Master's Thesis: 'The Misuse of the Studies and Observation Group'. p. 69

SR Paradoxes:

<u>Avoid Contact.</u> The mission of Recon Teams is typically to collect intelligence, and if the collection of this information can be accomplished without enemy detection, all the better.

- If undetected, the enemy will not take countermeasures or other remedial actions in anticipation of US forces acting on the basis of obtained intelligence (e.g. moving a headquarters after discovery of its location).
- Upon contact, the enemy then becomes aware of the presence of the Team in the AO and of the Team's proximate location. Enemy forces will then take a variety of precautions, to include tracking/pursuing the Team and increasing security patrols, all of which may diminish the Team's ability to accomplish its mission – or even to survive.

<u>Versus</u>

<u>Maintain Contact.</u> Although Recon Teams may be admonished to avoid or break contact, once they locate an enemy unit, headquarters, base camp or key capability, they may be expected to maintain contact, if possible, even if the Team presence has been compromised. For instance, if a Team locates an occupied insurgent base camp, and is detected in the process, the Team may be instructed to maintain contact with the insurgents, pursuing or tracking them and calling in friendly fires where possible, assuming the enemy force evacuates its base location. A pursuit would likely require the Team to engage in a series of firefights with enemy elements pending the arrival of a reaction force or until pursuit is no longer possible. In these circumstances, the Team assumes less the posture of a reconnaissance patrol and more the posture of a combat patrol. Note that the heavy burden of gear carried by an SR Team during operations, severely impacts Team capability to pursue a less burdened or more mobile foe. Furthermore, the enemy force may employ ambush tactics, or hastily deployed mines/booby-traps, against the pursuing Team. The T/L must dip into his bag of tricks to overcome these problems. His actions may include: employing tactical air support on or in front of the withdrawing enemy force to slow down its withdrawal and inflict casualties; requesting the deployment of a reaction/exploitation unit to act as a blocking force; conduct the pursuit on parallel and/or less rugged terrain to avoid enemy countermeasures and enhance cross-country mobility; use of terrain to channelize enemy movement; etc.

The SAS Malaysian counterinsurgency experience illustrates the point. The Malaysian armed forces had very limited air mobility assets. Movement of reaction forces to remote areas of the rain-forested interior to attack enemy elements/facilities that SAS detachments had discovered was rarely possible. This left the small SAS/Malaysian SpecOps detachments with the remaining options of ambush, raid, immediate assault and subsequent pressure on the enemy by continuous pursuit and reengagement.

> 'We are the few and the enemys [*sic*] are the many: be able to use few in striking the many when those who battle against us are confined.'[28] Sun Tzu.

<u>Disengage/Break Contact upon a Meeting Engagement.</u> The Team, especially when deployed into a target area with a high density of enemy will prevail, more often than not, in encounters with enemy personnel/units, because the Team may possess the element of surprise, the Team is able to instantly mass its fires, and the firefights are brief. Recon Battle Drills are generally based on the initial imperatives of avoiding casualties, avoiding decisive engagements with numerically superior forces and breaking contact to continue the recon mission. But disengagement is not simply a matter of Battle Drills, it is also a matter of mobility – the ability to rapidly withdraw from an engagement where a skilled enemy is employing his own Battle

---

28. Huang, *Sun Tzu.*

Drills that are designed not to break contact, but to maneuver and close with the numerically inferior Team to compel a decisive engagement. In nearly all circumstances, the enemy will be more mobile and less encumbered than the Team, especially in counter-SR operations.

<u>Avoid Decisive Engagements.</u> As indicated above, avoiding a decisive engagement is an initial imperative. I employ the qualifier 'initial' because there are exceptions:

- During a chance encounter, the T/L may determine that the Team faces a small enemy element, where the Team has a numerical and/or tactical advantage. By opportunistically engaging a smaller element, the Team may take an enemy POW or may capture documents or equipment of high intelligence value. If the Team has been undetected by the opposing element prior to a firefight, then the T/L may decide to deploy into an Hasty Ambush formation. Alternatively, if the Team has been detected, but not yet engaged, by a numerically inferior element, the T/L may decide to maneuver on the enemy unit. Note that FM 31-20-5 does not provide TTPs or even emphasize taking POWs during SR operations.
- During a chance encounter, the T/L may not be able to immediately determine the size of an enemy element. The T/L must rapidly weigh the risks of possibly engaging a numerically superior enemy force and may decide to break contact. If the T/L determines that the Team faces a large enemy element (perhaps initially encountering a small advance element of a larger unit) and cannot break contact, the Team must resort to CAS assets, or even artillery support (as in a COIN environment), to assist in breaking contact. The Team should continue to move/maneuver as long as possible, at least until fire support becomes available; nightfall may facilitate Team evasion. If the Team cannot move further (e.g. due to casualties), the Team may take a defensive position and call in fire support on the enemy until the arrival of a reaction force or until a break out can be executed.

<u>Avoid Compromise/Emphasis on Stealth.</u> While avoiding compromise is considered a pre-requisite for successful reconnaissance operations, the signatures associated with aerial insertion will often betray the presence and even the initial location of the Team. Despite actual or potential compromise, the Team will most often be expected to continue the mission. To do so, the Team must frequently employ a broad spectrum of TTPs throughout the mission.

Implications of Evolving Strategic Doctrine and Selected Technologies on SR Operations:

- As of this writing, the two major adversaries to the United States and the Western democracies, namely Russia and the Peoples Republic of China (PRC), have been making a concerted effort to harness all elements of their national power and to direct this power against the West – and particularly the US. The US and the West have not effectively responded and have consequently invited a dangerous decline in their respective elements of national power. Both Russia and China have each developed an implementing strategy, sometimes referred to as 'Hybrid Warfare', where both military and non-military tools are combined to further the national security strategies of the respective nations against both general and specific target sets.
  - Elements of national power include: geography/location/territory, population/demographics, natural resources, economic, military, political (strategic purpose), psychological (national will), and informational. In the context of Russia's use of Hybrid Warfare in the Crimea and the Donbass region of Ukraine, the Kremlin calculated the elements of power/Correlation of Forces respective to the major players and found that their risks were low and their potential gain was high.[29]

---

29. Derivative/extract: David Jablonsky, 'National Power', (Parameters, *US Army War College Quarterly* – Spring 1997), pp. 34-54.

- ○ Hybrid Warfare characteristics include. Conservation of Force; continual and global operations; and use of other assets in lieu of military operations. Its objectives include: expanding territorial presence without resort to overt conventional force, if possible, or preemptively creating the conditions for overt, conventional conflict; and using other national elements of power to manipulate Western democracies.
- ○ Hybrid Warfare uses information operations (disinformation, propaganda, etc) to manipulate Western institutions, media and public opinion; various forms of cyber attacks to damage Western information systems and steal intellectual property and national security information; indirect attack through the use of proxies; economic extortion through control of energy supplies; covert (intelligence and Spetsnaz) operations to execute Hybrid Warfare efforts on foreign soil; and through threats of overt force.
- ○ In keeping with one of the major themes of this book, the reader will note that there is nothing new about Hybrid Warfare or in leveraging the national elements of power in conflict among nations. Formal doctrine on the use of the elements power in armed conflict dates back to Sun Tzu. And there is nowhere a better example of its application than in WWII.
- ○ It is not the intent of the Author to provide a tutorial on the elements of power and how they are being applied (or misapplied). But it is vital to consider the implications of Hybrid Warfare for US/Western SpecOps organizations. Belatedly and after of years of discussion, the US is developing its own version of Hybrid Warfare, after noting repeatedly its successful implementation by Russia. It is likely that Special Operations Forces will be the lead military element in this new Multi-Domain Battle (Hybrid Warfare) strategy. Subsequently, adoption of this doctrine would constitute a revolution in US military strategy and a new and <u>very substantial burden on US Special Operations Forces</u>.

  > 'The Army is preparing to unveil a new approach for fighting future wars that combines space, cyberspace and traditional combat, in preparation for conflicts short of all-out war that require attacks or counterstrikes in uncertain situations,… The new Army approach was developed with advice from experts in special operations, space and cyberspace… The new doctrine identifies one key threat of a hybrid attack: that an adversary can effectively occupy territory, before the US or its allies have time to react, which means the US needs to be able to launch offensive operations before a shooting war begins… , which can be seen as effectively preempting armed attacks.'[30]

- WMD proliferation will continue with emerging nations acquiring WMDs and the means to deliver them. This will produce something of a stalemate between the US and its allies versus some hostile states. Under the threat of a mutual WMD exchange, hostile states may undertake limited war or OOTW (Operations Other Than War) against their neighbors. This environment guarantees that SpecOps will carry an ever increasing burden in future conflicts.
- Unmanned Aerial Vehicles (UAVs)/Drones: Older series helicopters (e.g. UH-1D or Kiowa) can now be modified with software and components that can convert them to autonomous UAV logistics delivery aircraft at no risk to aircrews in otherwise high-risk situations. Notwithstanding the obvious attractions of UAV logistics, such large aircraft are vulnerable to enemy ground and air interdiction and furthermore could compromise Team operations/location within a target area. Alternate uses for autonomous aircraft are inevitable over time.

---

30. Ben Kesling and Julian E. Barnes, 'US Army Rethinks Strategy As Threats Shift', *Wall Street Journal*, https:\\www.WSJ.com/articles/u-s-army-outlines-new-approach-to-shifting-battlefields-1507521602. October 9, 2017.

Small hover-type drones (e.g. quadrocopters) are especially useful in both SR and counter-reconnaissance operations. They are:

- Relatively inexpensive, and available off-the-shelf. Advanced systems (to include drone swarms) are under development by the US DoD and major threat nations. This technology is already being used offensively by hostile non-state actors.
- Capable of maneuvering nap-of-the-earth over/under tree canopy, along vegetative corridors (e.g. trails, paths, roads, streams, ravines, around defensive/facility perimeters, etc.).
- Detection: Relatively silent; Cannot be detected by radar, Very Low I.R. signature
- Capable of mounting live-video high-definition cameras (with multi-spectral capabilities), navigation systems (e.g. Global Positioning System (GPS), programmable autopilot), collision avoidance, laser distance/targeting devices, target auto-tracking capability and carrying weapons/ordnance (e.g. mini-grenades, Point Detonation Bomblets).
- Can be used to Find, Fix, Destroy (targeting/attack) or pursue an enemy element (or an SR Team).
- If it can be determined that these systems are in use by enemy combatants, the SR Team may resort to a variety of passive and active defensive measures that are described elsewhere in this book.
    - Essentially, passive measures would require strict execution of the most covert TTPs.
    - Active measures (e.g. downing the drone) may/may not be taken at increased risk of detection of the Team. These measures may include use of fishnets to capture the drone, then waiting to ambush the recovery team; signal disruption (navigation or transmission); blinding it with a laser or shooting the drone (suppressed weapon preferred).
- Priority of use by enemy units, assuming some limits on proliferation of drones, would include: security of essential base areas/installations, security of WMD units/asserts, key Main Supply Routes (MSRs), including railways, and key infrastructure and enemy SpecOps. Enemy use of drones will be a tip-off to the SR Team of vital enemy targets in the vicinity.
- Mini-drones, and the components mounted/carried upon them, will have a range that is limited by geography, distance, radio frequency, camera transmission range, cargo and weather conditions. At the higher price (military/professional grade) category, such drones can have a range of 1.5 hours/30 miles (line-of-sight); lower priced civilian/hobby drones may have a range of approximately ½ hour/5 miles. However, the live video transmission range (e.g. 1-4 miles) associated with Remote Person View (RPV)/First Person View (FPV) systems is a more limiting factor than the ground control range. Without the RPV/FPV capability, the user would be unable to receive real-time/near-real-time observation of the target; nevertheless, the limited RPV/FPV transmission distance still has substantial utility in the recon/counter-recon applications.
- Other capabilities include radio relay and logistics provisioning. Radio relay, especially in deep penetration operations, could be critical. Small, high-capacity drones can carry small packages (e.g. spare radio battery), which could be delivered from a covert radio relay site, UW base camp, or Advanced Operating Base (AOB) within range.
- Drone capability will be impaired by fog, snow, rain, canopy/vegetation density and hours of darkness.
- Drone and component capability, will continue to improve into the future.

- Quantum Computing capabilities, now emerging, will be able to more rapidly crack ciphers/codes and brevity phrases.

- Chemical and Biological Weapons (CBWs), Improvised Agents, Toxic Industrial Chemicals (TICs)/Materials (TIMs). These agents may be basic to industrial/manufacturing processes or may be specifically produced to be weaponized.
  - Fentanyl/Designer Agents:
    - Fentanyl is hundreds of times more potent than street heroin. As of this writing, it can be bought for $4,000-5,000/kilo. Some Fentanyl analogues can be 10,000 times more potent than morphine.
    - 5-7 grains of Fentanyl can cause respiratory depression, cardiac arrest and/or death and can be easily weaponized for inhalation or skin absorption.
    - Designer agents, such as Fentanyl analogues, are produced in relatively small scale laboratories for the drug trade. But these agents can be produced on an industrial scale and are apparently to be found in the military inventory of technologically advanced nations. Russian *Spetsnaz* security forces used a 'Fentanyl gas' in an attempt to incapacitate terrorists in the Moscow theater hostage crisis of 2002 – killing 130 of the 850 hostages.
  - TICs/TIMs are toxic industrial chemicals/materials that are produced, transported, and stored for the manufacture of various commercial and military products. TICs/TIMs may be deliberately used as improvised WMD, most likely in low-intensity conflicts.
  - Note that military grade CBW agents (lethal/non-lethal) and Improvised Agents can plausibly be used as area weapons to destroy/incapacitate an SR Team, especially where Team location cannot be pinpointed. If any agent use has been suspected/reported in the AO, the Team should consider personnel vaccinations (BW agents) and carrying protective masks and antidotes.

Effects of Fatigue:

'Now a soldier's spirit is keenest in the morning; by noonday it has begun to flag; and in the evening, his mind is bent only on returning to camp.' Sun Tzu

SR personnel generally operate in the most primitive conditions and in the most austere and difficult environments. Additionally, SR personnel carry substantial mission loads into their Target Areas; resupply is often infeasible due to the close proximity of the Team to enemy forces, the presence of extensive enemy air defenses and the potential for compromise of the Team location.

SOG RTs operating in the mountainous rainforests of Laos, typically endured environmental conditions where temperatures were consistently above 100°F and where humidity hovered around 100 per cent; the terrain was heavily dissected and was covered with vast expanses of triple canopy (or greater) accompanied by thick ground vegetation.

In the SOG AO, a stealthy advance toward the Team objective, or to evade enemy trackers, required a deliberate pace that generally did not exceed one kilometer a day. Furthermore, operations in extreme-threat situations invoked justifiable paranoia, frequent rushes of adrenalin and continuous mental stress, all

Figure 4. Typical Multi-Canopy Rain Forest.

contributing to fatigue. Consequently, an SR mission was always a grueling experience; fatigue frequently had a detrimental effect on operations.

Fatigue makes you listless and slipshod and seduces you to take shortcuts that elevate Team risk. Therefore, US Special Operations personnel must exercise the necessary fortitude and discipline to consistently resist fatigue-driven impulses or to take ill-conceived shortcuts in the field. US personnel who are not acclimatized to the environment, who lack sufficient endurance, or who lack the will to fight off fatigue are much more likely to make fatal mistakes during missions. Special Operations history is replete with examples:

> 'A fatigued Team Leader fails to do a perimeter check prior to settling in to a Night Defensive Perimeter (NDP). He subsequently does not detect an adjacent high-speed trail or route-of-approach from the vicinity of an enemy battalion bivouac area, etc. near the Team position. This negligence results in the Team being detected and overrun as they prepare to move out from their NDP position the following morning.'
>
> – Various SOG SR Operations (Lessons-Learned)

> 'Fatigued partisans manning a security outpost fell asleep. They were discovered by the enemy during a routine sweep and were captured, imprisoned, tortured and executed. The partisan group to which they belonged was subsequently aggressively pursued by German forces.'
>
> – WWII Unconventional Warfare Operations in the Balkans

Countering Fatigue TTPs:

- To mitigate the effects of fatigue, SpecOps personnel must maintain a high state of physical fitness. Special Forces Groups are regionally focused; so a rigorous physical fitness program, tailored as much as possible to the prospective deployment environment, will increase Team Member endurance and forestall fatigue.

  > 'We must remember that one man is much the same as another and that he is best who is trained in the severest school.' – Thucydides

- If the Team has a foreshadowing of where it might next deploy, the T/L should proactively arrange to routinely train in similar (but friendly) environments as much as possible, to attain and maintain acclimatization and to establish training realism. If possible, the training area should have similar weather conditions to the anticipated AO.
- The same proactive mindset should apply to individual Team Members as they maintain their own regional focus and their individual fitness programs.
- To illustrate the need for tailored, regionally focused physical fitness, note that during movement in heavily vegetated areas (e.g. rainforest/jungle) Team Members will most often be moving bent over in a crouch to negotiate overhanging vegetation and to constantly peer under leafy cover at ground level. This causes substantial fatigue of abdominal and lower back muscles. Team Members must ensure their fitness regime focuses on and strengthens appropriate muscle groups.

During the Malaysian conflict with Indonesia (1963–66), the,

> 'SAS in Borneo was characterized by the highest standards of self-discipline and field craft, resistance to mental stress, relentless pursuit of excellence in operations, dogged perseverance in going to one step further than required, and great confidence in itself. Some approached the standards of the aborigines in jungle craft and tracking. Endurance was essential for the SAS troopers, due to the distance they covered during patrols, along

with meticulous attention to detail. They were always isolated and exposed, under constant nervous stress from the danger of detection, and had to be keen observers, anticipating, making minute decisions, choosing the best route, and measuring options in event of emergency. Additionally, during the Malayan Emergency, the SAS experienced long-range reconnaissance, improved their language, hearts and minds, and raiding qualities, and trained in special operations, signaling, medicine, and linguistics. This made the regiment well suited to its assigned tasks in Borneo.'[31]

Team Leadership TTPs:

T/Ls should be selected on the basis of merit demonstrated by relevant experience; rank should be immaterial. An Operational Detachment Alpha (ODA) commander is a Captain. The typical time in grade for a Captain is 3 years before promotion to the rank of Major; during this period, the officer attends military education courses (to include the Special Forces Qualification Course), and may serve in line and staff assignments. An ODA command would probably be limited in duration and only part of that time would be spent on deployment, and not all deployments are to conduct combat operations. Of course, the officer may have served on combat tours as a Lieutenant or even a Captain prior to SF qualification, but the reservoir of combat experience actually resides with SF warrant and enlisted personnel who generally have served in multiple tours (including several SF combat deployments).

Leader Training TTPs:

The SR unit headquarters leadership should schedule and host periodic (monthly) T/L round-table discussions and should spend part of that time on selected specific topics (e.g. movement, ambush, raid techniques, etc.) to focus the discussion for each session. Each T/L should be prepared to discuss his approach to the selected topic. The schedule should leave sufficient time to discuss recent notable operations, share Lessons-Learned and elevate support issues. The sharing of this information will increase the professional expertise of the T/Ls, stimulate synergy and group problem solving – and likely save lives. Attendees should leave their egos at the door.

Team Leader TTPs:

- One of the most essential duties of a T/L is to identify and assess risks and to apply measures to mitigate those risks.
- As a Special Operations professional, it is assumed that the T/L and Team Members will make few major errors in the conduct of their missions. But be assured, entirely avoidable major errors in judgment and tradecraft do occur. It is one of the T/L's key tasks to ensure that the frequency/occurrence of major errors is brought to near zero. The T/L must uncover and correct problems, erroneous practices and deficiencies through rigorous, realistic training.
- Notwithstanding major errors, it is often the small things that get you killed in high-risk environments and operations. Every ill-considered violation of operational protocols, special tactics and techniques, tradecraft or field-craft, increases the chances of mission failure and, unfortunately, casualties. In a 'virgin' operational environment, a Team may be able to get away with these violations for a time – until the enemy becomes operationally wise to Team mistakes; but mistakes in a mature operational environment or where elite or highly skilled enemy counter-reconnaissance forces are deployed more often prove fatal.

---

31. Shamsul Afkar bin Abd Rahman, *History*, p.73

- However, if the Team conforms to operational protocols and routinely employs appropriate tradecraft, field-craft and advanced patrolling and SR techniques, the Team will mitigate specific, overall and cumulative risk. The Team can reduce risk in the small things through attention to detail. And by attending resolutely to the details, the Team will reduce the odds of mission failure and casualties.

> <u>True Account:</u> An experienced SOG T/L, operating in Laos during the rainy season, established a rule that Team members should change socks daily during the noon break – BUT, that (1) only two Team members at-a-time should do so, and (2) sock replacement should be one foot at a time, leaving the other foot fully shod. This procedure paid off when a tracker squad came up on the SR Team back-trail during its midday meal and scheduled communications transmission. The Team fired its deployed Claymore mines, killing and wounding most of the enemy. As the Team evacuated its security perimeter, only one Team Member had an unshod foot; he was able to retain his boot and sock in his free hand, to don them later.

- Risk reduction in routine matters, may allow you to assume other risks. You can take <u>prudent</u> risks that may violate operational protocol, based on operational 'wisdom' (knowledge + experience = wisdom). For example:

> <u>True Account:</u> An operational rule in SOG recon was to avoid walking on high-speed trails and roads that comprised the Ho Chi Minh Trail. The operational logic for this rule was that SR Teams were overwhelmingly out-gunned in their Areas of Operation and that Team survival depended heavily on stealth and avoiding casualties. A highly experienced and detail-oriented T/L reasoned that this rule was so pervasive and enduring that the NVA would never expect to encounter an SR Team in the open, walking on their thoroughfares and that the SR Team would therefore enjoy the advantage of surprise in a meeting engagement. Further, the T/L believed that walking on an enemy road would help reveal camouflaged branch roads leading to enemy bases, truck parks, and other high-value targets. The T/L made his risk-benefit assessment and decided to take his Team onto an enemy Major Supply Route in Laos and travel west for more than a kilometer in broad daylight. Before taking this risky initiative, he communicated his intentions to headquarters (via a mountain-top radio relay site) and then requested a Forward Air Controller to orbit just outside the Team's Target Area, thus facilitating rapid response air support should the Team become engaged with the enemy. This gambit proved fruitful, as the Team expeditiously discovered the locations of a well camouflaged branch road/turn-off and several intersecting high-speed trail junctions. The T/L's use of operational 'wisdom' led to intelligence discoveries that might not have otherwise been made or that may have required several other missions to uncover.

> "'Reconnaissance' groups, each consisting of four or five men, were to infiltrate the areas north and south of the Meuse and identify the enemy's armor and artillery units there. These 'reconnaissance' units were also to confuse the American troops by relating false information and issuing conflicting orders. They would also change street and direction signs, remove minefield markers and use them to indicate false fields."[32]

---

32. James Lucas, *KOMMANDO, German Special Forces of World War II*, (Cassell Military Classics, London, 1985), p. 129.

These Special Operations personnel were members of the elite Brandenburg Battalion, comprised of highly skilled, well-trained and experienced personnel, who were assigned a mission to disrupt enemy movements and operations and create havoc behind their lines in advance of German offensive operations. Additionally, the unit knew, through observation and study, American small unit tactics and techniques well enough to successfully imitate them. The elements were supplied with captured uniforms and equipment, and were trained and rehearsed extensively. The units then covertly infiltrated deep into enemy-controlled territory disguised as American troop elements. Their disguises helped them to capture enemy vehicles and to travel unchallenged along enemy roads; they were subsequently able to rapidly pinpoint command and control and other key units and logistics facilities and were able to transmit this vital information to friendly headquarters. – WWII Battle of the Bulge.

- It is essential that T/Ls involve US Team Members in planning and preparation for operations.
  - The T/L is responsible for training of his subordinates; operational planning and preparation is included in this requirement.
  - In respect to ODA SR Teams, much of the operational experience resides with warrant officer and enlisted ranks and not with a Team commander.
  - Even an inexperienced Team Member may offer good ideas. If the Team Member offers truly awful ideas, this will reveal training or critical thinking deficiencies that are worth knowing … , and correcting.
  - More importantly, the T/L must mentor subordinate Team Members to prepare them to take charge of the Team/element (for example: in the event of casualties, split Team operations, etc.). Further, one or more of the Team Members may ascend to a T/L or senior position through attrition or assignment.
- If the T/L demonstrates confidence, then the Team overall will feel confident. But do not confuse arrogance or cocky posturing with a confident attitude.
- Have respect for the enemy, but do not fear him. YOU MUST KNOW THE ENEMY, his capabilities, tendencies, tactics and techniques so that you can develop effective counters and contingency plans in anticipation of his actions. The T/L must communicate this knowledge to the other Team Members; this display of intellectual preparedness will engender respect and confidence in the Team's leaders.
- The Team may be assigned an operation with challenges that may seem at the outset to be insurmountable and the risks to survival exceedingly high. The T/L must focus the Team on planning and preparing to overcome the challenges and mitigating the risks.

Modern artillery tactics emphasize flexibility and mobility. Self-propelled artillery systems (friendly and adversary) have the capability of rapidly shooting a fire mission and then rapidly displacing to avoid radar-directed counter-fire. When they displace, they typically move into hide positions and/or preselected firing points that are sometimes prepared/revetted in advance; the systems are quickly surveyed in to execute their next fire mission; this 'shoot and scoot' sequence is then repeated again and again. As firing point selection and preparation consumes valuable time and resources, the artillery systems displace to a discrete set of firing positions cyclically, especially during defensive operations. Ammunition resupply is also highly mobile and guns are often supplied at/or near their firing points by ammunition supply vehicles; but then they too move into preselected, prepared/revetted hide positions.

During the Persian Gulf War, US and UK SR teams were tasked to locate and interdict Iraqi SCUD missile Transporter, Erector, Launchers (TELs). And the US and coalition air forces devoted significant aviation and sensor capabilities to accomplish the same objective. The Iraqi's used deception operations and 'shoot and scoot' tactics so effectively that no TELs were verified as destroyed by coalition forces throughout that conflict. TELs 'came in two forms: the Soviet-made, eight-wheeled MAZ-543 and the Al Waleed, a modified civilian Saab-Scania tractor-trailer',[33] which was tough for sensors to differentiate from civilian vehicles. 'In addition, a large number of vehicles, including fuel trucks and missile supply vehicles disguised as civilian buses, supported the mobile launchers.'[34] High-fidelity decoys, some of East German origin, were widely employed.[35] Additionally, prior to hostilities, 'the Iraqis prepared protective, hidden holding pens for the TELs along highways in western Iraq.'[36]

One might expect similar outcomes in future conflicts in terrain and vegetation that may be even more suitable to firing units (to include ICBM capabilities). How could TEL interdiction efforts have been made more successful? Some approaches:

- SR Teams could have moved to previously used firing points and hide locations for TELs and ammunition supply vehicles and then placed mines (equipped with self-destruct capabilities) to destroy the enemy systems when they reoccupied the positions.
- As it became evident that the Iraqis were repetitively using culverts, overpasses and wadis as firing and/or hide positions, SR Teams could have called for fire/air support to seed these areas with dispensed mines (with self-destruct), without risking the Team during ordnance hand emplacement.
- Once you have a thorough understanding of the enemy, you must then develop methods and techniques to counter his strengths and exploit his weaknesses. A comparative (Friend vs Foe) SWOT analysis (Strengths, Weaknesses, Opportunities and Threats), used to evaluate the enemy and also to assess the Team is a worthwhile operational planning and review technique.
- Remember that Special Operations personnel are almost uniformly 'Type A' personalities possessing natural aggressiveness and distinct leadership traits. These proclivities may create a competitive or contentious environment among Team Members that could become dysfunctional. If such a situation cannot be remedied collaboratively, dysfunctional personnel must be reassigned … , sooner rather than later.
- <u>Don't be afraid to take advice from your Team Members, including indigenous members – but then be decisive.</u> Lack of decisiveness suggests lack of confidence and tactical or leadership incompetence.
- While being decisive, remember to leave your ego behind; <u>do not become so wedded to a decision or plan that you discount fresh or countervailing information</u> and thereby unduly risk the mission and the lives of your Team Members.

> <u>True Account:</u> All available Forward Operating Base (FOB) T/Ls, had been ordered to attend a mission briefing (POW rescue) presented by a moderately experienced SOG T/L to a senior officer who was visiting from SOG Headquarters. After the briefing concluded, both the FOB commander and the senior SOG officer effusively commended the T/L on the quality of his briefing and mission preparation; they opined that they anticipated the mission would be highly successful.

33. Garry R. Mace, *Dynamic Targeting and the Mobile Missile Threat*, Department of Joint Military Operations, (US Naval War College, Newport, RI, May 17, 1999), p. 4.
34. Centre for Defence and International Security Studies (CDISS), Lancaster University, '1990: The Iraqi Scud Threat,' http://www.cdiss.org/scudnt3.htm.
35. Mace, *Dynamic Targeting* p. 4.
36. Williamson Murray, *Air War in the Persian Gulf*, (Nautical & Aviation Pub Co of America; July 1, 1995), p.168

During mission preparation, the T/L had declined an aerial Visual Reconnaissance (VR) of his Target Area, for fear of tipping off the enemy to his pending insertion; instead, he relied entirely on the S-2 (Intelligence staff) target folder and a 1:50,000 map to select his Primary and Alternate Landing Zones and for planning the Team approach to the Target (a suspected POW internment station).

One of his fellow T/Ls, who attended the briefing, had previous experience in the same Target Area; he noticed that the LZs being briefed were located adjacent to a concealed/camouflaged enemy road, but that the enemy road had not been plotted on the wall map used in the briefing. After the briefing audience was dismissed, the knowledgeable T/L approached the T/L/Briefer to discuss this problem. The Briefer initially argued the presence of the road, referring to the target folder that S-2 had provided, which was then sitting on the podium; the knowledgeable T/L examined the target folder and pointed out to the T/L/Briefer that it was an archival folder that was two years out of date; the wrong folder had been provided by S-2 for mission planning. Although embarrassed by this blunder, the T/L/Briefer was unwilling to change the plan that he had just briefed to senior officers, especially after he had been so effusively complimented by them regarding his thoroughness. The knowledgeable T/L elevated his concerns to the Recon Company Commander, who then spoke to the T/L/Briefer later that evening; but the T/L/Briefer stood by his plan. Result: Upon insertion, the Team was immediately pursued from its Primary LZ and was driven into an engagement with a superior enemy blocking-force lying in ambush; the T/L and one of his indigenous personnel were lost Missing in Action (MIA) in a firefight during their tactical withdrawal and the mission was a complete failure.

<u>True Account</u>: A newly appointed T/L, a 1st Lieutenant and West Point graduate, who had just completed SOG's in-country Strategic Reconnaissance Course (also known as One–Zero School), was preparing for his first mission in a Target Area that was not considered a 'hot target'. The other two Americans on the Team had minimal experience; perhaps three prior missions combined. Normally, T/Ls were selected on merit, but this case was an exception. The First Sergeant of Recon Company asked a veteran T/L, a Staff Sergeant who possessed recent experience in a proximate Target Area (where he had discovered a significant enemy presence), to offer advice to the Lieutenant. A half-hour later that evening, the Staff Sergeant spoke to the Lieutenant in his team room and told him that he had been asked to offer advice to the Lieutenant regarding his pending mission; he advised him on LZ selection … specifically to avoid using any 'slash-and-burn' agricultural clearing for an insertion LZ and to rappel in if necessary. The Lieutenant received this information with jaundice; as if the offer of advice to him by an enlisted man was an insult. The Team launched two days later and after landing … on a 'slash-and-burn' clearing … was immediately pursued by a superior enemy force, was pinned-down with their backs to the insertion LZ and was then assaulted by a platoon of NVA. They were saved from total annihilation by the just-in-time arrival of air support. The Team lost an American and an indigenous commando KIA (Killed in Action); the Lieutenant lost an eye and a testicle.

- Display respect and tolerance for foreign coalition partners and counterparts, even if your actual respect or tolerance is superficial or not genuine. On some future occasion, one of these counterparts may be in a position to assist you – or harm you. A damaged relationship may result in the latter circumstance. However, take care in what you say or do in the presence of coalition partners/counterparts; like your indigenous troops, they may have divided loyalties.

- Understand cultural differences. Montagnards would often avoid direct eye contact with an American Team Member; direct eye contact in certain cultures (e.g. Montagnard) is considered disrespectful, so it may be a mistake to construe this cultural tendency as deceptive behavior. Alternatively, Germans are prone to stare and do not consider it rude to do so.
- Test swimming skills of indigenous Team Members; then teach them to swim, if available time permits. Beware of swimming in contaminated water. In most third world countries, raw sewage flows into rivers/streams, lakes and coastal waters.
- In the Author's opinion, SR Team leadership should be based on merit/experience, not on rank. To assure T/L selection is based on merit, the SR Team would likely have to be comprised of enlisted SpecOps personnel leading indigenous commandos. If the SR Team is comprised of a SFOD, Team leadership will not be based on merit/experience, but rather on rank. SR experience and small unit tactical skills are generally conceded to enlisted personnel, who have spent most of their careers in the small unit tactical environment. This is generally not true of officer ranks; the exception being if the officer had previously served enlisted time. This Author concedes that there may be exceptions to this perspective.

Conditioning Team Members to SR (TTPs)

- Bodily/Biochemical and Psychological responses to Close Combat.
  - Team Members may experience extraordinary mental focus, emotional detachment, high visual clarity, and a perception of time unfolding in slow motion in response to the stress of close combat. These responses may be considered positive and normal.
  - Salivation will be substantially reduced while the body is pumped with adrenaline, leading to an overpowering thirst once the engagement is over. Adrenaline will also cause vasoconstriction, which will reduce the supply of oxygen to the brain and impairing cognitive, visual (reduced field of view and depth perception) and auditory (sounds will be muted) functions, motor skills and muscle strength. The pain of battle wounds will often not register while adrenaline is surging.
  - Team Members may react to combat engagements differently, based on the level of violence experienced and the amount and quality of Team Member preparation for close combat situations. The Team Leadership and medical personnel <u>must</u> monitor Team Members for post-engagement/post-mission effects; this protocol must be reinforced in training. After an engagement, a Team Member may often experience euphoria; feelings of euphoria may be intensified in SR missions, where the Team Members are continually subjected to high levels of stress through the duration of the mission. The Author recommends that Team Member feelings of euphoria be embraced and reinforced (e.g. using 'gallows humor' and subsequent 'war stories', recognition, etc.) and used for Team bonding, to assist Team Members in coping.
- The Team Leadership should consider various ways to prepare or condition Team Members for the shock of combat. Some remedies/mitigating measures follow:
  - <u>Escalating Levels of Violence</u>. Without doubt, the best way to prepare for the shock of combat is, you guessed it, actual combat. There are, of course, varying levels of violence in combat encounters. And there's a world of difference in violence seen from afar as opposed to seeing combat up close and personal. In the SOG experience, some target areas were deemed relatively 'cold' (where the enemy presence was limited or enemy units did not include combat veterans/cadres), moderately 'hostile', or extremely 'hot'. A Team reconstituted with new personnel (e.g. after taking casualties on a mission) or with a new/inexperienced leadership, should first be assigned a cold target. As the

Team and Team Leadership gain experience, they should be graduated to more hostile target areas. Assuming they survived these escalating encounters, and given the Team Leadership demonstrated a steady hand in combat situations, they should ultimately be assigned missions into hot target areas. This logic escalation did not always work in the SOG environment: sometimes a mission to a cold target area turned out to be unexpectedly deadly; sometimes a Team that was insufficiently trained or experienced would be inserted into a hot target area prematurely. But a plan to graduate a Team based on combat experience is certainly prudent nonetheless.

- ○ <u>Realistic Training</u>. Training must be infused with realism, and as close to combat as possible, to include Red Team participation; 'Crawl, Walk, Run' levels of training intensity; simulated combat stimuli (noise, casualties, chaotic situations) in difficult terrain/environments, etc. Live fire exercises, especially those that incorporate joint Service participation and integrated with other units (especially where it is prudent to develop a habitual relationship), can be especially valuable.

- ○ <u>Intelligence Preparation</u>. Where intelligence indicates that an enemy is infamous for barbarism/extremes of cruelty, Team Members should be fully informed. This especially pertains where missions are covert (sterile).

  > 'Depend upon it, sir, when a man knows he is to be hanged in a fortnight, it concentrates his mind wonderfully.' – Samuel Johnson

- ○ <u>Psychological Preparation</u>. American troops were often shown photos and video clips of the 9/11 attacks during early deployments to Afghanistan and Iraq; this served as a reminder of their purpose and as an inspiration to the troops throughout their deployments. PSYOPS preparation can also be directed at indigenous troops, especially if they are not sufficiently aggressive.

The 1969 film *The Wild Bunch*, directed by Sam Peckinpah, was an extremely violent American Western that featured 170 killings, many of them rendered in slow motion, which culminated in heroic deaths for the protagonists.

> 'During the civil war in Nigeria, the Nigerian troops had been sitting on their asses for weeks, not advancing against the Biafrans. Then they showed The Wild Bunch to the troops. The Nigerians went out of their minds. They shot their guns in the movie. The soldiers shot their guns at the movie. And the next day they went off to battle, shouting that they wanted to die like William Holden.'[37]

Thereafter, the results of combat operations against the Ibo tribesman in Biafra were decisive.

Study of Special Operations Missions (TTPs)

- During the pre-mission period, glean Lessons-Learned and tradecraft from prior operations and from studies conducted of prior wars and campaigns. The US Army Combined Arms Center (USACAC) and its Center for Army Lessons-Learned (CALL) at Fort Leavenworth, Kansas; the Army Intelligence School Center for Lessons-Learned and the USMC Center for Lessons-Learned have produced some excellent studies since the end of the Second World War. The CAC has delved into such relevant topics as WW II Partisan Operations in the Balkans; WWII Soviet Partisan Operations; WWII Soviet Airborne Operations; WWII Jedburgh Operations; French Resistance Operations; German Night Combat;

---

37. P.F Kluge, 'Director Sam Peckinpah: What Price Violence', (*Life Magazine*, Vol. 73, No. 6, August 11, 1972), p. 53

Military Improvisations During the Russian Campaign; Small Unit Actions During the German Campaign in Russia and WWII German Counter-Guerrilla Operations. Many WWII studies have been out of print for years, but some are available on-line, and hopefully, archival copies can still be found in the CAC or CALL libraries. While most WWII-vintage Lessons-Learned studies have been gathering dust for many years, be assured that many of the Lessons-Learned are relevant even today. For instance, an appendix in the WWII 'Night Combat' Historical Study[38] contains a valuable curriculum for night combat training. Subsequently, consider how these solution sets and Lessons-Learned may be used in SR Team training and operations.

- Post-mission, commit Lessons-Learned to paper and provide these Lessons-Learned to the next organizational echelon so that they may be incorporated into a compilation and/or circulated to other units. As importantly, conduct a post-mission Lessons-Learned briefing to other T/Ls and Exploitation/Reaction Force leaders, incorporating maps and other graphics. This information may be vital in assisting other leaders in the planning, preparation and execution of subsequent missions in the same target area – and may promote mission success and casualty avoidance.

Impacts and Utilization of SR in Future Conflicts

SR is not a blunt instrument. SR Teams are comprised of highly trained and experienced personnel whose value should not be squandered on commonplace, primarily combat patrolling missions, especially where other units or capabilities can be used. SR shootouts against enemy troops/combatants will always be an anticipated occurrence, but SR units should always be deployed on missions with prospectively high payoff or strategic purpose.

General/Limited War:
- Economy of Force. SR may have an outsize impact on a Theater or AO if their missions and OPTEMPO cause an enemy to invest large numbers of troops in rear area security. SOG's SR operations in Southeast Asia had such an impact (as did the operations of the SOE and OSS in WWII).
- Concentration/Mass:
  - SOG SR units, and especially Exploitation Forces (EF), when positioned along/astride an enemy MSR, would cause a larger enemy force to emerge from its hidden location(s) to pursue or/or to mass for an assault of the Team/EF unit. At some significant risk, use of Team or EFs as 'Bait', would lure lucrative enemy targets to be destroyed via intense bombardment by air assets. The extraordinarily high kill-ratio attributable to SOG operations can be partly attributed to an enemy who took the bait.
  - A broader use of this tactic was employed by the 5th Special Forces Group in South Vietnam. The bait tactic became the main purpose for border/'A' camps established by the 5th SFGA. Originally established to conduct patrolling of enemy infiltration routes into South Vietnam, it became apparent that 'A' camp patrols were easily avoided and ineffective. But the enemy saw the remote bases as lucrative targets and would periodically mass substantial munitions and up to a division of troops to overrun a camp. The combined firepower of 'A' camp weapons, CAS and mountaintop artillery firebases, plus aerial assault by Mike Force units, inflicted enormous casualties on the NVA.
- Force Multiplier: When SR units use indigenous or irregular troops in the conduct of their missions, SR operations and capabilities can be expanded greatly, with a lower ratio of

---

38. Toppe, Alfred, Brigadier General, Wehrmacht, 'Night Combat' CMH Pub, 104-3 (formerly DA Pam 20-236, June 1953), (Center of Military History, United States Army, Washington, DC), Facsimile Edition, 1982, 1986

SpecOps troops committed. This was the model used by SOG (and the 5th Special Forces Group overall), and the SAS in Malaysia.

- Enemy Support/Logistics Functions as a Center of Gravity/Key Target Focus: SR will have an outsize impact on a Theater or AO if it can destroy vital supplies or logistics support capabilities upon which the enemy depends. Military history is replete with examples to demonstrate this point; examples are cited elsewhere in this book. This effect is contributory to 'Economy of Force' cited above.

> For want of a nail, the shoe was lost,
> For want of a shoe, the horse was lost,
> For want of a horse, the rider was lost,
> For want of a rider, the battle was lost,
> For want of a battle, the Kingdom was lost,
> And all for want of a horseshoe nail – Benjamin Franklin

- Other Key/'World Series' Targets: When an SR Team identifies and destroys a C^3 hub or a similar high-value target, enemy operations may be severely impacted. During SOG SR operations, a 'World-Series Target' was considered to be a battalion-sized unit or higher, armored unit, and occupied major vehicle parks. SR T/Ls had the authority to summon extraordinary support to engage such targets, to include diverting of a B-52 sortie – perhaps an unprecedented level of power for an SR T/L to address opportunistic targets.
- Hunter-Killer Operations: SR units may be deployed purposefully to attack/interdict enemy forces or capabilities, or an opportunity to 'Find, Fix, Destroy and Exploit' enemy units/ capabilities may present itself during other (e.g. reconnaissance) missions. The conversion from purely recon to combat footing was routine in SOG SR operations.
- Intelligence Gathering/Coups: Whereas a stand-alone intelligence coup is generally rare, smaller, seemingly irrelevant information gathered from SR missions can be woven into an intelligence picture that represents a coup. The piecing together of disparate bits of information to complete the intelligence picture is the task of competent and focused intelligence analysts; however, in the experience of the Author, the absence of competent analyses is a significant failure mode in the intelligence apparatus.

Counter-Insurgency (CI), Irregular Warfare, Operations Other Than War (OOTW):

- Force Multiplier: As above.
- Enemy Logistics Functions as a Center of Gravity/Key Target Focus: Enemy logistics support is difficult in irregular operations. If SR Teams can identify and/or destroy logistics depots/caches or can interdict local provisioning of enemy irregular forces, the enemy may be forced to curtail operations, move to other areas, or even be defeated in detail. Some success stories are found elsewhere in this book.
- Concentration/Mass: The tactical use of 'bait', as explained above; examples are cited elsewhere in this book,
- Hunter-Killer: Unrelenting pursuit of enemy irregular forces is the key to their destruction. SR units are trained, and through similar experience, best able to 'Find, Fix, Destroy and Exploit' enemy irregular units/activity. Note examples cited later in this book.
- Intelligence Gathering/Coups: Exploitation of intelligence gathered, to include intelligence coups, may be even more challenging than those obtained in General/Limited war settings. Note examples cited later in this book.

Chapter 2

Pre-Mission Activities

'Therefore, the deliberations of the ingenious ones must refer to both the advantageous and disadvantageous. By referring to the advantageous, tasks can be fulfilled. By referring to the disadvantageous, calamities can be removed.'[1] Sun Tzu.

General Pre-Mission TTPs:

Often the keys to successful SR engagements are (1) to gain fire superiority early on and (2) to disengage before a numerically superior enemy force can effectively react or gain the initiative. The longer an engagement lasts, the more fateful the outcome will be for the Team. To meet this paradigm, Team planning and preparation must emphasize simplicity; effective use of all-source intelligence and Intelligence Preparation of the Battlefield (IPB); resolute discipline in matters of counter-intelligence security and employment of deception; repetition in general and specific training (Battle Drills, TTPs); operational and tactical surprise and a combination of both stealth and speed of operational execution.

In the face of uncertain intelligence (after all, the SR Team is in the business of collecting what is unknown about the enemy), the Team must exercise wisdom (as defined: knowledge, informed by experience), what is reasonable to expect (common sense) of enemy behavior and what anomalies may be detected.

Repetition in training must be tailored to the specific mission and may require innovations incorporating special equipment, specialized TTPs, and organizational modifications (e.g. specialized personnel attachments).

The SR team must be able to achieve surprise against a prepared or alert enemy through deception, exploitation of key enemy vulnerabilities, operational timing, speed of execution, and shock through the application of extreme violence.

Speed may be employed in execution of Team insertions and extractions, battle drills, rapid response to tactical situations, supporting fires, Time-on-Target (ToT); tactical withdrawals and similar tasks. For instance:

- Immediately after an engagement, speed is generally necessary for the Team to gain separation from pursuing enemy forces; but stealthy movement then resumes, especially, while the Team is within the target area box
- Time-on-Target during execution. In Special Operations, time operates to the benefit of the enemy. An operation that is not simple, that is complex, will take more time to execute and allow an enemy response.

Security TTPs:

- SOE, OSS and SOG, all compartmented operations, were all penetrated by enemy agents. This should be reason enough to have a healthy paranoia about security and to have alternate plans that were not made available to headquarters.

---

1. Huang, *Sun Tzu.*

Operations Security (OPSEC): All information associated with Team insertion, Team missions/objectives, Team locations throughout the operation, Survival, Evasion, Resistance and Escape (SERE) plans, fire and air support planning are especially sensitive and access should be based on 'need to know' only. Exercise extraordinary caution in the conduct of VRs.

- Inspect indigenous quarters and possessions for contraband or indications of criminal activity (or espionage).
- It pays to be paranoid! Protect yourselves against moles/traitors/doubles and other types of human intelligence compromise!

Author's Solution:
Always behave as if Indigenous Troops, allied/coalition personnel and even some US personnel have been compromised or doubled.

Always have Team contingency plans that are shared ONLY among your US Team Members and with confidential custodians or a trusted agent at the FOB or headquarters.

If possible, substitute a new LZ/DZ just before launch, rather than using the designated primary points indicated in your plan; this to ensure that a 'welcome party' is not awaiting you.

'The main trainer of al-Qaeda in the years before 9/11 was Ali Mohamed, an Egyptian American army sergeant who had served at Ft. Bragg, the headquarters of JSOC. In the 1980s, Mohamed taught courses on the Middle East and Islam at the Special Warfare Center at the army base. During his leave from the Army, he trained al-Qaeda operatives in Afghanistan using Special Forces manuals he had pilfered from Ft. Bragg.'[2]

'The Son Tay raid, officially known as operation Ivory Coast, was a mission led by Bull Simmons to recover 61 American POWs held in North Vietnam. With Military Assistance Command-Vietnam (MACV) hopelessly infiltrated by communist spies, the US military put together a [sic] ad hoc force of Green Berets to carry out the rescue.'[3] Note that while many of the Son Tay raiders were SOG veterans, none of them were recruited from SOG units, but were selected from those who had been subsequently reassigned to other Special Forces units upon end of tour.

'In 1977, the post-Vietnam drawdown had not been kind to the Army. 'It was not popular to be in or to stay in,' one Special Forces soldier remarked as he recalled this era. Due to personnel shortages in other Special Forces groups, the 5th Group was really the only group that could possibly be tapped to establish a dedicated POW rescue team which also drew inspiration from the Vietnam era MACVSOG and Bright Light Missions.'[4]

Intelligence Preparation TTPs:

- Beginning with mission preparation and continuing through mission execution and post-mission activities, Team Members are simultaneously collectors, analysts, generators and consumers of intelligence. The Team's intelligence products include both descriptive and inferential analyses. The intelligence products acquired during an operation will support mission preparation for subsequent SR or exploitation operations.

---

2. Peter L. Bergen, *Manhunt, The Ten Year Search for Bin Laden from 9/11 to Abbottabad*, (Crown Publishers, New York, 2012).
3. Jack Murphy, 'Blue Light', p. 36.
4. Murphy, 'Blue Light', p. 36.

Table 1. Descriptive vs Inferential Intelligence Products.

Intelligence Products				
Descriptive Analysis: • Recording; Statistical. • No evaluation.	**Inferential** Analysis: Evaluation. Inference to Draw Conclusions.			
	Infer the Past	**Infer the Present**		**Infer the Future**
Describe Structure (Attributes)	Explain Past Events	Describe Structure (Attributes)	Describe Behavior (States)	Predict Future Events
• Census. • Production figures.	Investigation.	Command structure.	Tracking. Processes.	Forecasting. Foreknowledge.

Source: Figure 1.2 Taxonomy of intelligence products by analytic methods. Edward Waltz, 'Knowledge Management in the Intelligence Enterprise, Artech House, Inc. Norwood, MA (2003).

- Upon return from a mission, principal US Team Members will undergo a mission debriefing with intelligence section personnel. It is during this process, that the Team Members deliver an AAR detailing actions and observations taken during the mission. These actions/observations would include the Size, Activity, Location, Unit/Uniform, Time and Equipment (SALUTE/Spot Report information) of the enemy, reported during the conduct of the mission; route of march; terrain features, vegetation and obstacles encountered; BDA narrative; Chemical, Biological, Radiological, Nuclear (CBRN) reporting; POW information; Roads/Trails/Communications Lines located; mine/booby-trap locations; etc. These data are largely Descriptive Analyses; narrative information is typically for purposes of clarification; rather little in Inferential Analyses may be sought during the typical debriefing. It is a serious mistake to ignore Inferential Analyses during debriefings.
- Team Members must never assume that the S-2 staff or a higher headquarters intelligence analyst will understand the importance or relevance of the Descriptive Analyses presented during the debriefing. During the debriefing, Team Members must insist on including their Inferential Analyses to complement the descriptive data provided, along with recommendations for exploitation or additional reconnaissance. In the following two examples, intelligence analysts failed to grasp the significance of hard-earned reconnaissance collection and apparently lacked the operational experience or knowledge to convert Descriptive Analyses to Inferential Analyses.

> **True Account:** A novice SOG T/L located an underground tunnel during a Bomb Damage Assessment (BDA) mission. The tunnel was situated on a hilltop in Laos and had been revealed by a bomb crater from a recent B-52 strike on an enemy mobility corridor. The T/L reported the tunnel during the post-mission debriefing. While Viet Cong/North Vietnamese Army underground tunnel complexes were relatively commonplace in South Vietnam, the T/L thought that it was highly unusual for a tunnel complex to be located in an enemy sanctuary/base area, deep inside Laos. He opined that the tunnel signified the presence of a sensitive enemy installation; he assumed that the analysis conducted at higher headquarters would surmise the same notion, and that a subsequent reconnaissance mission would be mounted to investigate. The T/L's assumption proved to be incorrect; no subsequent reconnaissance or exploitation mission was mounted to exploit his information.

> **True Account:** An experienced SOG T/L located a network of heavily-trafficked, high-speed trails, running along the eastern bank of a north-south oriented river in southeastern Laos; he then discovered, on the western bank of the river, a camouflaged road (Old Route 96) once abandoned but now in use, also with north-

south orientation. The high-speed trails and road evidently merged, in the northern terminus, into a camouflaged intersection with Route 110, the west-east MSR that passed through Base Area 609 and then on towards the South Vietnam border, with a major branch ('New' Route 96) proceeding southeast from Route 110 into Cambodia. Route 110 was the target of constant air interdiction and of considerable interest to 7th Air Force, and therefore to MACV-SOG. The major southeast branch (New Route 96) was considered the chief thoroughfare supplying North Vietnamese logistics activities located in northeastern Cambodia in support of NVA combat units operating in the Central Highlands of South Vietnam and further south. The T/L therefore believed that the trail and road corridor he had discovered represented another MSR into Cambodia. Report of this information during the post-mission debriefing did not result in subsequent SR missions targeted against the new corridor.

- One remedy to these analytical shortcomings is in the integration of analysts and operators to rapidly produce actionable intelligence. Optimally, all Team Members, not just the T/L, must synthesize and integrate intelligence inputs and other relevant mission information during mission preparation and then continually, in near real-time, during mission execution. Thereafter, Team Members, but especially the T/L, must work in concert with S-2/intelligence personnel, and subsequently with exploitation elements, to generate intelligence that is timely and actionable, especially so in a tactically fluid environment or where targeting windows are volatile.

  'The CIA Bin Laden Unit, code named Alec Station, fused analysts and operators… Events move and change in ways you can't see… A single fact may come in which may dramatically change your understanding and projection of where something is going… Information arrives in incremental fragments.'[5]

  'When the CIA analysts were transformed from Bin Laden intelligence gathering to al Qaeda targeting activities, their operations became focused on Time, Target and Method (When, Who, How). In the targeting domain, CIA intelligence analysis and operations (e.g. agent handlers) personnel combined with higher-headquarters analysis and Special Operators, that 'created a lethal synergy that was highly reactive and opportunistic to real-time operational dynamics.'[6]

  'Target analysis, mission analysis, integration of intelligence and an understanding from the beginning that you had to have a stand-alone intel and analysis capability that can deploy to a crisis site… I had learned that from the SAS. They taught me if I was going to do something unique, something very dangerous, then I better have all my own horses. When your life and those of your people are the stakes, you don't want to have to depend on strangers.' – Colonel Charles Beckwith.[7]

- Team Members, and especially the T/L, should meet with the FOB S-2 Staff Officer, or his staff NCOIC, on a periodic basis, to obtain the latest intelligence on the AO and the enemy and to study reference material. The T/L should also visit the Intel Staff Officer, or his staff NCOIC, of the next higher headquarters, for the same purpose, as current intelligence sometimes is not disseminated. This may occur because the S-2 staff may be incompetent or they may not understand the importance or relevance of certain information.

---

5. Source: 'Manhunt: the Search for Bin Laden', © 2013, Home Box Office, USA.
6. Source: 'Manhunt'.
7. Murphy, 'Blue Light', p. 49

<u>True Account:</u> One of the most experienced T/Ls at FOB-2 encountered a strange ambush formation that the enemy used on his Team during an operation, which cost the life of his point man. Shortly after his return, the T/L flew down to SOG headquarters and personally reported this tactic to the intelligence staff, only to discover that a report of the formation had been received several weeks prior. The information had never been passed down.

<u>True Account:</u> On another occasion, another T/L at FOB-2 dropped by to visit a friend at SOG headquarters and discovered that an outpost of the Army Security Agency had intercepted NVA radio traffic containing specific information on pending operations. The message information included insertion dates, the Teams assigned, and ominously, the coordinates of primary and alternate DZs/LZs. DZ/LZ selection was normally reported to SOG only two to three days before launch. Communications personnel from each FOB were being rotated to SOG headquarters to take polygraphs. While FOB commanders and key FOB staff personnel were informed of the disclosures and of the likelihood of a mole or moles in SOG, this information had been deliberately withheld from T/Ls. The leadership at SOG may have been more concerned about morale and catching the traitor, than alerting T/Ls so that they could take their own countermeasures.

Mission, Enemy, Terrain and Weather, Troops and Support Available, Time Available and Civil Considerations (METT-TC) TTPs:

- During mission preparation, the T/L, with input from his Team Members, analyzes All Source intelligence. This analysis begins with a review of planning factors including: the Mission, the Enemy, the Terrain and Weather, Troops and Support Available, Time Available, and Civil Considerations (METT-TC).

    'The terrains act as reinforcements for the forces.
        Anticipation of the enemy, the determination of victory, and the calculation of obstacles and distances are the duties of the spearhead forces commander.
        Those who perceive these and then operate battles have definite victory; those who do not perceive these and then operate battles have definite defeat....
        Therefore we say: by perceiving the enemy and perceiving ourselves, victory thereby has no unforeseen risks. By perceiving the geographical factors and perceiving the cyclic natural occurrences, victory thereby is complete.'[8] Sun Tzu.

- The METT-TC analysis may be supplemented with a Strengths, Weaknesses, Opportunities and Threats (SWOT) analysis. And if time allows, the T/L might also prepare a formal Risk Assessment (see Table 2). If the T/L is highly experienced and if Team missions are recurring in the same Target Areas, the SWOT and/or Risk Assessment may become more a mental process; consequently a written analysis may not be necessary or a previous, still current analysis may be used.

    'Simply put, in combat each side makes mistakes. The side that protects its weaknesses and exploits the enemy's wins.'[9]

---

8. Huang, *Sun Tzu*.
9. Richard E. Killblane, 'Convoy Ambush Case Studies', (Transportation Corps, Fort Eustis, VA, 2006), p. 2.

Sample Single Task Risk Management Worksheet*					
Mission: Conduct an area reconnaissance of Target T-7	Date/Time Group: Begin: End:			C. Date Prepared/Revised:	
D. Prepared by:					
E. Task	F: Risk ID	G: Risk Assessment	H. Risk Mitigation	I. Revised Risk Assessment	J. Mitigation Implementation
Insert into target	Few LZs, easily monitored	High	Rappel in	Moderate	Team rappelling training; Unit SOP
	Anti-Aircraft (AA) threat	Moderate	Nap of the Earth approach; Screen A/C with terrain	Low	Coordinate w/ supporting aviation unit
	Enemy observation/ fires	High	Rappel in; Screen A/C with terrain	Moderate	Team-aviation rappelling training; Unit SOP
	Enemy counter-recon units in vic.	High	False landing; Deploy Nightingale	Moderate	Coordinate w/ supporting aviation unit; Unit SOP
K. Overall Mission Risk (circle one): Low **Moderate** High Extremely High					

* Adapted from FM 3-21.91

- Typical Weaknesses of SR Teams:
  - Generally foot mobile; impeded by rough terrain and dense vegetation; poor mobility.
  - Limited fire support, except perhaps in a counter-insurgency setting.
  - Lack of support by other units, except perhaps in a counter-insurgency setting.
  - Numerically inferior to the enemy they are facing.
  - Enemy typically possesses far greater organic firepower in the aggregate.
  - Typically inferior knowledge of the AO and of the enemy dispositions.
  - Significant supply limitations, except perhaps in a counter-insurgency setting.
  - Limited staying power; short mission duration.
  - Very limited aviation support, except perhaps in a counter-insurgency setting or where the US enjoys air superiority.
- Develop and study decision trees to think through and train for tactical situations:
  - Modify for variations in operating environments, weather conditions, equipment, Team composition, Team mobility and other METT-TC variables.
  - Decide what tactical solutions make sense in different scenarios according to the above conditions.
  - Develop Tasks, Conditions and Standards based on these variables to establish a training plan.
  - Options for an SR mission, upon encountering an enemy or a threat are typically run, hide or fight.
  - By going through this mental exercise and then by conducting situational training the T/L and Team Members will react faster.

Mission Analysis (METT-TC) TTPs:

- 'Recon only' missions may be imposed by higher headquarters on a Team to exclude independent action and explicitly preclude opportunistic and subsidiary missions that might otherwise be at the discretion of the T/L. There are a few reasons why recon-only missions may be directed; but generally this is not a wise practice – and it sometimes presupposes that the higher headquarters believes it has better tactical insights than the ground (Team) commander. Regardless of the 'Recon only' tasking, Teams should plan and train for the entire range of discretionary missions, time permitting.

- If higher headquarters does not have a plan for rapid action against/exploitation of SR discoveries, then something is wrong with higher headquarters. The T/L should make and coordinate his own plans in this circumstance.

- Especially if the Team mission is to integrate Direct Action, COIN or UW operations, a CARVER analysis may be useful. The CARVER 'method is a SOF methodology used to prioritize targets. The CARVER methodology can be used to rank-order Critical Vulnerabilities (CVs), thereby prioritizing the targeting process against limited organizational capability/resources in a target-rich environment. The six criteria listed below are applied against each CV to determine the impact on the threat organization:
  - *Criticality:* Estimate of the CV's importance to the enemy. To what extent will the vulnerability influence the enemy's ability to conduct or support its operations?
  - *Accessibility:* Determination of whether the CV is accessible to the friendly force in time and place. In other words, does the friendly force have the resources and capability to accomplish destruction or neutralization of the CV?
  - *Recuperability:* Evaluation of how much effort, time, and resources the enemy must expend if the CV is successfully affected.
  - *Vulnerability:* Determination of whether the friendly force has the means or capability to affect the CV.
  - *Effect:* Determination of the extent of the effect achieved if the CV is successfully exploited.
  - *Recognizability:* Determination of whether the CV, once selected for exploitation, can be identified during the operation by the friendly force and be assessed for the impact of the exploitation.'[10]

- The CARVER methodology (Appendix D, FM 34-36) assigns numerical values to factors evaluated under the criteria described above. For instance: 'Criticality depends on several factors:
  - Time: How rapidly will the impact of the target attack affect enemy operations?
  - Quality: What percentage of output, production, or service will be curtailed by target damage?
  - Surrogates: What will be the effect on the output, production, and service?
  - Relativity: How many targets are there? What are their positions? How is their relative value determined? What will be affected in the system or complex 'stream'?'

Table D-1, FM 34-36 (below) shows an example of how CVs are assigned on CARVER matrixes.

Table D-1. Assigning CARVER **Criticality** Values	
CRITERIA – EFFECT	**SCALE**
Immediate halt in output, production, or service; target cannot function without it.	9-10
Halt within 1 day, or 66% curtailment in output, production, or service.	7-8
Halt within 1 week, or 33% curtailment in output, production, or service.	5-6
Halt within 10 days, or 10% curtailment in output, production, or service.	3-4
No significant effect on output, production, or service.	1-2'[11]

---

10. FM 3-05, 'Army Special Operations Forces', (Headquarters, Department of the Army, Washington, DC, 20 September 2006)

11. Field Manual 34-36, 'Special Operations Forces Intelligence and Electronic Warfare Operations', Department of the Army, Washington, DC, 30 September 1991, App D.

- The numerical values assigned to all CARVER criteria are then summed up to provide a ranking of target priorities. The CARVER methodology can be applied in the field by the T/L/Assistant T/L (or the Operations & Intelligence NCOs), if he is contemplating the execution of a secondary/alternative mission. He can also record target-specific CARVER data in his notebook, so that this information may later be concisely and coherently reported in an AAR/debriefing for subsequent exploitation. Note: An enemy may generate a similar methodology to generate a 'Red Team' vulnerability assessment, which may be used to allocate security resources/protective measures to counter US targeting.
- In all, the METT-TC, SWOT, Risk and CARVER analyses may be synthesized in a kind of 'Cost-Benefit' assessment.
- If these analyses are conducted by headquarters/FOB staff, the T/L should not automatically accept the validity of the analyses. At a minimum, he should submit them to a common sense test and challenge the analyses if necessary. Staff analyses may be performed by persons with no/little SR experience or who may be incompetent or overworked; resulting in cursory analysis or based on invalid or erroneous intelligence/information.

<u>True Account:</u> A SOG SR Team was assigned a mission to Target Area Sierra-7 near the northern-most boundary of the FOB's AO in 1970. The T/L took a VR (aerial Visual Reconnaissance) of the target area and returned with some serious concerns.

- Mission: Road Watch (surveillance) of 7-days duration along an enemy road believed to be a MSR.
- Enemy:
  - This was the first mission to be conducted in this Target Area; information on the Target Area and enemy was largely unknown, except for the results of the VR.
  - Location of the highway was on a mountain ridge top where crop cultivation (including plantain groves) was evident.
  - At the base of the mountain ridge, the VR revealed structures inside the forest periphery overlooking cultivated fields.
  - Based on operational wisdom, the T/L deduced that the crops were grown to feed enemy troops.
- Terrain & Weather:
  - The highway was oriented North-South along a narrow, high mountain ridge approx. 1,200M AGL), with cultivated areas along both sides of the road.
  - The VR was taken in clear weather, but the mountain ridge was normally under nearly complete cloud cover during that time of the year.
  - The T/L selected an LZ on top of the mountain ridge, rather than at the base of the mountain, because the near vertical climb from the base would have taken three days; and because insertion without compromise would not be possible regardless of insertion LZ location. The T/L reasoned that a forewarned enemy force atop the ridge would easily shoot down upon an ascending Team; landing on top of the ridge would give the Team a fighting chance from a defensive posture.
- Troops and Support Available:
  - Elevation on and around the target challenged the operational ceiling of the available insertion aircraft (UH-1D helicopters), mandating that no more than 4 Team Members were to be carried per insertion aircraft (2) for a total of eight Team Members (two thirds of the Team's normal complement). Other helicopters (CH-34s) were operationally committed to support Vietnamese SR Teams in the southern AO (Cambodia).
  - Given aircraft load, engine power, altitude and distance from FOB-2's northern-most Launch Site at Dak Pek, the helicopter support would only have 15 minutes

of loiter time over the target area. CAS would therefore largely be provided by A-1 Skyraider aircraft, given the limitations (altitude, range, loiter time, etc.) on AH-1 Cobra gunships.

- ○ Forward Air Controller (FAC) availability would be intermittent throughout the mission, due to distant mission commitments, with only two fly-overs to be expected per day for scheduled communications. FM communications was otherwise not available; only an emergency radio (AN/PRC-90) could offer a possible communications means capable of contacting the Airborne Command and Control Center (ABC3).

- Time Available: Planning and preparation time was the standard 7 days prior to mission launch.
- Civil Considerations: Not relevant. Rules of Engagement (RoEs) were unlimited.

Execution: The Team travelled from the FOB to the launch site at Dak Pek every day for nearly two weeks, with no break in the weather over the Target Area. The mission was shelved when one of the American Team Members was wounded during a mortar attack on the Launch Site – which also destroyed and/or damaged several insertion aircraft and killed and wounded several aircrew members. The mission was later reassigned to another RT, which was inserted at the base of the mountain ridge. The 2½-day climb was exhausting; Team Members lashed themselves to precarious perches on trees growing on the mountainside for two nights during the ascent. As the RT was nearing the completion of the climb, the Team was ambushed, with one US Team Member and one indigenous Team Member shot and both falling off the mountainside to their deaths. The remaining Team Members slid/scrambled down the mountainside as rapidly as possible; they could not find the bodies of their Team Mates and were subsequently extracted. Mission failure; the RT was fortunate to have sustained so few casualties.

What went wrong? The mission, as planned, was doomed to failure.

- Regardless of the insertion LZ, the presence and proximate location of the Team would be immediately compromised.
- Given compromise, the Team was certain to be engaged by a superior enemy force.
- Given the fairly narrow ridge and vertical ascent, Team maneuver, and the inability of the Team to break contact, would be extremely limited.
- Given cloud cover and altitude of the ridge, timely air support and extraction would nearly have been impossible – unless the Team could descend to the base of the mountain. Such descent would have been extremely difficult and dangerous, especially if the Team was transporting wounded. And the enemy would likely be waiting for the Team at the bottom of the ridge. Attempts at support and/or rescue of a Team pinned down at the top of the ridge, assuming good weather, would likely have resulted in lost aircraft (with their crews) and a commitment of even more aircraft in Search and Rescue (SAR) operations.
- Given Team surveillance would necessarily be established near the road, an enemy engagement would very well result in the Team being overrun with survivors, if any, forced down the mountainside into the waiting arms of an enemy blocking force.
- Any intelligence gathered by the Team would be minimal.

Lessons-Learned:

- T/Ls of both Teams should have challenged the wisdom of the mission, and/or its supportability, on the grounds stated above and a risks-benefits calculation.
- T/Ls should have requested the mission be executed only with a forecast of good weather with alleviated cloud cover at the higher elevations and with adequate support.

- If the mission remained slated after objections, T/Ls of either Team should have requested more insertion aircraft or aircraft of greater capabilities, so that a full Team complement could be landed on top of the ridge.
- The mission duration might have been limited to a single day of collection, with an extraction before nightfall, with no difference in the intelligence collected.
- The Team should have had priority on AC-130 gunship support every night for the duration of the mission.
- Author's preference: Better, the Team should have requested AC-130 gunship strafing up and down the road periphery during the first night, so that the following day, the Team could then conduct its reconnaissance as a 'BDA', with the possibility of taking a wounded POW, with limited enemy response.

- Resource limitations and global commitments may require that deployments will be conducted from Continental United States (CONUS) or from regional bases. This creates a number of problems:
  - The SR Team must conduct training for a range of environments and theaters of operations, impacting training resources and quality. Desired training specificity, so necessary to an elite capability, will be shortchanged; this may result in mission failures and casualties.
  - Inconsistent or short term relationships with indigenous troops will diminish unit cohesion and likely require much more training prior to mission tasking.
  - More reliance on national assets and higher-level intelligence sources for pre-deployment preparation with limited availability of vital tactical and operational levels of intelligence and information.
  - Few habitual relationships established with supporting units, to CAS/fire support. Supporting units will not be acquainted with the unconventional SR TTPs so essential to mission success and will be less inclined to take risks or to deviate from SOP or common practices.
- If the Team (or Team replacement personnel) deploy from home base in the Continental United States (CONUS) to subsequently support operations in a mature combat environment, SpecOps personnel may undergo pre-deployment training delivered at a CONUS facility. If the 'home base' is located Outside CONUS (OCONUS), this training may be delivered at a training station/facility in-theater. On no account should SR personnel/Teams be considered fully trained without unit-level training relevant to the AO. Once these SpecOps personnel arrive at their unit of assignment, further specialized and mission-specific, unit-level training should be delivered to supplement pre-deployment/deployment training. Alternatively, if the Team is to deploy directly on a mission from its home station, the Team should enter a secure isolation facility located on a secure government installation to conduct mission preparation, or the isolation facility may be located at another (CONUS/OCONUS) station en route. FM 31-20-5 contains some useful information on Home Station or CONUS/Theater-based pre-mission activities.
  - Ensure that Team-level and/or SR headquarters training curricula contains blocks of instruction on enemy weapons and equipment. This is especially important if the Team may use such items during future/planned operations.
  - Note that specialized SR missions (e.g. Nuclear, Biological, Chemical, Radiological (NBCR); Counter-Reconnaissance (CR); Wire-Tapping; Counter Insurgency (COIN); etc., require additional analyses and preparation.

Enemy Analysis (METT-TC) TTPs:

Commanders can have five pitfalls: in having a death wish, they can be killed; in having self-preservation, they can be captured; in having short tempers, they can be humiliated;

in having moral sensibility, they can be insulted; in having fondness for the people, they can be worried.

All of these five are commanders' flaws and are disastrous during the employment of forces.

An Army's failure and the commander's extermination are definitely due to these five pitfalls.

These must never be taken lightly.[12] Sun Tzu.

- Consider what would be normal behavior of enemy troops in a field or garrison environment. Unconventional/paramilitary troops will likely behave differently than regulars; elite troops will behave differently than grunts; Combat Arms troops will act differently to Combat Support or Combat Service Support troops; troops in garrison will behave differently than those who are expeditionary or in a hostile environment. How troops behave in different settings can be a tip-off as to their morale, aggressiveness, purpose, effectiveness, mission status, etc. and then how they will respond to an engagement with the Team.

  A 'great deal may be learned in advance about one's potential enemy. Differences between his armed forces and one's own are usually not fortuitous, but rather reflect a discrepancy in the military policies of the two nations. The observations of any striking deviations from standard procedures should therefore give rise to speculation about their inherent causes…. Military history is another source of valuable information. It is never too late to determine the reasons for the success or the failure of past operations. Many of the decisive factors have retained their validity throughout the years and their effect on military operations in our time would be very much the same as in the past.'[13]

- If a belligerent nation has been planning an invasion or offensive operations, it may establish, in advance, a network of camouflaged, protected positions for provisions and for occupancy by Command and Control (C2) and other key units. This same practice could be continued post-invasion as combat operations continue deeper into occupied territory. Tactical/operational logic will be applied to the selection of these positions.
- If the target area was contested in previous wars, campaigns or battles, there may be an account of the facilities used previously by belligerents. These same facilities (caves, quarries, bunker complexes, etc.) may be used again in current or future operations. Obtain this information from intelligence agencies.
- In planning and preparation for a mission, Team Members should familiarize themselves with up-to-date information on enemy organizational structure, order of battle; individual weapons, clothing (including branch and rank insignia), organic equipment recognition; major weapons systems/items of equipment identification; tactical/bumper markings; enemy tactics, battle drills and operational tendencies; and if available, biographical information on enemy unit commanders and other enemy key personnel. Supplied with this information, Team Members may better and more immediately grasp the significance of mission observations; this information may also prove essential to the survival of the Team itself. Special Operations FOBs should maintain a library containing this information so that Team personnel can familiarize with and stay current on enemy information.

---

12. Huang, *Sun Tzu*.

13. Unnamed German Generals and General Staff, 'Military Improvisations' During the Russian Campaign', CMH Pub, 104-1 (formerly DA Pam 20-201, August 1951), (Center of Military History, United States Army, Washington, DC), pp. 99-100.

'A tactician has to think like a hunter. A successful hunter thinks like his prey. A tactician has to think like his enemy. Only by understanding how his enemy thinks can the tactician predict his enemy's next moves. By anticipating what the enemy will do next, the tactician can then plan to exploit the vulnerabilities of his enemy.'[14]

- During pre-deployment training, ensure that Team Members are trained in basic, advanced and specialized enemy capabilities, formations, tactics and techniques, tradecraft and operational tendencies prior to deployment to a Forward Operations Base.
  - Are enemy officers or NCOs rewarded for initiative – or penalized?
  - SR Team Members, as the Germans did in facing unorthodox Soviet military tactics, must prepare 'for an encounter with an opponent whose pattern of behavior and thinking was so fundamentally different from their own that was often beyond comprehension.'[15]
  - It is vital to know how the enemy uses terrain tactically and operationally so that Team Members may prognosticate where key enemy units and facilities may be located on the battlefield.

<u>True Account:</u> **A highly experienced SOG Reconnaissance T/L was leading his Team on an area reconnaissance mission in northeastern Cambodia in 1969. On the fifth day of the operation, the team discovered a sizable logistics base concealed under dense canopy. Among many other items, the base contained stocks of 57mm and 37mm anti-aircraft ammunition (indicating the presence of a divisional anti-aircraft battalion); two flame throwers (indicating the presence of a Sapper/Special Operations unit); a few boxes each of 76.2mm OF-350 (Fragmentation/HE) and BK-350m (High Explosive Anti-Tank) ammunition (for the PT-76 tank) – and ominously, several new Soviet gas masks. This latter discovery, in conjunction with other information regarding enemy use of chemical agents near the Demilitarized Zone, resulted in a mandatory requirement that Recon Team personnel be equipped with M-17 protective masks. This requirement was subsequently ignored by SR Teams as T/Ls deemed the threat inconsequential to their operations.**

- All military forces, both enemy and friendly, exhibit certain routines and tendencies, which are meaningful for SR operational consideration. It is important to develop a profile of enemy behavior and tendencies. Consider for example, how a T/L might use some of the following points to plan his reconnaissance mission.
  - Enemy forces require access to water. Additionally, insurgents need access to local food sources.
  - Enemy forces will almost always occupy defensible terrain and/or terrain that can protect them from long-range fires or air attack.
  - Enemy motorized or armored forces, and the logistics elements that support them, will require easy access to the MSR or other roads. These forces will also require concealed or camouflaged locations (e.g. vehicle parks) along these routes to support the security of their movements. Use IPB methods to identify these locations.
  - Unlike US troops, many foreign military units cannot function without officer direction. But this proclivity may not apply to enemy SpecOps formations.

    'If your officers dead and sergeants look white / remember its ruin to run from a fight / so take open order, lie down, sit tight / and wait for supports like a soldier.' (Kipling 'Ghant Pagan')

---

14. Killblane, *Convoy Ambush Case Studies*, p. 2.
15. Halder, Franz. General Wehrmacht, ed., 'Small Unit Actions during the German Campaign in Russia, DA Pam 20 – 269', (Department of the Army, Washington DC, July 1953), p. 1

This, in fact, is the basic philosophy of both British and continental soldiers, 'In the absence of orders, take a defensive position', indeed, virtually every army in the world.[16]

- ○ The Team may find unit road markers, placed on MSRs, to mark turn-offs for unit elements, resupply vehicles, couriers, etc.
  - ▪ The Team should be trained, prior to deployment, in the enemy's protocol for road markers. Alternatively, the information should be made available in FOB S-2 files. Or, if the Team is using a tactical tablet, the information may be maintained in the device memory. This would be especially recommended if the markings are in a language that cannot be translated by Team Members.
  - ▪ The Team should take an annotated (date, location, direction of shot) photo of the marker and transmit a Situation Report (SITREP) with that photo at the next scheduled transmission time, if not sooner.
  - ▪ Investigate/follow the marker.
  - ▪ The Team leader should consider moving the enemy marker to another location to misdirect an enemy element, courier, transport, etc. to an incorrect destination or to direct the enemy element/vehicle to an isolated turn-off and onto a mine – or into an ambush kill zone.
- ○ Enemy forces or insurgents will establish exfiltration or escape routes from occupied positions.
- ○ Enemy/Insurgent forces will have established border/sanctuary crossing points. Indicators to look for might include:
  - ▪ Rope bridges over deep or swift water.
  - ▪ Fordable crossings in water not to exceed chest high.
  - ▪ Slow flowing and relatively shallow water where a covert underwater bridge or ford might be constructed.
  - ▪ River or stream crossing points for wheeled and tracked vehicular traffic may be different than those required for infantry/dismounted units. Such crossing points will require low and/or sloping banks.
  - ▪ Snorkels mounted or carried on vehicles.
  - ▪ Overhead canopy to avoid detection while crossing.
- • In Enemy Bivouac/Camp Settings:
  - ○ Enemy units/personnel may cease work and rest during daily peak temperature periods in high temperature environments. The T/L may decide to follow suit, if he believes that Team movement, during enemy quiet-time, may betray the Team's presence. Alternatively, he may decide that conducting reconnaissance during enemy rest periods may be optimum, as a resting enemy may be less alert.
  - ○ Enemy military personnel will be accustomed to being fed on a routine schedule (2–3 times a day). This may be an appropriate time to approach an enemy encampment.
    - ▪ Cooking or food aromas may betray the presence of the enemy; these aromas may travel especially well down ravines, from higher elevations to lower and during periods of weather inversions.
    - ▪ Additionally, enemy troops may be less alert while they are consuming their meals. The enemy's tactical sense of smell will be dominated by food aromas; and the mere act of masticating his food will impair the soldiers' hearing.
    - ▪ Furthermore, troops posted on watch at observation or listening posts, may be on a shift change/rotation schedule that is coincident with mealtimes. The Team may

---

16. Anonymous French soldier, 'HTTPS://InMilitary.com/A French Soldier's View of US Soldiers in Afghanistan (aka A Nos Freres D'Armes Americain), 6 January 17. Reprint from WarriorLodge.com

initially assume a watch condition during meal times so that they are better able to detect movement, and identify enemy positions, while this rotation is being conducted.
- Shortly after mealtime, troops will become sleepy; especially in warm temperatures.
- An unsophisticated enemy may not shield heat or smoke signatures from detection; but an enemy facing a technologically advanced opponent will soon learn to suppress these signatures.
- Enemy personnel will awaken and begin the workday by taking time, at or shortly after dawn, for personal hygiene. This may be an optimum time, while enemy personnel are shrugging off the effects of sleep, to seize an unwary soldier or to inflict casualties.
- Defensive camps of a sophisticated foe, to include enemy insurgents, will always have several characteristics in common. Consider these factors in planning an approach or mounting a surveillance:
  - 'Defense in depth
  - Extensive use of camouflage
  - Mutually supporting defensive networks
  - Restricted avenues of approach
  - Escape routes
  - Use of tunnels, bunkers, communication trenches and foxholes'[17]
- Expect also that the enemy may employ:
  - Active patrolling around the camp.
  - Deployment of night ambushes outside the perimeter, especially along routes of approach.
  - Watchers deployed in tree tops or on high ground to detect SpecOps insertions and/or observe danger zones and routes of approach.
  - Mines/booby-traps and/or noise-makers, especially on nearby trails and on other high-speed routes of approach.
  - The enemy may resort to bait tactics to lure friendly air assets into anti-aircraft traps – where heavy weapons are located – to maximize over-watch of prospective LZs or where attacking aircraft are channelized by terrain form into interlocking fields of fire.
- Keep the initiative and make the enemy react. If you know or can determine how the enemy reacts, you can maintain the initiative and make the enemy pay dearly.

ACTION – REACTION ANALYSES (Sample)		
ACTION(S)	**REACTION(S)**	**COUNTERACTION(S)**
Deliberate near ambush of an enemy infantry road column	• Counter fire by armored vehicles. • Base of fire by dismounted infantry (e.g. crew served weapons). • Immediate assault by dismounted infantry. • Flanking maneuver with mounted/dismounted infantry.	• Move to secondary, covered or to defilade positions. • Mine/booby-trap the kill zone and approaches of enemy assault elements. • Employ a 'Z' ambush formation to employ flanking fire on dismounted infantry and to counter enemy flanking maneuver. • Withdraw and move to next ambush position.

---

17. US Army, Combined Intelligence Center-Vietnam, Handbook: 'What A Platoon Leader Should Know about the Enemy's Jungle Tactics,' (US Army: October, 1967), pp. 1–44

ACTION(S)	REACTION(S)	COUNTERACTION(S)
Bury anti-vehicular/anti-tank mines on road; damaging/ destroying enemy vehicle(s).	• Unit on the scene attempts visual search for mines. • Mine clearance elements employ mine detectors, with protection from security elements. • Enemy vehicles rerouted to bypass mined road section.	• Initiate command/ remotely detonated anti-personnel mines to kill mine clearance crews and security personnel. • Employ command/remotely detonated off-route anti-armor mines to defeat armored mine clearing vehicles. • Secondary routes blocked with booby-trapped abatis and/or seeded with mines

Terrain and Weather Analysis (MET̲T-TC) TTPs:

Terrain-Based TTPs:

'The tactician must constantly be aware of the terrain and how it provides an advantage to the enemy and how he might use it to his advantage.' – Sun Tsu

• During the SOG experience in Southeast Asia, target areas were given an alpha-numerical designation and were established with a 6 kilometer X 6 kilometer boundary (No-Bomb-Line). The target area boundary could be shifted within (what the Author refers to as) a target area 'rosette' (see example). If the Target boundary/box lacked usable LZs or if the box did not offer terrain advantages, the boundary/box could be shifted within the

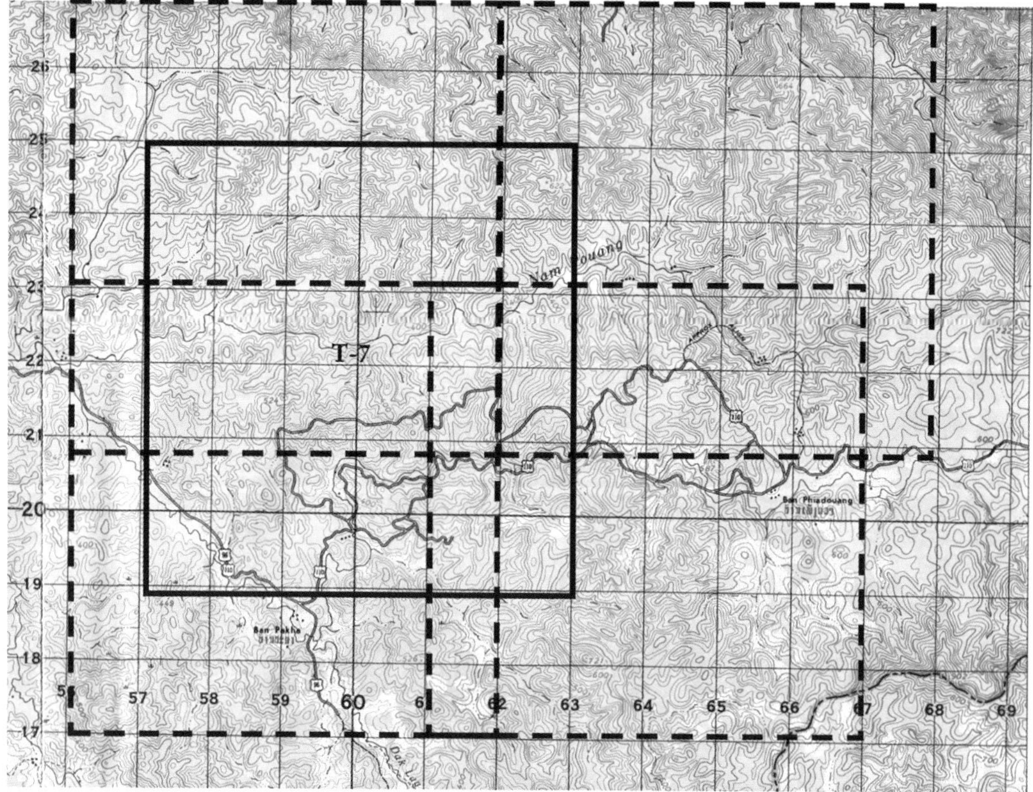

Figure 5. Example of a Target Area "Rosette".

rosette, to accommodate the mission CONOPS, as long as the target remained within the box.

- SOG SR Teams were given an impossibly large geographic area to conduct its missions with limited resources. The missions/target areas were assigned by command echelons well above the FOB level and were often illogical from a Team perspective. Whenever possible, the SR Team leadership should, within its assigned target area, focus on choke points or geographically suspect terrain features and known/suspected concentrations of enemy strength.
- OAKOC– Observation and Fields of Fire, Avenues of Approach, Key and Decisive Terrain, Obstacles, Cover and Concealment (aka OCOKA). OAKOC is used by unit leaders to analyze terrain and the effects of weather on unit operations. In its typical application, the analyses are used to support friendly unit operational planning, preparation and execution; however, it is also appropriate for terrain analyses from an enemy perspective. Its value for SR operations should not be underestimated. It is effectively a summary field tool comparable to the more elegant Intelligence Preparation of the Battlefield (IPB).

<u>True Account:</u> **In January 1968, SOG Recon Team Indiana, operating along the border in eastern Laos near Co Roc Mountain, detected tracks from multiple tanks, specifically those of PT-76 amphibious tanks. Soviet tanks had distinctive, common track design features that convinced the Team, the FOB commander and the SOG commander that PT-76 tanks were present at the South Vietnamese border and poised for an attack. This information was provided to MACV J-2; the MACV Chief of Intelligence and his Chief Estimator disputed the report, offering that the tracks were from earthmoving equipment, and actually calling the T/L a liar during a post operation debriefing at J-2. Subsequently, J-2 did not circulate the information to subordinate commands. Shortly thereafter, PT-76s were used to overrun the Special Forces camp at Lang Vei (which also served as a SOG Launch Site), near Khe Sanh, South Vietnam on 6-7 February 1968, during the Tet Offensive of that year; even then, 'the J-2 flat out refused to believe it. He just said, "There are no tanks in Vietnam"' As a relevant side note, according to the Center for Military History publication, 'MACV – The Joint Command in the Years of Withdrawal, 1968-1973' the J-2 consistently rejected reports and analyses that a general, nationwide NVA attack was forthcoming during Tet of 1968. This was an historic, major intelligence failure; that Tet offensive is considered by many as the seminal cause for the ultimate withdrawal of US Forces from Vietnam.**

Subsequently, PT-76 tanks were used in an NVA assault of another Special Forces camp at Ben Het, a year later (3 March 1969) in the central highlands of Vietnam near the tri-border area. Here too, tank tracks were detected by a SOG Team; and, despite the preceding Lang Vei episode, here too, they were construed by higher authority to be tracks of earthmoving equipment.

In both circumstances, follow-up SR missions should have been directed – but were not. OAKOC would have been useful in narrowing down the possible tank mobility corridors (Go vs No-Go Terrain) for SR exploitation.

- Team Members (especially Operations and Intelligence personnel) should collaborate on conducting an OAKOC/IPB analysis based on an understanding of enemy tendencies and doctrine and on the limiting characteristics of terrain under varying weather conditions.
- The enemy will normally establish trails on ridge-tops, alongside streams, parallel to roads and in networks to interconnect units and facilities. With knowledge of enemy tactics and techniques, enemy METT-TC, and a current map study, an experienced T/L can predict where these facilities and units may be found.

<u>True Account:</u> An experienced SOG Reconnaissance Assistant T/L returned from an operation where his T/L had been badly wounded. The engagement had occurred within a major enemy base area close to an enemy MSR, and in a location where the enemy had an installation that was afforded protection from B-52 Arc Lights and other airstrikes by surrounding ridges enclosing the installation on three sides. The Assistant T/L was subsequently assigned the T/L position. He conducted his own AAR, including a map study of the battle location, and noticed that while the installation was less than 200 meters from the road, the peculiar terrain effectively protected it from frequent B-52 strikes along the highway. The T/L used this knowledge to identify terrain that was similar in form within other target areas along the MSR, enabling him to estimate the location of other key enemy installations.

- An enemy may establish one-way traffic on an MSR toward the combat sector or marshalling areas, especially if the number of road lanes is few; the enemy may use an alternate MSR for returning traffic. The SR Team should expect that a separate return route will not be as well guarded/patrolled. So, while the route forward may have more lucrative targets, a Team may prosper on targets (including priority targets, POWs, etc.) traveling on the return route. See information on high priority targets later in this book.
- If the Team is to use pack animals, route selection planning is critical. If at all possible, choose a route that will not require hacking a path through heavy vegetation and that avoids terrain features too steep for the animal to negotiate.
- The Team, or its higher headquarters, should consider using terrain profiling and line-of-sight profiling software, if it is available. These tools can assist in the following tasks:
  ○ Selection of helicopter nap-of-the-earth flying corridors to LZs or DZs.
  ○ Route selection from LZ/DZ to the Team's target, using terrain features to mask the Team's approach from observation and to suppress the Team's thermal signature.
  ○ Pre-selection of sniper, laser-designator or crew-served weapons systems locations. Note that a laser-designator is unsuitable for some environments (e.g. jungle/rainforest) and should not be carried unless line-of-sight can be obtained from appropriate terrain features.
  ○ Pre-selection of outposts or surveillance locations, or identification of possible enemy positions.
  ○ Assist in indirect fires planning.
  ○ Pre-selection of possible defensive positions.
  ○ Pre-selection of possible ambush positions.
  ○ Prediction of enemy AA positions.
  ○ Note: These software tools are generally suitable only for terrain profiling using map data and contour intervals. The software does not consider the effects/height of vegetation on top of terrain contour/elevation which may further obscure line-of-sight.
- Enemy units may have weaknesses in their defensive posture. For instance, the enemy might be less concerned with maintaining a 360 degree defensive perimeter, if the camp abuts an obstacle (water, minefield, etc.), or if they are adjacent to other enemy units (e.g. for purposes of mutual protection) – especially if they are located on their own turf, in a sanctuary or at some distance from the 'front lines'. In these situations, the enemy may deliberately use terrain features and obstacles to shorten his perimeter. SR Teams can exploit these tendencies to approach an enemy location from an unexpected direction.
- During rainy seasons, intermittent streams depicted on your map will become dependable water sources – and small, active streams may become raging torrents, creating an obstacle. During dry seasons, intermittent streams will often not readily provide water; the Team must then use caution in approaching active water sources, as the enemy may be concentrated there.

Figure 6. The VR reveals a slash-and-burn area and varying types of vegetation (Reed).

- Vegetation type and density may prove critical in the planning, preparation and execution of a SR operation. Overhead imagery will often be unsatisfactory for these purposes – as compared to a Visual Reconnaissance by the T/L.
- Tree/vegetation types (conifer, deciduous, etc.) will vary according to altitude, climate/environmental, geological conditions and other factors.
  - Some military maps (e.g. US) will not show tree/vegetation types or level of canopy, but some foreign maps (e.g. Russian) and/or high-resolution satellite/aerial imagery may.
  - The S-2 section should include map overlays of vegetation variations (e.g. deciduous, conifer, bamboo) for every target area folder.
  - Overhead imagery is also necessary to identify recently deforested areas and where fires have swept away normal vegetation to complement topographic maps.
  - The higher the elevation and the more prevalent the condition of high winds, the smaller the trees and more limited the foliage. Without the presence of natural cover, the more necessary the need for artificial camouflage.
  - Different kinds of vegetation will reveal the types of soil, moisture levels, animal habitats, presence of water (e.g. oases) and other important information that may be useful for Team route selection and for forecasting enemy locations.
- Some raw information on prevailing winds, vegetation, animal habitat, geology, hydrology, etc. may be found in area studies produced by the intelligence community; it is up to the Team Members to interpret and correlate the information and determine its significance for an operation. Underlying geology reveals additional information about the terrain than is portrayed on a map. Area studies may contain this information.
  - Expect limestone areas to have holes and caves.
  - Granite areas may have moors and bogs, as well as dissected terrain.
  - Smooth rounded pebbles suggest substantial water erosion and a susceptibility to flash floods.
  - In northern regions, southern facing slopes will have more vegetation and lone trees will bear more leaves on their south facing sides.
  - Determine the prevailing winds in the AO. Slopes facing prevailing winds will generally have thinner soil and shorter trees and greater tree-fall (due to storm winds). In sparsely forested areas and in the case of lone trees, tree roots of large trees will typically face into the prevailing wind. Additionally, walls, hedges, roads and some natural terrain features will

interrupt the linear flow of wind, causing back drafts, and static voids to one side where leaf and twig debris will collect; movement through such an area should be avoided.

- ○ In Northern regions, trees exposed to more sunlight (e.g. south-facing slopes) will have more branches and more developed foliage on the sunlit side. This can be useful in land navigation, in Escape and Evasion (E&E) situations and in charting Team routes. Tree bark is typically darker on the north-facing side. Ivy or vine climbers will generally be on the north-facing side of a tree; but the top of mature vines/ivy will face south.

- Place-names, some dating to antiquity, can reveal significant information about the locale and its surrounding area. The T/L should seek translations of place names from headquarters S-2 and this information should be maintained in the target folder. In keeping with OPSEC discipline, seeking the information from an indigenous Team or FOB interpreter prior to an operation bears some risk.

- Water Signs (Terrain) TTPs:
  - ○ In desert, look for birds at dusk. They will not be far from water. Flies and other insects abound close to water (e.g. near an oasis).
  - ○ Seed-eating birds often fly fast and low when headed to water; carrion birds and seabirds do not behave in this manner.
  - ○ Some trees (e.g. willow) and plants require a water soaked environment and others cannot tolerate it. Knowing which trees favor water within the Team's AO, will help Team Members identify water sources from afar.
  - ○ An algae bloom will indicate water with high phosphate/nitrate levels caused by dead animals, contamination by cattle or farm runoff. Bad drinking water.
  - ○ Typically, water flows four times faster in the center of a stream or river; but if the stream/river turns, the water at the outside of the bend will be fastest and deepest. The inner bend will often have sand, mud, rocky deposits and sand/mud bars.
  - ○ When stream/river water slows (where it is impeded by shallows), silt will deposit on the stream/river bed – beware of quick mud. As water speeds up, it carves a dip in the bed – beware of the dips when crossing.
  - ○ Water on either side of an island will be faster than the center current of a stream/river upstream or downstream of the island.
  - ○ Furrowed ground in cultivated areas will deaden sound, especially if wet.
  - ○ Author's Safety Tip: If caught in rapids, a Team Member will be unlikely to survive the experience if he strikes his head on a rock. Feet should always be pointed downstream.
  - ○ Bottom feeding birds (ducks and geese) will be found in shallow water, indicating potential crossing locations.
  - ○ Quicksand/Quickmud:
    - ▪ Found along river banks, in swampy areas and along shorelines.
    - ▪ Taller troops tend to sink to their knees; shorter statured troops may sink to the waist. Panic will cause further immersion.
    - ▪ If the Team Member is alone, he may have to abandon his rucksack, Load Bearing Equipment (LBE) and weapon. However, he will be better served by throwing his weapon(s) and other items onto the verge, then letting his rucksack drop behind him, and then further back (beyond the rucksack) to provide leverage points to work his way out of the muck. If the Team Member has presence of mind, he can toss a weighted 550-cord, attached to LBE, etc., to firm ground, to retrieve his equipment once he is free of the quicksand/mud; this line must be weighted at one end and easily accessible to the Team Member during movement.
    - ▪ The Team Member may remove himself by slow movements and rotations and/or then laying flat on his back to 'swim' to firm soil.
    - ▪ An area of quicksand/mud can be an ideal location to ambush trackers and take enemy combatants prisoner who are stuck in the muck.

Figure 7. (A) Hoang Cam Stove at Left. (B) PAVN Recruit Digging Hoang Cam Exhaust Trench on Right. (PAVN training photos)

- Other Water TTPs:
  - The appearance of vegetation/foliage can reveal the presence of a hidden water source not evident from a map study. A VR or color photography may reveal these foliage variations. <u>Identifying a hidden spring/water source may lead to an ideal Mission Support Site (MSS) location, OP/Surveillance site</u> or patrol base.
    - Some trees require abundant water (willow, cypress, cottonwood) other trees may die in the presence of excessive water; knowing which trees require a lot of water can lead to discovery of a water source.
    - Large trees require a great deal of water. A stand of large trees surrounded by smaller trees may indicate a water source. An abundance of foliage on a stand of trees in a forested area with less tree foliage is an indicator of a water source.
    - Color-infrared (CIR) aerial photography is useful in revealing differences in foliage attributable to water. Very intense reds in CIR photos indicate vegetation which is growing vigorously and is quite dense. If the season or prevailing weather is dry, an area of intensely red vegetation will be an indicator of a water source.
    - In the autumn, trees that are close to water will show brilliant colors as compared to trees that are not close to water. This is especially noticeable if prior seasons were in drought conditions.
- Heat Signature Suppression TTPs: A sophisticated enemy will use field expedients to suppress cooking fires, and associated smoke and heat signatures, by terrain selection (e.g. screening behind terrain features) and using fire pits, tarps and underground conduits or exhaust trenches allowing the smoke and heat signatures to dissipate at ground level or under canopy.
  - While heat signatures may be suppressed by these measures, cooking aromas can be detected by SR Team Members, especially under certain weather conditions described elsewhere in this book.
  - Some of these same measures can be adapted to long-duration SR operations, for instance, when conducted from patrol bases/MSSs in special environments (e.g. cold regions, alpine). Note the illustration of the Vietnamese Hoang Cam stove and an exhaust trench under construction. The exhaust trench is oriented downwind of the stove.

Weather-Based TTPs:

Weather forecasting at the regional level, such as may be provided to the Team by higher headquarters, can be wildly inaccurate at the local level or as compared to conditions on the ground. Team Members must use 'weather sensing' to predict weather during the conduct of a mission. Weather sensing should not be based on a singular observed factor, but should rather be <u>based on several observed, mutually-confirming factors</u> (e.g. cloud formations, wind conditions, snow type, temperature, barometric conditions, etc.) to predict local weather (see Appendix C). Weather predictions in mountainous terrain is more uncertain.

- Rainforest environments are normally high-temperature (90–100 degrees) and high-humidity (80 to 100 per cent) throughout much of the year. But at high altitudes these high temperatures can drop rapidly. Be prepared.
- Monsoons (or rainy seasons) are characterized by continual, unrelenting downpours for several hours of the day and continuing for weeks at a time, with only brief periods when the rain will pause. The downpour may be so severe that artillery cannot be fired in support, as fuses may detonate in flight when passing through sheets of rain. And air support is often not possible for days, even weeks at a time. During rainy seasons, periods of heavy overcast with rain, snow or fog, the only fire support capabilities available to the Team are its organic assets, unless the Team is operating in concert with guerilla or partisan forces. This reality supports the notion that an SR Team should conduct its operations heavily armed with highly lethal weapons and munitions in similar operational conditions.
- Rainy season creates muddy roads that hinder enemy movement and increases his reliance on improved roads, hard-top highway and railways. Heavy rain may also limit the enemy's use of water transportation and water crossings. These conditions are helpful to SR Teams and should be exploited.
- Monsoons will significantly hamper the availability and employment of air assets. This may require you to suspend the mission, change to an alternate mission, change your line-of-march and/or exercise extraordinary caution. Plan to bring extra rations; your extraction may be delayed for days. Prepare for the monsoon season by establishing a MSS during dry season periods.
- Rain, snowfall, fog and/or the accompanying overcast constitutes both a blessing and a curse to SR Teams. During these periods, enemy units may come out of hiding and move about the battlefield, making them more detectable to SR Team personnel. Alternatively, enemy combatants may seek shelter from the elements. If the enemy element is sheltering, any noise or movements outside its bivouac might be considered suspicious.
  ○ Sound carries during weather inversions that accompany these conditions.
  ○ Moist soil and vegetation reduces some of the noise of movement but leaves more visible trail signs.
- The Team should never underestimate the importance of weather on enemy operational planning and execution. Weather will affect both friendly and enemy behavior and tendencies.
  ○ Inversion – When a layer of warmer air traps a cooler layer beneath it near the surface.
    - Sound travels further
    - Refraction may create optical illusions
    - Fog and smog endures in evening and morning, smoke will drift close to the surface and will descend elevations to low areas following terrain features.
  ○ Both winter snows and spring rains and thaws (or monsoons) will severely impact enemy wheeled/tracked vehicle and aviation mobility. Bad weather that prevails during these seasons, in many ways, can be a blessing to SR Teams; for instance, if the weather has the effect of channelizing enemy movements. This channelization can be predictable.
  ○ Enemy units will adapt to bad weather by changing their behavior and tendencies.
    - Enemy bivouacs/camps and logistics stockpiles located in wet areas will be moved to dryer or more traversable terrain in these conditions. If the Team has detected the

location of an enemy unit/camp sited in an area that will be subject to flooding during rainy seasons, this may present an opportunity for the Team to conduct ambushes, or route mining in anticipation of an enemy relocation.

- Enemy offensive, defensive or retrograde operations will be staged from or located in proximity to traversable corridors.
- Traces of enemy vehicular movement can more easily be detected in soft, wet soil or after a snowfall. Unless the enemy is disciplined and effective in camouflage and concealment of vehicular tracks, these traces can reveal enemy unit locations.

- Dense, cold winds descend from ridge tops at night, making it colder at lower elevations. Later in the day, as temperatures climb, wind direction will reverse and warmer air will flow upslope. This information is important for airborne operations in mountainous terrain. In this terrain, the combination of Team Member weight differences (especially US SpecOps vs Indigenous), winds aloft, cross winds, up (day) and down (night) drafts, ruggedness of the dissected terrain, varying parachutist experience levels, and other weather factors will generally guarantee a split or scattered Team with Team Members separated by one or more ridges (mountainous terrain). Expect split Team Members to be disoriented; reassembly of the Team may take days. If this is to be avoided or mitigated, Team training is essential, but careful preparation and (if possible) selection of an insertion date with optimum weather conditions is the key factor.

- Determine the phase of the moon for the duration of an operation and when the moon will rise and set (according to hemisphere). Dusk in mountainous terrain comes earlier and abruptly. Plan to move into NPD earlier. See Appendix C.
  - The amount of moonlight, and the amount of time the illumination will be present, is important for night observation and navigation.
  - A 7-day-old moon means that the moon will be at 90 degrees to the sun as the sun sets.
  - A 12-day old moon will remain up until just prior to pre-dawn. A 13-day-old moon means that darkness will prevail at dusk.
  - Newer moons (1-14 days old) rise earlier than sunset; older moons (17 to 29 days old) rise after sunset. Moon shadows are cast accordingly; moon shadows are very dark, especially when vegetative/terrain cover is included, making movement difficult when the Team Member is unaided by night-vision optics. Moon shadows can have a bearing on route selection (Team movement should be concealed by shadow, whenever possible), hide selection, ambush or raid planning and site selection and the possibility of tracking across open areas at night.
  - A 15- to 16-day-old moon (full moon) will rise as the sun sets. A full moon provides up to 10-times the amount of illumination provided by a half-moon.
  - Moon illumination is, of course, impaired in cloud (including fog) cover; so planning to make use of moon illumination should be considered in concert with weather forecasts. Moon illumination should be optimum during high pressure conditions.
- Water Signs (Weather) TTPs:
  - In coastal areas, tides higher than normal will predict low pressure and bad weather.
  - Beware of wadis during rainy season. Rain can fall miles away, but because water is not well absorbed by sand or calliche, flash floods may result across a wide area.

Troops and Support Available Analysis (METT-TC) TTPs:

Troops Available Analysis TTPs:

'Through thorough knowledge of his own troops, the tactician can defend his weaknesses and apply his strengths to the enemy weaknesses.' – Sun Tsu

Troops that comprise an SR Team may vary considerably based on a number of factors and mission environment variables.

- Team Organization (US): Aside from (for instance) a basic 12-man, A-Team structure used in SR operations, the Team may be supplemented with personnel possessing special skills, such as: a sniper team, USAF air-ground coordinator, translator/interpreter, indigenous tracker or guide, technical expert, etc. Additionally, if the SR Team is short of personnel, volunteer/'strap-hanger' personnel may be sought to round out the Team. Strap-hangers must either be current or recent SR veterans to be acceptable as temporary Team Members. Further, strap-hangers must join the Team early enough during mission prep so that they are fully trained.
- There are substantial advantages to an SR Team comprised of a core of US SpecOps, supplemented with indigenous commandos/mercenaries.
  ○ Economy of Force, where fewer US SpecOps personnel are used to accomplish the mission. Force-multiplier, where fewer US SpecOps personnel direct the actions of more numerous indigenous troops and/or indigenous troops are used in lieu of US SpecOps.
  ○ Indigenous personnel on the Team will possess more intimate knowledge of the AO or in operating within similar environments.
  ○ Indigenous personnel on the Team may possess skills that US Team Members lack, to include facility with local languages and field craft.
  ○ Disadvantages: Indigenous Team Members will require specialized Team training; they will largely lack key SpecOps skill sets; third-country mercenaries will likely lack sufficient language/interpreter skills. Some indigenous troops may have divided loyalties.

> The Senoi Praaq was a special operations unit established during the Malayan Emergency; it was mostly comprised of Malay aborigines, who were recruited,

Donald Davidson, Joe Walker, Masterson, and Gayland Mussleman pictured with Team California, Kontum (FOB 2) CCC. (Courtesy of J.J. Walker.)

Figure 8. RT California FOB 2 (CCC) 1970. Note Team size and the heavy armament: RPD Lt MGs, RPG-2s, CAR-15s, M-203s & a 60mm Mortar. This highly regarded Team was geared to realities of SOG missions, where intense engagements were nearly certain.

trained and led by SAS personnel in deep penetration operations against Communist Terrorist (CT) elements.

The 'Senoi Praaq was a crack unit organized to fight the communists, an efficient military and intelligence machine resembling SAS troopers. …. Senoi Praaq's deep-jungle operations proved extremely successful in the suppression of CTs. … In 1958, the Senoi Praaq held the highest number of kills on record among any security force's units in Malaya. By 1959/60, their kill ratio stood at 16 to 1 for killed or wounded enemy personnel. The MRLA [CTs] quickly spread the news about the Senoi Praaq's success among them. Their reputation as ruthless killers forced the CT to abandon its activities and withdraw rather than engage this foe.

After the Emergency, a small group of Senoi Praaq helped establish the Montagnard Scouts in March 1963…. The Senoi Praaq mission was to teach the South Vietnamese forces what they knew, and establish an intelligence network among the Montagnards.'

'The Senoi Praaq sometimes used traditional weapons, blowguns and poisonous darts made from the Ipoh tree, to pick off CTs one at a time in a leisurely hunt that lasted for days. This shows that the Senoi Praaq were excellent stalkers and hunters, because this sort of killing was best accomplished when the stalker was safely concealed behind thick foliage. They were free to flee if contact became too heavy.'

Unconventional (Guerrilla/Partisan) Warfare Mission Environment TTPs:

- SR Teams may consist of US Team Members entirely or include indigenous personnel.
- The Team can use friendly guerrilla/partisan base camps or guerrilla/partisan controlled areas to launch even deeper into enemy territory on a continuing basis. The advantages/disadvantages of this arrangement are:
  - Deeper penetrations and longer duration operations are possible.
  - Reduction in air operations in support of repeated SR Team infiltrations and exfiltration operations, with a corresponding reduction in risk to aviation assets.
  - Guerrilla/partisan units would likely provide security, interpreters, guides and some logistics, communications and medical support while hosting the SR Team.
  - A separate UW Special Forces team will likely be integrated with the guerrilla/partisan organization. The two teams (SR and UW) can both benefit in this situation as they can share intelligence and participate mutually in various aspects of operational planning, preparation and even execution.
  - Guerrilla/partisan fighters can be used in combined operations in support of SR operations. For instance, a guerrilla/partisan unit can cover the withdrawal of an SR Team being pursued by the enemy.
  - Guerrilla/partisan organizations can provide exfiltration networks for SR Team personnel.
  - Guerrilla/partisan familiarity with local conditions, threats, geography, etc.
  - Disadvantage: OPSEC threat (possible compromise) – too many personnel know of the presence of the SR Team.
- A composite SR Team, operating out of a guerrilla/partisan controlled area or base camp, may consist of a core of 2–6 US Special Operators supplemented by indigenous UW troops, who are native to the AO, who were previously exfiltrated and trained to operate with Special Operations personnel and who were subsequently reinserted as part of the composite Team. The advantages/disadvantages of this arrangement are:
  - Advantages and disadvantages as above.
  - Economy of Force, where US SpecOps personnel are used to the same or similar effect but with fewer US personnel placed at risk. Force-multiplier (as above).

- ○ Indigenous personnel on the Team will possess more intimate knowledge of the AO.
- ○ Indigenous personnel on the Team may possess skills that US Team Members lack, to include facility with local dialects and local field craft.
- ○ Indigenous Team Members who are killed or wounded during operations could be replaced by selected guerrilla/partisan fighters.
- ○ Disadvantages: Indigenous Team Members are unlikely to be trained to the same standards; they will largely lack key SpecOps skill sets; many guerrillas/partisan may lack sufficient language/interpreter skills.
- Alternatively, a 2–6 man Special Operations SR core element could be infiltrated into the AO and then the core element could use guerrilla/partisan personnel to form the composite Team. The advantages/disadvantages of this arrangement are:
  - ○ Advantages and disadvantages as previously stated.
  - ○ Lower US logistical burden.
  - ○ Disadvantages: Indigenous recruits must be trained within the AO, with very limited resources, by US SR core Team Members, to a moderate standard of proficiency. This will result in partially trained recruits and will consume valuable time; proficiency may be a problem.

Counter-Insurgency (COIN) Mission Environment TTPs:

- Many of the SR considerations that apply to the UW mission environment also may apply to the COIN mission environment.
- As in the UW mission environment, SR Teams may consist entirely of US Team Members, but they may also be of composite manning with a mix of US and indigenous personnel. The composite Team may have 3rd country mercenaries in addition to, or in lieu of, local indigenous personnel.

Indigenous Troops TTPs:

- See more information in Chapter 4 (Personnel).
- The S-2 section has primary responsibility for security screening of indigenous troops. However, the T/L also bears some responsibility for security screening of his indigenous Team Members.
  - ○ Screening should discover affiliations, motivations, sympathies and core beliefs of the recruit and his family. These influences may be derivative of ethnic, tribal, religious and cultural factors existing in the country and even down to the locality. Team leadership should be acquainted with any doctrine (e.g. religious, ethnic) that may govern the loyalties and behavior of the recruit.
  - ○ Screening should seek this information through the host government or through local authority figures, if feasible, down to the neighborhood/village. If the enemy is invested in the village, this approach may not be possible. But other indigenous personnel from the local area may be located in areas under friendly control. If the enemy controls the recruit's home district, the recruit may be vulnerable to threats to family members.
  - ○ Screening should account for the adult history of the recruit. Extended unverifiable gaps should be suspect. Any attempts at hiding the recruit's past or any similar deception should be a disqualifier.
  - ○ 3rd country/mercenary recruits are normally less of a concern.
- Develop an understanding of your indigenous personnel; their culture, beliefs and motivations are important to operational success.
  - ○ If at all possible, recruit indigenous personnel from the same ethnic group, BUT first ensure that their cultural traits and system of beliefs/values are reasonably compatible with our own and with military requirements.

- ○ Not all people value the same things that we do; some cultural or ethnic subsets can be inherently hostile to our own beliefs and ethical standards.
  - ○ If your indigenous personnel are drawn from different ethnic subsets, understand that some ethnic groups may have a cultural or historical animus toward personnel of other ethnic backgrounds on your Team.
  - ○ Additionally, ensure that leaders among your indigenous Team Members have a role in the selection of new-hire indigenous personnel.
- Show respect and understanding for indigenous Team personnel under your control and develop a relationship of trust and confidence with them. They must have full trust and confidence in you; however, temper your own trust and confidence in them with 'prudence'.
  - ○ In consideration of Operational Security (OPSEC), tell them what they need to know to perform their duties and little else.
  - ○ Never forget that some indigenous personnel may have divided loyalties and/or their family members may be vulnerable to enemy influences or threats.
- Some Indigenous Troops will develop or display a code of loyalty that requires them to stand and die with their American Team Members, rather than leave a wounded or KIA American Team Mate behind. This is a precious warrior ethic that American Team Members should nurture and never squander.
  - ○ US Team Members should make every reasonable effort to evacuate/recover KIA indigenous Team Mates.
  - ○ One of the few occasions that justifies abandoning a wounded or KIA comrade (US or indigenous) is during a breakout operation from enemy encirclement; this is a decision that resides only with the T/L or senior surviving Team Member.
- Use tact when correcting your indigenous Team Members. If possible, take the man aside to correct him. Turn on-the-spot corrections into a 'coach-player' training opportunity. This helps the Team Member respond positively to the critique, since he will not feel ridiculed, 'lose face' or self-confidence, or lose the confidence of his Team mates.
- Most of your indigenous Team Members may have limited English skills. Conduct English classes for your indigenous personnel, especially for interpreters. Additionally, have interpreters conduct classes for your US personnel in the indigenous Team Members' common tongue or in the most common dialect. If possible, conduct these lessons while simultaneously learning field craft from your indigenous personnel.
- Remember that some, or perhaps most, of your indigenous troops may be illiterate. Keep drills, formations and tactics simple for indigenous troops and use the 'crawl-walk-run' training approach. Indigenous troops require more drilling, training and rehearsal than US troops.
- Indigenous Team Members may have their own system of rank or seniority, which may not always be based on merit.
  - ○ If indigenous Team Members are divided among varying ethic groups, the highest ranking indigenous Team Member, will often be from the most numerous ethnic group on the Team. However, be aware that other factors may bear on rank and standing (e.g. the interpreter may be the oldest or most influential indigenous Team Member).
  - ○ Some positions warrant a higher pay level than others, because they require greater skills/ training or the position may be inherently more hazardous. Indigenous Team Members generally understand and respect this, as their lives are dependent on the proficiency of Team Members with higher skill levels/experience. The point-man and interpreter are typically at the peak of the pay scale; but the interpreter is the key indigenous position on the Team. The tail-gunner is normally next in line (in terms of trade/field craft proficiency and pay) to the point-man, so he will earn below the point-man/interpreter.

Those trained on essential equipment or crew-served weapons, may earn proficiency pay; others with special qualifications (e.g. HALO) should earn incentive pay.
- Profile your indigenous or third country Team Members.
  ◦ Changes in attitude, behavior or demeanor, e.g. loquacious to reticent; happy/funny to solemn. Slovenly appearance.
  ◦ Unusual estrangement/distance from other indigenous/third country troops. Emotional outbursts.
  ◦ Absences from the Team, especially in advance of pending missions.
  ◦ Riding sickcall.
  ◦ Hanging out with unsavory personnel (civilian or other military).
  ◦ Drug/alcohol abuse or an increase in alcohol consumption.
  ◦ Possession of contraband (to include drugs) or material of intelligence value. Especially once a warning order is issued, observe the comings and goings of indigenous/third country Team Members, if possible. Contraband materials may be hidden outside their Team room.
  ◦ Location of family members and influence of the enemy in that vicinity.
  ◦ Names, addresses, occupations of relatives: check records with police and national security/intelligence organizations.

Snipers:

- Snipers have both DA and Human Intelligence (HUMINT) (reconnaissance/surveillance) functions.
- In the UW/FID environments, snipers may be consolidated at FOB/battalion level, dependent on resources available, deployment environment and sniper mission requirements. If the Team is to be deployed in areas with long range small arms fields of fire (desert, steppes, tundra, etc.), all Team Members should be given supplementary, even intensive, marksmanship training.
- The SFOD may have two organic sniper slots authorized on the unit Table of Organization and Equipment (TOE) (likely the weapons MOSs); as sniper teams are two man operations (sniper and sniper-observers), other Team Members must be designated to support the primary snipers as sniper-observers. So, at least two non-weapons MOS personnel should receive sniper training.

Support Available Analysis TTPs:

Mission Preparation Phase:

- There may be circumstances where, for reasons of plausible denial, certain supporting weapon systems or support may be denied to the Team. The Team leadership must be resourceful in finding timely, plausible alternatives to request of the FOB/higher headquarters.
- The FOB or the battalion headquarters should possess the necessary staff to assist the Team during mission preparation and execution. While the staff answers to the FOB/Battalion commander, the T/L should consider them his own staff as well and should not hesitate to task them for support. If they are unresponsive to reasonable requests, use the chain of command.
- The T/L may designate Team Members, with appropriate MOSs, as a 'shadow staff' responsible for coordinating with respective FOB or battalion staff members; for instance, the Assistant Detachment Commander (180A) Warrant Officer might logically coordinate the activities of the Team 'shadow staff' and be responsible for coordinating with S-1 staff, to include indigenous Team Member personnel actions; the Operations Sergeant (18Z) and

the Assistant Operations and Intelligence (O&I) Sergeant (18F) might interface with the S-3 (Plans and Operations) and S-2 staffs respectively; Communications NCOs (18E) might interface with S-6 staff; Engineer NCOs (18C) might interface with S-4 staff and supply/logistics support; Weapons NCOs might interface with S-3 (Training) and the S-4; and Team Medics (18D) might interface with the FOB dispensary. This assumes that the Team is manned with the MOSs described above and is not a composite of US SpecOps and indigenous Team Members. On composite Teams, these burdens fall on fewer US Team Members.

- The T/L, with the input of other Team Members, should brainstorm whatever provides an edge to the SR Team in preparation for, and conduct of its mission. This should include, specialized training, specialized and non-standard equipment, capabilities in augmentation of the base Team, etc.
- Ranges and training areas must be scheduled or made available for both day and night training events.
- Intelligence staff should provide the Team all Intelligence Requirements required for the mission tasking.
  - The Intelligence staff should place mission related intelligence information into a Target Folder for Team use.
  - Information contained in the Target Folder must be kept current; the T/L must double check the currency of information in the Target Folder. If the information is not current, the fact must immediately be reported to the SR Company Commander and/or to the S-2 staff officer. If the T/L later detects further lapses in the currency of Target Folder information, the lapses should be elevated through the chain of command.
  - The T/L should request any intelligence products that are absent from the Target Folder, and/or any that he believes are necessary and/or reasonably available.
  - The FOB S-2 should maintain translations of enemy field manuals, Tables of Organization and Equipment (TOEs), equipment identification photos and field notices/orders (from Lessons-Learned).
  - Team Members should familiarize themselves with this information, as the information contained therein will help the SR Team in pattern recognition, identifying red flags/intelligence indicators. Examples:
    - The enemy may clear vegetation from alongside rail lines or MSRs to provide fields of fire and impair the ability of guerillas/partisans to execute ambushes.
    - Enemy FMs may show TOEs of various units.
    - Field Orders may instruct rail transportation officers to use heavily laden dummy cars at the front of a train to trip mines/explosive devices.
    - Field Orders may instruct rail transportation officers to give priority of transloading to ammunition and Petroleum, Oil, Lubricant (POL) cars; and/or may instruct that they be positioned separately or separated from transloading of other commodities by physical distance (e.g. separate rail spur). Secluded cars with minimal security make an excellent target.

Aviation Support TTPs:

  - Including SAR, UAV, FAC, CAS, Infiltration/Exfiltration and SOF-specific aviation assets must be sufficient in numbers, capabilities and availability. It is essential that T/Ls be observant of aviation asset mission slating and aviation asset availability during the mission window. If aviation assets appear over-committed, or if their availability is in any way in doubt, the T/L should ensure contingency planning has been accomplished by the S-3 (Air) and that backup assets will be made rapidly available, if needed. If the aviation support is not dedicated or prioritized in direct support, Team mission planning must consider alternatives/contingencies/adjustments for purposes of mission accomplishment and Team survival.

- ◦ The T/L should become familiar with the supporting units SOPs and operational protocols, when those units are in a dedicated/habitual relationship with an FOB. Combined training represents opportunities for closer coordination and the Lessons-Learned from these experiences will result in adjustments to SOPs and protocols that will increase operational effectiveness and prospectively save lives. If the T/L identifies shortcomings in the training, SOPs or operational protocols of the supporting units, he should report these problems as soon as possible through the SR unit headquarters to the FOB operations staff (e.g. S-3 aviation) so that solutions may be found.
- ◦ SR Teams may expect limited, minimal or even no fire/close air support during deep-penetration operations. Extra efforts in advanced planning and preparation will be necessary for such missions. See information on Mission Support Sites and Caches elsewhere in this book.

Other Support TTPs:

- ◦ The T/L, or designated Team Members, must be observant of Bright Light Team (BLT)/Exploitation/Response Force (RF) readiness. Some response forces may be located at the Launch Site. Observe and evaluate them and speak to the force leader to familiarize them with key elements of your mission. It is also worthwhile to provide similar information to any response forces based at the FOB.
- ◦ Long-Range Systems: In deep penetration operations, where the enemy has air superiority, an SR Team may encounter a very significant target that should be taken out regardless of Team risk. The best fire support option to address such a deep target may be a long-range system (e.g. cruise missile). SR Teams should have a direct communications link to the appropriate fire control staff, skipping intermediate headquarters, to summon these fires. Given Time-of-Flight issues, this arrangement may have utility only for enemy 'World Series' targets, such as WMD systems at-the-halt (e.g. firing positions, staging areas, etc.) or brought to a halt by SR Team ambush.
- ◦ Team Communication specialists must ensure that assigned primary and alternate frequencies are usable for the Area of Operations/Target Area, that communications equipment is in fully functional condition and that communications relay capabilities are in place.
- ◦ The Team should develop on-call logistics/resupply packages for the mission. See information on this matter later in the book.
- ◦ If the Team is to collaborate with UW or counter-insurgency units, the T/L and Assistant T/L, at a minimum, should receive appropriate briefings, review relevant Operation Orders/Operation Plans (OPORDs/OPLANs) and should coordinate extensively as possible. For security reasons, be very cautious in sharing Team mission information with these units; share only that information that is essential for the other parties to provide their support.

Time and Time Available Analysis (METT-<u>T</u>C) TTPs:

Time Analysis TTPs:

- • The time factor affects, and is a component of, all other METT-TC analyses and mission planning efforts.
- • Be aware that certain national holidays and anniversaries are of military or psychological significance to belligerents; enemy offensive operations should be anticipated, and should

require more vigilance and security precautions than at other times. SR units should consider and plan for these occasions as prime OPTEMPO periods where demand and competition for support assets will be high. Calendar Date Concerns include:

- US or allied nations observe holidays or other dates, where friendly OPTEMPO is reduced. These dates may not be propitious for the Team if air or other essential support is not readily available. These dates may also invite attacks if the enemy expects lower US/allied OPTEMPO or if the enemy wants to affect troop morale.
- Holidays or other dates observed by the enemy, where his own OPTEMPO may be reduced, or where he may celebrate observances by attacking, or where his attacks may bear special significance (e.g. 9/11 attack anniversary).
- At harvest time during counterinsurgency operations, or in circumstances where enemy insurgents/troops may be diverted to planting or harvest tasks. This is explained more fully later in this book.
- Peak periods for bad weather conditions.

Time Available Analysis TTPs:

- Use the 1/3 – 2/3 Rule: Upon assignment of a mission, the T/L should use less than 1/3 of available time to plan and issue the warning/operations order so that the Team has 2/3 of available time to prepare. This same rule should apply up the chain to the FOB.
- Reduce Planning and Preparation time through process design streamlining, checklists and SOPs, with emphasis on simplicity and reduction of bureaucracy, to afford the Team more time to conduct its own planning and preparation activities.
- Habitual and/or dedicated support relationships will streamline planning and preparation.
- Headquarters (HQ) should ensure up-to-date target folders; preprinted forms, supply and equipment availability, proactive planning and preparation, etc. – and should coordinate intensively with supporting units (e.g. aviation) to do the same.
- Supporting units also have time constraints. These constraining factors will affect Team overall mission planning and preparation.
- Make a Time and Distance movement estimate. This estimate may impact LZ/DZ selection, cross-country route selection, and the mission window.
- After pre-mission training, the T/L should be able to make a Time-on-Target (ToT) estimate.
- Obtain or calculate an estimate of fire support, CAS and the Brightlight/Quick Reaction Force (QRF) response times.

Sample SOG Timetable:

SOG FOBs were assigned AOs and SR Teams were repeatedly deployed against a set of Target Areas, which were frequently in the vicinity of major enemy base areas. This allowed the FOB command and staff to streamline processes, cut redundancies and unnecessary steps. For a typical timeline for a 7-day mission, the process would be something like this:

- On Day 1, the T/L would receive a verbal Warning Order (Day 1) delivered by the SR company commander/1st Sergeant/Operations Sergeant. The Warning Order would contain the essentials, to include:
  - Target Area designation and the Lower Left No Bomb Line (LLNBL), defining the corner coordinates for the 6-kilometer square confines of the Target Area.
  - Primary Mission. Supplementary missions (e.g. POW snatch, Ambush, employment of CAS, etc.) were a given/by SOP, unless otherwise directed.
  - Insertion Date and Duration (if the mission varied from the typical seven day duration)
  - Launch Site location.

- ○ Special Coordinating Instructions (if any). Typically issue and insertion of propaganda materials, issue and use of wiretapping equipment, issue of medical kit, issue of communications gear and SOI, etc.
- ○ Offer of a VR. Date/time typically would already be scheduled for the following day; T/L may decline at his discretion.
- ○ Date/time of back brief.
- ○ Post-Warning Order, the T/L would immediately go to the S-2 staff to receive maps, to review the Target folder, to familiarize with other operations to be conducted in the same timeframe, to mark one (T/L's) map with current friendly and enemy intel (e.g. minefields, trail networks, road spurs, estimated enemy locations, communications wire orientation, AA threats, etc.) and any other relevant information that would bear on the mission (e.g. weather, illumination, etc.).
- ○ The T/L would summon US Team Members and issue maps and delegate tasks. The Team Members would assemble and/or cut down map sheets and mark them up from intel depicted on the T/L's map. Immediate tasks assigned would include:
  - ▪ Indigenous Team Members would be informed of a pending mission and would be restricted to the FOB.
  - ▪ A Team Member would obtain a supply checklist from S-4.
  - ▪ The T/L would do a map study to develop his concept of operations/tentative plan, including notional LZs, deceptions to be employed, notional routes, notional weapons and equipment requirements.
  - ▪ T/L consults with any other T/L who conducted an operation in the same Target Area. This may prove crucial, especially if the target folder is not up to date/incomplete.
- • Day 2 – Conduct the VR. After the VR was completed:
  - ○ T/L would firm up and brief the US Team Members on his concept of operations/plan.
  - ○ Complete the supply request, to include an ammunition allocation for range firing and battle drill rehearsals and coordinate supply pick-up.
  - ○ Coordinate with the company/HQ S-3 (training) for range time and for transport to the range.
  - ○ Coordinate for training on specialized equipment.
  - ○ Conduct walk-through battle drill refresher training.
- • Day 3:
  - ○ Pick up training supplies and equipment and distribute.
  - ○ Move to the range and conduct training.
  - ○ Clean weapons and equipment. Restore readiness of tactical gear.
  - ○ US Team Members receive training on special equipment.
- • Day 4:
  - ○ Move to the range and conduct live-fire training and rehearsals, to include night training.
- • Day 5:
  - ○ Upon return to the FOB, clean weapons and equipment. Restore readiness of tactical gear.
  - ○ Amend S-4 mission gear request as required.
  - ○ Prepare back brief.
  - ○ Brief or coordinate with standby Reaction Force (RF), if available.
- • Day 6:
  - ○ Pick up mission gear/supplies, medical kit, propaganda materials, communications gear and Communication Security (COMSEC) and other special items.
  - ○ Conduct commo and special equipment checks.
  - ○ Deliver mission brief back. Briefing would follow a standard Order Format.
  - ○ Pack for mission.
  - ○ Conduct Team inspection.

- Day 7 (Launch date for mission):
  - Attend pilot's briefing. Aviation assets were in a dedicated/habitual relationship with the FOBs; they had their own SOPs and abbreviated planning process and protocols.
  - Conduct final checks.
  - Assemble on FOB helipad.
  - Move to Launch Site.
  - Upon arrival, coordinate with Bright Light T/L or RF commander.
  - Launch.

Civil Considerations Analysis (METT-T<u>C</u>) TTPs:

- Consider the impacts that Rules of Engagement (RoEs) will have on the mission, and on mission planning. Identify target areas inhabited exclusively by hostiles and work with the operations staff to have these exempted from RoEs or established as RoE unlimited 'free fire zones' and/or distinguish other areas that warrant RoE relief.
- Seek maximum flexibility from higher headquarters, to include situations when RoEs can be unrestricted/minimized, to include:
  - Civilians engaged in support of enemy military efforts.
  - Areas where all civilians are deemed to be hostile.
  - Where civilians pose a real threat to the Team.
  - Where civilians perform hostile acts.
  - Where civilians are carrying weapons.
  - Civilian law enforcement and paramilitaries.
  - Where civilians can be taken as prisoners.
- Always assume that civilians in enemy dominated territory will align with the enemy and will report sightings of the Team or signs of Team presence. Use 'catch and release' of civilians to lure the enemy into an ambush, minefield, firetrap or preplanned concentrations of long-range fires.
- When working with guerillas or partisans, it may be advisable to defer to their leadership in handling of civilians during operations. Seek guidance of higher headquarters.
- Civilian prisoners will often be more valuable as prisoners than enemy soldiers. In the military services of some countries, enlisted men (especially draftees and lower ranks) are not taught land navigation and are not provided maps; land navigation/map reading is the province of officers, NCOs and troops within certain specialties. However, civilians will know much about their local area and about the locations and routines of enemy forces located nearby. Civilians are also not trained to resist interrogation, unless they were previously in military service.

Pre-Mission Tactical and Technical Training TTPs:

- Team Members should learn and practice key knots: bowline, slip knot, square knot, prussic knot; timber and clove hitches. These knots can be taught by the Team Engineer(s) as 'filler' training during Field Training Exercises (FTXs) or other field training.
- Team Members should receive familiarization training on enemy or future adversary combat vehicles. Priority of training on vehicle operation/driving should go to Team Engineer(s), as many armored vehicles have controls similar to heavy engineer equipment and priority for the weapons systems training should go to Team Weapons personnel (for similar reasons).
- The Recon company/unit commander should consider creating a video library encompassing a broad regime of TTPs. These can be used to train new SR Team Members or new Teams assigned to the SR operation/mission.

- Have indigenous Team Members teach hazards and precautions associated with flora and fauna of the AO and how to recognize and prepare edible plants. Make a record of such instruction, to include recording of videos, to be used in future training.
- If the Team is to use enemy/foreign uniforms and equipment during the mission, preparation and training must be done in a secluded environment. If necessary, the items should be packaged for concealment and then donned in the training environment. Team Members should not speak of the planned use of enemy/foreign uniforms and equipment outside of the Team environment.
- SR Teams should train in austere field environments, similar to that of their designated operational area, frequently and for several week-long durations. Remember that the enemy may have accustomed himself to living and operating in such conditions for months to years.
- By SOP, the 'tail-gunner' should always eradicate evidence of Team crossing of trails, streams or other danger areas, without being specifically directed to do so. The actions of the Tail-gunner must be ingrained during training.
- SR Teams should implement/tailor TTPs and training to correspond to some of those used by 'Blue Light' or similar units. Several of these practices were pioneered by SOG Teams.
  - Train with other SpecOps units, such as the SAS, to 'cross-fertilize' Lessons-Learned.
  - If the SR Team is to operate in urban or residential areas, the SR Team should train in an 'Hogan's Alley' facility operated by SpecOps organizations and/or Federal law enforcement (or other) agencies, if at all possible.
  - The Son Tay raid was executed by an ad hoc force of Green Berets, many of whom were former SOG veterans who had cycled back to other assignments; ad hoc because 'Military Assistance Command-Vietnam (MACV) [was] hopelessly infiltrated by communist spies.'
  - The 5th SFGA, tasked to form Blue Light, 'drew its inspiration from the Vietnam era MACV-SOG and [its] Bright Light missions.'
  - One component of the 75-member Blue Light was a '24-man element … which also had an intelligence collection mission.' Another element was 'a sniper/observer team.' The duality of intelligence/surveillance capability coupled to the Direct Action capability of Blue Light was found to be necessary.
  - Training. 'Because [the Blue Light] S&K range ran into an impact area, they could get away with things that simply were not done at Army ranges … such as mixing mortars and small arms fire, or frag grenades and smoke grenades. S&K was also unique in that they could conduct 180 degree live-fire exercises….' much like the SOE, OSS and SOG routinely did. Additionally, in 'the shoot house, they would shoot with 25mm BB guns and then transition them to .45s….'
  - The precursor to the isosceles stance, used by the SAS and US SpecOps, was the shooting technique developed by the SOE and taught to OSS operatives.
  - Many Blue Light personnel were former SOG personnel, who brought their TTPs and experience to bear.

> 'A lot of it was on the fly. Target analysis, mission analysis, integration of intelligence and an understanding from the beginning that you had to have a stand-alone intel and analysis capability that can deploy to a crisis site…. I had learned from the SAS. They taught me if I was going to do something very dangerous, then I had better have all my own horses. When your life and those of your people are at stake, you don't want to have to depend on strangers.'

Team Battle Drill and Combat Skill Training TTPs:

'To master the art of war, the tactician must train his mind and body just as a fighter. The tactician trains his mind through an academic study. For a warrior leader, the commander represents the head and the organization represents the body. He should train his organization as a fighter trains his body. Battle drills or immediate reaction drills represent the building blocks of tactics. Like a fighter trains to perfect the punch or kick, the professional warrior trains his organization to perform its drill with the same level of perfection instantaneously. With the battle drills in place, the tactician then spars with an opponent in war games to bring the mind and body together.'

- The standard immediate action drill used by most SOG SR Teams, for a point contact situation (for instance, in a chance encounter), involved a single Team Member firing a magazine of ammunition at the enemy and then peeling off to the rear of the Team formation; the next Team Member in the formation would follow suit, and so on, until the T/L signaled a withdrawal in Team movement order. In the Author's view, this was a seriously flawed drill tactic, even though it generally worked to break contact.
  - While the tactic makes some sense if the Team is negotiating very close vegetation and difficult terrain features, the Team has no opportunity to mass fires on an enemy who may be numerically superior.
  - If any Team Member becomes a casualty, an already chaotic situation can become much more complicated. This can mean a cascade of problems that could mean the annihilation of the Team – especially if the enemy is experienced and aggressive.
  - Author's Recommendations:
    - If the SR Team is comprised of twelve personnel (e.g. an SFOD), consider using three fire-teams of four Team Members each.
    - In a point contact, four Team Members or the lead fire-team may mass the fires of four weapons; this is much more effective in inflicting casualties on the enemy and in suppressing his response. See Figure 9 for deployment scheme.
    - There are few circumstances where four Team Members cannot bring their weapons to bear, even in very narrow corridors, within a couple of steps.

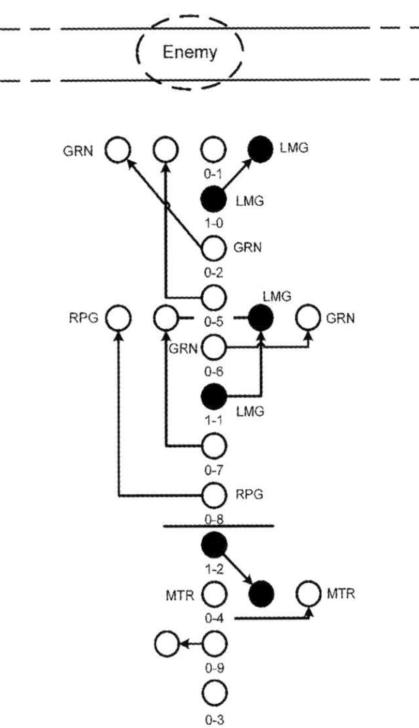

Figure 9. Battle Drill – Immediate Action (Deploying)

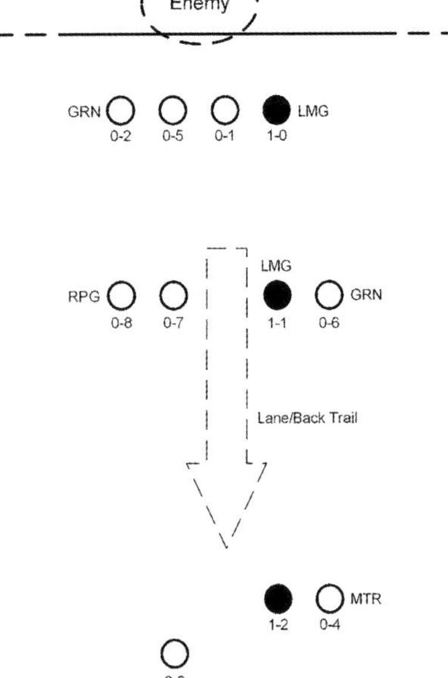

Figure 10. Battle Drill – Immediate Action (Deployed). Note withdrawal lane.

- If a Team Member becomes a casualty, two Team Members can retrieve and evacuate the casualty, while the remaining Team Member(s) covers the withdrawal.
- Meanwhile, the second fire-team has moved/maneuvered to cover the withdrawal of the first fire-team – with four weapons rather than one – allowing the first element to pass through to the rear of the second element. See Figure 10 for withdrawal scheme.
- If the third fire-team is a crew-served weapons element (e.g. with a mortar crew), they can move to a position to bring fire onto the enemy position; if it is not a crew-served weapons element, it can perform the same covering maneuver of the first two fire-teams. See Figure 11 for breaking contact scheme.
- A three fire-team unit has substantial flexibility to maneuver its elements against the enemy (e.g. to flank the enemy) at the T/L's discretion.

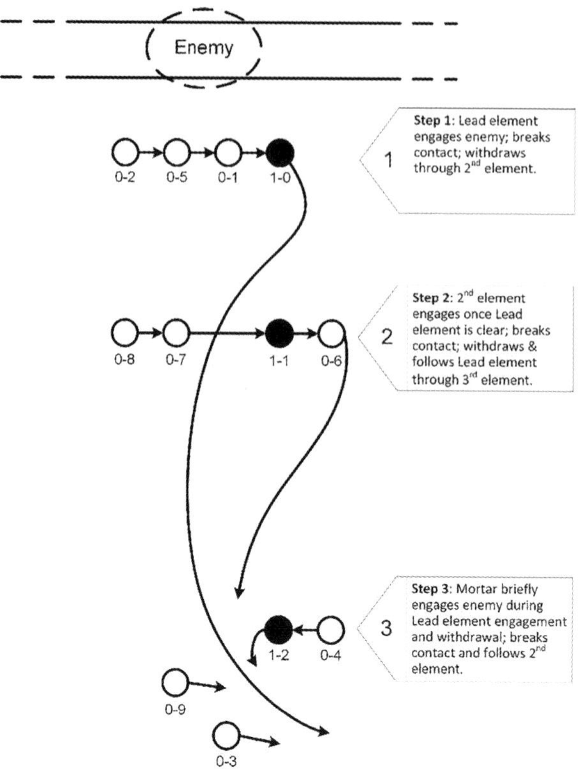

Figure 11. Battle Drill Immediate Action (Breaking Contact).

- This arrangement can also be used for a nine-man or six-man Team (SFOD split Team), with fire-teams of three or two men per respectively.
- A SR Team is almost always heavily outnumbered when conducting operations in the assigned Target Area. During a firefight, Team survival is often dependent <u>on reacting faster, upon contact, than the enemy</u> and on the ability of the Team <u>to mass and maintain its fires</u> until the Team can break contact or maneuver.
  - To speed reaction, the Team must routinely and intensively train on Battle Drills. This training should occur in, and be adapted to, all environmental and terrain conditions (light and dark; hot and cold; swamp; desert sand and winter snow, etc.), and should be exercised from various Team formations. <u>Speed of execution is very important and must be emphasized in realistic Team</u> training.
  - Team Members must train to achieve expert hand-dexterity in rapidly reloading individual and crew-served weapons; correspondingly, Team Members should become adept at rapidly changing magazines (optimally within 2 to 3 seconds) of primary and secondary weapons. A weapon (i.e. the M4 carbine) that possesses a magazine well, should enhance the speed of magazine exchange. Manual dexterity will be affected by cold weather and/or gloves and the stress of combat.
  - Ideally, while magazines or crew-served weapon ammunition are being swapped, the Team Member should be changing locations to avoid counter fire. Again, this training should occur in and be adapted to all environmental conditions.
  - <u>Work on increasing speed of execution and decreasing Time-on-Target in all Battle Drills and tactical training. The goal is to execute Battle Drills and other maneuvers faster than the enemy can execute his, e.g. beating the enemy to the flank.</u>

○ <u>Practice rapid deployment into a hasty ambush and the rapid deployment of a Claymore or booby-trap device while simulating reaction to pursuit.</u>

> <u>True Account:</u> A SOG T/L, on the first day after insertion into a target area in Laos, found a coaxial cable communication line crossing a ridge, indicating the presence of an enemy headquarters or major communications node at both ends of the line. Early on the second day, the Team traveled downhill (toward a river at the border with Vietnam), paralleling the line and encountered an ideal, isolated ambush location along a high-speed trail at the opening of a ravine near the base of the hill and close to the suspected communications node. The T/L left the bulk of the Team in its traveling formation (file) and moved forward 15–20 meters with two indigenous commandos to scout the ambush location; he spotted some bamboo shoots that had been dropped on the trail and paused to take photos of the trail and the bamboo shoots. As the T/L was taking photos, voices were heard, indicating at least four enemy approaching along the trail. The T/L saw this as an opportunity to take prisoners, but his commandos were unprepared and not properly deployed for such an action; they hastily withdrew back to the Team formation, leaving the T/L behind and making noise in the process of their withdrawal. The approaching enemy immediately stopped conversing, indicating that they had heard the noise, but they continued to advance along the trail. The T/L rejoined his Team just as four enemy soldiers crossed the Team's point. The Team opened fire from a dispersed file formation, killing all four enemy soldiers. <u>What went wrong:</u> The scouting element would likely have been able to pull-off a POW snatch on the trail, had the Team Members been better trained for a hasty take-down of an armed combatant, but to be fair, the T/L could not know how many enemy troops were approaching in this instance. Had the T/L deployed his Team into a hasty ambush, which the Team had been well trained to do, before moving forward to scout the trail, the enemy would likely not have detected the Team and a POW snatch would likely have succeeded. Because the Team initiated the ambush from a file formation, firing toward the point, one indigenous commando was struck in the chest by a 40mm grenade fragment (friendly fire) penetrating one of his lungs, thereby requiring the Team's extraction and denying the opportunity to continue the mission or following the coaxial cable.

○ Incorporate Tactics, Techniques and Tradecraft, as much as possible into unit SOPs and Battle Drills; then train on and rehearse SOPs and drills until they become second nature.

○ Standardize and 'routinize' such procedures as moving into a NDP position, MSS, ambush position, etc., requiring only a bare minimum of specific hand and arm signals from the T/L.

○ Standardize deploying the Team into a hasty ambush formation so that this drill requires no more than simple hand and arm signals from the T/L.

○ Simplify Battle Drills so that formations and actions are similar to/variations of one another.

○ If the unit will operate in split Team configuration, battle drills must be modified and additional training must be conducted.

○ Note that in actual combat, adrenaline surge will cause a number of physiological effects, to include tachypsyschia. In this condition, actions (e.g. during a firefight) appear to go into slow motion; additionally, the Team Member will experience some loss of motor skills. It is difficult to induce tachypsyschia in training; but it can be approximated if stressors are introduced. Stressors include: time pressure, surprise, noise/sound, test or competitive environment, exertion, etc.

- ○ Team Members should also be cross-trained (at least familiarity) on all weapons carried by other Team Members during an operation.
- ○ The Team weapons specialists should train Team Members on enemy weapons in current use within the Area of Operations, to include familiarization with enemy heavy and crew served weapons that may be present in the AO.

Author's Solution:

Team Members can improve their ability to maintain presence of mind while under stress by careful preparation, by training and rehearsals under stressful conditions – and by pro-actively anticipating situations. Team Members should constantly devise contingency plans or counteractions to problems they might encounter while conducting their mission, and incorporate these eventualities in training. Use 'virtual' or immersive mental processing to mentally devise, test and rehearse these actions and to perform 'what-if' analyses. This means that Team Members should mentally rehearse the execution of actions, counter-actions and other scenarios. For example, if an operation is to be conducted in or near built-up areas, use Google 'Street View' or similar information technology tools to assist the virtual rehearsal during the preparation phase. A 'Street View' mapping application, occasionally available for specific locations in urban settings, provides 360-degree, panoramic, and street-level imagery where team members can 'walk the terrain' during mission preparation.

In other situations, where information technology aids are not applicable or not available, Team Members must rely on maps, photographic imagery or sand-table terrain models. These tools or aids provide Team Members an overhead perspective, so they must use these aids to imagine, from a ground-level perspective, the landform, terrain features and danger areas that they will encounter upon insertion, along the planned route-of-march, and in the vicinity of the target. This mental mapping not only assists in land navigation during the mission, but can promote rapid decision-making in the heat of combat.

Mental rehearsals of actions and reactions to combat and other scenarios can instruct Team Members in what to do and how to do it when confronted with a tactical situation, e.g. what do we do if we are taken under indirect fire when we are crossing a particular danger area.

- ○ <u>Team work requires repetitive practice, rehearsals and training.</u> The best training injects realism into all levels and phases of the training experience, and Special Forces formal training, including Special Forces Qualification Training may often tend to under-emphasize situational realism. Without T/L emphasis, unit or team-level training may often fall short in training realism.
- ○ The Author strongly recommends the use of Situational Training Exercises (STXs) using a cadre of experienced and/or Subject Matter Expert personnel. During the STX, Team Members are permitted to ask TTP and Lessons-Learned questions of the cadre, and receive on-the-spot corrections and assistance with periodic/daily AARs.
- ○ To enhance subsequent team-level training, conduct training under conditions of evaluation or where experienced 'Red Team' participants are involved to inject the pressure of stress. Additionally, Team evaluators should be Subject Matter Experts or veterans who can identify and correct flawed performance on the spot.

The Malayan Scouts (22nd SAS) applied 'force on force' training down to an individual level during the Malayan Emergency. 'The concept of individualism was advanced by pitting one man against another to increase the efficiency of both. This 'hunter/finder' game … was routine as a nonlethal duel, a method of nurturing mutual regard and preparing men with jungle warfare skills.'

- ○ Practice and react to taking at least one casualty during Battle Drills or tactical training situations – to include instances where the T/L becomes a casualty. This realism will exercise the Team in a likely contingency and also may identify deficiencies during integrated chain-of-command, leadership and functional responsibility transition; in medical treatment, casualty reporting, leadership decision-making; and in battlefield tactics and techniques while under the pressure of stress.
  - ○ Practice engaging an enemy while the Team is in awkward or vulnerable positions, such as ascending from a ravine to a ridge top, descending from a ridge into a ravine and while crossing various danger areas.
  - ○ Elevate the realism of Team training with Red Team–Blue Team engagements, especially where Red Team personnel are competent in enemy formations and tactics. Equip the Red and Blue Team members with the Multiple Integrated Laser Engagement System (MILES) or equivalent, with area weapon add-ons (where available). Where MILES equipment is not available, consider using 'Paint-ball' engagements while practicing Battle Drills. Where Red Team assets are not available, have the Team use simulated force-on-force engagements using 'combat theater' and combat 'arcade' simulators.
- If ground mobility equipment is to be used on the operation, modified (mounted) battle drills must be established, and thorough training conducted. In the absence of aviation asset availability, Ground Mobility Equipment may be the Team's only salvation. Subsequently, the Team Members must be trained in operator level maintenance and service, e.g. Preventative Maintenance Checks and Services (PMCS) and emergency repairs of vehicles (and major components) in the field.
- Ultimately, culminate a training cycle with Joint/combined arms participation and/or live-fire Battle Drills where possible.

The normal or planned rotation for a SOG Recon Team was: Mission Execution – 1 week; Stand-Down – 1 week; Duty (e.g. training, manning security positions, conducting FOB patrolling and mounting night ambush positions) – 1 week; Mission Preparation – 1 week. During periods of high OPTEMPO, the Duty phase would occasionally be eliminated. Teams that were reconstituting from battlefield losses or Teams with new personnel might have an extended duty cycle to allow for Team training. During the Mission Preparation phase, the T/L would receive the Warning Order, review the Target Folder, conduct an aerial Visual Reconnaissance (VR), develop his concept of operations and conduct operational planning based on METT-TC and other considerations. The T/L would prescribe and obtain any special equipment and supplies necessary for mission training and execution. Throughout the Mission Preparation phase (and during the Duty phase, if possible), the T/L would lead training and Battle Drill rehearsals. A competent T/L would almost always conduct Battle Drill walk-throughs, at the FOB; conduct at least two days on the range (in daylight) consisting of marksmanship, Battle Drill walk-throughs and a substantial period performing live-fire Battle Drills. During the Mission Execution phase, the T/L would record battlefield Lessons-Learned during the course of the mission and would subsequently adjust Team training (in the next Duty and/or Mission Preparation cycles) based on these insights. Perhaps once every other rotation, the competent T/L would extend daylight range training to conduct night-time marksmanship, movement and live-fire Battle Drills. At least 50 per cent of Battle Drill training was conducted live fire.

Note: SOG SR Team Members carried/handled or operated fully loaded individual weapons routinely several hours each day (exceptions during leave, Rest and Recuperation, Stand-down). Individual weapons were always kept loaded and close by. Accidental discharges were rare.

'Training of the Malayan Scouts (22nd SAS) during the Malayan Emergency 'ensured that troopers had the ability to track, move secretly and silently, and react immediately. Training included grenade practice, immediate action drills, and live ammunition practice that sometimes disregarded the normal safety rules for field firing ranges…. Troopers also learned the use of explosives, setting booby-traps, and communications. Since there are many rivers in Malaya, boating was one of the most useful skills taught….

'In all, the purpose of the training was to make every man adept at surviving in jungle warfare, quick to act and react, and capable of getting a shot off a split second earlier than an opponent. All these skills were instilled and mastered through repetition. A squadron's cycle was two months in the jungle, two weeks' leave, two weeks' retraining, and back to the jungle.'

Checklist TTPs:

- Remember that it is often the little things that get you killed. Each T/L should have pre-mission and post-mission checklists to ensure no detail is forgotten. Prepare checklists in advance, as preprinted blank forms, prior to mission assignment or field training.
  - Pre-mission checklists should include preprinted supply lists, FRAGO formats, support request formats. Additionally, use a check list to inspect your troops the day prior to departure for the Launch Site.
  - Post-mission checklists should include supply requests for items consumed during the operation and for turn-in of COMSEC materials, special equipment, etc.
- If ground mobility equipment is to be used on an operation, ensure vehicle loading plans and checklists are prepared, to include:
  - Internal and External Loading Plans and Diagrams for air insertion, including air drop, air landing and sling load – for the variety of support aircraft that may be used.
  - Load plans for ground mobility equipment and trailers.

Aerial Visual Reconnaissance (VR) TTPs:

- A VR should be taken whenever possible – except for deep penetrations or areas where high AA threats exist. A VR may be the only way to obtain a low level slant view inside tree lines to identify high-speed trails, structures or other indications of enemy presence. If a VR is not feasible, then the Team must rely on satellite and/or UAV imagery. Future drone capabilities that might be released by a 'mother' aircraft (for instance) may offer both safe and high fidelity examination of LZs/DZs, prospective routes and prospective targets.
- Before conducting a VR, consult with the S3 (Air) and/or S2 to identify AA sites and AA activity in the AO, especially where enemy AA capability may affect the mission and the VR.
- When making an aerial VR, always identify every useable extraction LZ in the immediate vicinity of the Target and along the prospective routes from insertion LZs to the target. Plan the Team's route-of-march so that you will always know roughly how far and in what general direction the nearest LZ is located.
- VR of prospective LZs may sometimes reveal indicators of enemy and civilian activity in the immediate vicinity; but a VR, if improperly conducted, may also tip off the enemy to an imminent operation.

True Account: An experienced FOB-2 T/L decided to shun the offer of a VR for a pending mission. Upon insertion into an LZ that he selected from a map study, his Team was subjected to several ground assaults by a numerically superior enemy force launched from the periphery of the LZ. The assaults were intended to close with the Team as a defense against Tactical

Air (TACAIR) strikes, so the T/L employed CAS very close to the Team perimeter. The enemy employed bugle calls to signal coordinated assault maneuvers, which suggested the presence of an enemy regiment. The Team was fortunate indeed that the enemy assaults were mounted immediately after insertion, while USAF FAC, CAS and helicopter gunships were still loitering on station. The assaults and AA fires continued all day until the enemy broke contact. The T/L received minor wounds, but several of his indigenous troops had more serious wounds. The T/L declared that he would never decline a VR ever again.

- Identify prospective insertion LZs from a map study and examinations of aerial photography. Then plan the VR, finalizing the plan with the pilot the day of the flight. One highly experienced SOG T/L had a rigorous VR procedure; he instructed the pilot to:
  - Fly out to the Target Area indirectly and using any cloud cover available. Enemy air guards/observers would almost always look in one direction, expecting that VR, insertion/extraction and support assets would fly directly from the FOB or launch site. And some AA radars would also be oriented in the same direction, with their line of sight blocked in other directions by protective terrain features.
  - Approach the target area at a relatively high altitude, consistent with visibility; the pilot and T/L must be able to clearly see the ground. This allows the T/L to gain an aerial perspective of the target area in relation to the AO. If also allows the T/L to orient himself and his map. Once over the target area at this altitude, fly the boundaries of the 6-kilometer target area box, again to orient the T/L.
  - Fly out of the target area, descend to be a lower altitude that allows the T/L to see the complete target area box, and then fly a pass that follows one of the boundaries. This pass should begin several kilometers from the target box and end several kilometers after exiting the target box. Beginning and exiting the pass that far out from the target area box was to prevent the enemy from detecting the area of interest. Repeat for all remaining boundaries. This step further allows the T/L to orient and to take photos; at this altitude, enemy activity may be detected; and it allows the pilot to identify terrain markers for his next series of passes.
  - The pilot descends to nap-of-the earth altitude and will make passes slightly offset to the prospective primary and alternate insertion LZs, the prospective route of march and over prospective extraction LZs in the vicinity of the target. As in the above step, the pass commences and ends several kilometers from the target area box, and need not be straight azimuth passes. The pilot will alert the T/L as to which LZ/point of interest is in the approach, from which side of the aircraft he should see the objective LZ, and will give him a 'mark' as the aircraft approaches the objective LZ. The T/L should peer into the periphery of the LZ to detect enemy activity, trails, structures that may be inside the tree lines. Limit the number of passes (normally two) to reasonably determine if the LZ tree line is clear. If any enemy activity is detected, mark the location for air attack, and go to the next prospective LZ. Once the VR mission is concluded, and assuming that the VR aircraft is equipped with White Phosphorus (WP) rockets, the pilot may call in air support to attack the enemy locations detected during the VR. This may convince the enemy that the sortie was an interdiction mission and not a VR.
- Author's Recommendation: For OPSEC reasons, use only US pilots on VRs. Are any local national or host nation military pilots actually enemy agents? It could happen.
- The regular LBE is ill-suited for VRs aboard a fixed wing aircraft, where the T/L/Member must be seated on an aircraft seat. The T/L should have a second set of LBE that allows upright sitting.
- If the T/L is susceptible to motion sickness, or has an upset stomach prior to the flight, he should carry a plastic bag.

- Do not fly a VR directly over the prospective LZs or other points of interest/prospective routes within the target area. Fly offset to the LZ/points of interest, so that you can observe into the verge and possibly spot the presence of the enemy or of high-speed trails. Bring binoculars and a camera (or even a video camera); ensure the cameras are set for high speed to avoid blurring the images. The FAC should fly an extended, and if possible, non-linear route, commencing 2 or more kilometers before the LZ and two or more kilometers beyond the LZ, so that the enemy will find it difficult to identify the specific area of interest, target or prospective LZ.
- If the T/L has taken any film/photos during the VR, he should study them carefully upon return to the FOB.

Chapter 3

Employment/Execution

Pre-Launch TTPs:

- Conduct inspections on all Team Members, prior to departure for the Launch Site. Inspect each Team Member's uniform and equipment, especially radios (perform commo checks), protective masks and battery powered items (e.g. strobe lights), and check all Team Members' pockets prior to departing home-base for passes, ID cards, lighters with insignias, rings with insignias, etc. Additionally, check canteens prior to departure.
- The T/L should attend the daily Operations/Pilots' Briefing prior to the operation. Daily information on enemy anti-aircraft locations, weather, aviation assets, schedule of events and other information is delivered at that time.

<u>True Account:</u> An experienced SOG Reconnaissance T/L had a vacancy on the Team for an indigenous Team Member. Word of the vacancy circulated and the T/L was approached by an American Senior NCO, a member of one of the FOB's Exploitation Companies, who strongly recommended a veteran indigenous commando from his unit who wanted the additional pay and prestige that came with the reconnaissance mission. The T/L interviewed the candidate and found that he had very good English-speaking skills, better than those of the Team's designated interpreter. After the candidate left the interview, the T/L summoned the indigenous point man and the interpreter and inquired of them what they knew of the candidate and what they thought of him as a possible addition to the Team. The indigenous Team Members would not recommend the candidate, but would not state their reasons. The T/L believed that the interpreter may have felt threatened by a candidate who had better language skills than his own, so the T/L hired the candidate. Two weeks later the Team was inserted into a very hot Target Area on a high priority road-watch mission along an enemy Major Supply Route. On the morning of the second day of the operation, the Team discovered a large, flat open area beneath the shelter of continuous canopy; an unoccupied, perfectly maintained major vehicle park, complete with guard sheds and a large, recently used latrine facility. Despite this plum find and prospective ambush site, the Team was ordered to continue with the planned road-watch mission. The T/L decided that he would extend his mission in order to return to the vehicle park after completing the road-watch mission. The Team proceeded toward the MSR, but within an hour the new indigenous Team Member started quaking and sweating and complained of blurred vision. The T/L assumed that the new Team Member was suffering from malaria. Other Team Members had to carry his LBE, pack and weapon as the Team continued toward its objective. Due to the noise made by the ill commando, the Team had to settle for a listening watch that evening. The following day, an enemy tracker unit caught up with the Team at its hide location. The ensuing firefight caused an early extraction for the Team. Upon return to the FOB, the ill Team Member was taken to the dispensary for medical care. Later that day, the dispensary reported that the indigenous Team Member was suffering from

delirium tremens – and not malaria. The T/L, who had not checked Team canteens prior to the mission, discovered that the alcoholic Team Member had his filled with a strong indigenous alcoholic beverage; when he exhausted his supply, he became ill. The Exploitation Company Senior NCO knew that the commando was an alcoholic, but rather than fire him, he assisted in having him reassigned to the Reconnaissance Team. The SR T/L did not hesitate, and fired him immediately.

- The T/L should keep abreast of information at the Launch Site and should coordinate the SR Team's operational information with any Bright Light T/L who will be on duty during the duration of the mission.
- If the Team is to use enemy/foreign uniforms and equipment during the mission, the items should be packaged for concealment and then donned after reaching the Launch Site (OPSEC). Team Members should remain out of sight at the Launch Site once the items are donned. If the enemy has a particular manner of appearance (e.g. haircut, beard, skin coloration), Team Members may have to adopt a similar appearance.

Launch Site TTPs:

- The Launch Site Operations (LSO) Officer generally returns to the FOB after all daily insertions and extractions have been accomplished, assuming no emergency situations are in play; each morning he participates in the daily Operations/Pilots' Briefing at the FOB and will then displace to the Launch Site on mission aircraft, weather permitting. The LSO Officer is normally required to be present at the Launch Site prior to the launch of an SR Team, Exploitation Force, Bright Light Team (BLT) or Reaction Force (RF) operation. The LSO should have prior SR or RF experience, as this will help him anticipate Team/RF needs.
- LSO Officer Responsibilities at the Launch Site:
  - Every day, the LSO Officer should bring a map overlay for the operations shed map board, containing information on all SR Teams/Exploitation Forces on the ground, and those pending infiltration. Overlay information must include the most current information on SR Team/Exploitation Force locations and status.
  - The LSO Officer must also update the operations status board, which will include updated communications frequencies and call signs for all current and pending operations for the day.
  - The LSO Officer will also be the custodian of a spare radio and additional SOI (one-time pads, etc) or COMSEC devices that he may issue as required for operations at the Launch Site or by the BLT.
  - The LSO Officer will also bring complete map sets for all current operations and for those pending for the day in anticipation of on-call BLT or RF operations.
  - The LSO Officer will ensure that a sufficient quantity of administrative supplies and consumables (e.g. radio batteries, water, POL) is maintained at the Launch Site.
  - The LSO Officer ascertains the aircraft status after consultation with the flight leaders on site and determines sufficiency/readiness for planned and contingency operations. He coordinates resolution of any issues with the FOB G3 (Air)
  - The LSO Officer keeps all flight leaders, SR T/Ls and the BLT Leader apprised of emerging issues, as they affect current and pending operations, as they occur.
- As the stand-by BLT or RF cannot predict the requirements of a contingency insertion, the Launch Site should store an assortment of gear and munitions that a rescue might require. These may include: ropes, ladders, stretchers, body bags, extraction rigs, karabiners, claymore mines; grenades (fragmentation, White Phosphorus, colored smoke; CS); shoulder-fired rockets; mortar rounds (variety); demolition charges (including blasting caps, fusing, etc.).

Figure 12. Ladder Rigging on UD-1D Helicopter (Bosewell).

- Other items may be staged at the Launch Site for specific missions and resupply. Storage containers and/or additional revetments may be required. These items may include: other munitions (e.g. shaped and cratering charges); radio and equipment batteries; spare radios; pre-planned resupply packages. Munitions storage compatibility and Quantity-Distance (Q-D) precautions would be waived, but storage design should emphasize safety.

- A static Launch Site compound is an exclusion area which normally contains, at a minimum, an operations and communication shed, an antenna array, a shelter for the BLT and for SR Teams that have been extracted or that are pending insertion, ordnance magazines or ammunition bunkers (one for aviation gunships and one for Team or Rapid Response Force needs), a generator shed, defensive positions, a latrine and refueling bladders. All these facilities should be sandbagged or protected by revetments. The compound should be further secured with barbed wire and posted with warning signs. If the BLT deploys on a rescue or recovery operation, the FOB may backfill the BLT with another SR Team.

- While a BLT rescue/recovery operation is underway, all other insertion/extraction and support operations are suspended, except for SR Teams in contact. All Teams will be notified of the suspension and of impacts on SR Team support.

- A damaged aircraft might return from an operation and crash upon landing at the Launch Site. Additionally, the aircraft might land, but the pilot(s) could be wounded. All Launch Site personnel, including the LSO and US BLT Members must know how to shut down the aircraft engine and

Figure 13. DakTo SOG Launch Site (1970). Note Control Tower (Center); to its left, Team Quarters; to far left, commo/command building, generator bunker. To right of Tower: Ammo Bunker, Storage Conex, Defensive Bunkers and Latrines. (*Buckland*)

other aircraft systems/subsystems and they should be able to assist aircraft (mission capable) crew in re-arming weapon systems. Launch Site personnel must also be able to rig extraction helicopters for string/fast-rope and ladder insertion and for string or ladder extraction.

- The Launch Site should also have adequate fire extinguishers to fight an aircraft electrical fire and equipment fires within the exclusion area.

Bright Light (BLT)/Reaction Force (RF) TTPs:

- SR Teams may be periodically rotated to a Launch Site to serve as the BLT. The BLT is generally a small RF that is temporarily posted (normally a week) to the Launch Site. FM 31-20-5 does not discuss the uses or employment TTPs of a BLT or RF or of TTP's associated with operations at the Launch Site.
- Upon receipt of a Warning Order to deploy to a Launch Site for BLT duty:
  - Train on insertion and extraction TTPs, to include Fast-Rope, Rappelling and ladder use.
  - Confer with S-3 on planned operations for the duration of BLT duty. And conduct Team planning and preparation accordingly.
- BLT versus USAF Combat Search and Rescue (CSAR). USAF Pararescue personnel under AFSOC control and assigned to USAF Special Operations/Special Tactics units often operate in partnership with other Service SpecOps units and they have some functions in common with BLTs. The chief differences:
  - BLTs possess far greater ground combat capability than the Pararescue personnel. With the exception of medical personnel, skill sets and employment, TTPs of the two are far different. See BLT responsibilities below.
  - Bright Light Teams are on standby to launch on an <u>immediate</u> basis from launch sites to a target area. CSAR elements would typically have a longer response time; they are not likely to be found deployed to FOB launch sites.
  - CSAR elements will have specialized aircraft, equipment and capabilities. Bright Light Teams will use more austere SR mission aircraft.
  - CSAR elements will typically not be deployed to reinforce a Team, secure an LZ/DZ, recover KIAs, assist in security, logistics and launch activities, or conduct a BDA or other SR mission.
  - CSAR is normally limited to rescue of relatively few Wounded in Action (WIA), whereas the BLT may be inserted to rescue an entire SR Team, plus air crew personnel.
- BLT responsibilities include:
  - Continuous security and upkeep of the Launch Site. Operator level maintenance of Launch Site equipment.
  - Mission training and preparation.
  - Monitoring operational communications in the absence of the LSO Officer.
  - Assist LSO Officer in the conduct of infiltration and/or exfiltration operations. This may include assisting air crews in rigging aircraft and/or assisting in rearming of gunships.
  - Support of resupply operations and/or mission aircraft rearmament, as required.
  - Deployment on short duration operations to include: Bomb Damage Assessments (BDAs), SR Team/Team Member/Air Crew rescue and/or recovery. A BLT may be inserted on missions to, on call:
    - Rescue or recover downed aircrews.
    - Rescue and/or reinforce an SR Team or Team Members under emergency conditions, or to recover SR Team casualties.
    - Secure an LZ/DZ for an inserting Reaction Force/Rapid Response Force.
    - Act in concert with a Reaction Force/Rapid Response Force.
    - Conduct Bomb Damage Assessments (BDAs).

- Assist in resupply operations (e.g. bundle kickers)
- Other specialized or opportunistic missions, as assigned – on call (e.g. wiretap of communications wire discovered by an SR Team, where the SR Team is not trained or equipped to conduct the tap).

• The Launch Site is not immune to enemy attack. The BLT will maintain an Air Guard, in AOs where the US or Allied Forces do not have Air Superiority, at the Launch Site location.

• The BLT should spend some time improving the defenses and conducting maintenance of the Launch Site; but the BLT will otherwise have time on its hands; subsequently, the BLT Leader should take this slack-time opportunity to continue the training of his personnel, with emphasis on those tasks associated with Bright Light mission responsibilities. Include the Chase Medic in the training.

• The Bright Light Team must tailor its organization and equipment for rescue and recovery operations. This might include:

  ○ Two days' food.
  ○ Go 'Heavy' with a complete Team and additional armament and ammunition, as required.
  ○ Bring body bags.
  ○ Chainsaw and fuel to clear an extraction LZ.
  ○ Aircraft rescue equipment. For example, SR Team or aircrew personnel may be pinned inside the aircraft. Cutting and prying equipment/tools may be necessary to extract pinned personnel.
  ○ Aircraft transporting SR Team Members (and supporting air crew personnel) may have crashed on mountainsides or into canopy. BLT Members must be trained and equipped (e.g. with climbing gear) to rescue/recover personnel in these conditions.

• BLT Rescue/Recovery Operation Tips:

  ○ Scavenger birds and some animals are attracted to carrion/carcasses.
  ○ If no LZs are near to the Team location, the BLT Leader must create one. This may require that CAS aircraft drop a 500lb bomb nearby to clear a string/fast-rope LZ.
  ○ High grass LZs. See insertion TTPs below.

• BLT Members should be constantly ready to deploy on a rescue and recovery operation; therefore, the BLT Leader must tightly control his Team Members, and generally should not allow BLT Members to wander away from the Launch Site. The same may be said of the RF.

• If the BLT is tasked for a rescue or recovery operation or to conduct a BDA, the BLT or RF may be inserted into situations of extreme peril. For instance: an SR Team or a downed aircraft crew may be in contact with, or surrounded by, a numerically superior enemy force and will likely have casualties, rendering them immobile. If the SR Team or aircrew is surrounded, a heliborne BLT or RF operation may be anticipated by the enemy; the BLT may be inserted nearby to call in Close Air Support and interdict enemy forces along routes of approach to the defensive positions occupied by the SR Team/aircrew. If the SR Team is in danger of being overrun, the BLT may even land on/rappel into the SR Team/aircrew position to supplement their firepower and render medical care to wounded personnel.

• When the LSO Officer arrives at the Launch Site, the BLT Leader should confer with him on the operations slated for that day and on the status of Teams that are currently on the ground.

• The BLT Leader should also confer directly with SR T/Ls and/or an Exploitation Force/ RF Commander who arrive at the Launch Site pending launch. The SR T/L or Force Commander should provide a summarized briefing on the pending operation. This information may be crucial to the BLT Leader should he be called upon to deploy.

- Additionally, if an SR Team is extracted but Team Members (MIA, WIA, KIA) remain on the ground, the BLT Leader <u>must</u> immediately interview surviving members of the extracted SR Team Members, before they are transported back to the FOB. The BLT Leader must glean all relevant information pending the BLT mission insertion. Some surviving SR Team Members may be willing to accompany the BLT on the rescue or recovery operation. As these personnel have not been trained by the BLT Leader, only one or two surviving SR Team Members should be selected to accompany the BLT.
- The BLT will generally not be inserted in the waning hours of daylight or at night, but will typically launch as soon as possible the following morning. The BLT cannot expect effective CAS support at night, as aviation assets will fear fratricide of SR Team survivors. The exception to this is where SR Team survivors have been able to move to an LZ and clearly identify/mark themselves.
- The BLT Leader and the Assistant T/L should monitor the Launch Site radio, particularly during periods when the LSO Officer is absent from the communications shed/Launch Site, to stay abreast of developing situations on the ground and to stand-by for deployment alerts/warning orders from higher headquarters.

Infiltration/Exfiltration TTPs:

The method of infiltration or extraction depends upon METT-TC, means available (systems with the appropriate capabilities), weather and terrain, depth of penetration, and ever critical target area characteristics. One might think that the most desirable method for an SR Team infiltration/extraction would be one that reduces the possibility of detection. The argument would typically be that security and secrecy of movement should not be sacrificed for expediency. This is a simplistic view. The Team must <u>find a way</u> to maintain the advantage of security and secrecy of movement regardless of the insertion delivery method – this may require deception techniques, decoys, etc. Methods of Team infiltration include: stay-behind; fixed-wing insertion; heliborne insertion, to include fast-roping or rappelling; parachute delivery; water-borne insertion, and ground vehicle or foot infiltration. Exfiltration methods are obviously more limited. And in deep penetrations, the options are even more winnowed.

- <u>Stay Behind</u>. This method is normally employed during major retrograde operations of friendly forces. When properly executed, it is less risky to the Team than other means that require penetration of enemy airspace. In addition, supplies and special equipment can be stocked in caches or a MSS to provide for an extended operation with austere or non existent resupply options. The Team may also be able to glean supplies from abandoned friendly stockpiles. The Team may have access to abandoned ground mobility equipment, weapons, ordnance and other supplies. Team exfiltration methods are various.
- <u>Fixed-Wing Insertion</u>. This method of infiltration normally requires aircraft with short take-off and landing capabilities and a prepared landing strip, unless the delivery aircraft has Vertical Take-Off and Landing capability, which may require a large LZ. Mid-sized aircraft may have substantial cargo capacity and be capable of deploying Team vehicular ground mobility equipment and substantial amounts of supplies with the Team that can then be cached. This method would be especially preferred if the Team is to be infiltrated into a UW environment to operate out of, or deploy from, a guerilla or partisan base, using the base as a covert AOB. Use of this approach may be more easily detected than most other methods and may carry substantial risk if the enemy is operating anti-aircraft radars and weapon systems or possesses air superiority; the deeper the penetration, the higher the risk. Nap-of-the-earth flying is essential to a successful insertion. One overlooked fixed-wing insertion option is the 'flying boat' variant of seaplanes; some of these aircraft

are capable of delivering a complete SR Team, supplemental supplies and inflatable boats. This insertion method can land on lakes and rivers during good weather and smooth water surface conditions.

- <u>Heliborne Insertion</u>. This is normally the most desirable method of insertion for short to medium range penetrations; extended range may be achieved using a covert AOB that possesses re-arming and refueling capabilities. This approach bears some of the risks associated with fixed wing insertions. Heliborne insertion is rapid, responsive and flexible; depending on the type of aircraft used, it can deposit a Team with ground-mobility equipment and logistics stores in locations where fixed-wing assets have little or no operational prospects. As in fixed-wing deliveries, stealth is problematic due to aircraft signature. Nap-of-the-earth flying, night operations and deception techniques are requisite to successful insertion. The rotor size of the UH-60 requires a circular landing space of 50M and cargo helicopters substantially more. This infers that suitably-sized LZs, in many environments, will be limited, requiring SR Team Member insertion/extraction via rope/ladder and perhaps external cargo loads. Assume availability of 'stealth' helicopters to be limited.
- <u>Parachute Infiltration</u>:
  - On the positive side, parachute delivery is rapid, supportive of deep penetration operations and capable of delivering cargo along with the Team. It is appropriate to the UW or partisan environment where reasonably clear Drop Zones (DZs) are secure and available. As in other aviation based infiltration methods, some degree of secrecy may be obtained through night operations, nap-of-the-earth flying and deception techniques.
  - On the negative side, delivery aircraft become particularly detectable (e.g. to radar) and vulnerable when they pop-up from nap-of-the-earth flying to attain drop altitude. The Team may be required to jump at lower altitudes (especially in static line delivery) than is typical in training operations. If the enemy does not possess an air defense network or air superiority, parachute delivery becomes more appropriate.
  - But even in good weather a Team may expect to be scattered, especially if the Team is dropped in dissected terrain and where only small DZs are available or when High Altitude-Low Opening (HALO) parachute techniques are employed; Team assembly can consume valuable time while enemy forces may be closing on the Team. If the Team is a composite of well-trained, experienced US SpecOps parachutists, and less trained and experienced indigenous Team Members, insertion problems will certainly multiply.
  - If the Team must drop into rough terrain, and/or into forested areas where no DZs are available, tree landings will occur; the time to assemble may then be numbered in days and the likelihood of Team Member casualties are substantially greater.
  - Using High Altitude-High Opening (HAHO) techniques, a Team can be released in more secure airspace; and by using low-detectible steerable parachutes with a high glide ratio, an insertion can have a remarkably low signature. But depth of penetration may be limited. Most of the problems noted for static-line and HALO insertions also apply to HAHO insertions. And winds aloft, especially in mountainous terrain, may aggravate scattering of Team Members.
- <u>Water-borne Insertions.</u> Use of watercraft is a valuable insertion option that is too often ignored or underused by Army SF. The watercraft can be deployed at sea from a submarine, a littoral surface combatant ship, a covert vessel, or from an aircraft. The watercraft can silently carry Team Members and a limited amount of cargo to shore or up navigable rivers or streams. Watercraft can be delivered by air, can be deployed to the water surface from a hovering medium-lift helicopter, seaplane or personnel airdropped onto a lake. In the latter, Team Members should generally be Self-Contained Breathing Apparatus (SCBA) qualified/trained.

- <u>Ground Infiltration (vehicular or foot)</u>. Considered a stealthy method of infiltration, apart from 'stay-behind' or water-borne, this approach has limitations. Unless it is coupled with operations launched from a covert AOB (e.g. including guerilla or partisan bases), it is mostly appropriate for relatively shallow penetrations and COIN operations. The amount of time associated with cross-country movement to the objective and the number of danger areas crossed will likely expose the Team to more possibilities for detection, especially where substantial enemy ground forces are present. Ground infiltration lacks the speed, penetration depth and flexibility of air delivery methods, but most problematic, the stealth of the insertion will be obviated when the Team requires resupply attendant to its time-consuming cross-country movement. The only way that stealth can be preserved is when the Team is supported by a guerrilla/partisan unit, or when a network of MSS and/or caches have previously been established within its target area or along its planned route. MSS and/or caches can also become mobile when towed behind SR Team vehicles. Vehicular insertion requires terrain traversable for utility/tactical vehicles (desert, steppe, etc.) or where sufficient logging roads or trail networks exist. One purpose of a walk-in mission is to avoid detection caused by the tell-tale noise of a helicopter insertion. Walk-in missions (cross border) are feasible only for:
  - Shallow penetrations; otherwise, the Team must normally be resupplied by air – which will generally reveal the Team location and obviate the reason for the walk-in choice of insertion. If the Team is to be resupplied by air, resupply by air drop may be preferred to helicopter resupply.
  - Deeper penetrations; if pack animals or cross-country vehicles are provided and feasible.
  - If the Team is to rendezvous with a guerrilla partisan force within a reasonable distance.
  - If MSSs/caches have previously been established along the route.
- Never use a slash-and-burn agricultural clearing for an insertion LZ, unless it is overgrown and has obviously been long abandoned. These areas are often used to cultivate crops for

Figure 14. Example of a Slash and Burn Area in Laos (1970).

enemy consumption so an enemy cantonment/bivouac will almost always be nearby; as they may not be useful for insertions, active clearings may be of intelligence interest. Note that slash-and-burn clearings will generally be located on hillsides and may therefore be unsuitable for helicopter extraction landings. See Figure 14.

- Prior to launch, Team Members should insert their earplugs. The earplugs are removed after a successful insertion, after the aviation assets have departed, but prior to Team movement from the LZ location. This will accelerate hearing acuity recovery, post insertion.
- Consider using this technique to ambush trackers on the insertion LZ:
  ○ From the landing point, assemble the Team and move to the verge of the LZ.
  ○ Take a listening break to normalize the hearing of Team Members.
  ○ If necessary, re-cross the LZ to the opposite verge that is away from the likely enemy avenue of approach. This is to leave a clear trail in an open area, which may draw the attention of the enemy trackers, perhaps leaving them exposed to Team fires.
  ○ Follow the verge to a position where you can fish-hook and observe both the Team back-trail and LZ. Check the vicinity for enemy signs, including trails, and be prepared to use an alternative technique to set up enemy trackers.
  ○ Wait for the enemy to appear. If the enemy follows your trail across the LZ, you will have him in the open in a far ambush. If he won't take the bait and instead follows the verge to pick up your trail, you will have him in the kill zone of a near ambush.
- Don't move to an extraction LZ too early. If the trackers close on the Team while it is occupying the extraction LZ, they can ambush extraction aircraft.
- Air landing infiltration/exfiltration vs aircraft water landing (e.g. pontoons)
  ○ Water Landing:
    ▪ Minimum traces left.
    ▪ Beware of unknown underwater obstacles.
    ▪ Presence of enemy/civilians more likely near rivers/lakes.
    ▪ May be best option in areas with no/limited LZs/DZs.
    ▪ Likely will require a boat to transfer personnel/equipment/supplies.
    ▪ Water source is an obstacle/danger area; Team and reception party (if any) may be pinned against a water obstacle by an enemy force. May be hard to conceal landing activity. Aircraft and personnel may be exposed to observation and fires over long distances.
  ○ Ground Landing:
    ▪ More feasible where heavy cargo or ground mobility equipment is being delivered.
    ▪ Embarkation/debarkation is more rapid.
    ▪ A landing strip/large LZ most likely will be required.
- Some SOE TTPs:
  ○ If possible, obscure traces of land and beach crossing infiltration.
  ○ Use diversions and false trails.
  ○ Use a luminous/IR (InfraRed) ball swung on the end of a string for signaling. Different colored balls are available for signaling security.
  ○ During infiltration or exfiltration, ensure all clothing items are firmly attached/secured and that hats are carried inside individual apparel.
- Ensure air crewmen are trained in exfiltration techniques, cover stories, individual weapons, SERE and relevant TTPs, in the event their aircraft is downed.

<u>True Account:</u> **An injured helicopter crewman was recovered from a downed aircraft in Southeastern Laos using a McGuire Rig. The McGuire Rig somewhat resembled the 'horse-collar' rig used in Air-Sea Rescue, but unlike the 'horse-collar', the McGuire Rig was intended for the recovered individual to sit on the interior of the loop, not to loop the rig**

under the individual's armpits. Subsequent to extraction, the rig likely cut off the circulation to the arms of the recovered crewman; the door-gunner watched as the crewman slipped out of the rig and plummeted thousands of feet to his death.

Extracted SOE (and OSS) Fixed-Wing Insertion/Extraction/Resupply Aircraft TTPs:

- Ensure that SERE equipment is onboard and intact.
- Supplementary Emergency Equipment/Materials for the crew might include:
  - Local Currency
  - Individual Weapons
  - Land Navigation Maps
  - Packs/rucksacks with rations, water and other items that are appropriate to the operational environment
  - Pioneer tools (to help extract aircraft tires from mud)
- Aircraft crew should all wear boots that are suitable for cross-country travel.
- Train on aircraft landing procedures to include rapid aircraft ramp/landing strip loading/ unloading (if so equipped).
- Enemy Best Practice: The enemy would prefer to attack the aircraft while it is on the ground or as it is taking off – to capture the crew, cargo and/or passengers. This is when the crew and its passengers/cargo are most at risk. Ensure the reception party security is focused on threats rather than observing the landing and/or loading or unloading.
- During aerial resupply:
  - Conduct false airdrops on dummy Team positions, DZs.
  - Accurately place two observers (preferably) upwind on the drop zone, each possessing a clear line of sight to the DZ. Each observer should be equipped with a prismatic compass, and short-range radios, so that they can locate and report scattered bundles by azimuth intersection.
- Frozen streams and lakes can be used for fixed-wing and rotary wing LZs. If a fixed-wing landing is to be performed, consider the following:
  - A landing party must ensure no weak ice, ice ridges, or debris (e.g. logs, branches) are present where the landing is to take place.
  - Ensure there is sufficient wing clearance from banks and bounding tree lines and enough turn-around space and 'runway' is present.
  - Observation of the landing and loading/unloading area are limited, if possible, by stream bends or lake shore contours (e.g. consider an inlet).
- Trees and certain types of bamboo can reach astounding girth and height in the rainforest, with trees far exceeding 150ft in height. The height and numerous layers of canopy, makes it very difficult or impossible for string/fast rope or USAF canopy penetration extractions over large swaths of a rainforest.
  - The ground beneath the canopy might not be visible to the aircraft crew or the Team; hazards, such as a steep ravine, might exist unseen until the Team Members penetrate the first or even second layer of canopy. When aircraft altitude above ground level, vegetation height, and terrain form are all considered, there might be insufficient rope for a descent – with the possibility that Team Members might be suspended above the ground and may be unable to report this dilemma to the pilot.
  - Further, the potential for rope entanglements in canopy represents a serious threat to the aircraft and its crew.
  - Subsequently, the T/L must select string/rope LZs with minimal (or no) canopy; most often, a usable LZ may be found along a stream, where the path of the water (e.g. from post-monsoon deluges) will have carved a path through terrain and canopy.

- ○ Except for the limited availability of stream breaches to the canopy, identification of LZs in continuous rainforest canopy is very difficult using photo imagery, as depth perception and scale are lacking; a VR is therefore necessary.
- ○ It is not sufficient to train Team Members in rappelling/fast-roping off of a tower or even a helicopter (in open terrain). Team Members <u>and Air Crews</u> must be trained in these techniques/procedures within environments that closely parallel those of the AO. This means training in rope/string insertion and extractions into/from very small LZs. A ground safety NCO, with communications to the helicopter crew, is recommended during training in these conditions.
- When loading the aircraft for infiltration, ensure the Team is seated in an order that allows tactically expedited exits from each door. Load the Team in reverse-Team order with the tail-gunner being the first Team Member in the aircraft.
- The T/L is the first off the aircraft during infiltration and the last to leave the LZ on exfiltration. The radio man should sit next to the T/L on the same side of the aircraft. If T/L exits the aircraft under fire, the entire Team should also exit the aircraft.
- An extraction LZ may have to be cleared, which may include felling medium to large trees. But clearing trees requires tools/materials, time and labor that a Team simply does not possess. Clearing trees with bulk explosive requires a substantial amount of explosive, governed by size of trees, tamping and techniques used. Note that a claymore mine has virtually no utility in clearing trees from an LZ.
- The Team must make its best effort to determine if the LZ is under enemy observation, or if close to high-speed avenues of approach or in proximity to enemy combatant forces. If some of these elements are determined to be present, the T/L must consider aborting and using an alternative LZ, if the Team is able to do so, barring contraindications such as Team casualties, etc. The Team must inform the FAC and/or extraction assets of details regarding these threats/potential threats and have a notional plan on how to best deal with them to include a plan to employ close air support, post insertion.
- In heavily vegetated environments, especially where the Team cannot navigate to a reasonably proximate LZ, Team Members must be extracted by rope line (commonly called a 'string' extraction) or by ladder.
  - ○ LZs with high grass: The depth of high grass (e.g. elephant grass) on an LZ cannot be determined by aerial imagery and seldom detected by VR. This grass often reaches a height of 10ft or more and the stalks may not fully compress under the prop wash of the helicopter blades. Helicopter pilots may resist descent into the grass during insertions/extractions over concern that an unknown obstacle may lie beneath the high grass. T/Ls should discuss this with helicopter pilots to alleviate their concerns and should consider training the crews under these conditions.
  - ○ The lead pilot should be the most experienced in supporting SR missions, rather than the highest ranking.
  - ○ Ladders may be used when vegetation (e.g. vegetation or shrubs, elephant grass, etc.) or terrain form will not allow the helicopter to land or to hover within reasonable jumping distance above the ground. A ladder extraction will permit Team Members to ascend into the aircraft while the aircraft is in flight. Ladders are not appropriate to areas of multi-layered canopy or tall trees. See Figure 15.
  - ○ Lines are used where the vegetation or terrain form will not permit the use of ladders. The helicopter crew must be able to see at least one Team Member on the ground, before a string may be deployed in extractions.
  - ○ For line or ladder infiltration/exfiltration operations, a member of the Bright Light Team, or another American may be aboard the helicopter to assist the aircraft crew in the deployment of the lines or ladders.

Figure 15. A SOG SR Team being extracted from the Laotian rain forest. Note the helicopter crewman assisting a wounded Team Member.

- ○ Line/ladder extractions must be 'managed' by the T/L to ensure safe and balanced loading of the aircraft; optimally, the Team should be split to each side of the aircraft so that an approximately equal number of Team Members are assigned to ascend the ladders or to be suspended from lines.
- ○ Lines used for extraction should have loops (non-slip knots) established at the midpoint and free end of the rope to allow Team Members to hook on using karabiners. It is best to hook into the midpoint loop, rather than the end loop; if the line gets caught on a tree, a Team Member on a midpoint loop may be able to cut away the trailing end of the rope below him; this becomes much more problematic if another Team Member is hooked onto the lower loop (see more information on this below).
- ○ If a friendly casualty or a POW is being extracted by rope, one Team Member on another line should hook onto a loop at the same level and then secure himself to the casualty or POW with a karabiner. If this is not done, the casualty may slip out of his extraction rig, or may flip upside down in the harness, and a POW may attempt suicide by wriggling out of his harness and plummet to the ground.
- ○ Note that there are certain varieties of so-called 'fast ropes' that are manufactured with loops woven along their length, fashioned so that troops can hook onto the rope with karabiners. Loaded (personnel and/or equipment) 'fast rope' lines and strings cannot be recovered while the aircraft is in flight. Aircraft crews and Teams must practice landing techniques where fast ropes, strings or ladders might be used; training should be conducted both during daylight and during periods of limited visibility or darkness.
- ○ The Team Member may hook onto a line using a purpose-built extraction rig (e.g. Hanson Rig, STABO Rig, etc.), or commercially (e.g. mountaineering) acquired rigs. The field-expedient solution would be to use a length of nylon rope to form a 'Swiss Seat'. The Author recommends that Team Members carry both a Hanson Rig (uses a length of nylon webbing) and a 12 foot length of rope (or equivalent). These items can be used for a travois harnesses/litter slings, in negotiating difficult terrain and various other field-expedient purposes. If these items are to be used for load bearing, a shoulder pad or field-expedient pad should be used. The Special Patrol Insertion/Extraction System (SPIES) in current use is a legacy TTP from development of rigs used by SOG SR Teams.
- ○ Note that the ability of Team Members to endure in-flight suspension from ladders or lines will be governed by wind conditions/air temperatures aloft – wind chill factors

may be extreme due to the additional factors of aircraft airspeed and prop-wash. In this circumstance, the helicopter may have to lower Team Members to the ground and then land (which may still be in enemy territory) to allow Team Members to board the aircraft. Team Members who have been landed must assist the air crew in 'rope-management' tasks as the aircraft continues to lower suspended Team Members.

- If a ladder, rope or Team Member becomes entangled in trees, the crew (e.g. the crew chief and/or door gunner) must be in position to observe and identify this situation immediately. The helicopter pilot must then attempt to lower the suspended Team Member(s) to the ground and try to ascend once more, or the ladder(s)/rope(s) must be cut loose – even with the Team Member(s) attached. If the ladder suspension is fabricated using cable, the ladder must either be equipped with a disconnect mechanism or the crew must carry a cable/bolt cutter. If the pilot decides to cut loose a Team Member, he must first attempt to lower the Team Member to the ground. The crew chief and/or door gunner must try to signal the Team that they are about to be cut loose and then the notified Team Members must share this information with other suspended Team Members so that they can prepare themselves. The Team Members and crew must understand the hand and arm signals associated with this procedure. If the Team Members cannot be ascended on a rope, the preferred solution is for the entangled Team Member(s) to cut themselves loose; if the crew cannot communicate with Team Member(s), or if the Team Member(s) will not/cannot comply, then the aircraft commander must make the decision to sever the suspension system. Important Notes: If Team Members are suspended from the aircraft by nylon rope:
    - Remember that nylon rope is elastic; when the rope is severed from below by a Team Member, it will spring back like a rubber band, possibly posing a serious threat to helicopter rotor blades.
    - To save the lives of the aircraft crew and other Team Members, a Team Member situated above the entangled Team Member(s) may have to sever the rope below him. An entangled Team Member must NOT attempt to sever the rope above him if the rope poses a hazard to the rotor. This sacrifice will preserve the lives of the remaining Team Member(s) on the string and the aircraft crew.
- Karabiners should also be used to attach Team Member gear/LBE to lines or to vertical ladder suspension cables/rungs, rather than to continue carrying these items at the risk of heavily laden Team Members flipping upside-down during flight. Additional karabiners should be used to secure personnel to lines or ladders. If a casualty or POW cannot ascend into the aircraft, a Team Member must hook onto the ladder next to the casualty/POW to treat/secure the person.
- Ladders cannot be recovered while the aircraft is in flight unless all the Team Members (with their gear) can ascend into the aircraft. Additionally, ground crew, Bright Light personnel or Team Members must assist in aircraft landing by separating the ladders laterally from the aircraft so that the aircraft skids do not crush the ladders or any cargo/personnel carried on the ladder rungs.
- In environments with heavy canopy, ropes and ladders may also be necessary for Team insertions. When rappelling or fast-roping, make sure that the rope does not come in intimate contact with grenade pins or nylon gear (or rope friction will burn through the gear). When carrying a heavy load or descending a significant length of rope, Team Members might have to wear two sets of gloves (e.g. a 'gauntlet' outer shell or welder's glove, large enough to encompass the Team Member's hand and primary work glove). Team Members, once descended from the aircraft, must ensure that ropes do not become entangled in vegetation, canopy, etc.

- ○ If the aircraft cannot come to a hover over sloping terrain or within reasonable jumping distance (e.g. less than 6ft off ground) use a short ladder, fast rope or knotted rope to avoid injury to Team Members during insertions. Or, abort the landing on that LZ and proceed to an Alternate LZ. Plan and train for injuries during insertions.
- Team Members may be wounded or injured during insertion or extraction operations. An additional troop carrier helicopter may have a 'chase' medic on board to render care for wounded/injured Team Members. If any Team Members are wounded or injured, the chase aircraft may be the first into the LZ during an extraction.
- The T/L and lead pilot should discuss direction of approach to the infiltration LZ based on terrain form and current intelligence; this information should be covered at the pre-launch pilot's briefing. The T/L may also recommend/coordinate a direction of approach for the lead pilot for extraction operations based on his understanding of enemy locations and capabilities. The pilot may have concerns regarding aircraft and crew safety or may be more knowledgeable of anti-aircraft threats.
- The T/L should stay map-oriented once the insertion helicopters reaches the vicinity of the Target Area. The T/L should double check to ensure that the insertion is properly oriented and that the insertion will occur at the proper target area and on the primary LZ. Prior to the troop ships beginning their final approach to the LZ, CAS or helicopter gunships may 'prep' the LZ with rocket fire or gun runs. A pair of helicopter gunships may then descend to an altitude somewhat above tree top level and fly outside the LZ periphery in an oval orbit along the troopship approach axis. This race track orbit will allow each helicopter gunship to cover the 6 o'clock of its partner with its chin turret gun system. The troop-ship helicopters may then fly nap-of-the-earth beneath the gunship orbit during final approach to insert the Team – or the gunships may be pulled away to orbit in the near vicinity.
- Once the insertion helicopters begin final approach, the aircraft will fly nap-of-the-earth. As the helicopter (1st ship) doors are opened and the aircraft begins transition from final approach to landing/hover, all Team Members should observe the wood line to detect any enemy presence. If any Team Member spots enemy personnel during the descent, he should immediately open fire; all remaining Team Members and the aircraft door gunners should immediately follow suit while the pilot evacuates the aircraft from the LZ approach. If the aircraft and personnel are intact, the pilot should reorient and attempt an insertion at the secondary/alternate LZ.
- Team Members should normally unload simultaneously on both sides of the helicopter to help the pilot to stabilize his aircraft at hover, unless the aircraft has ramp exits, or if the aircraft has a single side door. After liftoff, Team Members should assemble at 0° from the aircraft landing orientation. Subsequent lifts of additional Team Members should follow suit.
- If the aircraft is taken under fire during an insertion, and some or all of the Team have already disembarked, the lead pilot should make his best attempt to recover those Team Members. Otherwise, a planned extraction operation will be necessary to recover those Team Members from the LZ; this effort should be mounted rapidly providing less time for the enemy to mass and making the extraction much more hazardous.
- If the aircraft is shot down during insertion or extraction, by SOP, the T/L or senior surviving Team Member should be in command on the ground, regardless of the rank of the aircrew survivors. He will accomplish the following:
  - ○ Account for his Team Members and aircraft crew and passengers, including KIAs, WIAs and enemy POWs.
  - ○ Direct the care for the wounded, and prepare wounded, KIAs, passengers and POWs for movement or extraction.
  - ○ Secure an area preferably near the front of the downed aircraft, and preferably in an area that offers cover and/or concealment. If the crash has occurred on the LZ, establish a

perimeter on the periphery of the LZ, if possible. If there is room on the LZ for another helicopter to land or hover, use cover and concealment on the periphery to move close to this prospective landing point.

- ○ With surviving and capable crew member(s), return to the aircraft to recover usable weapons, ammunition, medical items, classified materials (e.g. pilot's maps, notebooks, SOIs, etc.); then implement the aircraft destruction SOP (pilot decision) and destroy anything else that an enemy might recover and use. This should be practiced in training.
- ○ If enemy fire on the LZ cannot be suppressed, or if the LZ cannot accommodate the extraction aircraft, the FAC may require the Team and crew to relocate from the crash/landing site; however, it is often the case that some causalities will normally occur as a result of the crash – immobilizing the Team and crew. This will require the FAC to muster air assets, conduct air-ground suppressive fires and possibly deploy the Bright Light Team/RF.
- ○ If the LZ is secure and receiving no fire, and if the Team and/or crew members cannot be moved, and if no room is available on the LZ for a helicopter landing, string or ladder extraction may be required. Team Members must assist crew in rigging/preparing for string or ladder extraction.
- ○ Insofar as the Team and crew are able, the LZ should be secured prior to the arrival of the extraction aircraft.
- ○ The T/L should provide the FAC a recommended approach, and describe his LZ, especially regarding hazards.
  - ▪ Inform your personnel of the order they will be extracted prior to the arrival of the extraction aircraft.
  - ▪ Evacuate the aircraft crew, wounded/dead and any POWs on the first recovery aircraft. Caution the aircraft crew members to secure any POW(s) to the aircraft and to ensure that they cannot reach/obtain weapons of other evacuees (to include KIAs).
  - ▪ Evacuate the rest of the Team on another aircraft.
  - ▪ Team Members should approach the exfiltration aircraft from the front. This will enable the pilots and door gunners to better support the Team.
  - ▪ The Team could use both doors if the exfiltration LZ has enough space for a sit-down landing, but the Team should notify the pilot first.
  - ▪ The T/L or senior American Team Member is the last to board the aircraft and he will inform the pilot (with hand and arm signals) that he is the last evacuee as he boards.
- ○ Flight time to and from the Target Area, LZ altitude and the station time that supporting aircraft must have 'on target', should be factored into planning by the T/L when he selects the Team composition and helicopters required for insertion and extraction. Plan the altitude of the insertion and extraction LZs with the load carrying capacity of the aircraft in mind. A weigh-in of fully burdened Team Members may be required during mission preparation.

> <u>True Account:</u> An experienced senior SOG T/L was slated for a reconnaissance of a road located along a mountain ridge in central-eastern Laos at considerable distance from the most northern FOB-2 Launch Site. The flight distance and high-altitude would test the operational range, station time and operational ceiling of the UH-1D helicopters assigned to the operation. The Recon Company Operations Sergeant conducted a weigh-in and discovered that, if the mountain ridge insertion was to be done, the Team size needed to be pared down to 8 personnel from the standard 12-man unit. The T/L deemed this as insufficient Team strength for the mission and subsequently decided to select an LZ in the valley below the target, and that the Team would climb the mountainside to reach its recon objective. The

exhausted Team approached, but never did reach, the military crest of the ridge on the third day. Once the Team was no longer masked by terrain slope and vegetation, they were ambushed by an enemy force that had been waiting for them. The enemy fired down on the helpless Team, mortally wounding the Assistant T/L and an indigenous commando who both fell down the mountain side. The remainder of the Team slid and scrambled down the steep slope to the base of the mountain, but could not find the remains of their comrades.

Movement/Maneuver TTPs:

Land Navigation TTPs:

- While in garrison at home station, individual Team Members should consider joining an orienteering club or establishing orienteering as a recurring unit training event. Orienteering requires the participants run cross country through a course of control points across varied terrain, using map and compass. It therefore combines physical fitness with land navigation. As land navigation responsibility is typically deferred to the T/L, or to the lead Fire T/L, land navigation skills among the other Team Members may atrophy; orienteering training and events ensure individual land navigation skills are maintained and improved. To incorporate orienteering as a Physical Fitness Training alternative, or as a standardized, recurring event on unit training schedules, the physical course and control points should be established and maintained at the Battalion, Group, Command or Installation level.
- Conduct land navigation refresher training prior to deployment, with special emphasis on declination, peculiarities of the AO/Target Area/Terrain and differences related to another hemisphere.
- Compasses should be checked periodically for accuracy against a known (verified) azimuth. The verification azimuth checkpoint should be established wherever Team Members draw their land navigation equipment.
- During training and operations, always carry maps and notebooks in waterproof containers (e.g. plastic bag). If a plastic bag is used, mark at least one grid intersection on the plastic with permanent marker so that the map and its plastic cover can be matched up. A useful alternative is to coat the map with waterproofing such as Nikwax Map Proof or Aquaseal Map Seal; even so, keeping treated maps in plastic envelopes will help prevent map markings from rubbing off. Insect repellant may smear overlay markings.
- Map contours, when read by a skilled field soldier, reveals terrain features as one would expect to see them in 3-D imagery as viewed from ground level. Most soldiers are not sufficiently experienced to do this. The more experienced Team Members should teach others to 'see the shape of the terrain' from a map. Once the Team Member is able to 'see' the terrain shape, he can then walk through the terrain at ground level in his mind's eye. Orienteering clubs and exercises can also help substantially in this respect. See Figure 16.
- Creating individual Team Member maps for a given Target Area may require joining two to four separate map sheets. Particularly when a Target Area overlaps onto additional map sheets, Team Members may consider cropping the joined maps to cover only that which is considered

Figure 16. Mentally convert map contours to a 3-D ground level perspective. (US Army Field Manual 3-25.26)

necessary for the mission and to reduce bulk. However, Team Members should not cut off too much of the map, always retaining least an additional 5–10 kilometers space surrounding your Target Area boundary in the event the Team must evade or alter the mission.

- When drawing map sheets from the S-2 (intelligence staff section), ensure that ALL the sheets are of the same and/or most current edition number or map information date. Old map sheets will lack the most up-to-date geographical and infrastructure information so their planning and operational use could have a fatal result. Always take note of the contour interval; older military maps may give the interval in feet; only three countries do not use the metric system (USA, Liberia, Myanmar).
- Consider taping the map legend/marginal information to the reverse of the map sheet(s).
- Train on the use of enemy and/or foreign maps.
  - If an enemy map (and/or overlays) is obtained during an operation, the content may be of great value to intelligence personnel – if the information can be transmitted in a timely manner. A high-definition camera, capable of close-ups and collage shots, would be necessary to capture the content of the map and transmit the images to higher headquarters.
  - Once captured map information has been transmitted, the Team must discover if any of the information can be quickly exploited by the Team. To this end, at least one of the Team Members should be familiar with enemy map symbols, marginal information (legend), notations and coordinate conversion. Some foreign maps are more detailed than US/NATO maps. Some of this information may be contained in a tactical tablet.
  - After the Team is exfiltrated, the map should be rapidly studied at the FOB level, especially by Teams pending infiltration, and to populate S2/3 maps and target folders with enemy information – before being evacuated to higher headquarters.
  - The Team should be intensively trained if it is to use foreign/enemy maps (e.g. for deniability purposes) during the operation.
- To reduce clutter on your map, consider marking the map/map plastic cover sheet with alpha/numeric 'tags' that are indexed to a legend (e.g. on the back of the map or in a notebook). The iconography of the tags and the formatting of the legend and the information should be standardized (SOP) as much as possible. Remember that once a map is marked up with friendly or enemy information, it becomes a sensitive, even classified document and must be retrieved from Team Member casualties.
- When cross-country night movements are planned, select routes that are least difficult/hazardous and where terrain features, checkpoints and RPs can clearly be identified. When the T/L encounters a planned checkpoint or RP, he should pass the word to all Team Members. If tactically prudent, minimize the number of deviations/doglegs to be taken during night navigation; this advice may also apply to the 'terrain-hopping' technique.
- Mark selected/key information on the map or on the reverse of the map in luminous (glow-in-the-dark) markings for night navigation. Note that luminous markings must be exposed to light for 30 minutes prior to dark to yield 10 hours of luminosity. An alternative is to use markings that can be read with night-vision optics.
- The Team Member will mark a magnetic or grid declination line on the map, when orienting the compass and map. In using a folded map, remember to mark this line on each map side when preparing for the mission. All Team Members should use the same orientation method (magnetic azimuth vs. the grid azimuth).
- Once the Team has been extracted from its Target Area, US Team Members will normally be debriefed by S-2 immediately after returning to the FOB. If the Team is likely to return to the same Target Area in the near future, the T/L should consider retaining the mission map (if serviceable) and notebook (in secure custody), rather than turning these documents in to the S-2 for destruction. The mark-ups can also be used to populate a new map with information from the previous operation.

- A liquid-filled compass is preferred, even on button/wrist compasses. The liquid mitigates needle jittering and allows the needle to settle more quickly.
- All Team Members should know their estimated 'known distance' length of pace/pace count in varying terrain. The pace count during night operations will vary substantially, as length of pace is shorter at night; take 'known distance' night-time pace count in varying terrain as well.
- Every Team Member should carry a small protractor with degree and mil scales. Ensure that the protractor and its markings are durable.
- You will consult your compass frequently, so have it stowed for easy access; consider wearing your compass on a cord around your neck and tucked into a chest pocket. For terrain-following navigation, a wrist compass may be sufficient; but always carry a lensatic compass, with both degree and mil increments, as well. Do not carry/use a compass close to electronics (e.g. tactical tablet, GPS) or close to metal mass (e.g. pistol), as these objects will affect the magnetic field and may even permanently affect the compass over time.
- During daylight rest periods, meal-times, messaging breaks, the T/L should show his map to other Team Members and keep them informed of movement status to ensure they too capture the current location, enemy and terrain information and the trace of the route traveled thus far.
- In the northern hemisphere:
  - Moss/algae retains moisture and is more often found in north-facing shade.
  - Find the side of an isolated tree where branches are mostly horizontal to find South.
  - The side of an isolated tree that is more heavily populated with leafy (needles) branches will generally point South.
  - When using nature's indicators of direction, do not depend on a single indicator. Seek confirmation from other indicators.
- In the southern hemisphere:
  - Extend a line from the tips of the crescent moon to the horizon to find South.
  - Moisture retaining moss/algae is more often found in south-facing shade.
  - Find the side of an isolated tree where branches are mostly horizontal to find North.
  - The side of an isolated tree that is more heavily populated with leafy (needles) branches will generally point North.
  - When using nature's indicators of direction, do not depend on a single indicator. Seek confirmation from other indicators.
- In equatorial rainforest, the sun generally passes directly overhead. So moss/algae will <u>not</u> be a reliable indicator of direction. Other field expedient methods (watch or shadow methods) will also prove unsatisfactory, especially when the canopy does not permit sun observation or shadow formation.
- Winds will shift as the Team climbs a mountain ridge. Do not depend on prevailing winds for direction finding in this situation.

Reading Sand Dunes TTPs:

- The windward side of a dune will normally be firm, with a shallow slope; the downwind side will normally be soft, with a steep slope.
- These facts are useful in route planning based on the prevailing wind conditions; as a navigational guide based on prevailing wind direction and for selection of paths for vehicular movement.

Reading Snow TTPs:

- As with sand dunes, the windward side of a snow-bank will have hard-packed snow and the downwind side will have softer snow. If the prevailing wind direction is known, the snow pack can be a navigational aid.

- Snow stripes or snow stuck to the sides of trees will reveal the windward side and the direction of the prevailing wind.
- If falling snowflakes increase in size during a snowfall; that may be an indicator that a thaw is pending.
- If the ground is scoured of snow by winds, that area may be a poor place to establish a NDP, surveillance point or hide location.
- Conditions prone to avalanche:
  - Presence of an icy/cold snowpack.
  - Location on the downwind/lee side of a ridge.
  - Cracks in the snow surface.
  - Sun-balls increase to snow-wheels on a slope. This may suggest a wet snow avalanche.
- In snowbound areas, make trail markings (e.g. on trees) above the snow line, to indicate the trail route, hazards, etc. Markings may consist of putting stones and/or arrow sticks in the crotch of trees or marking/notching standing dead trees.

GPS TTPs:

- Wear or carry the GPS where you can reach it easily; you may be consulting it fairly frequently in certain environments.
- Note that GPS devices consume battery power rapidly and the device may be detectable from its antenna energy signature. It would be prudent to know where the navigation satellites are located. It may be possible to use a ridge to screen GPS signature if the instrument can still 'see' the satellite(s). GPS signature <u>may</u> be less of a concern in a COIN environment.
- Additionally, remember that satellite links may be impaired under heavy canopy or in low-lying areas (ravines) within heavily dissected terrain, especially where the satellite constellation is not robust. To obtain a GPS location fix, the Team may have to move to a clearing or atop an elevated terrain feature. Such limitations with GPS utility are not mentioned in FM 31-20-5.
- GPS-assisted desert navigation is usually unimpaired; traditional land navigation (map and compass) may require supplemental training in areas lacking terrain features.
- Due to terrain form, canopy, battery requirements, and the tactical situation, it may be best to rely on map and compass, and restrict GPS use to occasions where grid location confirmation is essential. These occasions might include: preliminary to transmitting Situation Reports or other messages, when coordinating supporting fires, when the Team has become disoriented, when planting mines/booby-traps, where the Team has discovered something of intelligence value or in proximity to a target.
- Prominent terrain features are rarely visible from a distance in Jungle/Rainforest; performing intersection or resection may be impossible. If you lack a functional GPS device, or if the GPS cannot receive satellite signals due to intervening canopy or terrain form, and you become disoriented in Jungle/Rainforest, you may have to:
  - Move to (or return to) and navigate from a known point (e.g. your insertion LZ). If you have been inserted on the wrong LZ (it happens!), a Forward Air Controller (FAC) must subsequently provide accurate coordinates.
  - Move to an open area or a prominent terrain feature discovered en route where a FAC can spot the Team or GPS reception can be achieved.
  - Move to a linear terrain feature (e.g. road, stream) and then find a reliable intersecting feature (e.g. stream/road junction) to orient to geographical location.

Compass TTPs:

- <u>Due to magnetic influence, keep your compass away from metal mass (e.g. the steel barrel/ receiver of a weapon; ammunition magazines; ground mobility equipment) and electrical</u>

<u>fields when taking azimuths.</u> Also, beware magnetic influence in lava beds/fields, as these areas will normally have high iron ore content. Similarly, consult terrain intelligence data for areas of iron ore concentrations. Areas that have sustained bombardment will also have much metal debris which may influence compass readings. Team Members/navigators should hold the compass away from an individual weapon by approximately ½ meter.

- If a Call-for-Fire requires transmission of an Observer-Target azimuth, the Team must indicate whether it is using magnetic or grid azimuth direction.
- For night navigation, the bezel click method of presetting an azimuth is inexact; each click represents 3° and is therefore a poor navigation technique (see below), unless prior preparation takes this into account. Prior preparation would include:
  - Click method should consider any GM declination (e.g. if maintaining compass-map orientation is required). During daylight, modify the planned azimuth(s) to the nearest 3° click increment(s) and identify/revise RPs and checkpoints accordingly. This modification step must be done for each night-time dogleg/pre-planned azimuth.
  - Or, navigate using the click preset method to a linear feature and then follow the linear feature to a landmark or intersecting feature (example: navigate to a trail, then parallel the trail until it intersects with a stream).
- There are approximately 6,400 mil radians (mils) in 360 degrees; so 1° is the equivalent of 17.8 mils. Bear in mind that one mil in deviation equates to 1 meter lateral distance (or height) at a range of 1,000 meters, and that one degree on your compass equals 17.8 *mils* – therefore a mere one degree in deviation will equal 17.8 meters lateral distance/height at 1,000 meters.

<u>Mil Relation</u> Formula

W = R x *m*

W = Shift in meters

R = Range in thousands of meters

m = Deviation in *mils*

Example: Assuming that the Team navigates to a point of interest <u>4km</u> from its insertion LZ, using the Mil Relation Formula (aka WeRM formula), an initial error of <u>3°</u> (=53 *mils*), an error which can easily occur when merely drawing an azimuth on a map, could ultimately place the Team some 214 meters from its intended target. This error would be compounded if the Team uses dog-legs or boxing around obstacles in its movement toward its objective, especially where reference points or prominent terrain features cannot be observed (e.g. in Jungle/Rainforest and dissected terrain environments).

- There are a variety of methods for Teams to compensate for likely land navigation deviation errors, including reorientation from a known point or linear feature. Upon insertion onto a DZ/LZ, the T/L should make his best effort to pin down his actual location as precisely as possible (GPS); this to verify and ensure that the Team has been inserted on the correct DZ/LZ. Once this location is established, the Team may use terrain following/hopping techniques that should serve for navigation in difficult environments. For instance: The T/L moves to the periphery of an insertion LZ bounded by hills/ridges, at a point where a stream flows from between elevations, past the LZ and exits between other elevations. The Team moves to where the stream debouches or departs the LZ location, ascends the ridge and follows the terrain form to the next terrain feature, and so on. As long as the T/L follows the terrain form on his map, he will be reliably guided along his route toward his objective. He will be able to confirm his location en route, as the Team crosses or encounters other features (checkpoints) during movement.

Local Weather Forecasting:

- It should be emphasized that Team Members should have the ability to forecast local weather conditions in the target area, independent of higher headquarters. Particularly in deep, long duration penetrations, the Team must be able to predict weather changes as an operational and survival necessity. For instance: the T/L will want to use terrain form to plot a course across desert, using a wadi or a ravine for concealed movement. If the Team cannot forecast local rain, the team could be wiped out in a flash flood. This forecasting ability bears on Team land navigation and operational movement, availability of friendly support (chiefly aviation), enemy behavior/practices, Team safety and general tactical decision-making. Team forecasting will often be more accurate than forecasts provided by higher headquarters for a broader operational area. As a weather front can move rapidly, Team Members should observe the weather during rest breaks or on other occasions when the opportunity to view the sky is unimpaired. To increase weather forecasting accuracy, the Team should not rely on a single weather indicator, but should collate indicators. These indicators can include: cloud direction and wind behavior, cloud formations, illumination phenomena and even the behavior of insects and animals. Note that weather forecasting in coastal and mountain areas are quite different from inland forecasting.

- The relationship between cloud direction and wind behavior is a good predictor of weather changes. Field procedure (northern temperate region):
  - Wait until fog has burned off.
  - Toss a leaf/blade of grass into the air to determine wind direction. Face in that direction.
  - Observe the highest clouds or contrails observable in the vicinity and determine their direction. This technique cannot be used in coastal areas or mountains or by observing, at a distance, clouds hovering above coastline or mountains.
  - Indicators:
    - If clouds drift from left to right: an advancing warm front with rain and low pressure may be expected. 'Left to Right; not quite right.'
    - If clouds drift from right to left: a cold front advancing (or has just passed). High pressure and good stable weather conditions are forecast. The higher the altitude of the lowest clouds observed, the dryer the conditions.
    - Clouds moving in the same direction as ground winds: no change in weather.
- Cloud formations/types are also indicators of changing weather: [See Appendix C.] If the Team is equipped with a tablet IT device, this information (and descriptive photos) may be stored in memory for use by the Team.
- Other phenomena:
  - The adage, 'Red at night, sailor's delight; Red in the morning, sailor take warning', is often a very good predictor of weather. Matthew XVI: 2-3 contains similar advice.
    - 'Red sky at night (sunset)' indicates that a westerly high-pressure system is moving in with good weather.
    - 'Red sky in the morning (sunrise)' indicates that a high-pressure system has passed and that a low-pressure storm system is moving in. A deep red would suggest that rain is on the way.
  - Halo around the moon or the sun may indicate rain or snow within the next couple of days. The ring is caused by ice crystals within thin cirrus clouds, made visible by sun or moon illumination. Cirrus clouds are often the precursor clouds to an inbound storm system.
  - High performance aircraft may leave contrails indicating the presence of moist air at high altitudes.
    - If the contrails are short and there are few or no clouds, then moisture is minimal in the upper atmosphere, predicting that the weather may be very good for the next 12–24 hours.
    - If the contrails are long and clouds are present, the moisture content aloft is high and cloudy to stormy weather may be inbound.

Stealth TTPs:
Stealth requires furtive movement, light and noise discipline and suppression of odors, and is essential to patrolling and reconnaissance operations.

- Stealth is essential to mission accomplishment and survival in reconnaissance operations. Rate-of-march in Rainforest/Jungle is approximately one kilometer per day. Move and halt at irregular intervals. Depending on terrain, vegetation, etc., move approximately 1 hour, then halt and listen for 5-10 minutes. Note that movement in temperate regions through wooded areas consisting of deciduous trees may also be challenging. The ground may not be as choked with vegetation as Rainforest/Jungle, but dead leaves, which may also conceal dead/dry branches or twigs, can create a lot of noise that can betray Team movement. In these circumstances, using the file formation, moving in a serpentine fashion, and where the point man chooses the most silent route, is preferred to open formations. Additionally, the rate-of-march may have to slow substantially, to suppress noise in movement.

- Avoid areas of fallen or dry bamboo (unless you are to use it as an impediment to enemy approach or to set up an enemy tracker team); it is virtually impossible to pass through a patch of dead bamboo without creating a great deal of noise. Breaking a stalk or a culm of dead bamboo can seem almost as loud as a discharging firearm. Additionally, bamboo will develop a leaf sheath that will drop off the growing stalk; these break with a distinct crack when stepped on, even when wet. See Figure 17 and notice the leaf sheaths and dead bamboo littering the enemy trail. It took great stealth and daring for the T/L to take this photo of a passing NVA soldier; the soldier's own noise passing through the cascading bamboo and stepping on debris was helpful. The photo was taken from a well-maintained intersecting high-speed trail.

Figure 17. NVA Soldier moving along a trail bounded by small diameter bamboo.

- Walk in small streams during the rainy season/monsoon. The topographic drainage and increased flow of streams will conceal any traces of disturbed sediment. But be aware that streams that are not shrouded with vegetation are danger areas. Do not walk in the middle of the stream, but keep near vegetated banks as much as possible.

- Avoid breaking limbs or branches on trees, bushes, or palms, or you will leave a very clear trail for the enemy to follow.

- Use of pack animals will impose certain restraints/constraints on Team movement and navigation. Pack animals, and their burdens, may cause noise and traces that are easy for an enemy tracker team to follow. This is especially true if the Team is moving through dense vegetation that requires Team Members to cut a trail with machetes. To mitigate this violation of noise discipline, the T/L may be forced to use well worn animal (e.g. deer) paths, trails (preferably old/unused) and ridge tops – all of which increase tactical risks to the Team.

- When approaching the target/Team objective, move when sounds of wind, rain, enemy movements/operations will screen the sounds of Team approach, if possible. If the prevailing wind/breeze is blowing toward the enemy location, sounds of Team approach will travel further in that direction. Sounds will be better conveyed downhill in the evening, as air cools, especially on ridge/hill sides opposite from the setting sun and down ravines. In the morning, when the sun warms ambient air in the valleys, sounds will be better conveyed upwards on ridge/hill sides and up ravines.

- Team Members may create substantial noise in moving through dry leaves found in deciduous forests. Techniques used in patrolling through these areas have some similarities (and some dissimilarities) to those used in crossing through areas of bamboo.
  - ○ Avoidance: During the mission planning phase, select routes that will avoid deciduous areas for evergreen/conifer areas. Movement over a pine needle carpet is much more silent. Notes:
    - ▪ Beware of twigs/dry limbs that are concealed beneath the pine needles.
    - ▪ Some military maps (Russian) will depict the types of vegetation in an area (e.g. deciduous versus conifer).
    - ▪ Areas that are windswept (e.g. ridge tops) are often free of dry deciduous leaves.
    - ▪ The Team may have to move along animal or human trails to suppress noise.
    - ▪ The Team should move in a file formation, following the path selected by the T/L or Point Man.
  - ○ Weather: movement during, or for a period after, drenching rains, will render dry leaves/ dead vegetation moist. Plan your mission with weather conditions in mind. Note: Beware of twigs/fallen limbs that are concealed beneath pine needles or dead leaves. Also, be aware that a layer of dry leaves may be found beneath the layer of wet leaves, especially in areas of heavy leaf drop.
  - ○ A layer of wet snow will also moisten leaves and suppress movement noise (as will powder snow). Note:
    - ▪ Crossing snow-covered areas may also cause noise, especially if there is a layer of crust or ice, or if the snow causes noise when compacted (e.g. pellet snow). Be aware that a crust layer may exist below a fresh layer of powder. To minimize sound, spread Team Member weight over a larger area by using skis or snow-shoes.
    - ▪ Movement in a file formation will reduce the noise of movement substantially over that of a more spread formation as trail is broken by the point man for the rest of the Team.
  - ○ Pace/Speed of Movement:
    - ▪ To maintain stealth, the Team must reduce its speed of movement in areas of dry vegetation, consistent with the mission timeline. Subsequently, a swifter pace may be required initially, after an insertion for instance; an increasingly slower, stealthier pace, then becomes necessary as the Team draws closer to its objective. The T/L should plan the speed of movement, LZ/DZ location and date/time. of insertion accordingly.
    - ▪ In areas where there is a high concentration of enemy troops/units, the Team must move at a uniformly slow and stealthy pace. In the heavily dissected rainforest of Laos and Northeastern Cambodia, the pace of SOG Recon Teams was typically a kilometer a day.
    - ▪ Frequent Team pauses, to listen for enemy movement, are essential.
- Load-bearing equipment, rucksacks, weapons and other cargo, will often make Team Members top-heavy and prone to losing their balance in broken terrain. When ascending or descending steep terrain, maintain balance by grasping nearby roots, bamboo, small trees or other plants, but only grasp at the base of the vegetation so that the vegetation does not rustle or sway from the contact.
- If the Team or its elements are in surveillance/hide, ambush or raid positions, geographic location, time of day and wind (and other weather) conditions will dictate whether Team Members can safely eat, urinate or defecate without alerting enemy combatants – or if special precautions are necessary. For instance, if the Team or its elements are located on high terrain valley breezes will often prevail during daytime, blowing winds upwards; during

night-time, mountain breezes will often prevail with winds blowing down towards lower elevations. The T/L must assess these conditions on the ground during the mission.

- Do not become overconfident, it leads to carelessness. If you have seen no sign of the enemy for three or four days, you should <u>never</u> assume that he isn't nearby or isn't aware of the Team's presence.
- Correct all Team and/or individual errors as they occur (if possible) or shortly after they happen.
- Movement of vegetation:
  ○ If a Team Member accidently disturbs vegetation, causing it to sway, he should grab it immediately to stop its movement.
  ○ If a Team Member needs to grab vegetation to steady himself or to assist in ascending or descending a hill, he should grasp the tree or shrub near its base, to minimize its movement.
  ○ Alternatively, attach a length of 550-cord to a shrub to deliberately make it sway and distract or deceive a tracker team – and perhaps draw enemy fire or cause the enemy to maneuver in the wrong direction, exposing his flank.
- Avoid silhouetting Team Members by:
  ○ Selecting clothing and equipment that is properly camouflaged (day and night) or blends in with target area terrain, weather conditions (e.g. snow) and vegetation.
  ○ Select routes that minimize/avoid the possibility of skyline silhouetting of Team Members, e.g. bald or thinly vegetated ridges/hilltops.
  ○ Select a route-of-march that avoids open areas whenever possible. Take time to go around these areas, keeping to vegetated fringe areas or clusters. Modify the mission timeline during the planning phase to allow for such route deviations.
  ○ If crossing open areas is unavoidable, use folds (e.g. ravines, gullies, gulches, arroyos) in the terrain to conceal movement and avoid silhouetting. Habitual use of the same fold in the same target area will invite enemy mines/booby-traps. See example at Figure 18.
  ○ When crossing high points, seek out elevation dips and/or clusters of vegetation and cling as close to the ground as possible.
- Ensure that straps on load-bearing equipment, rucksacks, weapons and other cargo, are secured so that they do not make noise or become entangled with vegetation.
- Team Member's senses of hearing and smell become very focused and acute during an operation, and the longer the duration of the operation, the more sensitive the senses become. Correspondingly, Team Members should never smoke throughout the entire duration of a mission, as this practice impairs the sense of smell and reveals the Team presence to the enemy. The aroma of American tobacco products smells distinctively different than those manufactured in foreign countries. Furthermore, Team Members should bury all used food containers and human waste as soon as possible.
- Low wind inversion conditions will cause aromas to linger close to the ground for extended periods of time and will cause the scent to drift downhill, along ravines,

Figure 18. Use Arroyos/Dry Stream Beds for concealed crossing of open desert/ landscapes. (*Depositphotos.com*)

towards low-lying areas. Well disciplined and experienced enemy personnel understand this and will often establish cook fires in low-lying areas, under canopy, employing measures to ensure that heat and smoke signatures are suppressed.

- While operating in a Target Area, SR Team Members behave much like rabbit or deer in the forest during hunting season. In heavily vegetated terrain (jungle, rainforest), or in close proximity to an enemy, the SR Team should typically move perhaps ten paces, and then should pause to listen and watch, before moving again another ten paces. This slow pace also allows each Team Member to place his foot or hand optimally to maximize stealth.
- When the SR Team is in a perimeter to eat its noon or evening meal, a Team Member may suddenly stop chewing to listen and watch. The entire Team will instantly notice this and should immediately follow suit. Only when the initial Team Member resumes chewing, should the remaining Team Members relax.
- Consider carrying selected items from a single food ration (soft pack) in a trousers bellows pocket so that the Team Member needn't remove his rucksack at a meal break.
  - If the ration is dehydrated, water can be added prior to tucking it away in the pocket; by mealtime, the ration will be completely saturated. Double seal a prepared ration to minimize any aroma and/or leakage.
  - No food should be consumed, to include candy and gum, except during designated mealtimes. Chewing interferes with hearing. Exceptions: caffeine candy may be sucked, but should not be chewed on ambush, Observation Post/Listening Post (OP/LP) or surveillance duty, but the candy should not be aromatic (e.g. spearmint, etc.).
  - A Team Member can make his own caffeine candy by using ingredients from a standard ration. Blend the contents of a sugar packet, an instant coffee packet, a coffee creamer packet and a little water in a small plastic bag and consume when needed.
- The senses of an indigenous Team Member will often be superior to those of the US Team Member, particularly if the indigenous personnel are recruited from primitive tribes, so whenever an indigenous Team Member freezes, all other Team Members should also freeze. Freezing in place will occur several times a day and it will often summon a surge of adrenaline.
- Team Members will inevitably and involuntarily feel the urge to cough or sneeze during the conduct of a patrol or a reconnaissance mission, while manning surveillance OPs/LPs, or while waiting in ambush or raid positions – when silence is most essential.
  - To suppress the sound of a cough or sneeze, muffle your nose and mouth inside your hat, in a scarf or using the inside of your shirt or coat to smother the noise. Force yourself to cough whenever a high performance aircraft passes over or artillery is firing. It will clear your throat, ease tension, and is more difficult to detect.

Author's Solution:
It is physically impossible to sneeze with your eyes open. Therefore, to suppress the impulse to sneeze, simply keep your eyes wide open, even if it requires holding your eyelids apart with your fingers.

- You will note that your sense of smell becomes more acute during operations in primitive environments. Enemy combatants and especially tribal indigenous personnel, who spend extended periods of time in these environs, can easily detect lingering foreign aromas. During a combat mission, your armpits and clothing will acquire a putrid odor within two days, derived from exertions in high-humidity, high-temperature conditions and from nervous sweat. Human odor and the scents of 'civilization' (e.g. soap, detergents, etc.) can carry further than one might expect and will linger during inversion weather conditions that prevail in jungle or rainforest environments. Enemy military 'service' dogs (e.g. security or trackers), native village dogs, working dogs or dogs within enemy facilities will come alert and may 'sound' on detecting a Team's odor.

> Dogs can hear four times further and can hear at a much higher pitch than a human; their hearing is so acute, they can hear a quartz crystal in a digital clock. The part of a dog's brain that controls the sense of smell is 40 times that of a human; a dog may possess 300 million scent glands in its nose, as compared to five million in a human.
>
> Source: Television Documentary, *The Secret Life of Dogs*,
> National Geographic Channel (Nat Geo Wild)

- Do not smoke, shave, or use scented soap, deodorant or shampoo just prior to a mission. Additionally, do not have your field uniform washed or laundered with scented detergent or additives prior to a mission. Unscented soaps, deodorants, shampoos and laundry detergents are commonly available.

Author's Solution:

Prior to deployment to the Launch Site, use only neutral (scentless) soaps and shampoos and deodorants that contain antimicrobials. Additionally, ammonium aluminum sulfate (aka ammonium alum or alum) is sold in crystal form, and often referred to as deodorant crystals, and has been used throughout history (and is still used extensively in third–world countries) as a personal deodorant, it does a superb job at suppressing body odor.

Stealth Formations TTPs:

- The file is the stealthiest formation, as it allows Team Members to follow the point-man or T/L's selected path as he navigates around or through difficult terrain and vegetation.
- In other formations, Team Members each forge their own path through the vegetation and terrain, substantially increasing the noise of Team passage, leaving a broader trail and increasing trail signs leading to detection. This is an especially hazardous practice where the Team must cross danger areas or is operating in areas populated by rural or primitive inhabitants. If the Team uses a spread formation (e.g. movement-to-contact) when crossing danger areas, the T/L may choose to re-form in file formation, after the Team crosses danger areas.
- The track left by a file formation is harder to detect from the air, in open terrain and snowbound conditions, than spread formations. The file formation also allows Team Members to aid each other in ascending or descending difficult terrain.

Camouflage TTPs:

- Wear uniforms that blend with the environment. If the issue uniform does not blend in, find an alternative that does, or use dyes or spray paint to conform the uniform to the environment.
- Some camouflage patterns may blend in poorly with local backgrounds or in different light conditions. Or the pattern may seem to blend in with backgrounds, but then become easily detectible by IR sensors or when viewed through night-vision optics.
  - Check the issued uniforms (US, allied and indigenous), and even Ghillie suits, against the prevailing backgrounds, under different lighting conditions and when viewed by night-vision equipment. If the camouflage does not hold up to scrutiny, swap out the uniforms for a different pattern or use clothing dyes and/or spray paint to mitigate camouflage pattern inadequacies.

- ○ Sometimes a mixed uniform approach is best. For instance:
  - ▪ In snowy conditions, wear lightweight white shell trousers to blend with the snow covered ground, but wear a woodland camouflage pattern above the waist to blend in with evergreen vegetation. This was common practice among German and Finnish ski troops during WWII.
  - ▪ Where earth or lower tier vegetation is different in colors, shades or patterns than colors/shades/patterns present in vegetation above the waist, dress accordingly or use dyes/spray paint to modify the existing pattern.
  - ▪ If the patterns vary within the target area, consider carrying lightweight shells, smocks, nets.
  - ○ In heavy vegetation and/or under multiple canopies that affords a shaded environment, a cheesecloth-type fabric, appropriately dyed, can provide a decent expedient camouflage cover.
- Consider wearing luminous spots/patches under your collar; flip up the collar to be seen during night movement.
- Skin Camouflage.
  - ○ All personnel should periodically refresh skin camouflage throughout the day. Have a Team Member check your camouflage and assist you in its application in hard to see or hard to reach areas.
  - ○ Use insect repellant to soften the tip of a camouflage stick or of camouflage appliqué, especially in cold weather.
  - ○ The easiest way to remove camouflage stick after a mission is to apply baby shampoo and rub or wipe off the residue.
- Blond, redheaded or bald Team Members should wear a boonie or subdued native hat, or wrap their heads in a subdued kerchief or triangular bandage. Blond or redheaded Team Members might consider dying their hair a dark color. Team Members who wear glasses or protective goggles should wear a boonie hat to reduce or eliminate reflective glare.
- All personnel should wear loose fitting and untailored clothing on field operations. Tight fighting clothing often tears or rips, allowing easy access to exposed parts of the body for mosquitoes, leeches and other parasites/disease bearing insects. Additionally, ensure that the uniform is 'seasoned'; made of a material that will not rustle when you move. Ensure that the trousers have a zipper fly, rather than a button fly; otherwise, leeches or other parasites will find a way inside. If a zipper fly is not obtainable, ensure that the fly is thoroughly doused with long-lasting insect repellent.
- Ensure that all bellows pockets (shirt and trousers) are sufficiently vented at the bottom to allow quick drainage. Otherwise, the pockets will fill with water during a stream/river crossing and will significantly impair movement as the Team Member ascends a stream bank, etc. This is even more essential for load-bearing equipment and rucksacks.
- Use a wide belt (1½ to 1¾in), such as a so-called 'Rigger's Belt', rather than the narrower issue cotton web belt, the wider belt is strong enough to hold up your pants when the cargo pockets are loaded. The issue belt will cut into your hips, and a resulting abrasion may become infected. A Rigger's Belt will have an anchor point on its buckle that should be robust enough to support your weight in an emergency. A Rigger's Belt can be purchased, fabricated from a military A7A cargo strap or can be fabricated from commercial materials. Ensure that the Rigger's Belt has at least an additional 1½-2ft of length, which is tied off next to the buckle. This will allow:
  - ○ The Team Member to drop his trousers just by loosening the belt, rather than by unbuckling or un-threading the belt entirely; this provides the capability to rapidly re-secure the trousers in an emergency merely by pulling up the trousers and yanking on the

free end of the belt. As the belt is not unbuckled, items attached to the belt by a belt loop will not fall off either.

- ○ Adds sufficient length to use the belt to bind a pressure bandage(s) to a chest wound.
- ○ Provides additional security to the buckle if the belt must be used for climbing, during ladder or string extraction, etc.
- Consider wearing enemy uniforms and equipment to confuse the enemy, should a Team Member be spotted. If your appearance causes an enemy to hesitate upon a chance encounter, you will have an advantage.

> **<u>True Account:</u> A SOG Reconnaissance Team routinely carried or wore a mix of US and NVA weapons, load-bearing equipment, and uniform items. On two separate missions, the Team passed completely through an enemy ambush kill zone without the enemy initiating fire.**

- The Author does not recommend 'baseball' hats for reconnaissance operations; the relatively long bill or brim of the hat is frequently knocked askew by vegetation; this is not only frustrating, but it can be fatal if it obscures vision at the wrong moment. Further, as the Team Member often moves bent over through dense vegetation, the bill of the hat will obscure forward vision. A better alternative is a short brimmed Patrol or Ranger hat. And a full-brimmed 'boonie' hat (e.g. Hat, Sun, Camouflage Pattern) provides better protection from the sun and from insects and debris falling down the back of the shirt collar; the brim on a boonie hat should be approximately 2in deep. Again, larger brims may obscure vision. Wearing a hat is almost essential if the Team Member wears glasses or protective goggles (to suppress glare from the lenses), is bald or has light hair color.
- Wear a subdued triangular bandage around your neck. The bandage can be quickly applied as a tourniquet. It can also be used in combination with the patrol hat to shield the neck from debris, insects and from the sun.
- See Chapter 4 for more information.

Movement Technique TTPs:

- Each Team Member should adopt the following procedures for individual movement:
  - ○ While bearing a top-heavy load, posture, stable foot placement and center of balance should be optimized to ensure the Team Member does not fall or stumble while shifting weight or even pausing during movement.
  - ○ If the Team Member must grasp rocks shrubs, trees, bamboo, or other objects, the Team Member must look to determine if the object can bear his weight or if a hazard is present near the object and should test the object before committing his full weight. Further, vegetation must be grasped near the base to ensure the plant does not sway and give away the location to the enemy.
  - ○ Each Team member must 'LOOK' (not glance) to determine his next foot placement. He will look for a spot that will be free of dry, noise making debris; a spot that does not contain a mine or booby-trap initiator and that will offer stable footing. He will also look to ensure that his subsequent steps will optimize stealth and proper direction. He will then look in the immediate area, on the ground and among surrounding plants, for venomous snakes or other hazards. Only until these brief examinations are done, will he shift his weight and take his step. This process is especially important for the point man or for Team Members who step away from the formation (e.g. Leader's Recon, etc.) or defensive perimeter. If a Team Member has been preceded by other Team Members in the Team formation, some of these steps will have already been taken so that he may place his steps in the footprints of the other Team Members before him.

- ○ Each Team Member will guard/observe his area of responsibility, and overlapping areas to detect the presence or signs of enemy combatants and other items of tactical importance. This will often require the Team Member to frequently lean over to peer under leafy vegetation. The Team Member should frequently glance behind him to determine the status of other Team Members in trail.
- The T/L should take account of his fatigue and the fatigue of Team Members to gauge when to pause for breaks during movement. Breaks in rough terrain should typically be 10 minutes every hour; 15 minutes on occasion. During these breaks, Team Members should take a knee or sit and the Team should then observe utter silence for several minutes to detect any sounds of enemy movement, engine noise, water flow, etc. A listening break requires slow breathing; no chewing or drinking/swallowing; no rustling of clothing and equipment. Only after the T/L determines that sufficient time has elapsed for the listening break, he may take the opportunity to orient other Team Member on location/land navigation status, etc., before resuming Team movement.
- Always keep an eye out for defensible ground while moving.
- Treat all trails (old and new), rivers/streams, and open areas as danger areas. When crossing a linear danger area (rivers/streams, roads/high-speed trail networks, rail lines, enemy obstacles, fire breaks, etc.) the Team may not know what is on the other side of the danger area, particularly where the area is heavily vegetated or bounded by vertical terrain. Rivers and moderate to large streams will often have a network of trails or even a road running in parallel on either side or even both sides. In this circumstance, crossing may be highly hazardous, as enemy personnel may come upon the Team flank unexpectedly during the crossing or may spot the Team from a distance from either bank. The T/L may want to conduct a Leader's Recon, or dispatch one or two Team Members to recon the opposite side before crossing the main body of the Team. The Author recommends the following procedure (Refer to Figure 19):

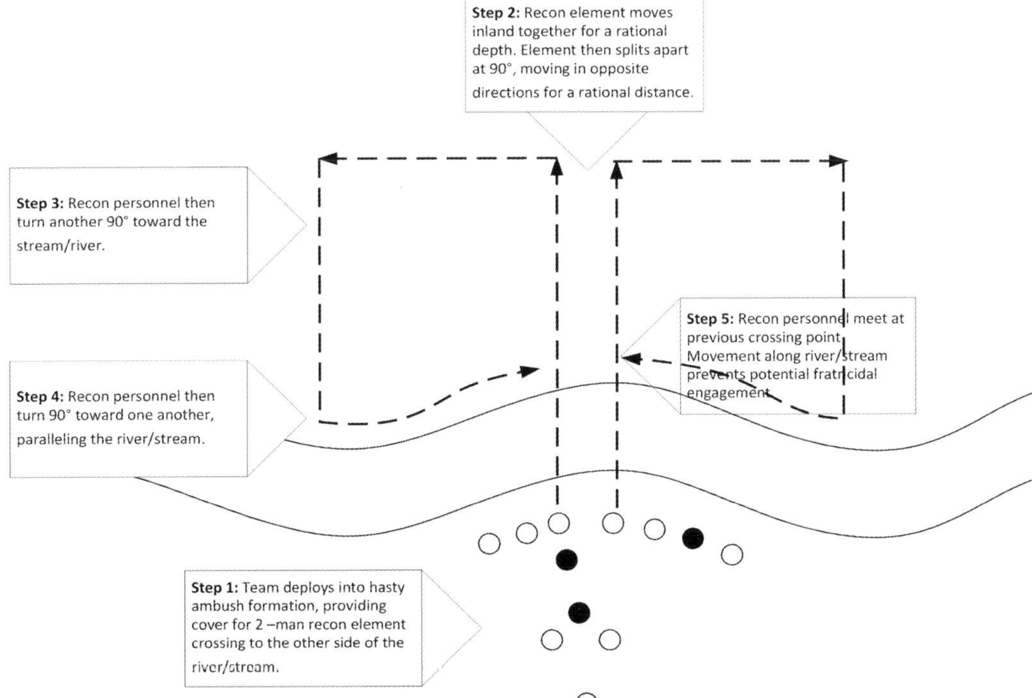

Figure 19. River Crossing Technique.

- ○ Select a crossing point with terrain and vegetation that limits enemy long range observation (at a dip, ravine, saddle, bend).
- ○ Post a grenadier to both flanks as security. It may even be worthwhile to set booby-traps on the back-trail and along high-speed routes of approach from the flanks (e.g. on high-speed trails); booby-traps will alert the Team to an approaching enemy, as well as inflict casualties, during the crossing.
- ○ Send a pair of Team Members across at the crossing point. These Team Members use a boxing method to reconnoiter. The boxing method consists of the Team Members moving as a pair at sufficient distance to observe beyond obstructing terrain and vegetation, or as directed by the T/L, then split 90° to the flanks to continue the recon, then box again 90° returning to the linear terrain feature, where the Team Members again box 90° toward each other along fringe of the linear terrain feature until they meet at the original crossing point. A recon element of two personnel speeds up the recon. Using the box technique allows the Team Members to meet during the return in a manner that mitigates the possibility of a fratricidal encounter.

- When crossing large streams, normal practice is to first ascertain a crossing point; next to observe the far bank for activity; then to send one or two Team Members across to check the opposite side; then cross the rest of the Team.
- To gather water at a stream, the bulk of the Team should move into the stream and travel upstream for a short distance, then ascend the opposite bank and establish a perimeter and security; 1–2 Team Members may remain on the near side to observe the back-trail; 2–3 Team Members should drop their rucksacks and gather empty water containers from the other Team Members; the water containers (canteens, bladders) can be carried by using 550-cord inserted through the container cap loops. The water team should take care not to rattle canteens while moving. The rest of the Team provides over-watch while the water team completes their tasks.
  - ○ Note that a stream is a linear danger area and that the water team will be exposed as water is collected. This can be mitigated by gathering water at rivulet/cascade or where vegetation shields detection.
  - ○ An alternate method may be for the Team to use a 'bucket' consisting of a heavy-gauge plastic bag (e.g. 4 gallon, 6 mil), encased in a sandbag (to provide strength and protection to the plastic bags) to gather water. Dark colored 6-mil plastic bags can be acquired with string/scrim reinforcement which would not require sandbag protective covers, but the Author recommends the reinforced plastic and sandbag combination. The mouth of the plastic bag should be folded over the mouth of the sandbag to form a 2in collar.
    - ▪ If the water is deep enough, the water can be gathered while the Team crosses the stream (or a river), rather than using a water team. The Team Member should only fill the bag partially, as each gallon would weigh approximately 8lbs.
    - ▪ This same approach can be used to gather water where the stream/river bank is steep or along the shore of a lake, essentially using the encased bag like a well bucket. This will require 550-cord tied to opposite ends of the plastic/sandbag collar. Use ½-in

Figure 20. A "loaded" field expedient bucket, using a plastic bag encased in a sandbag (winter white).

pebbles, placed underneath the collar at the tie-off points, so that the 550-cord will have firm anchors. A strap, attached to the tie-off points, can be used as a bucket handle.

- Place a rock at the bottom of the plastic bag so that the bag will sink with the mouth oriented upwards and so that air is not trapped in the bag.
- The bag will be semi-rigid, facilitating the collection of water in shallow streams.
- Retrieving water by this means facilitates the use of a water filtration/purification device while transferring water to individual bladders/canteens.

- During rest halts don't remove your pack or move out of arm's reach of your primary weapon.
- During communications breaks, the radioman should not remove his pack until the perimeter has been checked and Claymores deployed.
- Frequently practice mounted and/or dismounted movements and Battle Drills during night training. Also, practice Battle Drills that react to enemy contact from the point, rear and each flank in varying terrain situations; and ensure that casualty simulations are incorporated into these drills as well.
- The tail-gunner, or rearmost Team Member in a file formation, is responsible for concealing, observing and defending the back-trail during movement. This applies only if the Team or organizational element is moving in the file formation.
- There are five basic techniques of movement that can be employed by SR Teams to avoid being detected or attacked by enemy forces. Each of these are explained and illustrated below:
  - <u>Terrain Hopping Technique (Author's preference)</u>: The Team will move to a terrain feature (ridge/ravine/streambed) and follow it for a time, then move to another terrain feature – each terrain feature will take the Team, in circuitous fashion, closer and closer to the target. This technique makes it fairly difficult for the enemy to predict the line-of-march or the likely target location. Ridgelines are typically followed along the military crest. This technique is indirect and more time-consuming.
  - <u>Dog-Leg Technique</u>: Another effective Team movement method is the Angle or Dog-Leg technique. The patrol will change the direction of movement, in dog-legs, generally oriented toward the target in a series of angular directions. As with the Terrain Following

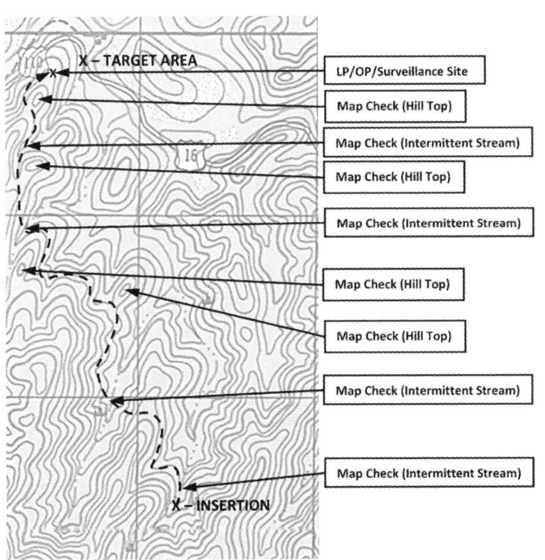

Figure 21. Terrain Hopping Movement Technique

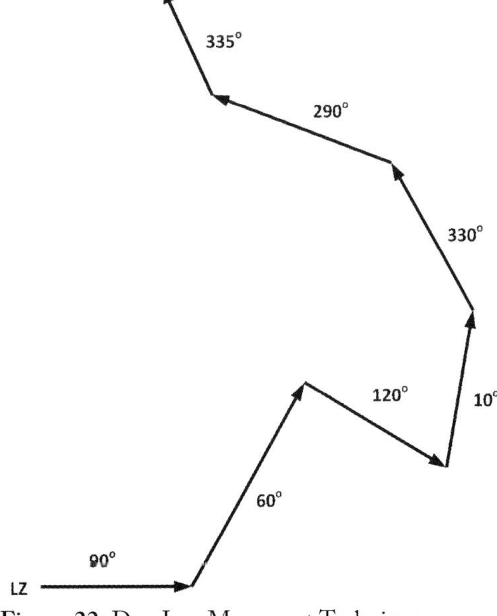

Figure 22. Dog Leg Movement Technique.

approach (above), this technique makes it difficult for the enemy to predict the likely target and line-of-march. If many doglegs are used during a single day's movement, the line of march may become laborious to the Team and time-consuming as well. Do not allow this technique to become too complicated. An example is shown at adjacent figure.

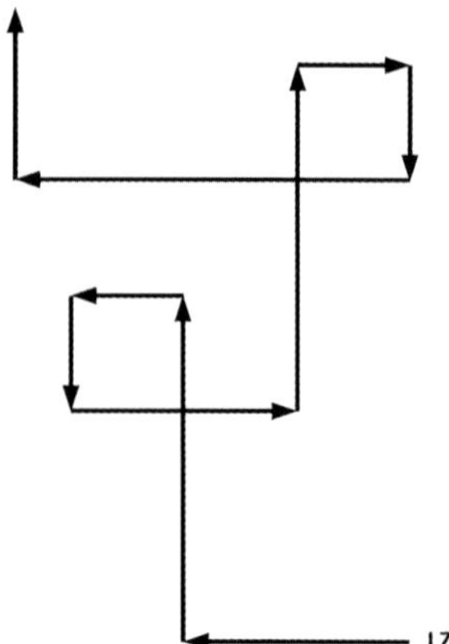

Figure 23. Box Movement Technique.

○ <u>Box Technique</u>: This method is intended to shake trackers (actual/suspected). This is not to be confused with the Fishhook Technique which is used prior to communication breaks, NDP or establishing a surveillance hide. The box technique is used in series to cut the Team's own back-trail. From a given point the Team moves out on a set azimuth for a distance; the Team then makes a 90° turn and moves a distance before completing another 90° turn, etc. making three 90° turns in form a 'box'. At this point the T/L can direct any of several actions. The Team can wait in ambush for trackers or pursuers; walk backwards across the back-trail, if the vegetation and soil is such that it is impossible to hide your tracks; or continue toward its target. The Team can repeat this technique, if time allows.

■ By forming these boxes, the Team can set up the enemy for an ambush; at a minimum, the Team will confuse any trackers as to your direction of movement and objective.

■ The enemy may find such erratic Team movement especially dangerous for them to track. It will cause the enemy unit to track the Team much more slowly or the enemy may change tactics entirely and attempt to anticipate Team's route in order to set up ambushes, or a 'Hammer and Anvil' operation.

■ The chief disadvantage with the box technique is that repeated box maneuvers will consume valuable time (especially if the Team sets up an ambush on the back-trail); which may allow the enemy to mass additional troops to the area. The second disadvantage is that little headway toward the objective is made while these maneuvers are undertaken. The third disadvantage is that if the Boxing maneuver cuts across enemy trails in the process, the Team may be exposed while it crosses these danger areas and the trail signs of the crossing may obviate the purpose of the maneuver altogether. Fourth, crossing the Team's own back-trail may provide the enemy a shortcut, by detecting and following the trail as it crosses the back-trail, to catch up with the Team. Fifth, changes in direction may bring the Team up against obstacles or other impediments (e.g. heavy thickets) which may bungle the box configuration. Sixth, unless the Team is equipped with an operating GPS capability (not always the case in heavily dissected terrain or heavy canopy), the Team may become disoriented.

■ The Author suggests that use of the box technique should be used sparingly, if used at all. For instance, it might be used immediately after an insertion.

○ <u>Cloverleaf Technique</u>: The Author recommends an alternative to the box technique. The Cloverleaf Technique loops back toward its own back-trail; then the Team, rather than crossing its back-trail, retraces part of its original path. At the point where the original path would've made its first loop deviation, the Team makes a deviation in the opposite direction creating another loop. When the Team re-approaches its original back-trail, it may again retrace its original path, and so on. It can also be combined with the terrain-hopping or dogleg techniques; and as the enemy tracker team will pause a while to decipher the trail, the technique provides opportunity for the Team to ambush the pursuit. When the T/L is ready to resume the movement to the objective, the Team should do its best to conceal the location where it breaks from the Cloverleaf looping. The Cloverleaf Technique has the same advantages of the box technique, plus it can be executed more swiftly (without taking compass bearings) and with less disorientation. Disadvantages (chiefly in consuming valuable time) associated with the box technique may also apply to the Cloverleaf Technique.

○ Figure Eight Technique: The figure eight method is similar to the box techniques, but the Team will be making circles instead of squares. The circles can be used to curl around certain terrain features such as hilltops, thickets and is more flexible in bypassing obstacles or other impediments. Terrain hopping would be the best land navigation technique to use in conjunction with the Figure Eight. An example is shown.

> **True Account: One survivor of an ambushed Reconnaissance Team succeeded for three days in ambushing and killing six enemy pursuers by employing the figure eight method before being spotted and recovered by searching aircraft.**

○ Step Technique: The simple method of changing the route-of-march in 90° turns for a distance of a 100 or so meters, prior to taking a heading towards the objective. Related to but different from the Dog-Leg Technique, as it uses 90° turns rather than smaller angles, to navigate towards the objective from start to finish. An example of the Step Technique is shown.

○ Skip Method: This method is employed if the Team has been moving along a trail or a stream/streambed. The Team stops in place and on command will move laterally off the route, stepping widely off the path, taking care not to leave signs. Alternatively (and a better approach), the point-man alone may step laterally off the trail, followed by the rest of the Team in file. Best practice is to step on rocks, fallen trees or tree roots to conceal footprints. The Team point man may lay a false trail for 30 to 50 meters further along the trail or streambed before the skip method is employed. If well executed, and if enough time has passed (e.g. for moisture to dry after exiting a stream; or for disturbed vegetation to recover), this method may deceive a tracker team into continuing its pursuit along the trail or streambed for a while. See the example.

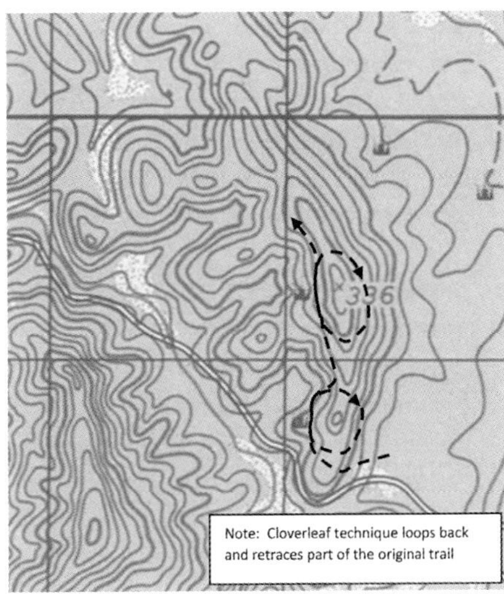

Figure 24. Cloverleaf Technique.

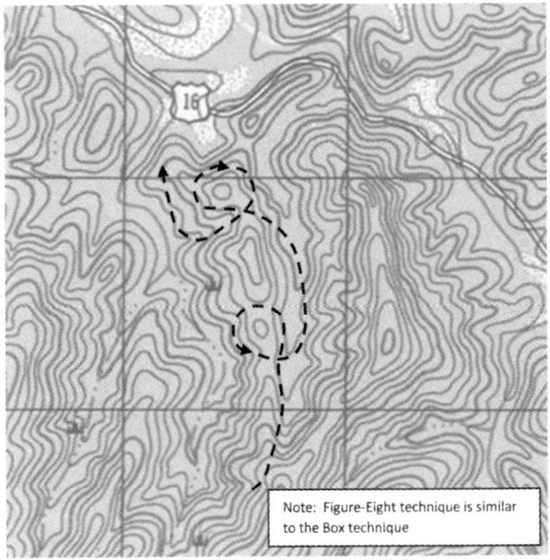

Figure 25. Figure 8 Movement Technique.

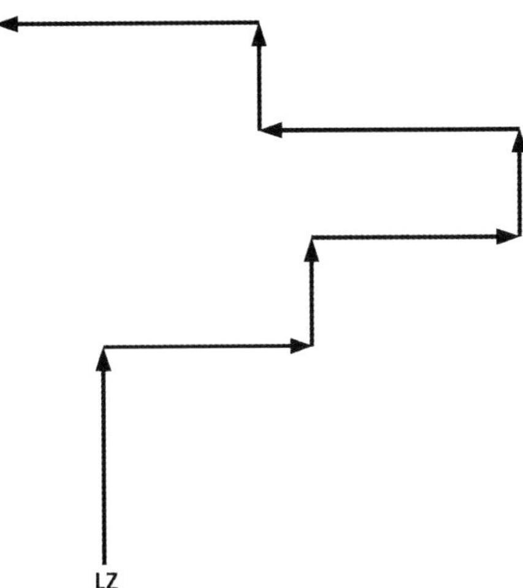

Figure 26. Step Movement Technique.

- ° Additional Tips:
  - ▪ Avoid setting a pattern, unless you intend to use a pattern to set-up the enemy for an ambush.
  - ▪ During the dry season, CS powder spread over your back-trail is extremely helpful in stopping dogs.
- There is a danger that Team Members may get 'tunnel vision' upon seeing, observing, sensing, or engaging the enemy. Team Members must have the discipline to fully attend to their areas of responsibility including visual and auditory focus.
- Be aware of what is lurking <u>above</u> the Team:
  - ° Team Members tend to look down, ahead or laterally.
  - ° In mountains, beware avalanche and falling rock hazards.
  - ° Tree canopy may contain:
    - ▪ Enemy observers.
    - ▪ Dead branches (widow makers)
    - ▪ Rotten or dead trees.
    - ▪ Suspended ordnance.
    - ▪ Booby-traps (e.g. deadfalls)
- Vegetation in dense jungle and rainforest concentrates its foliage mostly at the top of the plant or tree. The lower portion of the plant is often sparsely leafed. Note that an enemy lying in ambush will be close to the ground to obtain a good field of view/fire and to seek whatever cover the ground can afford. This is where your Team Members should scan or observe for the enemy during Team movement. Having a point man of short stature is helpful in this endeavor. Note that frequent bending over to peer under leaf cover can test back and abdominal muscles – train accordingly.
- The rearmost Team Member of a Team/element (the tail-gunner) is responsible for concealing the back-trail leading to the NDP, hide position, etc.
- Each man on a Team must observe the man in front of him and the man behind him, and be watchful for other Team Members' arm and hand signals.
- River currents speed up at the outside of a bend and slow down at the inside of a bend. Stay away from holes in the riverbed caused by these flows.
- Beware of using culverts or ground adjacent to culverts in certain geographic areas. These may be a favorite habitat for predatory animals or snakes, especially in hot weather environments.
- Movement along a high-speed trail or road is a high-risk endeavor. Only do so if the benefits outweigh the risks, and do not make a habit of it.
- If the Team can expropriate small boats (e.g. native dugouts), these water craft can be used to ferry the Team cross lakes/inlets, large streams or rivers to conduct an ambush or raid or lose enemy tracker teams; the Team can use the same watercraft to withdraw, leaving little or no trail for the enemy to follow.
- If the Team is to operate with mobility equipment, train and practice mounted battle drills.
- If the Team is to operate with skis, snowshoes and/or *ahkios*, train and practice battle drills while encumbered with this equipment.

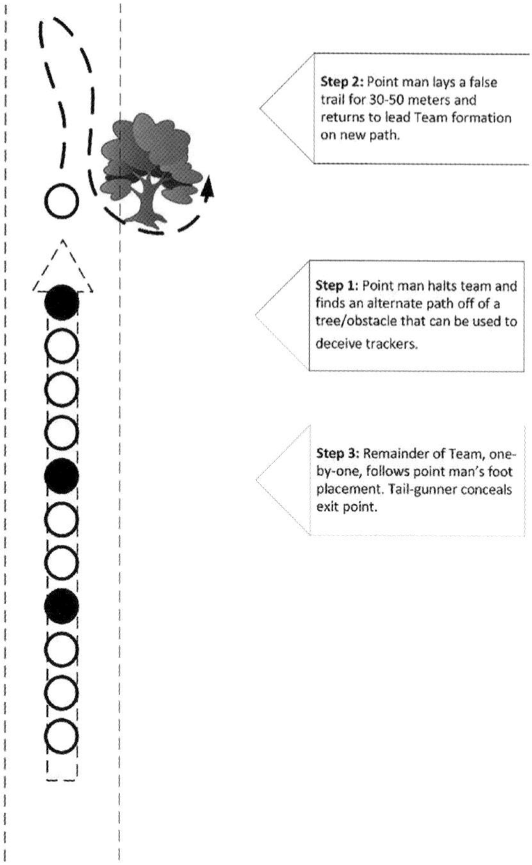

Figure 27. Example - Skip Technique.

Mounted Movement TTPs:

- Consult FM 31-23, Special Forces Mounted Operations, Tactics, Techniques and Procedures,[1] a compendium of Lessons-Learned by personnel at Fort Bliss, Fort Campbell, Fort Bragg, and overseas, to include Operations Desert Shield/Desert Storm, Restore Hope, and Provide Democracy. FM 31-23 is filled with generally good advice, with some exceptions.
- Cross-country movement in ground mobility equipment is much slower than one might imagine.
  - Mounted cross-country movement will be impeded by vegetation and over landform that vegetation may conceal and through heavily dissected terrain requiring frequent navigation deviations (dog-legs, terrain hopping). Often, a mounted Team will have to navigate around areas of steep terrain and/or dense vegetation.
  - Weather, dust and lighting conditions will require reduced speeds.
  - Logging trails, high-speed and rudimentary trails, animal paths as well as paved roads, may be seeded with mines/booby-traps, or they may be under observation and fields of fire. Such movement must be governed by caution that will reduce speed of traverse.
  - The need for stealth will require reduced speeds.
  - An odometer can get a Team into trouble. Note that the vehicle odometer during cross-country movement (especially in sand, snow, mud and vertical terrain), where frequent deviations are required, or during movement on roads through mud and snow will not be an accurate measure of distance (due to tire slippage); therefore, odometer accuracy must be periodically checked against maps and/or GPS, especially in desert or grassland landscapes where landmarks/terrain features are not present, and dead reckoning must be used to navigate.
- Extensive training in the operation of motorcycles or ATVs/UTVs is essential. Cross-country operation of these vehicles can be hazardous. Additionally, battle drills executed from these vehicles will necessarily be entirely different from those executed by dismounted troops or by troops mounted in larger vehicles.
- ATVs/UTVs or other four-wheeled vehicles (including foreign/enemy light tactical vehicles) can be fitted with valuable accessories/attachments that can increase Team capability and survivability. Examples:
  - A light front-end loader, excavator or towed plow could save the Team Members considerable time and effort in establishing MSSs/caches, defensive positions, surveillance hides/Ops and mine placement.
  - Accessories/attachments should be removable, so that they can be dropped when specialized tasks are completed, and allowing them to resume their primary purpose of Team mobility.
- Vehicular movement during rain or snowfall will suppress sound. Be aware of this as you approach, or cross roads, during these conditions.
- Vehicular movement on gravel roads, even for slow-moving vehicles with suppressed engines and exhausts, can be heard for a half-mile or more in rural areas.
- FM 31-23 recommends that during short breaks (<15 minutes), vehicles should stay in traveling formation and that they should turn off their engines to conduct a listening break. Two points to make on this:

---

1. Field Manual 31-23 (ID), 'Special Forces Mounted Operations, Tactics, Techniques and Procedures', (Department of the Army, United States Army John F. Kennedy Special Warfare Center and School , Fort Bragg, 3 March 1998). Note that this document contains a recommended reading list from which some of its TTPs are drawn.

- ○ Team Members' hearing will barely recover from vehicular noise within a 15 minute span.
- ○ If vehicles pause in the traveling formation while they are exposed, and while Team Member hearing is impaired, they can more easily be detected by airborne assets even at night, as the Team may not hear the approach of enemy aircraft. Furthermore, if the tactical formation happens to be the traveling column (file), they are not well disposed for chance encounters. The most likely chance encounter would come from the front or the rear of the column, where the team is most vulnerable; if Team tactical discipline is lax, the enemy can use a flanking movement upon spotting the Team.
- Mounted movement is generally conducted at night; therefore, mounted SR Teams do not generally occupy a NDP, but rather a RAD (Remain All Day) or a hide.
- ○ As with NDPs, SR Teams should avoid communicating status reports from RADs. Mounted SR team should communicate prior to dawn, before it occupies its RAD. In long-duration hides, the SR Team should move off from the RAD to transmit off-site. If possible, use terrain and directional antennas to shield the transmission location from RDF.
- ○ As with the NDPs, the SR T/L must reconnoiter outside the RAD perimeter. He may delegate this task (e.g. to Team Members with mobility equipment).
- If motorcycles are being used for ground mobility, their track is narrow enough to use on animal trails during cross-country movement. Team Members equipped with motorcycles cannot use map and compass while they are moving, if they're using the terrain-hopping technique, they should memorize 2–3 navigation checkpoints/RPs to avoid having to stop frequently to make reference to map and compass.
- Do not leave equipment or supplies (e.g. Satellite Communications (SATCOM), water cans, etc.) out after use. Remove items from the vehicles when needed; use the items immediately after off-loading and then immediately place the items back on the vehicles and secure the load to be ready for immediate movement. Ensure key items (ammunition/munitions, medical kit, etc.), while secured in the vehicle, are rapidly accessible.
- IR headlights on ground mobility equipment should rarely be used, as enemy troops equipped with night-vision equipment will spot the illumination from far away.
- Driving fast across the desert may kick up a plume of sand and debris that will obscure the vision of other vehicle drivers in column. This plume will screen rocks, wadis and other impediments from the driver's view which may result in fatal accidents. The plume will also reveal the presence of the Team to enemy aviation and pursuit and is also detectable by ground surveillance radar. Unless a brisk wind crosses the direction of travel, use another formation – or drive slowly enough to prevent plume generation.
- If ground mobility equipment is to be inserted using air assets, vehicle fuel tanks and gas cans should only be half-full to account for expansion at altitude.
- Carry 'sand channels' to extricate the vehicle from sand.
- Ensure a device (e.g. thermite grenade) for the destruction of the vehicle is kept on-board.

Tactical Dismounted Formations TTPs:

- Whenever the SR Team has indigenous Team Members, it is wise to limit the number of Tactical Movement Formations (and Battle Drills) used by the Team.
- As noted in the Stealth Formation paragraph above, the file formation is preferred in dismounted SR patrols. The file formation is the standard formation for reconnaissance operations in most terrain, weather and vegetation conditions. The file formation maximizes stealth, maximizes fires to the flanks, minimizes a detectable trail, enhances navigation control and hand and arm communication, and facilitates scaling/climbing steep or dissected terrain and in the penetration of dense vegetation and obstacles. It is best used in traveling

through jungle, rainforest, areas of heavy underbrush, cross country over deep snow and in restrictive terrain (e.g. mountains, along streambeds, down ravines, urban environments, etc.) and when visibility is poor. The file formation is not optimum for fires to the front or rear; hence the importance of Battle Drills that emphasize reaction to front or rear contact (discussed elsewhere in this book).

- Other formations are also recommended, in special circumstances or when the SR Team converts to a combat patrol purpose.
- If the SR Team is to deceive the enemy by imitating them, The SR T/L should consider using enemy formations, especially when crossing danger areas or in operating in open terrain.
- If the SR Team is to cross a large danger area, the T/L <u>may</u> opt for a more open formation, but after crossing the danger area, the Team will normally reform its file. Even so, the file may sometimes be more appropriate if terrain folds permit concealment and cover.
- If the SR Team is to cross a sequence of danger areas, or when the SR Team approaches its objective or a prospective ambush site, the T/L may opt for a suitable tactical formation. In crossing a series of ridges, the Team is at increased risk as it descends from a ridge into a ravine and ascends from the ravine to surmount another ridge. Being attacked from an enemy force from atop either ridge is one of the worst tactical nightmares an SR Team can face. Even a single enemy soldier can take out the entire Team with grenades tossed into the ravine without exposing himself to fire. In the Author's opinion, the 'Diamond' formation may be the most suitable for this situation. This approach allows one or more elements of the Team to ascend a ridge in a Diamond Formation, while other elements remain in overwatch. The Diamond provides enhanced fires to the front, rear and to the sides during the crossing. The overwatch element is responsible for guarding against an enemy tracker element, guarding the lateral high speed approaches of its ridge; and covering the opposite ridge top while the first element is crossing. Once the lead element has reached the bottom of the ravine and has begun its ascent, the second element begins its descent to the ravine. Once the first element is secure on the opposite ridge, it covers the crossing of other elements that follow. The Diamond is similar to the Bounding Overwatch (Wedge), but is faster in execution, more compressed and easier to control. This formation is also useful while being pursued by an enemy force in more open areas, as it provides better, more responsive firepower coverage (than the file) to the front and rear of the Team during movement. This formation may also be the best for use during breakout operations.
- When the SR Team must execute a breakout, the Diamond formation would be the best formation, with WIA, litter-bearers and medic occupying the center of the formation.
- If the T/L is anticipating a meeting engagement, the T/L should opt for an appropriate tactical formation. The Bounding Overwatch (Wedge) formation may be the best formation for this purpose, but the Diamond would be even better if the SR T/L is not convinced that he has shed the likely enemy tracker team. If the Bounding Overwatch is used, the elements ought to be in tight wedges, or even Diamonds, where all Team Members of the element can be seen and controlled by the element leader (in open terrain) or the entire Team seen and controlled by the T/L

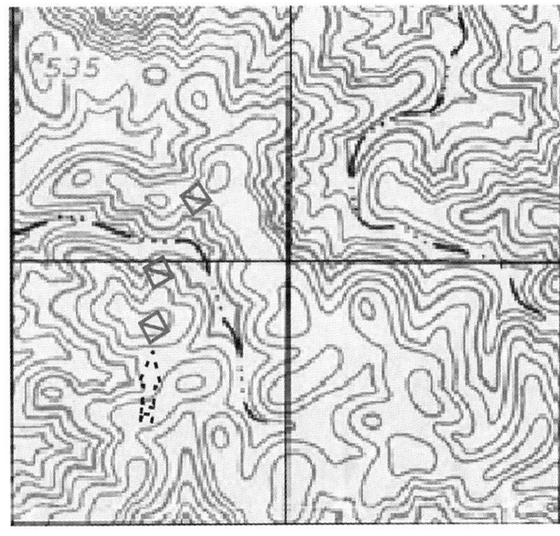

Figure 28. A 12-man Team in 3-Element Diamonds Crossing a Ravine.

(in closed terrain). Moving into this formation would be very much like executing a 'hasty ambush' formation, which simplifies training and execution for indigenous Team Members. The Team can use the same formations (elements in Wedge/Diamond) if it operates in three elements with a weapons support or special purpose element in trail.

Movement in Steppe, Savannah, Grassland, Tundra TTPs.

- Teams deployed to these areas should be equipped with mobility equipment or horses, due to the vast area of uncovered, unbroken terrain to be traversed. Captured enemy vehicles may be preferred, as enemy observation of such equipment in such open terrain may not raise suspicions.
- Crossing Grasslands:
  - Navigating across Steppe/Savannah/Grassland/Tundra is similar to crossing of desert or flat, snow-covered terrain, in that the Team can be observed from a long distance in daylight and during periods of increased moonlight. And at night, the Team can be detected at long distances using thermal optics. If the Team is using enemy equipment, especially mobility systems, the enemy may be fooled by the thermal signature.
  - If possible, use sun and shadow to provide cover of Team Movement. Cross near dusk or dawn when the sun will be in the eyes of any enemy observers. Use shadows cast by folds in the earth whenever possible. Also, observe for reflections from enemy equipment/ optics while the sun is glaring in the enemy direction.
  - Where possible, navigate across such terrain using whatever gullies, washes, ravines or other terrain form may exist. Even so, use folds in the earth to the extent that they are available. High grass may appear uniform across the grassland but may conceal actual folds in the earth; they might not be revealed by contour lines on a map.
  - Movement during early morning will leave an easily detected path in the morning dew that will endure, the path may be especially noticeable to low flying aircraft. If sun or shadow is not in the Team's favor, delay crossing, if possible, until time or weather favors movement. If animal paths are available, use them. If neither of these options are available, make a zig-zag path to the next available terrain fold.
  - Bad weather and periods of low visibility generally favor stealthy, covert movement across grasslands. Periods of fog, falling snow, rain and high wind and/or overcast night sky are optimum conditions for crossing such areas; rapidly cross while such conditions prevail. Consider weather forecasts in mission planning and execution.
- Notice the Steppe/Savannah/Grassland/Tundra similarities in the following photos. Also notice the folds in the terrain, some barely noticeable. Grass or crops grown on these types of terrain will often conceal folds in the earth; but relief/contour lines on a map may reveal their locations. Day and night optics capability is a must for operations in these expanses.

Figure 29. African Savannah and Russian Steppe. (*Depositphotos.com*)

Night Movement TTPs:

'Once a soldier has learned how to move and fight at night, he will be all the more effective in daytime when good visibility facilitates his tasks.'[2]

'Night movements and night combat require the most exacting preparation by officers and men, including detailed map and terrain study.'[3]

'To maintain control and intra-unit contact and communication is difficult during the hours of darkness, and unit commanders therefore prepare every detail of the operation plan with meticulous care. Any contingency, however far-fetched, must be taken into consideration.'[4]

'Success of night operations depends primarily on careful planning, detailed preparation, simplicity of the operation order and tactical procedure, achievement of surprise, and the leaders' calmness and circumspection.'[5]

'In some instances individual Russian reconnaissance patrols, led by capable and energetic officers, managed to slip through gaps or weakly held positions in the German front under cover of darkness. They either restricted their activities to obtaining information or expanded the scope of their mission by disrupting wire communications, laying mines, and carrying out commando-type raids on CP's.'[6]

'[T]he advance guard of a German infantry division was attacked during the night in a large village where the reinforced battalion had stopped on the way to Kharkov. After the Russian attack had been beaten off, the German battalion commander found that a Russian rifle platoon had been left behind in the village after all other troops had withdrawn and that the men had concealed themselves in groups of two or three in the dung hills near the farm buildings. Their mission was to observe the Germans after their entry into the village and to communicate the information to their parent unit....'[7]

- Train Team Members in individual and Team/element movement, land navigation, weapons firing (including cross-training, field stripping and clearing of malfunctions), battle drills and tactical tasks (e.g. deploying into hasty ambush positions), first aid, at night and in all weather conditions. Train Team personnel by first navigating over ground during daylight, and then follow the same path during periods of darkness.
- The likelihood of making noise during movement increases substantially at night. Therefore, the pace of movement at night is slowed, with frequent listening stops.
- Extended hours of daylight (during peak summer months) and extended hours of darkness (during peak winter months) occur in northern regions. These phenomena also occur in extreme southern regions. Developing a capability to operate during extended periods of darkness is essential for operations in these circumstances.
- Militaries located in northern (e.g. Russia)/southern tier nations will typically have substantial training and experience in night operations.
- The possibility of chance encounters, and therefore close-quarters combat, increases during periods of limited visibility while the Team is moving. Train Team Members accordingly.

---

2. Toppe, 'Night Combat', p. 1.
3. Ibid, p. 2.
4. Ibid, p. 4.
5. Ibid, p. 5.
6. Ibid, p. 21.
7. Ibid, p. 22.

- Sample WWII German Army training schedules for night training can be found in the appendices of CMH Pub 104-3.[8] These schedules and the subject matter contained therein (with some tailoring) are still relevant today and are recommended by the Author of this book.
- Indigenous Team Members, partisan attachments, etc. may not be equipped with night-vision equipment. Train the entire unit to operate with partial night-vision capability and train the entire unit without the benefit of night optics.

'German field commanders with many years of practical experience advocate that up to 50 per cent of all training be conducted at night.'[9]

- Night operations, with and without Night-Vision Devices (NVDs), is a training imperative. This includes practice of Battle Drills, weapons firing, land navigation, reconnaissance and surveillance and actions on target.
- As nearly all policemen know, a very bright flashlight or a strobe light (with limited life batteries) can ruin a person's night vision for up to ten minutes. If the Team is being pursued at night, lay a booby-trap with Self-Destruct (SD) next to a flashing strobe light. The strobe light will act as a decoy, attracting the attention of the enemy and luring him into an approach where the Team may have placed booby-traps/mines.
- Sounds travel especially well at night. Use the sounds of enemy activity to mask the sound of Team movement.
- The cross-country pace for a Team during night movement is normally reduced from that of daylight movement due to requirements for stealth and avoidance of detection. However, night movement using trail networks might actually exceed the typical pace of daylight cross-country movement.
- Use friendly aircraft fire/dropped ordnance to mask the noise of Team night movement. Monsoon rains will also mask the noise of Team night movement.
- Enemy combatants may use flashlights in rear areas or under multiple canopy forest. The T/L must weigh the pros and cons of following the enemy protocol. If the Team is detected moving without lights in an area where the enemy freely uses lights, this movement may draw attention. If the Team uses a flashlight during movement (e.g. along an enemy trail), the flashlight should be fastened to the end of a walking stick/pole.
- Night movements will generally require the Team Members to use NVDs.
  - While US Special Forces personnel are equipped with NVDs, indigenous Team Members may not be so equipped. Night movements and night operations will therefore be constrained by those personnel who lack night-vision equipment. This impairment can be mitigated through training and the implementation of specialized techniques or field expedient measures.
  - The use of the Night-Vision Goggles (NVG) built-in infrared LED should be avoided if possible, especially where the enemy is close, as enemy troops in proximity who are equipped with their own NVDs may spot the LED illumination.
- The T/L must weigh the advantages and disadvantages of night movement. Considerations might include:
  - Tactical exigency, such as breakout, hot enemy pursuit, movement into tactical positions (e.g. ambush), etc.
  - Conducive weather/environments. Phase of moon; snow cover, etc.
  - Making up lost time.
  - Availability/non-availability of an enemy NVD capability.

---

8. Toppe, *Night Combat*, Appendices.
9. Ibid, p. 42.

'Before any night operation, the responsible commanders must familiarize themselves with the theater of operations, become thoroughly acquainted with the enemy materiel and his methods of employing them, and observe carefully his tactics in different situations....

'Reconnaissance must be an uninterrupted effort; frequently the most useful information is gathered through night reconnaissance. During the hours of darkness, friendly patrols are able to penetrate deep into enemy territory to points from which they can observe enemy movements during daytime.... In darkness, reconnaissance patrols can usually determine only whether or not a specific area is occupied by the enemy.'[10]

— Lessons-Learned on the WW II Eastern Front.

'German field commanders with many years of practical experience advocate that up to 50 per cent of all training be conducted at night, starting from the very first day of basic training.... They advocate that the most important features on the weekly training schedule take place at night and that the Lessons-Learned in daytime be repeated and driven home during the hours of darkness. By shifting part of the regular schedule from day to night, what may achieve the dual-purpose of top running the soldier and making him a night fighter.'[11]

- Terrain hopping, from terrain feature to terrain feature, may be the best technique for land navigation at night for a variety of operational environments. See Figure 17.

Rally Points (RP) TTPs:

- The T/L should periodically designate primary and alternate rally points while on the march; however, in rainforest/jungle, noticeable/remarkable features that might be suitable for RP designation are often not easily identifiable. Subsequently, the T/L should periodically (especially at the noon halt and prior to occupying NDP positions) notify Team Members of the direction to, or locations of, the nearest LZs — or stream valleys, ravines, etc. that will lead to LZs. Whenever possible, two LZs, each in a different direction, should be designated. Alternatively, during Team movement, RP designation should be in accordance with SOP (e.g. Move in the opposite direction from enemy contact 'X' number of meters.)
- Ideally, rally 'points should easily be identifiable during both daylight and limited visibility, show no signs of recent enemy activity, offer cover and concealment, be defendable for short periods of time and be located away from … high-speed avenues of approach.'[12]
- If possible, RPs should be down slope to facilitate accelerated movement away from enemy contact, to ease carriage of WIA/KIA and should be off of the Team back-trail, if possible (given the possibility of enemy trackers in close proximity).
- Night Movement RPs:
  - If the Team is executing night operations, RPs may be established along the back-trail, for instance at a dip in the trail, or where the back-trail crosses a streambed, etc. This is notionally appropriate if the Team has contact from the front and if the Team back-trail is easily retraced. Enemy trackers will be much less a concern during night movements.
  - More appropriately, the night RP should be downhill, into a ravine or toward another linear terrain feature to the flank of the Team's direction of travel and to the rear from the direction of travel for a standard distance (by SOP). Move quickly, before the enemy can hurl grenades down into the ravine. This technique can be used if enemy contact is from

---

10. Toppe, *Night* Combat.
11. Ibid.
12. John D. Hurth, *Combat Tracking Guide*, (Stackpole Books, Mechanicsburg, PA), p. 79.

the front or the flank. This technique should work in most cases, barring an enemy area ambush that anticipates the SR Team reaction.
- ○ The T/L should account for Team Members at the RP. If any Team Members are missing, the T/L should determine where the Team Member was lost. The T/L must make a decision to either try to locate the missing Team Member(s) or move to the secondary RP/LZ, anticipating that missing personnel will join the Team there.
- How to find a Rally Point Tracing Back From a Terrain Feature:
  - ○ In open terrain, select a recognizable feature (e.g. a lone tree) in the distance and measure it using the knuckles of the hand (or a marked stick), held at arm-length, to estimate the distance.
  - ○ If the Team engages the enemy and is scattered, Team Members may use the knuckle/stick method to estimate the same distance measured from the singular feature (from where the first measurement was taken) and then move in an arc to intersect the back-trail and subsequently move to the location from which the original measurement was taken. From that spot, separated Team Members can then navigate to the designated RP. This field expedient is especially useful in desert or steppe.

At the Halt TTPs:

- Teams will normally take a 10 minute break each hour to rest and listen for enemy movement. Team Members should sit as quickly and silently as possible and then listen intently. A tracker team, that is closing on the Team, will move in concert with the Team movement to obscure the noise of its own movement. If the Team suddenly stops, Team Members may be able to hear the enemy movement before they too take a halt. When the Team resumes movement, they should do so as quietly as possible.
- SITREPs may normally be scheduled (approximately) around the noon break and prior to establishing the NDP or a Rest All Day (RAD) 'day position'.
- Be aware that the enemy may have RDF capability; brief, rapid transmissions are essential.
  - ○ This problem is largely alleviated if the Team is equipped with a radio that features frequency hopping encryption and/or burst transmission capabilities. Otherwise, it will take time to compose and transmit an encrypted message using a one-time pad.
  - ○ Make allowances for the time consumed in this process so that there remains enough ambient light near dusk to subsequently move to and set up the NDP. This time period is referred to as EENT (End Evening Nautical Twilight) or nautical dusk.
- If feasible, the T/L confers with each American, to share location coordinates and other information, during occasional rest breaks, at the noon and pre-NDP breaks, prior to approaching the target, or upon discovery of enemy combatants/activity.

Night Defensive Perimeter (NDP) TTPs:

- Get in the habit of practicing proper defensive perimeter and NDP procedures whenever your Team is training. This is especially relevant when training in the zone of conflict (e.g. in the vicinity of the FOB). Take advantage of all training opportunities.
- The evening meal is consumed in a perimeter established prior to dusk; food is not consumed in the NDP. The last SITREP is composed and transmitted from this same perimeter and is not transmitted from the NDP. The T/L should consult his map to identify prospective NDP sites that are nearby and that offer a defensive advantage.
- The T/L must be conscious of approaching dusk. He must allow sufficient time to accomplish meal consumption, message preparation and transmission and movement to the NDP location. Furthermore, the T/L must build in additional time if the prospective NDP

location proves untenable. Note that nightfall can occur very rapidly in mountainous terrain. Lastly, there must be sufficient light for the Team to occupy positions and deploy Claymores.

- After passing a suitable NDP site, 'fish-hook' and move into your selected position so that you can observe the Team's back-trail.

- When in the NDP, Team Members should keep their equipment on and remain alert until the perimeter has been checked 360 degrees for at least 20 meters beyond the perimeter line, depending on terrain and vegetation. Equipment, LBE, rucksacks should not be removed until it is dark, and then only when necessary. Prior to putting out Claymores at the NDP, the T/L must check beyond the perimeter at least 20 meters

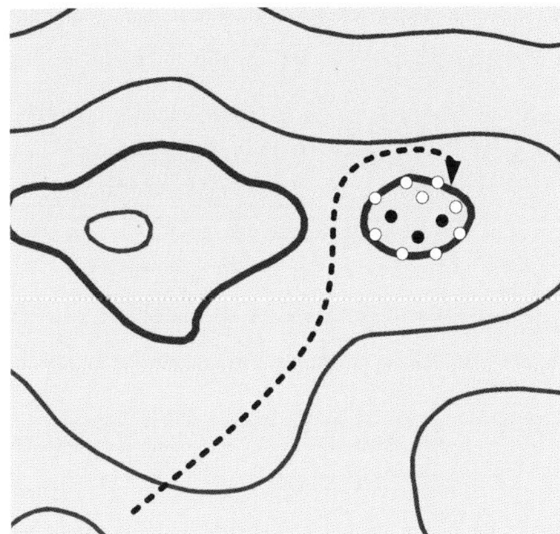

Figure 30. Fish Hook for Team Commo Break, Surveillance/OP and NDP Locations.

(depending upon terrain and vegetation). This check is to determine an enemy presence or to detect if trails or high speed routes of approach are in close proximity to the NDP. Alternatively, some Team Members can fan out from their Team element defensive positions to check the perimeter; take care that this is done when there is sufficient light so that these Team Members are not misconstrued for enemy combatants upon their return. The Team Member(s) doing the check should remain visible to their Team Mates as much as possible. Prior to departing the perimeter, the T/L must coordinate his Rally Point (RP) instructions with the rest of the Team, and coordinate his Leader's Recon, as it is always possible that the Team may be attacked while the perimeter reconnoiter is being conducted [refer to the subsequent paragraph below pertaining to **GOTWA**].

- The jungle/rainforest floor, under multiple canopies, is pitch-black at night, even when the moon is full. The only natural illumination will be small pieces of rotting luminescent twigs or bark. You can collect these pieces and consolidate them next to your Claymore firing device to mark its location, to mark firing lanes or to mark the direction to an LZ.

- Claymore mines should be deployed prior to darkness, while there is still enough light to see, and the Claymores should be located as close to the Team positions as is safe for employment. Normally, the Claymore should be situated on the opposite side of a tree to the Team Member. Close proximity of the Claymore to the user, allows a better kill zone for protection of a small perimeter and enables faster, more convenient deployment and recovery. After dark, collect luminous debris and use it to mark the location of the Claymore firing device; alternatively use some other method to mark the location of the firing device if luminous debris is not available. Marking the firing device with luminous paint markings may not last throughout the night, unless the paint is exposed to daylight for a period of time. If no luminous debris or paint is available, place the firing device next to a stake or plant stem or at the base of a tree/sapling where it can be within easy arms reach from the Team Member/defensive position. When firing the Claymore in close proximity, cover one ear with a free hand and press the other ear against the bicep, forearm or shoulder to mitigate eardrum damage.

- When placing Claymores around your NDP, emplace each using two-man teams; one man emplacing the mine while the other stands guard. Claymores may be employed quite close to the perimeter, if placed behind a tree or in front of an earth mound.

- In open terrain/vegetation, Claymores may be emplaced further out from the perimeter, but the mine and firing wire should be within line of sight of the defensive/NDP position.

- Determine in advance who will fire each Claymore and who will give the firing command or signal (typically the T/L). Normally, common sense will dictate firing the Claymore (e.g. a tracker team approaching the perimeter requires detonation).
- The Author strongly recommends that Claymore mines be dual-primed; one fuse-well primed with an electrical blasting cap, the other primed with a non-electrical blasting cap. If the Team displaces under enemy pressure, a Claymore may be delay-initiated using time fuse (non-electrical) to detonate after the Team has departed the NDP, prospectively inflicting casualties on the enemy and focusing the enemy forces and enemy fires on a false location.
- Some SOG T/Ls chose not to deploy Claymores during commo breaks and while occupying NDPs, opting instead to use CS grenades to break contact. The Author recommends that the combined effects of both Claymores and CS should be used in this situation for the following reasons.
  - A daylight engagement at the Team perimeter should literally take <u>seconds</u> (the time it takes to expend a single magazine of ammunition, or less) before the Team makes its withdrawal. But it takes <u>minutes</u> for a sufficient cloud to emit from the CS grenade; and the cloud may be insufficient in volume to affect enemy combatants in all directions from the perimeter during crucial moments.
  - If the Team perimeter has been discovered by the enemy, a larger enemy force may have been summoned to the site. The Team is unlikely to know the enemy dispositions and may withdraw from the Team perimeter directly into a concealed enemy ambush or blocking force. Claymore detonations may clear the enemy from the Team line of withdrawal and the CS cloud may deter pursuit.
  - If a tracker team has closed on the Team position, it will often approach the perimeter with stealth and attack by fire. If the tracker team is well led, the enemy leader will use a flanking maneuver to create crossfire conditions, to further limit Team lines of withdrawal and to subject the Team to fire during its withdrawal. Claymore detonations may inflict casualties 360° around the perimeter, enhancing the Team's chances of a successful withdrawal.
  - CS is a temporary impediment to the enemy; it will not stop him from moving away from the cloud and pursuing the Team along a different angle. Inflicting casualties on enemy is a more lasting impediment.
  - If the Team Members must don masks, their vision will be impaired; even worse if the Team must also move in pitch-black conditions.
  - Once Claymores have been emplaced, and the enemy is discovered to be moving in on the Team, the Team may have a tendency to stay in place too long, waiting for the enemy to get closer or fully within the kill zone. It is often better to blow the Claymores a little early and deploy CS gas, than to risk the enemy getting off the first shot(s) at close range and inflicting casualties on Team Members.
- However, in some instances, it may be better not to deploy (or detonate emplaced) Claymores around the NDP, but rather to rely on the use of CS grenades for the following reasons:
  - If the Team is expecting to move at night, deployment and recovery of Claymores will impede rapid and silent Team displacement.
  - Enemy combatants may be sweeping 'on line', not knowing the exact position of the Team. A Claymore detonation will reveal the proximate location of the Team. If Claymores are not yet emplaced, US Team Members may throw CS grenades in the direction of the enemy combatants; preferably upwind from the enemy. Team Members should be equipped with lightweight protective masks, which would allow the Team to withdraw while enveloped in the cloud.

- ○ When enemy combatants are enveloped by the CS at night, they may panic. If the enemy does not have protective masks he may run away and may even fire his weapon indiscriminately, causing overall confusion and panic. Most certainly, the vision of enemy combatants will be blurred with tears. If the enemy is equipped with, and dons, protective masks in reaction to CS employment, combatant vision will be restricted by the mask. In either case, the Team has an improved chance to escape unharmed.
  - ○ If the Team has no CS grenades, throw a smoke grenade and have the interpreter (or language qualified team member) cry out in the enemy language 'gas'.
  - ○ Remember that grenades (including CS) are difficult to employ, especially at night. Additionally, a small flame or some sparks may emit from the CS grenade when it ignites and the hiss of the burning ignition compound may be audible to the enemy.
  - ○ Do not employ projected CS munitions (e.g. 40mm CS grenades) unless the Team is engaged or unless the Team is attempting a breakout.

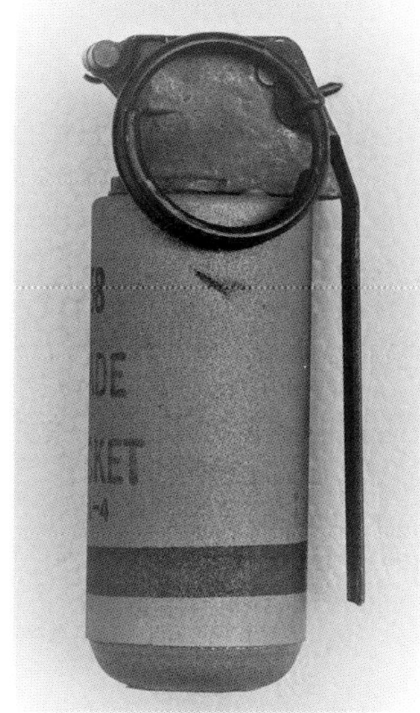

Figure 31. A Mini-CS Grenade. Preferred over the Larger/ Heavier Version for SR Teams.

  - ○ Consider using a double-bagged quantity of CS powder placed in front of a Claymore, for combined effect.
- SR Teams should be equipped with Mini-CS grenades rather than the much larger/heavier M-7 series grenades, for obvious reasons. Regardless of the type of grenade used, the Team Member should always 'roll the spoon' prior to throwing, to avoid the tell-tale 'pling' sound.
- Throwing a fragmentation grenade in total dark is a prescription for disaster (the grenade may strike a tree, bamboo, vine or other obstacle and bounce back toward the Team), unless the following measures are applied.
  - ○ The hand-grenadier should be designated based on his skills and experience (coolness in combat). The hand-grenadier is more likely to be US than an indigenous Team Member who may lack upper body/arm strength.
  - ○ Before dark, each Team Member should memorize the direction and distance to the trees and larger bushes around his position. Stakes, luminous debris or other marking methods may be used to designate grenade throwing lanes (to avoid fratricide), routes of approach and Final Protective Fire (FPF) zones. Note that the throwing lane may be 'up and over' to clear low vegetation. Note that you may lose your ability to see luminous marks once muzzle flashes or detonations create 'night blindness'.
- Prior to dark, the T/L insures that each man is informed of the orientation and distance to primary and alternate RPs/LZs. If the enemy comes from the direction of the primary RP/LZ, the Team will move toward the alternate. Resume Team march-order as soon as possible upon evacuation of the NDP and <u>account for Team</u> personnel.
- When deploying the Team for NDP, place the point man in a position oriented toward the primary escape route/rendezvous/rally point.
- If the Team is within range of friendly artillery (unlikely in most SR operations, except COIN), and has preplanned concentrations, azimuths should be taken (Observer-Target line) to concentrations, noting distances, prior to nightfall. Under dense canopy, use stakes or other marking devices to aid in calls for fire at night. Ensure that the Team is not located near the Gun-Target azimuth (especially in long range fires) to avoid fratricide.

- Do not send radio transmissions from your NDP site unless it is essential. (Such eventualities might include calls for fire or employment of air assets against high-priority fleeting targets, Team emergencies, enemy engagement, etc.). Be prepared to move if you do send radio transmissions.
- After a night engagement, and during subsequent movement, ensure weapons are on safe. Team members will stumble on uncertain footing and accidental discharge becomes more likely.
- Extra karabiners can be essential in porting additional equipment.
- During movement in utter darkness (e.g. under multiple canopies), daisy-chain (hand on pack/LBE) with Team Members who do not have NVGs. When time permits, 550-cord or similar materiel might be used to keep the Team intact. Take count of Team Members at every rest break.
- It is the T/L's decision as to whether a wakeful watch will be established. Consider using three Team Members per NDP defensive position. This allows one Team Member to sleep during each watch, while the other two Team Members keep each other alert. The three-man position automatically establishes a buddy system should one Team Member in the position become a casualty at night. One man will take care of/carry the casualty; the other will carry the casualty's equipment. Note that such a burden can only be carried for a short distance.
- Caffeine gum, hard candy or stimulants dispensed by the Team medic, can help keep Team Members awake and alert during surveillance operations, in emergency defensive situations, etc. Note chewing of gum/candy impairs hearing.
- If the rucksack is removed during NDP, ensure that the carrying straps are in the 'up' position for easy insertion of the arms to rapidly don. Most SOG SR Team Members slept in a semi-reclined position with the rucksack on.
- It is permissible to unhook the LBE harness buckle, but the LBE should rarely be removed at night or at any time through the duration of the operation (exceptions: defecation, medical care, donning or shedding additional clothing layers, etc.).
- If a person coughs, snores or talks in his sleep, placing a gag in his mouth; covering his face with a heavy scarf may also help.
- US Team Members should not 'bunch up' or sleep back-to-back in a cluster. One grenade or automatic burst from a weapon could get them all.
- An entrenching tool may be used to quietly scrape a shallow depression to sleep in. This depression may be sufficient to save the Team Member from being wounded.
- In NDP, boots stay on, but laces may be loosened (not untied). Web gear stays on, but may be unbuckled. Pack stays on, with waist strap (if provided) unbuckled; the pack may be removed briefly while setting up Claymore mines, firing sector markers and/or mines and booby-traps; while removing sleeping items or warm clothing; while taking a 'nature break'; while erecting antennae and other tasks that require full freedom of movement. Once these tasks are accomplished, the pack is re-donned.
- Regardless of intent to move from the NDP position the following morning, gear should always be stowed immediately after use. Always be ready to move at a moment's notice, with minimum groping around for individual or Team equipment. Repack most items into rucksacks prior to dawn, whether movement is intended or not. Recover and repack firing sector markers and Claymore mines at dawn. Don and readjust LBE by dawn.
- All Team Members should be awake, alert, and nearly ready to move prior to first light. Claymores are retrieved as soon as there is enough light to see.
- Remember that NDP departure is a window of vulnerability for the Team. As the Team may have been detected when it established the NDP, the position may be swept by the

enemy at first light or the enemy may have established point or area ambushes in the vicinity. Choose a departure point on the perimeter that least exposes the Team to these threats.

- The T/L or his assistant checks the NDP site to ensure that nothing is left behind and that the entire site is sufficiently sterile. Team Members check and clear their own NDP positions. The tail-gunner is the last to leave the NDP; he will eradicate footprints and other signs. An enemy tracker team can learn much from signs and debris left at an NDP; this information may include:
  - The number of Team Members.
  - Team composition (e.g. indigenous vs US)
  - Team discipline.
  - How the Team establishes its defense/defensive positions and arrays its Claymores or other defensive weapons.
  - Food/diet of Team Members.
  - Health of Team Members (e.g. by examining feces, bandages, etc.).
  - Other practices/trends.
- If the NDP is in a dense swamp the Team might use hammocks (perhaps with netting) or may build sleeping platforms. The NDP should be established on whatever earth is available and amid concealing vegetation. Claymores may be strapped above water level to tree trunks using bungee cord.
- Avoid establishing the NDP near ant/termite hills. If feasible, establish the NDP at a location which faces in the direction of, or has more exposure to, the sun, as this will reduce infestation of mosquitoes and leeches. Alternatively, establishing the NDP (or surveillance position) in an insect infested area may be tactically sound, as enemy combatants tend to avoid such areas.
- Never eat in your NDP position; food odors are dead giveaways, as US rations will likely have an aroma distinctly different from rations of enemy combatants. Also the aroma of food at times when the enemy is not preparing his own rations, will tip-off the enemy to the Team presence.
- T/Ls should ensure that the Team does not form fatal habits, such as:
  - Always fish-hooking to the 90° or 270°.
  - Team Member removal of boots and socks at the same time.
  - Eating and sending messages from the NDP.
- Sound travels much further during temperature inversions (often present at dawn).

Special Reconnaissance (SR) and Advanced Patrolling TTPs:

General SR/Patrolling TTPs:

- Tactics and techniques are dynamic, not static. Be agile, adaptive and innovative. Understand that the enemy will constantly attempt various ways to counter you … be alert and be prepared.
- The closer the enemy, the less reaction time there is for the SR Team to execute. Preparation, training and mastery of TTPs will lead to speed in decision making and execution in the heat of close combat.
- It is essential for the T/L to constantly maintain situational awareness and the locations of RPs, defensible terrain and nearest exfiltration LZs. It is equally important for the Team to understand and adopt TTPs that will enable the unit to avoid enemy contact during the execution of its reconnaissance mission, break contact upon enemy contact, possess highly lethal capabilities to inflict substantial damage and casualties if an engagement is necessary, evade a pursuing enemy, bitterly defend or breakout to continue the mission and/or exfiltrate

to fight another day. In other words, SR Team Members must master all relevant TTPs to stay alive and accomplish the mission in the face of overwhelming odds.

- Beware of an increase of UAVs and/or aircraft over-flights in or near your target area. This may be a tip off that the Team presence is known. The closer the over-flight to the Team location, the more cause for alarm and the more urgent the necessity to take steps, as explained in this book, to mitigate detection and further compromise.

- It is often the case that the Team will suddenly spot enemy units, locations/facilities, danger areas, etc, while en route to its area of interest or target. Further, due to imprecision of target or enemy intelligence, vagaries of terrain and vegetation, etc., the Team may unexpectedly reach its area of interest/target. In more seldom occurrences, the Team may actually reach its objective as expected and on schedule. In these, and similar, circumstances, the Team may have to take up an Objective Rally Point (ORP) to perform contingency/adjusted planning and to conduct a Leader's Recon. The ORP should be established at a location where enemy sweeps and security patrolling do not occur, where local national civilians do not frequent, where the standard criteria for RPs are met (described elsewhere in this book). Some doctrinal guidance suggests that the ORP should be established 200–400 meters from the objective, or at least one major terrain feature away. This distance criteria is largely irrelevant; METT-TC factors (e.g. terrain, vegetation, weather, illumination, tracker elimination, routes of approach, etc.), should govern ORP selection.

- Whenever the Team encounters the enemy, the T/L should deploy the Team in the most tactically advantageous position possible. Whether the Team is in an ambush position, or is using a battle drill to counter an enemy ambush; whether the Team is in a firefight during a meeting engagement or executing a break contact battle drill or whether the Team is in a defensive position, the T/L should maneuver to a position to inflict the maximum casualties on the enemy. Typically, this means that the T/L should seek to align Team Members, or optimize the lethal focus of its firepower, in enfilade, upon the long axis of the enemy formation. A skilled enemy should be expected to do the same.

- Additionally, whenever the Team encounters the enemy, the T/L should attempt to use defilade; using terrain, obstacles and/or vegetation to protect and/or to conceal the Team from enemy observation and fires – and especially to protect the flanks of the Team.

- If the T/L can hear people speaking in the target area, move close enough for the interpreter to get the gist of what they are saying. The Team Interpreter or language proficient Team Members should translate for the T/L <u>in real time</u>. The reasons for real time translations are many and obvious; chief among them is to discover if the enemy is aware of the Team's presence.

- If the Team is successfully inserted and is able to complete its primary reconnaissance/ surveillance mission within the mission time window, the Team may generally take on discretionary (unassigned) missions and/or address opportunistic targets, by SOP. Teams must plan and train for discretionary missions. This common SOP applies, unless: Higher headquarters establishes hierarchal mission rules. For example, mission tasks that may surpass the primary assigned mission might include:
  ◦ Rescue of friendly POWs.
  ◦ Taking enemy POWs
  ◦ Friendly long-range fire support is not available to strike targets of opportunity.
  ◦ Fleeting high-priority targets (e.g. WMD systems).

- While on patrol, don't always take the obvious course of action and don't set a pattern in your activities, such as, always turning to the right when fish-hooking to ambush your own back-trail. However, establishing a pattern can prove fruitful in deceiving and setting up the enemy for tactical surprise.

- It is essential for T/Ls/Members to act/react quickly to situations as they occur, often based on scant information. Analyses-based decision-making is generally only feasible when the Team has the luxury of time and is almost never feasible during Team movement. The T/Ls/Members must often act/react based on cues/anomalies that are juxtaposed against a known pattern of norms, some of which may be gained through all-source intelligence, but most of which are gathered through Lessons-Learned, personal experience (including the exercise of common sense) and focused training – nearly all of which must be acquired in advance of deployment.

'Analytical decision-making is neither practical nor useful in the high stress situations encountered while on patrol: whether to shoot or not in a matter of milliseconds, whether to travel a particular stretch of road…. Combat environments … require people to be able to recognize threats and patterns quickly and then act immediately based on that information.'[13]

- ○ The Team may prepare for rapid decision making by:
  1. Studying enemy doctrine; tactical situational tendencies and battle drills; order of battle information, including commander's proclivities – to establish an enemy pattern baseline.
  2. Study Lessons-Learned from well-regarded SME veterans with recent exposure to combat in the AO. These may be written, recorded or live presentations by US SpecOps personnel, ralliers/deserters from the AO, US and allied intelligence units/ agencies, SpecOps personnel from allied nations. News and internet posted video clips may reveal enemy TTPs.
  3. Study and train (FTXs/STXs) in accordance with the TTPs established in this book and as informed by items 1 and 2 above. SMEs should participate in training, if possible, perhaps as lane-graders, mentors, trainers or members of a Red Team force.
- ○ An anomaly is detected when conditions, actions or inactions are observed that do not fit established patterns in a given situation. SR examples:
  - ▪ An absence of an indigenous Team Member prior to mission deployment.
  - ▪ Increased guard forces or patrols in the target area.
  - ▪ Crops are grown in remote areas, where the scale of agriculture exceeds the local market and/or where no thoroughfares exist to get crops to a market. Or crops are grown in a manner that is inconsistent with local practices.
  - ▪ Enemy indiscipline (e.g. heat, light, sound, etc.) in the target area.
  - ▪ Condition of enemy combatant uniform, which may indicate construction of field fortifications/tunnels.
  - ▪ <u>See the table below on some behavioral</u> cues.
- ○ Does enemy behavior fit the environment and situation? Enemy combatants act differently based on level and immediacy of a threat. When the enemy is in his comfort zone, the signs are clear. Weapons are carried at sling arms or set aside. Personnel are in a relaxed posture, and may be seen sitting or leaning against trees or equipment. Crew-served weapons may be unmanned, or pointed in a non-tactical manner.
- ○ Note that combat units, will behave differently than rear area support troops. Security force personnel, who may include former frontline troops (who may be recovering from wounds) may comport themselves more like combat units. Note that some behavior may be transient due to the presence of enemy senior officers or when enemy combatants are participating in training.
- ○ More Anomalies – Enemy Threat Behavioral Cues (Examples):

---

13. Patrick van Horne and Jason Riley, 'Left of Bang', Digital Version, no pagination.

'Red-Flag' Anomalous Posture and Comportment of Enemy Combatants in Rear Areas		
• Carrying weapons at port arms.	• Running under arms.	• Dropping to cover.
• Tactical hand and arm gestures.	• Binoculars up and scanning.	• Active sweeps/patrols and tactical movements.
• Shouting; signal shots.	• Intense facial expressions.	• Sudden changes in direction.
• Changes in vehicular movements (e.g. escorts).	• Personnel moving at a crouch.	• Weapons safeties off.
• Wearing of camouflage facial paint.	• Wearing of protective vests and tactical LBE.	• Wearing of NBC protective uniform. [Danger!]
• Tactical versus administrative unit movements.	• Uniform and equipment indicative of combat echelon and/or elite units.	• Increase in guards posted and/or new security stations established.
• Presence of Reaction Force(s).	• Evacuation or flight of civilians.	• Uncommon armament spotted.
• Crew-served weapons manned and/or aimed to defend a sector of fire.	• Insignia/markings associated with elites.	• Increased aviation patrolling.
• Artillery/mortar registration firing.	• Night illumination rounds fired.	• Test or H&I firing.

Observation of one or more of these anomalous cues may require rapid Team action. The more cues observed, the more urgent the Team's situation may become. A T/L/Member should not require more than three related anomalies before making a decision and acting; often one will suffice.

- The Team may have mere milliseconds to react to cues/anomalies. Range of reactions may include the following.
  - <u>Freezing</u>: Will frequently result in death, wounding or capture. This does not refer to a deliberate tactical freeze of motion in close proximity to an enemy; this is when a Team Member hesitates/does not react to a threat due to shock, surprise, confusion – or too much thinking/analysis
  - <u>Hiding</u>: An option only if there is a reasonable chance of not being discovered.
  - <u>Flight</u>: Have a plan/SOP and a well-trained battle drill to break contact or the Team may be trapped/cornered or ambushed.
  - <u>Fight</u> if cornered; ambushed; engaged by chance or if the risk is worth it (e.g. discovery of a WMD system, POW opportunity, opportunistic ambush, etc.). Be attentive during daylight while manning ambush or surveillance positions, etc.
- Whenever you are in position to execute an ambush/POW snatch, raid or target surveillance, or if an engagement with the enemy is imminent and Team Members notice that an enemy in close proximity is carrying their weapons either at <u>high port or some other weapon-ready position</u>, you must assume that the Team has been detected and/or the Team position is either specifically or generally known.
  - The T/L may consider a silent withdrawal, only if time and situation permits.
  - The alert status of the enemy may signify that a tracker unit has communicated the Team whereabouts or proximity to an 'anvil' element. If so, then the Team must take every precaution to avoid the tracker unit 'hammer' during its withdrawal.
  - However, since some movement noise will be unavoidable and detectable to an alert enemy combatant, the best practice is to <u>shoot the alerted enemy combatant(s) without hesitation</u> (preferably with a suppressed weapon) <u>before</u> the enemy initiates contact.

- Tactical Enemy Behavior Red Flags:
  - <u>Weapons Ready Posture</u> of an Enemy Combatant, e.g. Weapon at High Port, may signify that the Team presence in the vicinity has been detected.
  - <u>Signal Shots</u> may signify local hunting activity (or hunting by enemy combatants); but a sequence/series of shots may indicate enemy discovery of the Team presence, a tracker team signaling Team direction changes to other enemy elements or enemy pursuers 'beating the brush' in an attempt to flush the Team in the direction of a blocking force.
  - <u>Signal/Electronic Countermeasures</u> indicate that the Team presence in the Target Area has been detected, that the enemy is committed to interdiction/destruction of the Team and that the enemy may have high-value targets nearby that they must protect. Implementation of these countermeasures may also signify compromise/betrayal of the mission.
  - In a chance contact/firefight, during an enemy pursuit or in several other situations, the Team may become disoriented or lost. If Team Members cannot resort to GPS or other Position Location System (PLS) technology, a FAC may have to locate and reorient the Team Member(s).

    'A contact would normally result in a mad moment of gunfire, a shot of adrenaline and a hasty retreat (if possible) away from the enemy. The combination of adrenaline and running would invariably result in the team becoming disoriented and lost. To get reoriented, the team would contact the FAC and request an updated fix. This would be accomplished through the team vectoring the aircraft by sound and then using either a 'shiny' (a mirror) or a colored panel to identify the team location. Performing this task, meant the team had to find an open area, significantly increasing the risk of another compromise.'[14]

  - <u>Enemy Aircraft in a Search Pattern</u> signifies that the Team presence has probably been detected, that the enemy is committed to interdiction/destruction of the Team and/or that the enemy may have high value targets nearby. This activity, if focused, may also signify mission compromise/betrayal.
  - <u>Increase in Enemy Sweeps/Patrols</u> will likely occur in the vicinity of vulnerable sections of MSRs, marshalling areas, Command, Control, Communications, Computers, Intelligence, Surveillance and Reconnaissance (C⁴ISR) facilities, WMD capabilities and other key facilities/capabilities.
  - <u>Enemy Use of Dogs</u>; trained tracker dogs are typically not available in great numbers. If deployed near key facilities/capabilities, dog employment may be routine, suggesting that the SR Team should use counter-dog TTPs (noted elsewhere in this book) to investigate. If dogs are deployed to patrol along enemy MSRs, this may indicate that the Team presence is suspected. If deployed with tracker teams or similar security elements, the Team presence has definitely been detected.
  - Arrival of Enemy Special Forces, Infantry and/or Security Units (including Reinforced Guard Posts). This tip-off may signify that the enemy is preparing military operations within the target area and the arrival of enemy Special Forces, Infantry and/or Security Units may signify that the enemy is stepping up security precautions or is attempting to counter guerilla/partisan activity. Alternatively, it may suggest that the Team presence has been detected and the enemy is taking exceptional measures to interdict or destroy

14. COL Al Greenup (Ret), 'No Glamour No Glory – A Ground Pounder's Perspective of Project Delta Forward Air Controllers in the South East Asia War', (The Drop Magazine, Winter 2017 Edition, Special Forces Association, Fayetteville, NC) p. 107

the Team. Identification of enemy Special Forces would be especially concerning. This activity may also signify mission betrayal/compromise.

- ◦ <u>Enemy Presence/Rapid Response to Insertion or Resupply</u>; if the enemy is present or in proximity to a Team LZ or DZ, such activity may signify mission betrayal/compromise.
- ◦ <u>Absence of Civilians</u> from a target area may convey that major enemy operations are imminent, that the populace is considered disloyal/uncooperative, or that the exclusion is designed to strip away logistics support of guerilla/partisan forces.
- Enemy soldiers generally do not sit around idle in the field.
  - ◦ After a movement, enemy troops will normally dig in, even in rear areas and even if the stay is relatively brief. This includes Combat, Combat Support (CS) and Combat Service Support (CSS) – but the level of effort will vary according to branch/type of unit. Armored units may plow revetments. The Team may move under cover of this noise and activity. Units that are constantly moving on the battlefield will likely be exhausted at the end of the day when they move into bivouac. This is an opportunity for the Team to move into an OP/LP surveillance position or to take a prisoner.
  - ◦ Some units will have a set routine while bivouacked. It is operationally important to acquire an understanding of these routines. If the enemy breaks standard routine it may may be significant. A North Vietnamese infantry company, for instance, would have the following regimen:
    - 0500–0600: Reveille, Brief Period of Calisthenics and Cooking. This time frame (dawn) was optimum to avoid fire/heat signature detection.
    - 0600–0700: Eat and Rest while Occupying Field Fortifications.
    - 0700–1130: Improving Fighting Positions/Fortifications; Study and Tactical Training; Local Food Gathering (UW/Insurgents) – an opportunity for the Team to take a prisoner.
    - 1130–1330: Rest and Cleaning/Maintenance of Equipment. This period may be extended during the summer during peak heat of the day; to include rest periods; this may be an opportunity to take a prisoner.
    - 1330–1830: Operational Training and Rehearsals.
    - 1830–1930: This time frame (dusk) was again optimum for cooking to avoid fire/heat signature detection.
    - 1930–2000: Rest while Occupying Field Fortifications.
    - 2000–2200: Establishing/Improving Fighting Positions/Fortifications under Tactical Conditions (Complete Silence).
    - 2200–0500: Sleep.
    - Notes: This regimen may have been relaxed while the unit is in a sanctuary location or a rear area. Support units would have had a less rigorous regimen. HQ staff had an entirely different schedule.
- During an engagement or when adrenaline is pumping, Team Members may effectively lose peripheral vision and fixate on obvious enemy activity and they may not be attentive to their own sectors of responsibility. Team Members must learn to keep their 'heads on a swivel'.
- Beware of enemy minefields/booby-traps around key enemy unit, infrastructure (e.g. bridges) and logistics positions, especially in areas not under enemy direct fire or observation (ravines and other low areas).
- Observe enemy patrols and other enemy behavior as their units/personnel skirt or negotiate their own minefields.
- Heavy bracken (tall fern) will signify that low wind conditions prevail and where gnats and ticks abide in substantial numbers. This is not a great spot for an NPD, OP/LP or hide location, but the enemy may not occupy the same ground for the same reasons.

- If a chainsaw is to be used on an operation, muffle the engine noise with cloth for <u>brief periods</u>, but do not continuously obstruct the exhaust. Use the chainsaw when the noise of enemy operations is at its peak or when weather conditions will help suppress its noise. A chainsaw can be used to partially cut through trees in preparation for rapidly creating an abatis later. Except in UW or COIN scenarios, enemy forces at strategic depths might not consider the sound of a chainsaw as especially unusual. Some scenarios for chainsaw use: (1) The Team is able to flush an enemy unit from its marshalling/assembly area leaving it vulnerable to air attack; so creating an abatis can delay or channelize its movement; (2) The Team needs to create an LZ or DZ. A chainsaw would normally be stored in a MSS/cache and drawn for use when required.

- Cause an enemy unit to evacuate its hide positions, thereby exposing them to satellite/aerial observation and/or attack. This may be accomplished by simple measures such as setting fire to the vegetation in which they are hidden or by using a time delayed/remotely detonated white phosphorus grenade to mark their location for CAS engagement.

- Drop/attach tags or time delayed incendiaries/demolition devices on enemy vehicles as they pass by using magnets, dangling fishhook lines or other expedients. These techniques may later reveal the vehicle location to airborne assets or may cause substantial damage to other enemy assets/capabilities that may be nearby.

- Only use the IR diode (light) on NVGs, when it is necessary. The IR light may be detectable to an enemy who is also equipped with night-vision optics, depending on terrain and vegetation.

- Zone/Area Recon using a Patrol Base. Conducting SR operations in the blind is not an optimum use of scarce SR assets and support capabilities. But this approach may be advisable in the following circumstances:
  - Higher headquarters has little or no information/intelligence regarding the enemy presence and/or intensions in the target area, but the presence of high priority enemy units/capabilities is suspected.
  - Friendly UW operations are being conducted. Guerilla bases can serve as a patrol base deep within enemy territory.
  - Prior reconnaissance missions, or other intelligence, reveals presence of significant enemy elements/capabilities making it essential to discover the locations and activities of remaining elements. Example: vehicles tracks consistent with enemy TELs or specialized vehicles or equipment associated with enemy WMD capabilities have been spotted within the target area.
  - Higher Headquarters places value, and even priority, in other effects consequent to reconnaissance patrolling (Economy of Force) on the enemy.

- In certain circumstances, an SR Team equipped with enemy small utility vehicles, may drive cross country and along secondary roads, even in daylight, if they can deceive enemy troops into believing that the vehicle and its occupants are friendly. The advantages and disadvantages to this include:
  - The SR Team may traverse substantial distances to close on its target area. This will allow the Team to use LZs further away from its target area/area of interest and possibly further away from enemy anti-aircraft systems.
  - The SR Team may surprise enemy soldiers/units and take them under fire. This may also allow the SR Team to more easily capture enemy POWs and rapidly convey them to an exfiltration LZ.
  - The SR Team may more rapidly identify key targets that are hidden along or in proximity to secondary roads. These targets may include vehicle parks, assembly areas, logistics areas, rail sidings, ballistic missile systems, etc., that would otherwise go undiscovered.

- ○ At night, such a vehicle may surreptitiously join enemy tactical and logistics convoys and follow them to their destinations.
- ○ The chief advantages to this type of operation are lost once the enemy discovers the deception. But once the enemy discovers these operations are occurring, the enemy would be obliged to take extraordinary security measures that should be substantially burdensome to his own operations – an Economy-of-Force outcome.
- A dead enemy's uniform and contents of his pockets, LBE and/or pack, if he has one, almost always renders more valuable intelligence than his weapon. If the Team has booby-trapped ammunition in the caliber of the enemy weapon, insert the bad round into the KIA's weapon chamber or into a magazine, and leave it behind with the corpse.
- Generally, treat enemy KIA with some respect. Do not allow indigenous Team Members to mutilate or otherwise mistreat corpses and only photograph enemy KIA for intelligence purposes. Do not allow Team Members to pose with enemy KIA. An enemy seeking vengeance is one more problem a Team does not need.
- The Team should stay alert at all times, even in an area considered a 'dry hole', with no or little sign of recent enemy activity. The Team is never considered safe until it has returned 'home'.
- Team Members may be able to detect the approach or presence of an enemy element from the behavior of insects, birds, and other animals. Bird and animal calls that signal danger may be previewed on various internet websites. Insurgents/guerillas/partisans and other combatants (elite unit personnel), who spend long durations in the field, may be tuned into these signs.
  - ○ Startled birds may fly off in the same direction. And they may fly in the direction of the threat. Observe their behavior in the AO as they react to Team Member presence as a predictor of how they will behave in the presence of an enemy element.
  - ○ Bird alarm calls will typically be short, simple and high-pitched. Nesting birds may not abandon eggs or their young; when a nesting bird sounds an alarm and then goes silent, the threat is probably close-by.
  - ○ Squirrels make a distinctive chattering sound when they are alarmed or spot a predator.
  - ○ Frogs, crickets, toads, and some other nocturnal animals are often noisy throughout the night; they will go silent in the presence of a threat.
- Bury all trash unless you intend to use it as bait for a booby-trap/mine or ambush.
- You may be able to get an indication of where (along which route or direction) a major enemy facility or camp is located by examining the ground at a trail, trail-road or road junction. The trail that is more heavily trafficked is likely to take you towards the troop concentration.
  - ○ One side of the road or trail will have more signs of travel than the other, emphasizing traffic flow. Also look for where enemy troops may cut corners, or ruts where vehicles make their turns at/near a junction; this is another sign of an enemy presence nearby, and the direction in which it abides.
  - ○ When these signs are detected, the wise T/L will slow the Team's forward rate of movement and increase the Team's level of stealth. The high-speed trails leading into an encampment will normally be under OP/LP surveillance and may have command-detonated mines pointing down the trail. It may therefore be prudent for the T/L to:
    - Check the map for likely locations for the encampment (e.g. near a water source, defensible terrain, etc.) and for likely positions for OPs/LPs (fields of observation, etc.)
    - Move the Team far enough off the trail so as to avoid detection by enemy combatants who may move along the trail or who may be occupying an OP/LP. Parallel the trail, with increased stealth, to minimize noise of movement, tracing it to its destination.

- Consider cautiously circling around the suspect encampment location to verify Team suspicions and to determine other key information (e.g. camp size, evidence of patrol routes, other trails, other avenues into the encampment, escape routes, barriers, communications wire, etc.).
- As the Team approaches an enemy encampment, look for patrol paths. Patrol trails will normally be outside the OP/LP positions; patrols would normally be focused on seeking signs of enemy approach, especially in areas that are not under observation by OP/LPs. Consider deploying into a hasty ambush formation before the T/L advances to examine the trail (to determine frequency of use, etc.). Ensure traces of Team presence are concealed. Increase stealth and be on the lookout for OP/LPs.
- The presence of high-use primary and converging paths, will also indicate proximity to an encampment.
- Look for cuttings (e.g. for construction of field fortifications, clearing vegetation for camouflage, fields of observation or fire, etc.) as evidence of an enemy encampment. Also look for the ground surface to be cleared of dead vegetation (used for campfires).
- Depending on time of day and enemy routine behavior patterns (changing of OP/LP personnel, meal times, etc.), go into a hide location and be attentive to signs of enemy activity (e.g. sounds, movements, aromas). If the Team presence in the vicinity of an encampment coincides with key calendar occasions (e.g. harvest times, anniversaries of national, religious or historical events), enemy combatant activity may be substantial.

- The enemy may send periodic (routine/random) patrols to sweep along roads and major trails. Take the time to establish the enemy's routine (if any) for these sweeps, especially if the Team is to linger nearby (e.g. to establish a hide/surveillance outpost, ambush or raid position).
  - Enemy troop/leader behavior will indicate if the sweep is cursory, or if the enemy is aware of Team operations in the target area. Observation of security patrols may be difficult in densely vegetated areas.
  - Most of an enemy's vehicular and major troop movements will be conducted at night, unless there is a solid protective canopy or dense cloud cover overhead allowing concealed daily operations. Based on the enemy sweep schedule, a Team that has a road-watch or ambush mission may stand off (depending on terrain and vegetation) until the sweep has been conducted and then move into position.
  - If the enemy conducts sweeps near dusk or during periods of reduced visibility, he may be less likely to detect the Team/element position. Sweeps conducted during or after a rain will often be more cursory, as troops do not appreciate getting wet.
- Blousing your trousers inside your boots or inside gaiters/leggings will help deter leeches, ticks, chiggers, stinging insects and disease bearing pests. Note that, during stream crossings, bloused/gaitered trouser legs will fill with water, and the water weight will make climbing from the stream onto a bank surprisingly difficult. But in crossing shallow water, leggings can mitigate this bladder effect.
- Smoking should be forbidden throughout the duration of the mission. Team Members should not use chewing tobacco or snuff during an operation; this practice represents a life-threatening aspiration hazard should the chewing Team Member become wounded.
- Insect (bees/wasps) mating swarms can kill a human. Team Members should not move when a swarm passes over; if someone moves, the entire Team will likely be stung multiple times; to survive the swarm, the Team may then have to flee.
- The fuzz on certain types of bamboo (and some other types of vegetation) may act on the skin like itching powder. If you cannot avoid these areas; ensure that skin is covered/clothed. In any event, shirt sleeves should be rolled down, with the cuffs fastened at nearly all times in the field.

- Time-delay devices <u>must</u> be used for SD of mines/booby-traps whenever a Team Member emplaces them; this ensures that civilians are not harmed or Team Members do not become casualties on subsequent operations in the same Target Area/AO. Using a SD capability also eliminates the typical necessity of recording 10-digit minefield coordinates. A time-delay device can also be used to create a diversion.

Author's Rule of Three: When possible, clusters of three Team Members should be used on surveillance outposts, road watch positions, and in defensive or NDP positions. One man sleeps, while the other two keep each other alert. This has particular utility if the Team is exhausted.

- Inversion and high humidity conditions are the norm under heavy canopy, and generally no wind or breezes are present in this circumstance. Under these conditions, smoke (except White Phosphorus) will lie close to the ground, will dissipate slowly and will generally not penetrate the canopy (except perhaps in a wisp through spaces in the canopy layer). Under canopy, these constitute good characteristics for screening smoke and poor characteristics for signal smoke. If the smoke drifts, it will drift down into low lying areas/ravines. CS will behave in the same manner as signal smoke during inversion conditions, keeping a higher concentration of the agent lingering close to the ground; CS powder, squeezed into the air, can also carry far under canopy. Odors (human waste, food, scent) will also linger and carry a longer distance into low lying areas.
- To attract the attention of a Team Mate, toss a small twig toward the Team Member or make a non-descript noise that is common to the habitat.
- Consider carrying an extra Extraction Rig (e.g. Hansen Rig), for string extraction of a POW. Ensure Team Member Extraction Rigs are adjusted for rapid use before leaving on a mission.
- <u>Always</u> assume that your Team is being tracked. Be careful to plot your line-of-march so as not to become trapped against an obstacle or where you can be boxed in (e.g. river, lake, cliff, mountain side, road, etc.). See counter-tracker techniques in subsequent paragraphs.
- An enemy will often pursue or track a Team with a squad-sized unit, supplemented with an indigenous tracker and/or a dog team. This tracker unit may attempt to predict your objective on the basis of the azimuth the Team is following; they may then, using a high-speed trail network, shortcut or road system, race ahead to establish an ambush along the Team's anticipated line-of-march.
  - Alternatively, an enemy tracker team may openly pursue a Team (making noise or firing shots) in an attempt to drive it toward a blocking force (Hammer and Anvil tactic) or ambush.
  - Further, an enemy tracker team may use a tactical radio or signal shots to communicate your line-of-march to a blocking force.
- The 'Hammer and Anvil' tactic, using a blocking element (Anvil), coupled with a pursuing or driving element (Hammer) is standard procedure in counter-reconnaissance/COIN operations.
- <u>Never</u> move in a straight line for long periods; frequently deviate in your direction-of-march. As the Team navigates and moves toward its objective, the following navigation techniques should be considered:
  - <u>Best</u>: From the LZ or last confirmed location, move from terrain form to terrain form ('Terrain-Hopping' – e.g. ridge to ridge) to reach successive points of reference, leading ultimately to the target destination. (See text in Land Navigation)
  - <u>Good</u>: From the LZ or last confirmed location, use compass azimuths to dog-leg to the next point of reference leading, ultimately to the target destination.
  - <u>Fair – Poor</u>: From the LZ or last confirmed location, parallel or follow natural or manmade linear terrain features (e.g. trails, roads, electrical/telephone wires, ridge lines) or other

features (e.g. tactical communications wire, streams) to the next point of reference, leading ultimately to the target destination.

  ○ <u>Worst</u>: From the LZ or last confirmed location, take an azimuth directly to the target destination.

- US Team Members should consider wearing valuable jewelry that can be used during evasion operations. Some SOG US SR Team Members would wear quality wristwatches, solid gold wristwatch straps and neck or wrist gold chains that they purchased in Thailand; the strap/chain links could be separated to be used as currency if the Team had to evade to a safe haven. A few others carried Laotian currency for the same purpose.

- Indigenous troops may often have weak upper body strength. This would be a serious problem if they are to retrieve and evacuate an American, who will generally be substantially heavier. This muscle weakness will be aggravated by adrenaline which will course through the human system during a firefight, restricting blood flow to the limbs. Ensure that the Team's physical training regimen includes a focus on upper body strength.

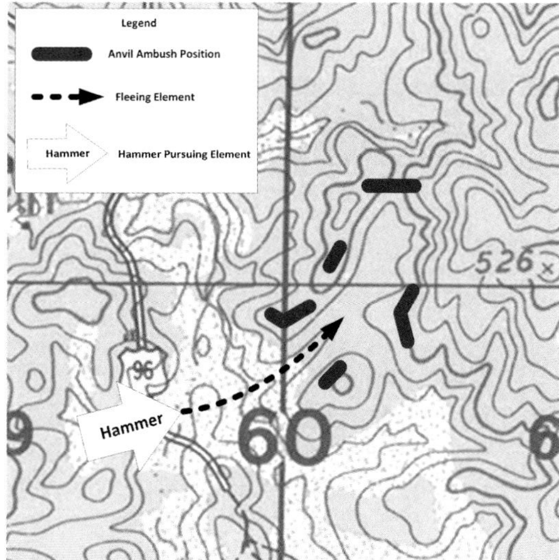

Figure 32. Example of a "Hammer & Anvil" trap.

- <u>GOTWA</u>: Spot planning occurs constantly throughout an operation. And the T/L (and selected Team Members) may have to separate from the main body of the Team to conduct a Leader's Reconnoiter of prospective NDP sites, Objective Rally Points (ORPs), ambush sites/positions, raid objectives, approaches to MSSs/caches, etc. The acronym, GOTWA, is one method that can be used to plan, coordinate and communicate these excursions to other Team Members. Simple plans can often be communicated using hand and arm signals; GOTWA elements are performed in accordance with (IAW) SOP. GOTWA elements include:
  1. Where the leader or designated Team member is <u>Going</u>.
  2. What <u>Others</u> accompany the T/L or designated Team Member.
  3. <u>Time</u> that he or they will be gone.
  4. <u>What</u> to do if the T/L or designated Team Member does not return on time. [by SOP]
  5. What <u>Actions</u> will be taken on enemy contact [by SOP]:
     a. If the T/L or designated Team Member becomes engaged with the enemy, the. . .
        i. T/L or designated Team Member will . . .
        ii. Team will . . .
     b. If the Team becomes engaged with the enemy, the. . .
        i. T/L or designated Team Member will . . .
        ii. Team will . . .

Human Waste TTPs:

- A skilled enemy can gain worthwhile information from human waste. Examination of feces can reveal Team Member diet, health, and the time and duration of occupancy at that location. If one or more Team Members evidence symptoms of diarrhea or intestinal parasites, the enemy can surmise the effects of such maladies on Team tendencies and mission conduct.

- Use a buddy system when a Team Mate must void. A Team Mate stands guard while his buddy defecates in a hole that is dug outside the perimeter and generally beyond the Claymore location. The Team will fish-hook prior to occupying a perimeter; therefore, voiding must occur outside the perimeter on the <u>opposite side from the fish-hook</u> curl.
- Always carry your weapon with you; carry an entrenching tool or a large knife to excavate the hole.
- Choose a depression for the location, if possible, so that the Team Member can sprawl behind a fold in the ground if taken under fire or approached by an enemy combatant.
- Trees in the rainforest or jungle sometimes have substantial roots that are often partly exposed above ground; this offers an alternative concealed and covered site to defecate. Examine around the root system to avoid biting insects, etc.; and cover the feces with earth as quickly as possible to mitigate the odor.
- Urinate from your knees, or even lying on your side (if warranted by proximity to the enemy), into a hole. Again, cover the hole with earth as quickly as possible.
- If you are in a concealed and covered security or surveillance outpost or listening post (OP/LP), defecate and urinate into an ammo can or a heavy duty plastic or barrier-type container (e.g. ration container or resealable, zip-lock plastic food bag) and seal the container as rapidly as possible to mitigate the odor. Take the container with you when replacements come to relieve you at the OP/LP or hide location; replacements should bring their own sanitation container(s). Once you have returned to the main unit location, bury the container or its contents.
- There are WAG (Waste Alleviation and Gelling) Bags obtainable from military inventory or commercial sources. The gelling function suppresses odor. These bags can be fabricated using heavy duty zip-lock type plastic bags and plain (odor-free) 'kitty litter'.
- In swamps/marshes, defecate into a pool and stir/mix it with the water.
- Alternatives to toilet paper.
  - Smooth stones or sea shells.
  - Immersion in a stream (downstream from drinking water) or tidal pool, swamp pool – or rinse with water from a puddle (the bidet approach).
  - Seaweed or other aquatic vegetation.
  - Mosses or peat.
  - When using any plant leaves, make sure that the leaves do not have milky sap or possess hair or prickles.
  - Lastly, manually (preferably with the left hand). Cleanse the hands as soon as possible afterward.

Operations During Night/Periods of Limited Visibility TTPs

- An enemy may be expected to routinely and extensively use periods of impaired or limited visibility to screen their movements and activities and to frustrate US superiority in combat aviation support, aerial intelligence platforms, satellite imagery, long-range fires and other capabilities. When operating against a technologically sophisticated enemy in conditions of impaired or limited visibility, the US SR Team must be properly equipped with most recent generation night-vision and thermal-imagery equipment, and other sensor devices (acoustic, seismic, motion, etc.).
- A technologically sophisticated enemy may deploy its own night-vision and thermal-imagery equipment in counter-reconnaissance efforts, particularly in the vicinity of high-value targets (command and control installations, WMD systems, key logistics installations, etc.). In these circumstances, if you can see the enemy, he may be able to see you. The T/L

must select surveillance positions that use terrain form to shield Team personnel from night-vision and/or thermal observation, yet are close enough to permit Team observation of the enemy, and that are outside the detection range of other enemy perimeter security sensors.

Weapons Employment TTPs:

Weapons General TTPs:

- Place magazines upside down in the magazine pouches, preferably with bullets pointed away from your body, to help keep the magazines dirt-free and facilitate draining of moisture.
- Magazines should be modified with 'tabs', allowing the Team Member to more easily extract magazines from LBE under stress of combat. 'One-Hundred-Mile-An-Hour' tape can be used as a field expedient for this purpose. So-called 'Ranger Plates', used to replace magazine base plates, have built in finger loops; they may improve manual dexterity (especially in cold weather) and speed magazine exchange; they are commercially available.
- Change positions of crew-served weapons during an engagement. Once the enemy detects a crew-served weapons position they will engage it, as a priority target, with heavy fire in an attempt to take it out.
- Call out 'reloading' to fellow Team Members, especially if you are manning crew-served weapons. This would also be the optimum time to change positions. The other Team Members could then pick up their rates of fire until the reloaded weapon is back online. If a Team Member is out of ammunition, he should call 'out' and advise of the type of depleted ammunition (e.g. 40mm HE).
- Always carry your primary weapon with the selector switch on 'safe'. During tactical carry, the Team Member should always rest his thumb (left hand safety) or forefinger (right hand safety) on/close to the selector switch, if at all possible (depending on the weapon). This facilitates rapid selector switch operation, with no fumbling, saving as much as a second in putting the weapon into action.
- The M-16/M-4 series of firearms has a trigger guard that can be folded down against the weapon pistol-grip, normally to accommodate cold weather gloves/mittens; DO NOT use this trigger-guard option except in extreme cold environments.
- Tracers have inferior ballistics. Use of tracers to mark enemy positions should not be necessary for well-trained and experienced Team Members, as priority of fire should be directed at enemy crew-served weapons, communicators and leaders; this priority of fire should be prescribed by SOP and therefore automatic. Note that Russian and Chinese tracers are typically green; NATO tracers are red.
- The weapon muzzle should generally track with your line of sight during Team movement.
- Firefights in jungle or rainforest environments are frequently at extremely close quarters and the SR Team will often be outmanned and/or outgunned by enemy forces. Speed and

Figure 33. Look closely. An enemy combatant moving along a trail (right to left). If this enemy spots you, he will either hit the ground directly in front of him, or he will take off running. Plan your shots accordingly. Note: An enemy wearing skin camouflage in a rear area could be a critical anomaly; beware if he is.

violence of action in raids, ambushes and meeting engagements and in attaining and maintaining fire superiority, is essential to tactical success – and survival – in these environments. The dense vegetation often does not afford aimed shots during the first few seconds of a close-quarters firefight; aimed shots may become possible once enemy firing positions can be defined.

Author's Solution:

In a meeting engagement, both the Team and the enemy element will drop to the ground to take cover and to have a better view beneath the ground foliage. Team members should hit the ground … and roll laterally, so that the enemy fire is directed at the members' last seen location. Team members will then be able to see dust and debris kicked-up in the air in front of and beneath enemy muzzles; Team Members should focus their aim immediately behind this cloud. Of primary importance is to immediately concentrate fires to knock out crew-served weapons (including rocket-propelled grenade launchers). Team grenadiers should displace, under covering fire if necessary, to get a clear shot through intervening vegetation. Note that a well-trained enemy force will immediately employ fire and maneuver in a close engagement to gain the Team's flank. The Team must react faster than the enemy, to flank the enemy's primary position first, to intercept the flanking element or to break contact.

Figure 34. Look closely. A kneeling enemy with weapon aimed and scanning for a target. If you see this, do not hesitate; shoot him immediately. He may fall wounded, so anticipate where he will fall and shoot him again.

- When the SR Team initiates contact on enemy combatants, anticipate where the enemy combatants will be once the shooting commences. This is especially important in heavy vegetation or other situations of reduced visibility, where only glimpses may be had of the enemy combatant(s).
  - The initial shots should certainly be fired where the enemy combatants' location is clear.
  - Priority to enemy combatants who are in a weapon-ready posture.
  - If only a glimpse was gained of the enemy combatant while he was moving, the Team Member must estimate his initial shot placement where the enemy's center of mass may be, based on his speed and direction of movement.
  - Once the shooting begins, anticipate that the enemy combatant(s) will drop to the ground. Team Members should no longer shoot where the enemy was, but where he is expected to be (nearer to the ground). If the enemy combatant is not put down by the initial shots and drops to the ground, a lateral cut at boot height should finish him. If using a shotgun, two shots, the first fired at center of mass chest, the second at boot height, should do the trick.
  - Note: If the enemy combatant initiates contact, Team Member shots should be directed laterally at boot height as the enemy may be engaging from field fortifications/foxholes or from the prone position.

Fire, Flame and Incendiary Weapons TTPs:

- Fire has been used in warfare at least since the dawn of recorded history. Notwithstanding its history of use, modern militaries don't always calculate the potential of fire in tactical or operational planning. Fire has both anti-personnel and anti-materiel application, destroying both if they cannot be withdrawn from its path.

- Flame weapons and munitions are available in US and foreign inventories and are largely used to defeat enemy field fortifications. The use of fire or fire weapons has a devastating effect on enemy morale.
- Fire can denude an area of vegetative concealment, exposing enemy forces, facilities and infrastructure and limiting the ability of the enemy to maneuver undetected. A fire front can flush an enemy from bivouacs, assembly areas, installations, emplacements and field fortifications that lie in its path and expose them to attack.

In the aftermath of the Soviet victory over the German forces and their allies in the Baltic States tens of thousands of Estonians, Latvians and Lithuanians took to Baltic forests to conduct partisan activities. 'The Soviets … , conducted widespread deforestation campaigns, burning vast tracts of forest, to flush out resisters.'[15]

- A fire can be used as an obstacle to block enemy movements or maneuvers, to channelize an enemy unit into a kill zone or a cul-de-sac, to cut off the enemy unit from sister/supporting units and to block enemy reinforcements. Smoke from a fire will also inhibit enemy observation of the battlefield. The ash generated by such a fire will facilitate observation of subsequent enemy movements in the burned-out areas.
- The Team may be able to block an enemy's aggressive pursuit by igniting a fire on the back-trail using a thermite munition or other incendiary device. For this to be effective, there must be sufficient distance between the Team and the pursuers (allowing time for the fire to surge), conditions must be dry and the wind brisk and in the right direction. Further, ensure that the enemy is downwind (and uphill if in appropriate terrain); that the fire will not block the Team from its mission objective or extraction LZs; and that the Team will not create its own cul-de-sac, should an enemy force appear to its front, e.g. in enemy Hammer and Anvil operations. It is also helpful if the Team can withdraw across a waterway after the fire is set.
- Forest fires can generate incredible heat, exceeding 1,472°F on the forest floor. Given abundant fuel, such temperatures are capable of sucking oxygen from underground tunnels that lie beneath the fire, suffocating the occupants. The speed at which the fire can spread is dependent on terrain; available fuel; presence of natural and man-made fire breaks; wind speed, moisture and other weather factors; the speed of a raging fire can reach approximately 6.7mph in forest and 14mph in grasslands, faster than dismounted troops can move.
- Wind conditions must direct the fire toward the enemy and not toward friendly forces. If winds reverse, the Team will be in substantial danger. Note that wind conditions in mountainous terrain and shore lines will shift near dusk and dawn; bear this in mind when setting a fire.
- Smoke may obscure both enemy and friendly forces, inhibiting the use of enemy/friendly aerial and artillery support.
- Fire can be used to flush out or destroy guerrilla/insurgent units, their base areas and their material resources and expose them to attack.
- Despite its obvious advantages, using fire as a weapon has its limitations. Fire generally does not always have great utility in verdant jungle or rainforest environments. Jungle or rainforest vegetation is almost entirely green and is moist through much of the year; dead vegetation quickly rots or is eaten by insects, generally leaving a relatively 'clean' rainforest floor under canopy; wind conditions on the floor of the jungle/rainforest are almost always

---

15. Švābe, Arveds, ed. (1950-55). *Latvju Enciklopēdija* (in Latvian). Stockholm: Trīszvaigznes. p. 3. OCLC 11845651

still; monsoon seasons soak everything in frequent cascades of rain. However, fire can be used in rainforest environments during dry seasons.

- Ambush, Raid or other Attack Operations with Flame Weapons/Munitions:
  - Use air dropped incendiaries/flame weapons (if friendly forces have air superiority); or use artillery support within range, firing WP (e.g. in COIN operations). Otherwise, use remote initiation or delay timers with incendiary devices whenever possible, to get safe separation, especially in areas that will promote rapid burning.
  - The Team may be able to flush out WMD/missile systems by mining roads that are patrolled by missile unit security forces. Or by setting fire to suspect wooded hide areas.
  - Use in rugged, mountainous and forested terrain under dry conditions and where wind conditions are stable and predictable. Avoid fires in zero/near-zero wind conditions or when winds are shifting. Mountain winds flow upwards in daylight and downwards at night.
  - Study terrain and prevailing weather and forecasts, including (night-day) wind variations, for the target area during mission preparation. And be observant of <u>actual</u> local weather, conditions, as these conditions may vary substantially from the predictive norms. <u>Warning</u>: Ensure that winds are brisk and constantly toward the enemy. Be mindful of night to day wind shifts, especially in mountainous elevations.
  - Set incendiaries upwind of the enemy, in ravines and/or in areas with plenty of dead brush and fast burning conifers. And employ where the enemy is channelized or blocked by natural/man-made obstacles.
  - Block vehicular movement at both ends of a road (e.g. abatis, road crater, etc.) and/or near an enemy encampment, to trap the enemy within the path of the fire in an attempt to annihilate the entire column/camp. Block all exits from a logistics installation to trap logistics personnel, supplies and equipment in the path of the fire. Use organic weapons or remotely initiated anti-personnel devices to inflict casualties on enemy personnel attempting to escape.
  - Ensure initiation occurs far enough away from the MSR or logistics installation so that the fire super-heats and builds up speed.
  - Withdraw upwind, but be conscious of post-dusk/dawn wind shift. <u>Ensure</u> that the Team route of withdrawal will permit rapid movement, preferably with mobility equipment. The Team should withdraw to a location that will provide overwatch of fleeing/dispersing enemy units. Pre-plan for on-call CAS/fire support to use against exposed enemy units.
  - Have a safe haven available if the wind shifts dramatically. Examples: an island/sandbar in the middle of a river; a swamp; a bare, rocky outcrop, a previously burnt-out area, etc.
  - Notes:
    - Attacking with fire (flame) is especially useful in destroying logistics stockpiles that the Team is typically ill-equipped to deal with.
    - C2, WMD systems and logistics installations will likely occupy well hidden sites under canopy, amid concealing heavy vegetation and behind terrain features.
    - C2 and logistics installations/complexes may be otherwise inaccessible to reconnaissance due to enemy patrolling and troop concentration/collective protection.
    - A fire will force evacuation of enemy units from hidden sites in the vicinity, providing many targeting opportunities. Enemy units may attempt to flee upwind, if the road/ trail network is oriented to facilitate movement in that direction. Prearrange supporting fires so that artillery/CAS have prepositioned and/or stockpiled ordnance; have fire support/CAS standing by for abundant target opportunities.
    - A fire can remove vegetation concealment from an area for extended periods, depending on season, rendering the area unusable for enemy units or installations.

- A fire will reveal concealed/camouflaged routes to aerial or satellite observation, as hard-packed paths will stand out from surrounding ash.
- Consider the offensive and defensive use of fire in operational planning. In Unconventional Warfare planning, consider what protective measures and contingency courses of action can be used to mitigate enemy use of fire against friendly UW forces co-located with SR Teams.
- Determine prevailing, current and forecast weather conditions in the operational planning phase and check for changes to these conditions prior to launch and throughout the operation.

Decoy or Deception Techniques TTPs:

> 'Be near but appear far, or be far but appear near' – Sun Tzu

- It is SR 'High Art' to influence the enemy to do what he might otherwise not do. Study enemy behavior to understand how to influence him. This may include:
  - Cause the enemy to mass so that he may be taken under long range fires/CAS, etc. and be destroyed.
  - Cause the enemy to abandon a place of relative safety, by making him feel threatened where he is – exposing him to destruction.
  - Cause the enemy to relocate to a position where he is trapped or bounded by obstacles/danger areas or where his mobility is impaired.
  - Use 'bait', deceptive communications, weapons fire, mines/booby-traps or remotely initiated explosive devices, marking devices (smoke, star-cluster munitions), ambushes, propaganda leaflets/PSYOPS materials, etc. that will cause the enemy to feel vulnerable.
- Deception Techniques:
  - Helicopter noise during insertions will alert the enemy to the location of the Team LZ. Consider using a false landing or decoy to confuse the enemy regarding your actual LZ. A Nightingale Device, used by SOG SR Teams, consisted of waterproofed, clustered firecrackers, fuses and fuse lighters attached to a wire mesh panel. The device would be thrown from a helicopter near a false LZ, and after the lapse of a time delay, would create a firefight simulation.
  - Use air assets to attack false locations to deceive the enemy during an insertion/extraction.
  - Two small parties of the British SAS were dropped in the Cherbourg Peninsula on a deception operation in support of the Normandy invasion. They were equipped with 'very [*sic*] pistols and gramophones; gramophones played suitable records of small arms fire interspersed with soldiers' oaths, while the very pistols lit the sky for miles around…'[16]

'Dust raised by motor or horse-drawn vehicles behind the lines also deceived the enemy. The vehicles dragged tree trunks or brush wood along the roads in order to raise more dust.'[17]

> 'Show gains to lure them; show disorder to make them take a chance.' – Sun Tzu

- Consider using a fake/rubber venomous snake as a decoy:
  - If the enemy notices the decoy snake, they may attempt to avoid it. This may direct the enemy onto a mine/booby-trap; or cause him to move into the kill zone of an ambush.
  - If the decoy snake itself is booby-trapped/mined, and the enemy attempts to 'kill' the decoy snake, the enemy will set off the explosive device.
  - Tracking dogs can be attracted to the decoy snake using bottled animal/human scent, which will focus the handler's attention on the decoy.

---

16. M.R.D. Foote, SOE in France (digital rendering without pagination).
17. Unnamed German Generals and General Staff, 'Military Improvisations' p. 57

- Enemy combatants are most susceptible to PSYOPS/propaganda immediately after experiencing casualties.
- Carry a small sealed (airtight) container of human or animal (e.g. pig) blood to leave a false blood trail for the enemy to find.
  - Fake blood may not fool a tracker dog, use actual blood. An airtight container will keep the blood in a usable form.
  - A blood trail or bloody bandages will encourage the enemy to pursue the Team, if they hadn't already decided to do so. If the enemy is already in pursuit, the blood sign will encourage him to move much faster in pursuit, believing that the Team is burdened with a casualty. This presents the Team with the opportunity to:
    - Cause the rapidly pursuing enemy trackers to blunder onto a mine/booby-trap.
    - Influence the enemy to deploy into a line to conduct a sweep, the Team may take up a position to inflict enfilading fires on the enemy unit.
- Deploy false enemy minefield signs to tactical advantage.
  - One application may be to deter an enemy unit from conducting a sweep toward the Team position or from approaching the Team flank.
  - The signs must be accurate, in the enemy's native tongue, must be written in the enemy's military terminology and consistent with enemy warning sign fabrication standards and protocols.

Author's Solution:

If the Team is in a defensive perimeter (perhaps with a POW or wounded comrade) it may be necessary to attract enemy fire away from the Team by having one or more Team Members, move off a several meters and rattle bamboo/ vegetation, cry out in feigned pain or mislead the enemy in his native tongue. This may cause the enemy to assault towards the distraction and expose his flank to friendly fire.

- Consider using a short-time-delayed grenade or demolition charge during your withdrawal to confuse the enemy as to Team location. Such a device should be pre-assembled for rapid employment.

Author's Solution:

- *If under attack at night, consider cracking a light-stick (white or yellow) and throwing it a distance away from the perimeter to lure the enemy away from the team and/or toward a field of fire. A light stick could also be thrown in a false direction by the tail-gunner as the Team withdraws from contact or conducts a breakout operation. Even better, if the light-stick can be attached onto a plant or branch, the light-stick may show some deceptive movement. Use of a light-lure, aside from a diversion, may have the added value of ruining the night vision of some of the attackers/pursuers.*
- *A noisemaking device, such as a small battery-powered toy can either disturb vegetation or make deceptive sounds to deceive the enemy as to your location and would deter or interrupt pursuit. This device may also serve as a decoy to lure the enemy onto a booby-trap or mine; lure the enemy into an ambush kill zone or lure a lone enemy into a POW snatch trap. These decoys may be accompanied by deployment of a Pursuit Deterrent Munition (PDM) to inflict casualties on a pursuing enemy.*
- *Light and sound decoys may cause the enemy to assault towards the distraction and expose his flank to friendly fire.*
- *Use something to distract/attract the enemy; perhaps something fascinating or out of place that they might covet, or even find irresistible. For example: a seemingly dropped pack of cigarettes can be an irresistible lure to enemy combatants.*
  - *The lure is used to cause an individual enemy soldier or element to pause along a road or trail within a POW snatch or ambush site. A mounted patrol may even be susceptible to the* lure.

- ○ *Ensure the cigarette brand is common among enemy soldiers or available* locally.
- ○ *To allay suspicion, use a used, partial* pack.
- ○ *Tuck a partially used pack of matches into the cellophane wrapper. This may cause the soldier(s) to pause to light* up.
- At night, a lit cigarette can be used to lure trackers to a false NDP and into a kill zone. The cigarette should be attached to or hung from vegetation, so that it moves and so that it can be seen by the approaching enemy.
- If the Team is to use a far-ambush to attack an enemy during daylight, consider dangling a mirror/shiny object from a bush at another location to attract enemy fire. This might be supplemented with a remotely initiated explosive charge.
- Note: The Author has conceived of some 'infernal' deceptive devices/decoys that are too sensitive to be included in these pages.

> Author's Tip: To mark enemy trackers or pursuers for air strikes, consider using 40mm or rifle grenade smoke (WP) munitions. Move a several meters away from the Team location in order to fire. This temporary repositioning will mislead the enemy as to the actual Team location and will influence the enemy unit or team to deploy toward the firing location instead, leaving its flank is exposed.

Collection TTPs:

- Attention to Detail. Do not overlook the small stuff.
  - ○ For instance take photos of abandoned dunnage, used to elevate supplies off the ground (in field storage), to provide intelligence insights: (1) <u>Transportation</u> dunnage debris, to include transportation pallets and banding material, can reveal the type of cargo and the proximate location of logistics field storage/installations; (2) <u>Storage</u> dunnage is essential for the protection of food, medical, clothing and personal equipment, personal demand items, etc. These items must be kept off the ground especially in wet conditions.
  - ○ Pay attention to any appurtenances, markings/insignia, weapons, special uniform items, etc., which may signify military branch, unit, special status, vehicle bumper markings, membership of an elite force, etc. Team Members should familiarize themselves with such information during pre-deployment training, with refresher training thereafter.
  - ○ Oncoming vehicular traffic can often be detected by observing vibration/ripples on the surface of puddles.
  - ○ Moss covered rock will reveal tracks. Look for moss discoloration as another indication of foot traffic.
- Team Members should carry small plastic bags to collect items for subsequent intelligence analysis.
  - ○ Bags should be appropriately marked when items are collected.
  - ○ Even bloody bandages may have significant intelligence value, once read for DNA, antibodies and other markers.
- If an enemy radio/radar antenna is discovered, photograph/record it, especially its form, construction, and orientation.
- Spent shell casings and other combat debris found at enemy locations will reveal the types of arms and ammunition with which they are equipped.
  - ○ Firing-pin imprints on casing primers can differentiate specific weapons that use a common caliber (e.g. AK-47 vs RPD).
  - ○ If the casings reveal that the weapon used is of an advanced design, this could be important intelligence.
  - ○ If Team Members do not know how to read primer imprints or are not carrying a tactical tablet with this information, casings of different calibers or with different imprints should

    be collected into a small plastic bag (marked with the Date Time Group (DTG) and location) for subsequent analysis.

- Cross-country vehicular movement along unimproved roads creates a plume of organic material that is distinctive (cloud form/contours) and that is easily detected by long-range radars and lasers (LIDAR), even at night. The enemy may attempt to deceive friendly forces by dragging treetops/limbs behind vehicles to create deceptive plumes.
- All US Team Members should take notes while on an operation and share significant observations with each other when an opportunity presents itself.
- <u>How to identify an enemy leader</u>: Uniform, insignia or equipment may not always differentiate a leader, but his behavior and how others behave in his presence will reveal his presence and authority.
  - Expect a senior leader/commander to observe training and operations rather than participate in it.
  - A senior leader/commander may sit in the presence of subordinates. He may have a security escort.
  - A senior leader/commander may stand aloof from the rest of the troops and he may have a small cluster of subordinates/staff in his orbit. Other subordinates will visit the cluster or commander to communicate and receive guidance. He may eat alone or with a few subordinates.
  - A senior leader/commander may often be in close proximity to a Command and Control (C^2) vehicle or personal transportation and will be seen entering and exiting the vehicle frequently. The senior leader/commander's vehicle may have multiple antennae.
  - If a C^2 vehicle is not close-by, the C^2 operation may be conducted from a requisitioned building or large tent (tent cluster). Look for heavy camouflage and the presence of antennae nearby.
  - A senior leader/commander will often be accompanied by a radioman equipped with a tactical radio. He may be observed frequently speaking on a manpack or vehicular radio. If the suspected senior leader/commander is observed speaking on a satellite phone, he may be considered a key leader/staff member indeed.
  - The senior leader/commander will often be well groomed with a clean, fresh uniform. He may be armed with a pistol and may carry binoculars. If he is armed with a standard rifle, it will generally be slung out-of-the-way; he will not carry many spare magazines. He will rarely be seen carrying grenades on his LBE. His LBE will generally not be heavily laden and he will not be bearing a rucksack. If possible, observe his boots to see if they are clean.
- There will always be a reason for the presence of enemy patrols or guard posts. Assuming that the presence of the Team itself (or a friendly UW force) is not cause for the added security measures, the Team should try to determine the reason for the added security.
- When the enemy conducts sweeps/patrols does he channelize himself to avoid his own minefields, difficult terrain features, etc.? If so, these areas may be suitable to establish ambush positions, Team hide locations or caches.
- Detection of Enemy Deception Operations[18] (TTPs):
  - An enemy will not invest his valuable resources in deception operations without a purpose. SR Team detection of enemy deception is extremely important to framing the intelligence picture. The SR Team must understand enemy deception operations or the Team may return from a mission with bad information that supports the deception.

---

18. Derivative/extract from Richard N. Armstrong, Lieutenant Colonel, US Army, *Soviet Operational Deception: The Red Cloak*, US Army Combat Studies Institute, US Army Command and General Staff College, Ft. Leavenworth, KS, 1986.

- ○ An enemy may establish false assembly areas, that he hopes will be detected (e.g. from the air or from sensors), to mislead US/allied forces as to his true intentions. These intentions may be that he actually (for instance) intends to defend or that he intends to strike elsewhere. If an assembly area is detected, an SR Team may be tasked to gather Priority Intelligence Requirement (PIR) information and confirm the operational viability of the assembly area. The enemy will aggressively seek to deter SR discovery of a false assembly area.
  - ○ Identification of smoke generation units (dummy or actual), can be crucial, as the purpose of these organizations is to blind observation and conceal crossing points, but may also be used in feigned attacks.
  - ○ Train movements may be easier to conceal than road or cross-country movements, as heat and debris plume signatures are mitigated. Look for stretches of rail that are concealed by overarching foliage and rail cuts that are closely bounded by conifers. The areas may conceal covert rail sidings and intermodal logistics transfer points, which may be instrumental to supporting enemy operations.
  - ○ When the enemy is in a defensive posture, expect enemy units to assemble or stage in the vicinity of road intersections that maximize lateral movement flexibility.
- Possible Red Flags of an Enemy Deception Operation:
  - ○ Radio silence in an area (with perhaps the exception of communications checks); but normal or increased radio traffic in other/adjacent areas. Reliance on alternative communication methods.
  - ○ Night movements in the area of main effort; daylight movements in other areas to attract intelligence interest.
  - ○ Dummy equipment positions.
  - ○ High visibility (false) transload operations.
  - ○ Loudspeaker simulations.
  - ○ Smoke obscuration.
  - ○ Tracer fire converging on US/allied aircraft.
  - ○ Increased engineer operations (e.g. construction of dummy positions; road improvements; crossing points.).
  - ○ Thermal plume from field kitchens/campfires.
  - ○ Light discipline not enforced.
  - ○ Dragging the crown of trees to cause dust clouds; discarded tree crowns as an indicator.
  - ○ Be suspicious of movements timed to coincide with known US aerial/satellite over-flight periods.
- Enemy C^3I Systems:
  - ○ Presence of MPs or manned traffic control points at a crossroads may presage a major tactical movement.
  - ○ Enemy communication and radar equipment (especially antennae) will likely be sited on high ground to achieve line-of-site/maximum range and coverage.
  - ○ Enemy SATCOM antennae, associated with important facilities, systems and headquarters, will be oriented toward communication satellites in synchronous orbit.
  - ○ AA will generally require an elevated and clear field for radar (early warning/tracking/guidance) or ground optics and good fields of fire along corridors of expected attack to protect enemy units and critical facilities.
  - ○ Anticipate likely antennae locations with terrain analysis.
  - ○ Some communication systems may avoid or mitigate the threat of detection, by using vehicular-mounted, telescoping (pop-up) antenna masts.
- In enemy occupied areas/terrain that are subject to brisk, cold winds, look for the enemy to locate in the lee of terrain that will shield their troops from the elements.

- Depending on soil and weather conditions, a Team Member may detect some sounds by putting his ear to the ground, which would otherwise be inaudible.
- Maintain a map and notebook record of your route and observations. Use a pencil, or a quick drying permanent ink, to make notes during an operation. Some inks will smear in wet conditions or will freeze in cold regions, whereas lead generally does not. Paper must also be moisture resistant. The ink in marker pens will often dry out at the worst moment. Grease pencils/china markers will not write in cold temperatures; a Team Member may have to insert the tip of the grease pencil into his mouth for a period of time to restore its capability to mark.
- Ideally, a digital camera will have a tagging capability to record GPS location and camera/ photo shot direction. If this capability is not present, if reliable GPS coordinates are not available, or if a film-based camera is used, tagging data must be recorded in a photo log. Photo log information for any photo taken on a roll or on a memory device should include: roll number, grid coordinates, photo subject or description, orientation of the shot, camera-object distance, F-stop, dimensional data and remarks. If possible, include a size-reference object (e.g. ruler or common object) positioned next to the item and within the frame of the shot. Never take pictures of Team Members while on patrol; if the enemy captures the camera, they will have gained useful intelligence/propaganda.
- Use flash and sound (flash-bang) ranging (if the flash is observable) to estimate the distance to and location of enemy artillery. This same technique can be used by Team Members to orient them (including from the detonation of friendly dropped ordnance) in the AO. From the flash, count the seconds to the sound of ordnance impact near your position; this will estimate the approximate number of kilometers to the firing locations.
- An SR Team may observe an enemy penal unit on the battlefield. These units are comprised of criminals, soldiers under punishment (hard labor or even death penalty) or political undesirables, whose sentences may be lifted as a condition of serving in a penal unit. Penal troops may be mostly used as infantry (including motorized or mechanized units) and may be integrated into a taskforce with armor. A penal unit will often be tasked with suicide missions such as breaching a minefield under fire, frontal assaults of strongly held defensive positions/field fortifications, assaulting through a choke point, attempting a contested river crossing, or assaulting key infrastructure (e.g. a bridge). Subsequently, the presence and location of such a unit on the battlefield is significant. These units may be detected or differentiated from other units, by:
  - A significant presence of political officers or military police.
  - A prisoner enclosure (e.g. barbed wire/concertina fencing) with an inwardly facing guard force.
  - Harsh treatment/corporal punishment of the soldiers.
  - Troops without individual weapons or without ammunition. These items would be issued at the marshaling/assembly area just before deployment for an attack.
  - The general conduct and composure one might expect of penal troops.
- How are refugees and local inhabitants being treated by enemy combatants?
  - In which directions are refugees flowing?
  - Is the enemy using population or refugee control stations/check points?
  - Where are the young civilian men? Are refugees/civilians being used as forced labor?
  - Are civilians being forced from their homes? Are enemy personnel moving in?
  - Are crops being maintained, harvested, seized?
  - Are civilians being used as human shields?
- Which structures are occupied by enemy soldiers and which are occupied by local civilians? Are religious buildings, schools, medical facilities, buildings of historical significance being occupied or used by the enemy? For what purposes?

- Local civilians may hunt, fish, or trap to provide or supplement available food. Discovery by local civilians represents a threat to the Team.
  - Stay alert for game traps, stands, blinds and fishing line.
  - A large military presence or military operations may drive away game, forcing hunters, fishers, trappers to move further afield. Enemy units may deplete local game.
  - Hunters, fishers, trappers are likely to have the skills needed to detect trail sign of the SR Team. They may also be accompanied by dogs.
  - If the civilian is hunting with a firearm and operating a vehicle to retrieve game, this may signify that he is aligned with the enemy in order to receive permits and ration cards issued by the enemy. Hunters using firearms will typically hunt outside the enemy security zone to avoid engagement with enemy forces.
  - Hunters, fishers, trappers may have significant value as captives; even more than a military POW. Not only will he thoroughly know his hunting area and the rural environs, but he may have comprehensive knowledge of the local enemy situation and other significant information.
  - If the Team kills, wounds or captures a local, it is likely that enemy will be informed and they may conduct a search, perhaps with local personnel acting as guides and/or accompanied by dogs. This will naturally compromise the Team's presence, and may require the Team to plan for mission revision or extraction.
- Locals may hide livestock some distance from MSRs and villages to prevent seizure by enemy combatants. Alternatively, livestock may be hidden because local sympathizers may be deliberately providing cattle/pack animals to guerillas, insurgents or partisans.
- Estimating the width of a river (or other obstacle) using geometry, if a laser or coincidental range finder is not available.
  - The Team Member picks a landmark (e.g. a distinctive tree) on the opposite bank of the river (Point 'A') directly opposite from his location on the near side bank. This would best be at a straight stretch of the river.
  - The Team Member marks his position as Point 'B'.
  - The Team Member walks to the left/right until he reaches a 45° angle to the landmark (Point 'A') and marks this spot (Point 'C'). He does the same on the other side of the initial mark; again marking the spot (Point 'D') as the second 45° angle; forming an isosceles triangle (Points 'A', 'C' & 'D')
  - Count the paces between the two 45° marks (Points 'D' and 'C') and convert paces to meters. The river width is approximately half this distance.

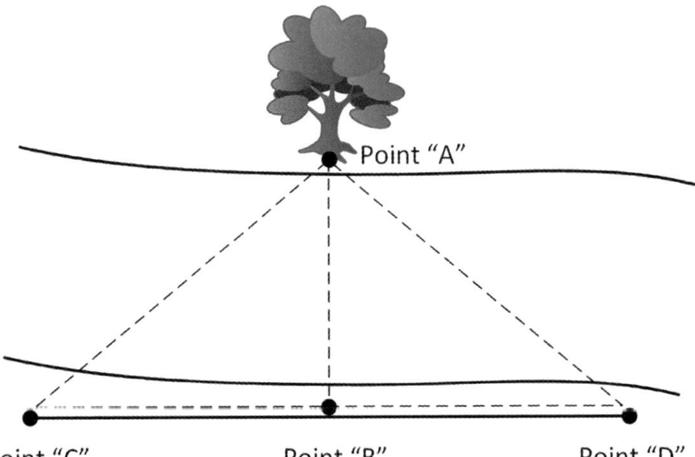

Figure 35. Estimating Stream/River Width.

- Use field expedient 'blinders' on binoculars and scopes to keep the objective lens in shadow to mitigate lens reflection in daylight; blinders are also useful to shield the observer's eyes during hours of darkness from bright moonlight and the light of enemy fires/illumination, thus preserving night vision.
- A prudent enemy will place his OP/LPs in defilade from his main perimeter camp fires and lighted activities to preserve the night vision of his observers. Consider this possibility when Team Members approach an enemy camp, guard post, security force, etc., especially in winter, when campfires will be more in use.
- In the vicinity of a village/abandoned village, fruit trees may exist in current/former crop areas.
  - If it appears that enemy combatants have been picking the fruit, as may be determined from footprints, debris and other signs, the T/L may decide to:
    - Track the enemy to their unit/element location.
    - Establish an ambush to await enemy return.
    - Lay a mine or booby-trap beneath a fruit laden bough, before moving on with the Team mission.
  - If it is local civilians that harvest the fruit, capture one or more of them.
    - These people may possess an abundance of useful intelligence and they may harbor ill sentiments toward the enemy combatants.
    - They may also be auxiliaries who are supplying enemy combatants (guerillas, partisans) in friendly territory. Likewise, they will possess a wealth of useful intelligence. These prisoners must be evacuated quickly, before the enemy realizes that they are missing – which would cause the enemy to relocate or to launch a pursuit.
    - If it is friendly paramilitary personnel (guerillas, partisans) operating in enemy controlled areas, who are harvesting the fruit, this may present an opportunity for the Team to establish contact.
  - Orchards/cultivated area may be a source of supplemental/survival rations for the Team.

<u>True Account:</u> A SOG SR Team was assigned a road watch mission in Target Area H-9 located in Southeastern Laos, south of where the primary MSR had descended from hilly/mountainous terrain into flatlands partially covered by single to double canopy rainforest, and then split, with one MSR highway branch headed east toward the Tri-Border Area (Laos, Vietnam, Cambodia) and the other MSR highway branch leading south into Cambodia. The Team successfully infiltrated into the target area and headed west toward its objective; within two days, the Team engaged an NVA tracker unit as the enemy closed on the Team NDP location at dawn. The Team continued on with its mission and navigated west towards the enemy southern MSR. The Team encountered an area of broken canopy populated with very tall trees. One of the indigenous Team Members happened to look upwards and found wooden slats embedded in the trunk of a tree leading to the high boughs of the tree. The slats did not descend all the way to the ground, probably to avoid detection of the observation post from the ground, and suggesting that a ladder would be used to ascend to the lowest slat. No enemy observer was present in the treetop OP. As the entire Team cast their eyes up, they then detected an entire line of trees, running north and south of their position along the road, that were similarly slatted. The upper reaches of the trees offered long range observation of the flat land to the east, allowing detection of approaching aircraft and observation of infiltration LZs for several kilometers. Lessons-Learned: (1) As the Team navigates in its target area, Team Members must also visually scan the vegetation above them, and (2) the Team can also use an elevated OP or tree stand to obtain extended fields of view.

Electrical Power Operational Indicators TTPs:

- A different type of pylon (not apparent on a topographical map) is used whenever commercial power lines take a sharp 'dog-leg'; this to support structural stress. This information can be useful in land navigation orientation.
- Be observant of branch power lines and transformers suspended from telephone poles; if the lines lead into areas where no habitations or other infrastructure are known, and where tree foliage has not been cleared, this may be an indication of an important, prepared, covert C^2 or logistics site hidden in the vicinity. Recently installed power lines running to abandoned buildings may also be a red flag. Also, check for power lines that may descend to the ground and run under roads via culverts, etc.
- Additionally, the enemy may tap power lines along a MSR to support the temporary power needs of units and facilities in the vicinity; new lines/poles, etc. could be an indicator of this. Tapping into commercial power would be a prudent enemy measure to conserve fuel that would otherwise be consumed by generators and would suppress the signature (heat, noise, etc.) associated with generators.

Bomb Damage Assessment (BDA) Mission TTPs:

- A BDA mission is often a high-risk operation used to assess the effects of friendly bombing of enemy targets. BDAs normally follow a major airstrike (e.g. B-52 sortie), but they can also follow smaller precision strikes on high value targets. After a major strike, it may be possible to scoop up enemy POWs who may be wounded, unconscious or dazed. It may also be possible to seize documents or materiel of significant intelligence value from enemy field fortifications or C^2 centers. The enemy may realize that a BDA operation routinely follows a major strike, so enemy forces may be prepared to react to a BDA Team and they will be especially vigilant in the vicinity of critical nodes.
- Teams sent on BDA missions should be prepared and equipped for opportunistic missions and intense combat situations.
- The RF/Bright Light Team should be on standby at the launch site.

  <u>True Account:</u> SOG Reconnaissance Teams operating in Laos would typically insert on a two-day BDA mission immediately after the dust cleared following a B-52 strike on an enemy base area along the Ho Chi Minh Trail. The Teams would typically insert via a LZ located outside the beaten zone and then move through the rainforest and through air strike debris and fallen trees to the area of interest. This movement was often difficult, time-consuming and allowed the enemy to organize its pursuit and defense of critical nodes. Teams were almost always driven away from their area of interest by enemy forces and subsequently thwarted in their mission; they were often extracted under emergency conditions the morning following the insertion. One T/L decided on another approach; he designated the Team insertion LZ within the strike zone immediately next to Laotian Route 110 and planned for a same-day extraction prior to nightfall. The Team landed shortly after the strike, as planned; ascended to Route 110 and patrolled in a westerly direction along the road for approximately 500 meters, taking photos en route. Though vulnerable to long range enemy fire across the denuded landscape, the Team was not engaged by a dazed enemy. The Team also had excellent fields of fire across the vista of the strike zone; but no enemy troops could be detected. Subsequently, the Team ascended a hill on the northern periphery of the road, which provided an excellent panoramic view of the strike zone. The Team then found an underground tunnel network lacing beneath the hill that

had been revealed by a bomb crater, indicating the presence of a secret and high-level C2 facility. Had the T/L followed the typical protocol, this tunnel complex would never have been detected.

Fixed Site Surveillance TTPs:

- Select surveillance positions that are not silhouetted against backgrounds (e.g. sky, hill crests or dissimilar backgrounds). Ensure that the site is not revealed by changing light conditions as the sun changes position during the day. Also, ensure that the site will not be detectable to night-vision/thermal optics.
- Select a surveillance position (and its access route) that is in an unexpected location, that will not attract enemy attention, or is in an area unlikely to be patrolled or swept by enemy units. Enemy troops will routinely avoid swampland, areas covered in thorny thickets or razor grass, etc.
- Select a surveillance position with concealed (and preferably covered) access, so that surveillance elements can be rotated from the Team hide position, or so surveillance elements can be covertly withdrawn, if required.
- If defiladed, the position itself will provide cover against direct fire. If any excavation is required, the spoil must be concealed or moved to another area and scattered.
- Dogs may be used by enemy patrols or sweeps. Consider using a light dusting of CS powder at some distance from the position and along the back-trail to confound detection by dogs. Additionally, consider using game animal attractants/scents (e.g. musk, urine, lures) that will draw game to the area, befuddle the dog or cause the dog to chase the game, which may cause the handler to conclude that an alerting dog was not reacting to Team presence but to game presence.
- The surveillance position should ideally provide natural concealment that requires little or no additional concealment measures. For moderate to long term surveillance, note that use of leafy vegetation cut from plants to conceal a position will wither and be easily detected by native enemy troops or patrols. If additional vegetation is needed, consider cutting sod or small plants taken from another area that the enemy will not detect; keep the roots intact, and re-plant these at the hide position if the position is to be used for a lengthy period. Additionally, use line (e.g. fishing line) to reposition plant foliage or to secure low tree boughs to camouflage the position.
- If sensors or remote cameras are to be used for the surveillance, select a sensor/camera position that is optimum for sensing or observation. For cameras, this includes ensuring that the lenses can view maximum angles and capture key information such as cargo, bumper/unit markings, etc. Use a dusting of CS powder around these positions.
- Use predator musk/urine in pastureland to keep dogs away from a hide location.

Tracking and Reading Sign TTPs:

> 'The greatest enemy will hide in the last place you will ever look.'
> – Julius Caesar 75 BC

- It is the Author's view that combat tracking is a very important skill set that is especially relevant and transferable to SR operations. Combat tracking skills applies, or transfers, to a spectrum of SR mission tasks, to include: intelligence gathering, COIN operations, SERE, counter-tracking, etc.
- Many commonalities exist between SR and tracker disciplines, to include a variety of TTPs associated with field craft, stealth, movement and mission purpose. The overall purpose of a

friendly tracker team may be similar to that of SR Teams (or a Mobile Guerilla Force), as has been discussed elsewhere in this book; that is: to 'Find, Fix, Finish and Exploit'[19] the enemy.

- Combat tracker missions relevant to SR include:
  - <u>Pursuit</u>: to gain, maintain or reestablish contact with a moving or fleeing enemy element.
  - <u>Reconnaissance</u>: to track and observe enemy elements, routes, locations, etc. and collect information pertaining thereto.
  - <u>Security</u>: patrolling to secure friendly units (e.g. Mobile Guerilla Force, reaction/exploitation forces) and locations.
  - <u>Other</u>:
    - In SR Bright Light/RF missions to rescue friendly Team personnel/aircrew, recover KIAs and rescue POWs, tracking is considered an essential function.
    - Tracking was routinely performed by SR Teams and was made possible because indigenous Team personnel had well-developed tracking and field-craft skills.
    - Additionally, BLTs were often assigned BDA missions. If a Team spots blood trail(s) leading from a strike zone, the Team should be prepared to follow that trail to capture a POW.
- A K-9 (dog) tracker unit can track and alert to a quarry at night. Some concerns:
  - The dogs will typically need rest. It may be difficult to keep a dog focused and alert during continuous day and night operations, especially in hot, hot/humid conditions.
  - Scent will dissipate more rapidly,
    - Along ridge tops or along other terrain features where brisk winds will scour the ground.
    - On hard packed road surfaces, especially where subsequent pedestrian or vehicular traffic has passed.
    - After rain/snowfall, in fog or during/after a wind storm.
- Relatively few combatants, outside of those found in aboriginal/tribal hunting cultures such as exist in Africa, South and Central America, etc., are exceptionally skilled in reading trail signs. Many US and foreign military service personnel tend to be young men with rural backgrounds, as opposed to urban/suburban males, and some of these may be expected to gravitate to SOF; these rural males represent a good recruiting pool for SOF. As one might expect, most SpecOps personnel have an affinity for the outdoor lifestyle and many may have experience in hunting, but this is not the same as having the requisite knowledge and skills associated with tracking. Despite the paucity of recruits or servicemen with tracker skills, even limited training in tracking TTPs can have a huge payoff.
- The Author recommends required reading of the *Combat Tracking Guide*, by John D. Hurth (former Special Forces soldier), and TC 31-34-4, Special Forces Tracking and Countertracking, as an introduction to combat tracking. This book describes how information can be derived from trail signs. The Author also strongly recommends that all SR personnel attend a combat trackers course. Note: In the tracking TTPs contained within this book, the Author supplements or tailors content found in the Combat Trackers Guide as it pertains to SR operations and how tracking experiences have been/are relevant to SR.
- In the AO, the Author also recommends that local trackers-hunters, from primitive tribes if possible, be recruited as training (tracking and survival) cadre and/or as indigenous Team Members. If local talent is not available, recommend that 'mercenary' recruitment be conducted from among third-country primitive or tribal hunting populations living in similar environments. The British SAS used native (Iban) trackers recruited from Borneo to great effect during the Malayan Emergency. Native trackers must be thoroughly trained

---

19. Hurth, *Combat Tracking Guide*, p. xii

in unit operations, in TTPs and in identification of battlefield hazardous items or they may become a significant liability to the Team.

- The Author also recommends using commercially obtainable and specialty dyes (both permanent and impermanent), in various colors, to serve the following purposes:
  - Visible or invisible permanent dyes, in powdered form, can be used on an enemy trail in a counterinsurgency environment near local villages. Local villagers who are part-time insurgents or who are logistically supporting enemy forces can later be identified by the dye.
  - Invisible luminescent dyes, in powdered or slurry form, can be used on enemy trails and roads in counterinsurgency, sanctuary/base area environments, covert marshalling/assembly locations in more remote areas. Enemy combatants or vehicles (for short distances) passing through a dye contaminated area will leave a marked path that can be traced by the SR Team to camps and other facilities. The invisible dye can also allow the SR Team to track an enemy at night and can be useful in avoiding enemy booby-traps/mines.
  - A bag of invisible dye can be attached beneath an enemy vehicle to leave 'bread crumbs' as it travels along a road to a hidden location. The dye can also be used to mark the vehicle exterior.
  - Invisible dyes can be detected at extended distances, even through intermittent canopy, by IR illumination/optics operated by Team Members or mounted on UAVs or other airborne platforms.
  - Tips:
    - Team Personnel should not handle tactical dyes during the mission preparation or execution phases whenever they are to be used to mark or track an enemy unless a fully dependable solvent is available to eradicate all traces. Blacklight or other illuminant should be used to verify that Team Members are not contaminated. Use of surgical gloves is recommended.
    - During mission execution, Team Members should not pass through areas that are contaminated with tactical dyes or be exposed to powdered dyes that are aerosolized by traffic.
    - Dye packages should be prepackaged (e.g. at the factory) in a range of configurations and quantities for SR Team use. Dissemination should be done remotely (or by time delay) and as silently (e.g. using a squib or compressed CO_2) as possible.
    - Dyes should preferably be deployed by the Team during lapse weather conditions or where dye powder or contaminated debris will not be blown back toward Team Members.
    - Certain situations suggest permanent dyes, other circumstances suggest impermanent dyes. And note that <u>very little</u> powdered dye can go a long way.
- All Team Members, not just the designated tracker, must use the senses of sight, hearing and smell, while tracking an enemy.
- While tracking, beware of leave-behind mines/booby-traps planted by the enemy at the site of an earlier enemy ambush or on the enemy back-trail. The designated tracker may find it handy to use a walking stick or a twig while following enemy traces. The twig can serve to reveal booby-trap/mine trip-wires; a walking stick can be used to uncover leaf cover (for instance) to find surface/buried traps/mines. The tracker's assistant provides security.
- Footprints can reveal much about an enemy element to include: 'direction, rate of movement, number, sex and whether the individual knows he is being tracked.'[20]

---

20. TC 31-34-4, 'Special Forces Tracking and Countertracking', Headquarters, Department of the Army, Special Operations Press, September 2009, p. 1-4.

- ○ Tread design may reveal the type of enemy combatant/unit (standard/elite military, guerilla, partisan, etc.) that the Team is tracking.
- ○ If the Team tracker detects that the quarry has uncharacteristically accelerated his/their pace, there is a reason. This urgency may be benign, or it may indicate that the Team has been detected. If the tracks reveal the enemy accelerating to a run, the Team may be very close to the quarry and the quarry is attempting to gain separation or the quarry is attempting to gain high ground or a defensive position. The Team should immediately take precautionary measures.
- ○ Determine the number of enemy in the pursued party. This can be done by counting the number of tracks that are found within a typical stride length (36in on level ground) on a narrow trail or on the enemy's trace through soft ground (off trail), then dividing the number of tracks by two.
- ○ Key prints are distinctive because they possess some identifying mark or feature. Take photos of key prints.
- ○ If tracks have become difficult to trace due to ground cover (e.g. leaves, conifer needles), recover the trail by lifting up ground cover to locate traces.
- 'Normally, a person or animal seeks the path of least resistance; therefore, when searching … trackers will find signs in open areas….' If the Team tracker detects that the quarry is deliberately choosing a difficult path, take heed and use caution.
- Team Members must take extreme care, especially once they have tracked the enemy combatants to their MSS perimeter or camp, not to leave signs of their presence or passage.
- If the tracker notices enemy countermeasures against the pursuing Team, be aware that such measures constitute a certain degree of professionalism and field craft. Increase security and resourcefulness in these circumstances.
- <u>Age of Signs</u>.
  - ○ Winds of Beaufort Scale #5 or greater, depending on rain/condensation, will cause debris to drift over quarry tracks. By recollecting when the winds occurred, age of track may be estimated.
  - ○ Fresh bloodstains are bright red, but will darken with exposure to air and sunlight.
  - ○ Light rain will affect track definition. By recalling when rain fell in the area, the age of track may be estimated.
  - ○ If sap is still running from damaged tree bark, the trace is fresh.
- Troops normally will not walk through a puddle (e.g. along a road or trail) if they can walk around it. This tendency may provide a 'trap' for footprints and an ideal place to plant mines/booby-traps.
- Tracking, reading and interpreting trail signs is time-consuming, especially as the Team must exercise stealth and caution to avoid mines/booby-traps and ambushes. Meanwhile the enemy is evading and moving more swiftly than the Team. An exception to an enemy's better speed-of-march is when the enemy element is carrying WIAs/KIAs, porting burdens, or is taking evasive/deceptive measures; these are opportunities for the Team to catch up. Eventually, the enemy will return to their base, where, again, the Team may close in. There are techniques to accelerate tracking or pursuing enemy combatants, some shown within these pages, but note that these techniques bear risks.

Closing the Distance:
- One method is the leap-ahead or bounding technique, requiring a split Team operation. The T/L estimates the amount of lead held by the enemy element, the direction of enemy travel, speed of enemy movement and restrictive or high-speed terrain and vegetation that may assist or impair pursuit. The T/L then identifies linear danger areas (e.g. streams) or

terrain features (e.g. ravines) that the quarry must cross. While the primary tracker element continues to follow the existing trail, the second Team element moves rapidly along a separate path to intersect the enemy element's trail at or near the linear danger area or terrain feature. If the second element is successful in intersecting the quarry's trail, it then becomes the primary tracking element and continues the tracking/pursuit. The other tracker element can then, based on fresh information from the new primary tracking element, bound forward in an attempt to intersect the enemy's trail even further ahead, again at a linear danger area/ terrain feature. However, if the leap-ahead element's movement is rapid enough, the leap-ahead Team element may establish an ambush at a linear danger area/terrain feature ahead of the enemy's arrival.

- In a COIN environment, helicopter support may be able to assist in repositioning the bounding element, if the distance between the pursuit and enemy is substantial.
- Risks: CAS assets will often decline to drop/fire ordnance when the Team is split and where the elements cannot positively be pinpointed. Additionally, the split Team element will obviously possess only 50 per cent of the Team's total firepower, making it much more vulnerable to an enemy meeting engagement/ambush.

• In pursuit of an enemy unit, the T/L may elect to pursue in parallel, with the lead element following the trail and the other element off the trail, moving in parallel to the lead element. This may be prudent if the terrain and vegetation permit relatively silent movement and sufficient maneuver space. This technique may have the following advantages:
- If the enemy is about to be run to ground, the element in parallel is poised to provide over-watch, to detect an enemy ambush position and to maneuver in flanking counter-ambush battle drill.
- If the enemy tries to covertly peel off its personnel to evade the lead element, the parallel element may be able to detect their trail or even intercept these combatants.

• <u>Tracking in Limited Visibility Conditions</u>: Under darkened conditions, even at night, the Team may successfully track an enemy element by using ambient or artificial light.
- Tracking at night is a risky enterprise. The T/L must consider risk versus reward in making a decision to track the enemy at night. Tracking at night may be driven by the need to pursue the enemy (to establish, maintain or reacquire contact) and/or to close the distance between the Team and the enemy element. The T/L must employ good judgment and use this procedure only to close the <u>distance</u> with the enemy, rather than to close <u>with</u> the enemy. The risks of approaching too close to the enemy element are obvious:
  - Team Members will make more noise during night pursuit.
  - The enemy element may have fish-hooked into a NDP location and may be in position to mass fires on the Team.
  - The enemy may have deployed a sniper or ambush.
  - The enemy may possess night-vision optics and may be able to detect the Team before the Team detects them.
  - Mines/booby-traps are more difficult to detect at night
  - Continuous movement of the Team will exhaust Team personnel and may lead to tactical lapses.
  - Detection of the Team may cause the enemy element to scatter and flee. It will be much more difficult to track single combatants than track an enemy unit.
- If trail signs indicate that the enemy may be nearby, (as noted above) night tracking should be suspended. If there are clear indications that the enemy may continue his movement at night, then night tracking may be necessary to close/maintain the distance on the prey.
- The night tracking element would normally include the tracker, a point/security man, and a tracker's assistant (to operate a light). Factors attendant to the use of artificial light (to include NVDs) may be acceptable under the following conditions:

- Heavy vegetation and/or heavily dissected terrain may minimize the range at which a light may be detected by the enemy.
- Lens filters and shrouds are used to limit loss of night vision, or to facilitate night-vision optics of the Team.
- The distance between the Team and the quarry must be closed to maintain the pursuit.
- Note that IR light can cast a revealing shadow on footprints, much like other forms of artificial illumination, as long as the light is cast at a proper angle.
- If the enemy is known to employ night-vision optics, the risks of night tracking with artificial illumination increase substantially.
- At night, the point/security man should be positioned to the right (especially if he bears his weapon left-handed) and immediately behind the tracker; he focuses his attention to the front and to right flank. The assistant tracker should be positioned to the left and immediately behind (no more than arm's length) the tracker; he focuses his attention on the needs of the tracker and to the left flank. All three will be on the lookout for mines/booby-traps. The obvious risk is that an ambush may wound or kill all three of the tracking element in the initial burst.
- Use of NVGs (ambient light device), aided by an IR light source, is much preferred to using a flashlight with the naked eye, for obvious reasons. If a flashlight is used, recommend that the beam be filtered (with a lens) and shrouded to preserve the night vision of the Team as much as possible. Tests have shown that the green lens generally provides better visual acuity and color differentiation than lenses of other colors; however, color differentiation between Team Members may naturally vary. Subsequently, some lenses may be better for certain individuals and purposes (e.g. tracking) than for other purposes (e.g. map reading). Team Members should test how different color lenses perform for varying purposes during training. Rather than change lenses according to purpose (inconvenient and time-consuming), consider carrying small, light-weight flashlights, pre-fitted with different lenses.
- The enemy may be tracked across open areas using moonlight, as long as the moon position is at an angle to the tracks sufficient to cast a shadow. If this angle is not sufficient, shrouded supplementary artificial light (e.g. IR) may be necessary.
- A chemical light (to include the IR variant) may also be useful in night tracking. Risk: an active chemical light can only be doused if placed in a pocket, inside clothing/equipment or buried.

- If the Team encounters abandoned materials or debris left by the enemy, the T/L must assess the situation immediately and with caution. The items may have been deliberately left behind as a decoy.
  - Immediately suspect mines/booby-traps or an ambush.
  - Consider dropping the Team to ground; withdrawing a few meters and then either scouting around to detect an enemy ambush or deploying for battle drill.
- During tracking, be careful of enemy mine or booby-trap placement in the following circumstances:
  - Areas requiring that Team Members crawl.
  - Where foot placement options are limited or constrained such as stepping over roots, fallen trees/limbs/logs.
  - Danger areas, especially if trail signs indicate that the quarry is professional, disciplined or elite.
- If the quarry selects the more/most difficult path or makes a dramatic course alteration, there's a reason. Be especially wary in these circumstances.

- ○ Troops will generally seek the easiest and most direct route from a point of departure to a destination. Only the presence of some hazard or leadership direction would deter this tendency, so if enemy troops are avoiding a direct/easy path, trail, road or area, be suspicious.
- ○ If the quarry's trail enters severely restricted terrain/vegetation, or <u>small</u> open/danger areas, the Team should circle around.
- ○ If the quarry's trail crosses a <u>large</u> open/danger area, employ the appropriate tactical formation and movement techniques to mitigate risk. This may mean the bulk of the Team moving along folds in the earth, while the trackers remain exposed following quarry signs.
- ○ If the enemy trail makes a suspicious 90 degree turn, go to ground immediately, modify the Team formation and modify movement techniques to maximize stealth (crawling if necessary); the 90 degree turn may signify the enemy element moving into a perimeter with overwatch of its back-trail. The T/L has other options: he may deploy the Team into assault line or, more appropriately, split the Team into two elements – with the trailing element to flank the possible enemy location.
- When the enemy is about to move into its base camp, the enemy leader may be expected to separate from his element to exchange sign-countersign with base security. But proximity of the enemy element to its base camp may not be obvious to the SR Team. The enemy leader may also separate from his element to reconnoiter a hide, NDP or ambush site. The Team should be attentive to these circumstances and take appropriate actions, to include:
- ○ Go to ground.
- ○ Exercise extreme stealth and/or tactically deploy.
- ○ If a base camp is ahead, carefully circle around the camp to identify exit/escape routes, patrol routes, OPs/LPs, watering points, etc.
  - ▪ If the camp turns out to be a MSS, then the T/L must decide if a raid is warranted or if the Team should continue to observe and/or track the quarry to its subsequent destination.
  - ▪ If the camp is large, the Team must summon support to 'Finish' the enemy. The Team should be prepared to continue pursuit contingent on the effects of supporting fires, heliborne assault by friendly ground elements, and enemy attempts to flee.
- At a trail junction, or at a turn at a dirt/gravel road junction, pedestrian/vehicular traffic will wear a curved path/pattern at a corner that will indicate the direction to occupied areas/troop concentrations. The more substantial the wear pattern, the clearer the indication of high traffic and occupied areas. This worn area will also collect water into puddles, where footprints/tire patterns may be detected. This puddle may be an ideal spot to bury an anti-vehicle/anti-tank mine.
- Team tracking formations (see Hurth's Combat Tracking Guide) to be used 'will depend on the mission, terrain [and vegetation] and likelihood of enemy contact…. Leaders should understand that movement formations need to be flexible and should adjust them according to mission, enemy situation, troops available, terrain [METT] and amount of time and distance between the trackers and the quarry.'[21]
- ○ The tracking formation will also relate or be tailored to the standard Team tactical reconnaissance formation selected by the T/L for the situational threat, terrain and vegetation environments.
- ○ If the Team is using indigenous Team Members, the tracker and the point-man should be paired (in a file formation, for instance), followed by the T/L.

---

21. Hurth, *Combat Tracking Guide*, p. 47

- ◦ A modified diamond formation (again, with the tracker and point-man paired), would be useful when crossing certain danger areas, in ascending ridges (where the enemy may be waiting), in other situations where contact may be expected and in open terrain. The diamond provides all around security and immediate fire in all directions.
- A high-lumen flashlight may be useful to better see footprints in shadows under canopy or ledges. An assistant tracker/Team Member may use the beam to cast shadows on enemy foot impressions for the primary tracker to better detect. This may be warranted if distance to the quarry is substantial.
- If the Team loses the trail:
  - ◦ Search the immediate area first. If flank security is being employed, use them as well to reacquire the trail on the flanks; this may save substantial time.
  - ◦ 'Read' the map to discover areas where the lost trail may be intersected, and where in the target area, the trail sign may be more easily detected (e.g. stream banks); this may require that the Team split, one element to try and reacquire the enemy trail and one element to intersect the trail further ahead. This same technique may be used to accelerate the tracking pursuit of the enemy. Also, an area reconnaissance technique (e.g. box) may be used to reacquire the trail.

  > 'If you know the enemy and know yourself, you need not fear the result of a hundred battles. If you know yourself and not the enemy, for every victory gained you will also suffer a defeat. If you know neither the enemy nor yourself, you will succumb in every battle.' Sun Tzu: The Art of War

- Trail signs, whether they are left by the Team or enemy combatants, can provide valuable information that may go well beyond such routine matters as numbers, direction, activities, etc. This information may offer substantive clues to enemy or friendly vulnerabilities and capabilities.
  - ◦ All Team Members, regardless of mission type, must develop habits to conceal signs of Team presence or passage – but responsibility resides especially with the tail-gunner.
  - ◦ When a Team Member/the tail-gunner brushes out tracks, the brush stokes should be logically applied. The brushing should not score the ground (a dead giveaway), disturb rocks, sweep away preexisting debris, etc.
  - ◦ All Team Members must avoid 'track traps' (e.g. puddles, mud, clay, sand or other soil media), which will provide a clear print impression of footwear. An alternative to this would be to wear footwear with the enemy's tread pattern. If the quarry is notably professional, he would be unlikely to leave prints in a 'track trap' (unless the quarry is feeling secure, is in proximity to its parent unit, or is setting up the tracker for an ambush) – be alert! Note that the footprints of US Team Members may be much larger than the footprints of indigenous enemy combatants (or civilians) of short stature/slight build.
  - ◦ Team Members will leave impressions where they sit/lie in a NDP; this will reveal the number of Team Members. When the Team leaves this location, every Team Member should do what he can, given available time and illumination, to restore his position to its previous condition. In grass or dead vegetation, 'lift' the vegetation back into place with a stick. This will accelerate the recovery of pressed vegetation.
- While tracking, observe enemy combatant TTPs to determine trends. Observations may indicate the state of training, discipline, leadership and expertise and may include:
  - ◦ How the enemy establishes his NDP.
  - ◦ If he routinely establishes OPs/LPs.
  - ◦ If he uses fish-hooking to observe his back-trail.
  - ◦ Where he sites his crew-served weapons.

- ○ If he routinely deploys mines/booby-traps.
- ○ How he uses tactical deception.
- Mine/booby-trap Precautions in Tracking:
  - ○ Team Members should know enemy trail or boundary warning signs and enemy sign-marking protocols for mines/booby-traps, so that these devices can be avoided. The FOB/battalion S-2 should maintain reference materials describing enemy signs/indicators and marking protocols. Team Members should become familiar with these prior to or immediately after deployment into an AO.
  - ○ Team Members can move or remove enemy warning signs to inflict casualties on the enemy; Team Members might also mark trails/boundaries with false enemy markers, to reroute enemy elements into an ambush site; to cause an enemy element to pause in a kill zone; or to generally harass and confuse enemy forces. These deceptive actions, and the locations where these actions are implemented, should be recorded, photographed and reported.
  - ○ Understanding enemy mine warfare/booby-trap techniques better prepares Team Members for operations in the presence of these hazards and for employing the proper countermeasures. If indigenous Team Members include former enemy combatants, they can be valuable assets to the Team's mine/counter-mine practices. These assets should train all Team Members in enemy mine warfare methods and may best serve in Team point and tail-gunner duties. Note: The Author has experiences and perspectives in both the offensive (SR, sabotage and demolition training and experience) and defensive (Explosive Ordnance Disposal (EOD) training and experience) operations regarding such devices.
  - ○ If the Team encounters a field of mines, punji stakes or other booby-traps in a COIN/FID environment, they are there for a reason. They are typically placed, at some expense of time, labor effort and commitment of logistics resources, to protect something of importance.
  - ○ An enemy element will attempt to camouflage its mines and booby-traps, but not every enemy soldier has sufficient expertise to successfully camouflage his work; this would especially pertain if the enemy combatants are not infantry or members of an elite unit. Further, if the enemy is under time pressure, his efforts at camouflage may be superficial. Beware though if enemy camouflage is sloppy; this may be a tip-off of a trap or deception. Mine tell-tales:
    - Small mounds and/or depressions in the ground.
    - Dry or dead grass.
    - Color difference of the earth, from turned over soil.
    - Smeared mud.
    - Cuttings of vegetation nearby.
  - ○ EOD units/personnel (to include those assigned to SpecOps EOD elements) may be available to train Team Members in recognition of mines/booby-traps and IEDs being used in the AO, the techniques used by the enemy in their employment, the hazards associated with the particular devices and methods to counter or disarm those encountered. The Author strongly recommends training in these skills.
- As dusk approaches, the Team should heighten stealth and security if the enemy element is suspected to be nearby, as the enemy element may have moved into its NDP location.
- If the enemy is carrying WIA, they may split the unit, with one element carrying the WIA to medical care and the other to reunite with the parent unit or camp. The Team may have to split as well to follow both elements. If the Team has only one skilled tracker, then the T/L must make a decision as to which trail to follow; the easiest trail will be the one carrying the WIA.

- Enemy Bloodstains: Consult TC 31-34-4, paragraph 1-22 to estimate wound location and severity.
- Know enemy vehicle tread-patterns and be able to differentiate between military tread patterns and those of commercial and agricultural vehicles.
  - Take photos of enemy vehicular tracks and always include an object (e.g. ruler) in the shot to give size comparison.
  - If the Team carries a tablet IT device, the device might carry a database of enemy vehicular tracks that can be matched to photos.
  - If no tablet IT device is available, Team Members should familiarize themselves with track patterns contained in S-2 reference books.
  - If the Team is mounted, its vehicles should, if possible, have the same tire-track pattern as a corresponding type enemy vehicle.
  - If the enemy is using commercial vehicles/equipment, the tire tracks will not be a reliable intelligence indicator of military use. Additionally, the tracks of military engineer equipment may be identical to those of commercial construction equipment.
- Enemy vehicle tracks are especially visible after rain/snowfall, in the morning dew and in sand. A disciplined enemy may use various means to eradicate vehicle tracks. The enemy may:
  - Station a guard at a road intersection, who may be responsible for traffic control or who may be used to obscure vehicle tracks at turn-offs.
  - Drag brush or tree branches behind a vehicle to conceal vehicle tracks. This may be broadly used in desert or snow-bound environments. But it may also signify exceptional precautions to protect high-value targets. Always consider why the enemy is taking such steps; it may be used to conceal tracks and locations associated with large rocket/missile TELs or C^2 units.
  - When stealth and maintaining a covert presence is key to Team survival and mission success, it is generally bad policy to use a flashlight on an operation (unless the beam is concealed from observation). However, if the enemy is freely using flashlights (e.g. in rear areas), then Team use may seem innocuous. Cautions:
    - Take care to use the same color lens as the enemy is using within the target area.
    - If the enemy possesses NVDs, the illumination provided by a flashlight, regardless of lens color, will light-up the user/Team. This is another good reason to wear enemy clothing and equipment.
    - NVDs used by the enemy will likely be of a less advanced generation than current US military models. The capability of these NVDs to adjust to sudden, increased light may therefore be insufficient to mitigate optics dazzle and the device may automatically shut off in bright light. A high lumen flashlight or strobe light may 'blind' the use of an enemy NVD. S-2 should have information on the capability of enemy NVDs.
    - As in routine tracking procedures, the trail/tracks of the quarry should be between the tracker and the light source. Attach a light source (e.g. IR or flashlight) to a stick or pole so that the light beam will be perpendicular to the pole. This technique has several important features:
      - If an enemy combatant fires on the light source, he may miss the user (tracker/tracker's assistant) entirely, and he will reveal his own location.
      - If an unfiltered flashlight is used, it will impair the night vision of Team Members, but it may also impair the night vision of enemy personnel.
      - NVGs used in tracking will likely require an IR beam, directed from a side angle, to create a sufficient shadow for night tracking.
      - Attachment to a pole/stick will reduce or eliminate the necessity for the tracker's assistant to bend over to shine the beam at a proper angle. The pole mount may even reduce/eliminate the necessity for a tracker's assistant, as the tracker himself can use the pole – reducing the tracking element size.

 * Note that the pole-mounted light cannot easily be doused in an emergency.
  ◦ A full moon may provide enough illumination to track an enemy unit across grassy or crop areas or even other open areas (e.g. desert). The best use of the full moon is when it is at an angle to the trail so that it casts a moon shadow accentuating the tracks.
- It is never wise to assume that the SR tracker Team is not, itself, being tracked by an enemy element. All SR Team counter-tracking TTPs should remain in force during tracking operations.

Counter-Tracking TTPs:

- There are a number of techniques that an SR Team (or the quarry) can use to evade, lose or delay trackers. The sophistication of these techniques will vary according to such factors as Time Available to the Team, the level of Team training, experience and preparation, and the circumstances of the environment. When followed by enemy trackers (always assume that you have trackers), set up the tracker team for an ambush or channelize the tracker team into a mine/booby-trap. There are several ways to achieve this.

<u>True Account:</u> A veteran T/L, having had experience in a particular Target Area in Southeastern Laos, knew that the Team would pick up trackers shortly after insertion. During operational preparation, he studied the elevated terrain in the vicinity of the LZ and established a plan to set the trackers up for an ambush. Immediately after insertion, he led the Team up and over a ridge. Ensuring that the crest itself was not occupied by a high-speed trail, he then moved the Team relatively quickly, following the ridge just below its military crest, for approximately 150 meters until he found a suitable kill zone. Below the crest was a steep ravine, and beyond that was another ridgeline running parallel to the ravine. He then led his Team down across the ravine and up to the top of the second ridge, which also lacked an enemy high-speed trail. He moved the Team a short distance along the ridge top in the opposite direction from the Team's previous direction of travel and then positioned the Team for a linear ambush with excellent line of sight to the back-trail along the opposite ridge. The Team waited in this position for an hour-and-a-half before the lead element (Laotian tracker, tracker squad leader, RPG-7 grenadier, and a riflemen) of an enemy tracker squad, entered the nearly ideal kill zone, bound by a steep hillside on one flank and a steep ravine on the other flank. The enemy element was quickly eliminated without the enemy able to return fire. The Team moved off the ridge towards its objective without further threat of trackers. The remainder of the tracker squad, now lacking its squad leader, grenadier and tracker, would not pursue.

- The T/L/quarry must weigh the time spent in implementing counter-tracking methods against the value of separation distance between pursued and pursuer. For instance, the quarry could do a number of sequential clover-leaf loops that could confuse a tracker; but while the quarry is looping, the tracker element is gaining ground. And if the tracker element is skilled, they will box around the clover-leaf to quickly pick up the true trail.
- Pay close attention to sanitizing Team approaches to Team MSSs/caches, OPs/LPs, surveillance hides, and crossings of danger areas (trails, roads, stream banks, NDPs, enemy patrol areas, etc.).
- Quick, simple methods may be best in most circumstances. For instance: leaving false trails at a stream costs the quarry little time; the tracker will lose more time as he searches up and/ or downstream to find the quarry's true exit point.
- The SR Team may determine that the tracker team is closing by the sound of signal shots, by increased over-flights of observation or combat aircraft, or by detection of the trackers at danger areas.

- If the enemy tracker team continues to close on the SR Team, despite false trails, it may be an indication that (1) the tracker team is skilled; (2) the SR Team is <u>not</u> skilled in counter-tracking techniques; (3) the SR Team has established a routine that the tracker team has identified or, in the case of a US – Indigenous SR Team, a Team Member may be an enemy agent.
- One of the quickest, simplest counter-tracking methods is to use footwear with the same tread pattern as the enemy.
  - Alternatively, each Team Member could have a set of <u>special overshoes</u> (e.g. rubbers) that has the enemy boot tread design. Overshoes should fit snugly over the Team Member's boot, heel-to-toe, so that they will not come off during movement, or they should be designed with straps that secure the covers over the boot bottoms. These may be donned or removed at the discretion of the T/L.
  - A simple, temporary expedient is for Team Members to wear oversized socks over their boots to suppress US pattern boot prints; this measure should be used sparingly, as it has limitations: (1) the socks will wear out quickly, and (2) unraveled, separated threads will leave traces of their own.
- There are occasions when Team Members might walk forward in the footsteps of another Team Member; a simple measure that may be especially appropriate when crossing linear danger areas (e.g. high-speed trails, stream banks, etc.); this practice will make the tail-gunner's job easier. This may also confuse a passing enemy combatant, should he happen upon the Team's sign, as to the size of the Team.
- Counter guerrilla and counter reconnaissance units are very fond of employing the hammer and anvil technique.
  - If the trackers are equipped with a radio, they will send messages ahead to have a blocking/ambush force positioned in front of the Team.
  - If the trackers are not equipped with a radio, or if they wish to act as 'beaters' to drive the Team towards a danger area/blocking force/ambush, they will fire signal shots to drive the Team in the desired direction and/or to alert 'anvil' forces as to Team directional changes. See Hammer and Anvil discussion elsewhere in this book.
- Frequent changes in direction and use of movement techniques explained elsewhere in this book, may cause the enemy to temporarily lose the trace, giving the Team separation distance/time. The enemy will also be wary of bold direction changes as an indicator of the Team moving to set up an ambush and will take time-consuming precautions.
- The tail-gunner's efforts to brush over the Team back-trail may skew enemy estimation of track age.
- To shake enemy trackers, consider a stream, river or standing-water crossing. Ambush the enemy from the opposite bank as they cross in pursuit of the Team. Warning:
  - A stream, river or lake is a danger area. Other enemy forces may be nearby. The opposite bank of a river or large stream may be occupied or patrolled by the enemy.
  - Note that both sides of the stream/river may have extensive trail networks running parallel. Lakes may also have circumnavigating trails.
  - The trackers, if experienced and well trained, may cross up or downstream of the Team crossing location, and subsequently pick up the Team trail where it ascended the steam bank. Be prepared for this and take countermeasures.
- Drop debris and other lures to deceive enemy trackers.
  - Carry dry, smoked cigarette butts (acquired at the FOB) in a small plastic bag and drop them to convince the trackers that the Team is poorly disciplined or unprofessional. If the Team is wearing enemy clothing and equipment, trying to pass for enemy troops, the butts should be of native/enemy manufacture.

     ◦ Carry a small squeeze bottle of animal blood to deceive the enemy that a Team Member is wounded. This may cause the enemy to accelerate his pursuit and fall victim to a Team ambush.

- Enemy Dog Teams:
      ◦ Drop aromatic debris and other lures as bait, to attract animals and to deceive enemy tracker dogs.
      ◦ Food articles (e.g. peanut butter) are a strong attractant to wild animals and tracker dogs. If the enemy is using a dog team, the scent of wild animals along the Team's trail may cause the dog to abandon the Team's trail in favor of an animal trail – especially if the dog and/or handler are not well trained.
      ◦ Use bottled animal scent or musk to lure animals onto the Team back-trail, to confuse tracker dogs. Alternatively, some scents (e.g. predator urine) may repel tracker dogs.
      ◦ Embed fish-hooks in food lures/bait to disable tracker dogs.
      ◦ Sebum, skin oils and debris that accumulates on the skin in a waxy, paste-like form can be scraped off the skin onto vegetation to be used as a lure to tracking dogs.
      ◦ High humidity, cloudy days and moist conditions will increase/extend odor traces and will aid an enemy dog team. Heavy rain/snow and direct, strong sunlight (in dry conditions) will help eradicate odor traces.
      ◦ Be mindful that a dog also has incredible hearing and eyesight (spotting of movement). The rushing sound on a radio handset, nearly silent to a Team Member, may easily be heard by a dog several meters away.
      ◦ If enemy forces can narrow down the search area for the SR Team, they may cordon off the area and employ the dog team with a hunter-killer team or may employ sweeps and Hammer and Anvil tactics. This situation is a very lethal combination that will test Team skills, resources and TTPs to its very limits.
      ◦ Teams should train against Red Team dog units, in realistic settings, to 'shake-out' its inventory of appropriate TTPs. This should be incorporated, if possible, in culminating training exercises, at the conclusion of a SR course curriculum.
      ◦ Killing/wounding the handler is often better than killing/wounding the dog.
      ◦ Deploying puffs of CS or capsicum powder, especially when wind and weather conditions are favorable, will often deter a scent tracking dog.
- Consider enemy tracker dog team TTPs to counter their capabilities.
      ◦ A dog will usually track 20–30 minutes at a time, resting for 10–20 minutes, before resuming work. This cycle may be repeated up to six times during a 24 hour period; or up to 3 hours a day. But given the very slow cross-country speed of an SR Team, this may be more than sufficient for the dog team and its security to close the distance on the Team. If the Team is aware that it is being tracked by a dog team, the Team must substantially increase its speed of movement to buy time for implementing counter-tracking TTPs. If the enemy is determined enough to expend its dog team resources to track down the Team, they may use a second dog team.
      ◦ Local area weather forecasting by the Team is very important to countering enemy dog tracker teams. [see Appendix C]
      ◦ Optimum time for dog team tracking is early morning and pre-dusk or whenever inversion conditions prevail.
      ◦ Optimum terrain for dog team tracking is north-facing slopes (northern hemisphere), across low areas, areas under shadow and in moist conditions, where scents will linger.
      ◦ If the Team travels along ridge-tops, strong breezes may diffuse the scent. During daylight hours, scents drift uphill; therefore a good handler will know to parallel track the Team

from a higher elevation, along the crest, and opposite to prevailing winds. In the evening, scents drift downhill, so the dog team will work below the military crest.

- Other favored dog handler TTPs include:
  - Ridge top saddles have increased air flow and are a good place to pick up scent.
  - Forest openings will act as a vent through which forest drafts will flow. This may help the dog team to catch an initial scent.
  - Dense forest will substantially slow a prevailing stiff breeze, which may otherwise inhibit the dog's capabilities, by up to 80 per cent – allowing continued operations.
  - Terrain form that creates an eddy will collect scents. These are easily identified by the handler, as leaves and other debris will collect there.
  - Downdrafts occur on ridge/hill sides under shadow (e.g. northern slopes in north temperate regions). The handler may search down slope and at the bottom of ravines. Sunlit ridge/hill sides will be searched up slope and at the top of ravines.
  - Note: Wherever/whenever these conditions/situations exist, consider using CS/capsicum powder to impair dog capabilities.
- The Team must consider wind direction/conditions in choosing a hide, if dog teams are known to be in use in the target area. A strong breeze will make an upwind Team or Team Member more detectable. 'A general rule is that a dog can smell a man-sized source downwind out to 50 meters and a group-sized source – a hide – out to 200 meters under ideal conditions.'[22] Given the strong body odors of Team Members, under conditions of exertion and stress and in high-humidity and temperatures, detection by a dog tracker team becomes more likely at these distances. If Team Members are <u>downwind</u> of the dog team during strong breeze conditions, their scents may not be detected even at much closer distances.
- When the SR Team is approaching its target, especially if it is an enemy facility, Team Members should immediately report any dog tracks. If these tracks are accompanied by human tracks (especially with military tread patterns), this may be an indication that the enemy is actively patrolling the target and that the target is consequential. In this situation, employ some CS/capsicum powder; withdraw and diligently conceal the Team back-trail; seek a hide location that is optimized for terrain and weather conditions consistent with advice in this section. Also consider using animal extracts to deceive the dog and/or the handler. For instance, a dog may alert on a Team Member scent, but extract from a skunk will convince the handler that the dog is alerting to a skunk. Employ all appropriate stealth measures and wait to approach the target when the weather conditions are optimum.
- The best TTPs to deter a dog tracking team are to injure/wound or kill the handler or the dog. This is best done proactively and as early as possible, whenever dog teams are assumed to be in use within the target area. Use long-range sniping, mines/booby-traps, poisons, lures and traps and scent deterrents, until the threat is subdued. Once these measures are employed, move out of the area for several days, before resuming the mission.

- Look for bottom-feeding waterfowl (e.g. geese, ducks, heron, etc.) at streams, rivers and lakes. Shallow water will be indicated wherever these birds are feeding.
- Lakes/shorelines will often have inlets, often fed by small streams, and often between two semi-peninsular arms, which may be shallow enough to cross. The Team can use inlets to an advantage.
  - Unlike a stream crossing, the enemy tracker element must either cross in the same manner as the Team or it must move around the closed end (head) of the inlet. With diminished enemy options, the Team can better position to ambush the trackers e.g. at the head of

---

22. TC 31-34-4, 'Special Forces Tracking and Countertracking', p. 3-3

Figure 36. A Small Lake Inlet. Marshy Conditions. Note the Waterfowl Indicating Shallow Water. (*Depositphotos.com*)

the inlet, overlooking the inlet and the prospective ground route around it; or the Team can fish-hook back to overwatch the back-trail and putting the enemy's back to the lake. Caution: in either circumstance, the Author recommends using a 'Z' ambush formation, in the event that the enemy takes tactical precautions. The Team should not attempt to ambush the enemy from the opposite side of the inlet crossing point, as the Team would then be caught in a cul-de-sac on an opposite peninsular arm.

- ○ The Team may be able to deceive an enemy tracker element into believing that the Team boarded boats, by making shallow excavations, resembling bow impressions, into the near-side bank. The Team must also ensure that the point where it emerges from the inlet (opposite bank) is well concealed. If sufficiently convincing, the enemy may terminate his pursuit.

- Time, terrain and vegetation permitting, a Team can emerge from a stream and travel parallel to it in the upstream direction for about 15 to 30 meters. The point man can carefully approach the stream and then use a long stick/bamboo pole to disturb rocks, sediment, etc. This technique can be repeated more times at 15 to 30 meter increments to mislead the enemy trackers as to the direction of travel and path taken. The Team may then fishhook back to occupy an optimum ambush position.

- As the Team must always assume that it has picked up trackers, the Team must always fish-hook or occupy a position allowing observation of the Team back-trail to observe for an approaching tracker unit or to conduct an ambush.

  - ○ It is generally best to fish-hook to the right, terrain permitting, because the enemy on your back-trail will typically carry their weapons with the muzzles pointing left – giving the Team a second or two of an advantage at initiation of an engagement.

  - ○ Always fish-hook to a position that has concealed observation of your back-trail, preferably from a higher elevation, as you move into your NDP, prior to the midday meal and commo break (if scheduled), into

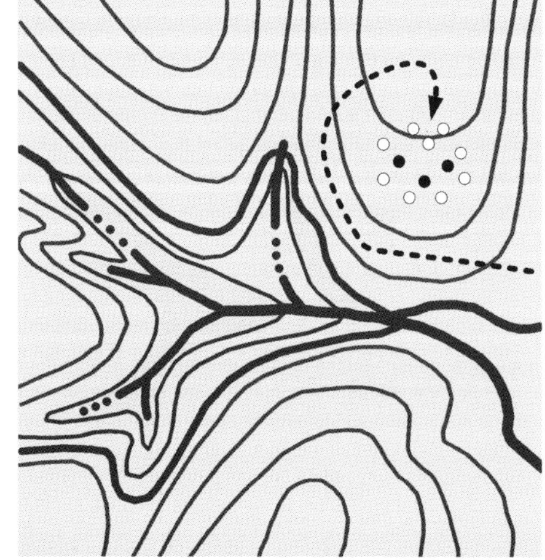

Figure 37. Fish Hook into Team Perimeter.

a surveillance position, into an ambush/raid release point, or if the T/L has reason to believe that an enemy tracker team is nearby.

- ○ Fish-hooking to establish an ambush on trackers is a great way to deter tracking/pursuit, but, it has some negatives: (1) the SR Team/quarry loses time and distance to the trackers and in the execution of its mission timetable and (2) the SR Team/quarry may take casualties during the ambush. Absent the ambush, the mere practice of fish-hooking will be detected by the tracker element and will slow the pace of pursuit as the enemy becomes wary of an ambush wherever the Team trail takes a bold turn.
- ○ In flat areas (e.g. desert, savannah, etc.), where a dominating terrain position is not to be found, the Team may have to use vegetation to establish a hide or NDP. The fish-hook technique is still appropriate in this situation, but the Team must ensure that it does not inadvertently cross over its own back-trail. Note that steps to conceal tracks leading to the hide must be detailed and effective.

- If the Team uses a booby-trap/mine to eliminate trackers, it should be placed at a choke point along the Team's back-trail and be very well concealed. Note:
  - ○ Trackers may be expert at detecting booby-trap/mine emplacement signs.
  - ○ Have the booby-trap/mine readily assembled <u>in advance</u> and available for rapid deployment.
  - ○ Ensure the booby-trap/mine is equipped with a Self-Destruct feature. The SD feature should be set for a minimum of 4 hours – unless the T/L foresees the near term need for the Team to backtrack. Much longer time delays (e.g. 90 days) may be selected by the T/L for various reasons, but then an accurate GPS location must be recorded and reported.

<u>True Account:</u> **An experienced SOG T/L always carried booby-traps and mines to deploy on his operations to inflict casualties on tracker teams and to deter pursuit. On an operation within a very hot Target Area the T/L discovered a perfect location to place a mine. The Team had come upon a large area of dead, fallen bamboo that could only be crossed by crawling on hands and knees. The T/L believed that the enemy tracker team would be compelled to crawl along the same path or risk losing time in reacquiring the trail once the Team emerged from the area of dead bamboo. If the tracker team followed, they would be channelized onto a buried mine in an area where they could not maneuver. Further, the wound inflicted by the mine would be to a hand, elbow or knee rather than to a foot. The psychological effect would be devastating, because the enemy unit would not be able to maneuver. The SR Team crawled through the bamboo until the T/L passed over a fallen tree limb that crossed the Team's direction of travel. Here, he moved aside to allow the rest of the Team to pass, then buried an M14 (toe-popper) anti-personnel mine just past the fallen limb, where the mine location could best be concealed from a crawling enemy and where an enemy soldier couldn't avoid initiating the mine. Beneath the mine, the T/L placed a Limpet detonator, with a 24-hour chemical delay, as a SD device. The Team was not bothered by trackers for the remainder of the operation.**

- Using hard-packed or paved roads that are infrequently used by the enemy, especially just prior to dusk, assumes increased risk, but will make it more difficult for trackers to pick-up signs and will allow the quarry to gain substantial time and/or distance on the pursuer; risk can be mitigated if the Team is wearing enemy uniform and equipment. It may be best to access the road where it crosses a stream, so the Team can clean footgear of mud. When the SR Team/quarry leaves the road; it must find a way to conceal its exit point.
  - ○ This may be achieved by cautiously stepping off the route onto a rock, tree roots, a fallen tree, etc. without disturbing vegetation or leaving sign. This method can be used when the Team is traveling along other trails (animal, footpath, high-speed trail).

- ◦ Ensure that the Team is not caught at a road cut, bridge or channelized roadway. The enemy is more likely to control or observe these chokepoints. If not controlled or observed, such channels still represent a significant hazard to the Team should the enemy appear by chance. If trackers are able to follow Team sign (muddy prints) on surfaced roads, the Team may use these same channelized points to ambush the trackers.
  - ◦ The headlights of an enemy patrol vehicle will illuminate Team signs (mud deposits, dew imprints, etc.) on a hard surface road if the Team does not take precautions.
- Prior to establishing a MSS/cache, the Team <u>must</u> ensure that it has no trackers, either by successful evasion or by killing them in advance. Once confident in losing the trackers, the Team may employ much more complex, time-consuming counter-tracker techniques, or a combination of techniques, to conceal the presence of the MSS/cache.
- The Team should not cross any open areas while dew is forming or present – unless a trap/ambush is to be employed. Crossing dew-laden grass/crops leaves a trail that is ridiculously easy to follow. The same caution applies to crossing burnt out areas, leaving a trail through the ash. Weather permitting, low-flying observation aircraft can detect tracks where the quarry crosses open wet grass, ash or snow-covered areas. Here again, the Team should either have no other options or intends to incorporate a trap or ambush.
- Bad weather (aka 'infantry weather') can help suppress Team trail sign.
- A drop-off technique, if properly executed by the Team, can be used to set up an ambush, to dispatch an enemy tracker element or a stalking sniper. It can also be used to drop off a sniper team to observe local civilians who may be insurgents or insurgent supporters.
- Combine techniques to be more effective in throwing off a tracker element. For instance, the Team can move along a fallen tree/log, leaving a false trail on top of and then leading away from the log; the Team can then back track to the log and carefully dismount the log at another location. In concert with this ruse, the Team can set a mine/booby-trap in a location where the enemy is channelized to the device.
- Lay a false trail at an appropriate spot (e.g. with rocky ground ahead), then step off the trail onto tree roots; Team Members, one-by-one, move to the blind side of the tree to shield from enemy tracker view where the Team changes direction or where Team Members drop-off. The tail-gunner must clear any boot-tread debris from the top of the roots. See example at Figure 27.
- Team Members, especially the tail-gunner (in a file formation), must sanitize areas previously occupied and eradicate Team signs at danger areas.
- Walking backwards (backtracking) is simple and rapid in implementation and it may be successful in briefly deceiving a tracker, but, in the bush, it is generally only feasible for a <u>small</u> party of SR Team Members. Attempting to walk backwards in larger parties is virtually useless in deceiving even a novice tracker.
- <u>Best</u>: Practice any deception or counter-tracking method that causes enemy causalities will probably result in the termination of the pursuit.
  - ◦ If the Team/quarry prepares mines/booby-traps (with SD) <u>in advance, for rapid deployment</u>, setting these devices will consume little time.
  - ◦ Caution: If some of the trackers are killed, the enemy may cache the KIA remains – and continue the mission using a secondary tracker element. This would be a determined, disciplined, and perhaps an elite unit tracker team. Such an enemy may leave behind a single team member to care for a seriously wounded comrade to continue the pursuit.
  - ◦ If the tracker unit has sufficient assets and the will to continue the pursuit, or if more tracker assets arrive, tracker pace will still be dramatically reduced as the trackers will be intimidated by the prospect of other devices ahead and will take precautions.

- It is very difficult to eliminate traces of where a Team climbs a stream/river bank. When emerging from a stream, grab onto the base of small trees for balance and cautiously step on roots to minimize traces on the bank. The first to ascend might use a rope/strap to help other Team Members mount the bank with minimal sign. Note that water traces will be evident on the tree roots for a variable period of time, depending on weather conditions. In wet weather, this concern 'evaporates'.
- SR Teams (and enemy combatants) can use streams to lose trackers. There is always a risk in doing so, as any stream is a danger area.
    - The tracker team can leap an element ahead (bounding technique) to intercept or ambush the quarry during a stream crossing.
    - The quarry can ambush the tracker team as it follows across the stream. A quarry can loop back (fish-hook style) along the stream to establish an ambush on the pursuer.
    - If warranted by the situation, and time permitting, either the tracker team or the quarry may use the box technique to clear the opposite stream bank. If the enemy uses a box technique up or down stream from a Team crossing point, the enemy combatants will likely intersect where the Team emerges from the stream. This may be a good spot to set a mine/booby-trap. See Figure 19.
    - In a close pursuit, disturbed stream mud and debris may still be suspended in the water after the Team crosses. This is an indication to the tracker team that it has closed the distance with the quarry; the tracker team should then immediately increase its stealth and/or transition to a combat formation or stalking role.
    - Consider laying a false trail from the stream leading back onto the same bank where the Team originally entered the stream; this will cause the leader of the tracker team some real concern and will consume time as he tries to figure out what is going on.
    - A well trained Team/quarry can rapidly leave multiple false trails at a stream. This would require designated Team Members to briefly split away (best in pairs), simultaneously set false trails up and down stream and/or at intersecting streams and then return to the core Team to continue its journey. The chief problem with this technique is that Team Members will spend more time in the danger area.
    - The best location along a stream for the quarry to lose a tracker team is at a stream junction, where the possibility for false leads is increased.
        - If 'stepping stones' are stable, they can be used to conceal the quarry's trail. If wet prints (without mud or other debris) are left on a dry rock surface (transfer signs), the passage of time (especially in dry weather) may evaporate the prints, making detection more difficult; but if the wet prints are still detectable, then the trackers will know that they have closed the distance on the quarry.
        - If the steam bed is a submerged rock shelf, this may be an ideal point to cross the stream while leaving little trace – especially if the water is murky or is obscured by sediment (e.g. after a rain) so that disturbed algae on the rock surface is not visible.
        - Note: If an SR Team must pause in a stream/river, it should cling to one of its banks – do not linger mid-stream – and the Team should move about its business rapidly.
    - A stream junction may also be a good place for an ambush, as the stream junction is generally more open than its feeder streams; additionally, the enemy tracker team may pause in mid-stream to sort false signs from the true trail. See section on ambush techniques.
        - Situation:
            * SR T/L knows or suspects that Team is being followed by an enemy tracker team.
            * T/L consults map, identifies a ravine with a stream junction, navigates to the main stream, follows the stream (upstream in this case), and approaches the junction from the left in the photo overleaf.

Figure 38. Good site for a Stream Junction Ambush. (*Depositphotos.com*)

* Enemy tracker element <u>must</u> either track the Team along the streambed or lose substantial time and distance in trying to cut the Team trail along both/either slopes of the ravine.
- Features:
  * A typical, heavily vegetated ravine with a main-tributary stream junction (tributary in center of photo; main stream flowing right to left at bottom of photo)
  * Steep banks/slopes to channel an enemy.
  * Open overhead canopy and illuminated kill zone.
  * Rocky bed with plants that can be disturbed to leave a trail to lure the enemy.
  * Areas of submerged gravel, to leave footprints.
- Opportunities and Comparisons (not in order of preference):
  * Two 'Point' Ambush positions. Ambush 'A' directly ahead in tributary on slightly elevated ground. Ambush 'B' to the right (out of frame) on main stream. Easy and quickly deployed. Purpose is to kill the tracker element and the follow-on element at the junction danger area. More difficult to control and to reunite Team elements.
  * 'L' Ambush with short arm crossing the ravine on slightly elevated ground (facing in center of photo), with long arm deployed on slope (to the left in the photo). Easy and quickly deployed. Purpose is to kill the entire tracker team in the tributary streambed. Entire tracker team must be lured into exposing itself in the danger area; this can be done by the SR Team deliberately being sloppy during movement. Easy to control; easy reunification of the Team.
  * 'V' Ambush with head and arms on elevations respectively: head straight ahead and arms to the left and right of tributary (middle distance in center of the photo). Most complex and time-consuming to deploy; requires a well-trained Team. Purpose is to kill the entire tracker team in the tributary streambed. Perhaps most lethal; employs crossfire in kill zone. Again, entire tracker team must be seduced into exposure. Less easy to control; Team reunification relatively simple, but more time-consuming.
  * Linear Ambush deployed onto ravine slope to the left in the photo. Easiest to control; least time-consuming to deploy. Purpose is to kill the entire tracker team in the tributary streambed. May not have optimal fields of fire. Team reunification not needed.

Nuclear, Biological and Chemical Reconnaissance TTPs:

- FM 31-20-5 provides scant information regarding this crucial strategic and operational-level reconnaissance mission. What information is provided is basic and common to many other US military units and other FMs, including such tasks as collecting samples, use of standard protective clothing and equipment, etc. The FM recommends that this task should be allocated to specialized NBC assets; more likely, some EOD personnel, from within the SpecOps community, who may be attached to an SR unit, if an enemy NBC capability is targeted.
- Terrorist organizations have shown considerable interest in acquiring or developing WMD, and in fact have succeeding in obtaining and deploying WMD in the past. Given terrorist desire to stage mass casualty events, it is likely that terrorist organizations will use WMD in the future. These weapons may be provided by a nation/or non-state actor hostile to the US and its allies; they may be purchased commercially (e.g. toxic chemicals) or they may be developed in a terrorist lab. The likelihood of their use increases where rogue states or non-state actors are engaged in hostilities (e.g. with western nations) and turn to terrorists as a means of delivery.
- Consider the possibility of an SR Team, while conducting a deep penetration operation, as it comes upon a suspicious site, such as a laboratory, a storage facility, NBC protective gear, processing equipment, etc. In these circumstances, the Team should not enter the site if it is possible that contamination exists. The Team may also be directed to gather the relevant information and materials without the insertion and support of an EOD/NBC attachment (a very bad idea); higher headquarters may drop Personal Protective Equipment (PPE) and other items for safe handling by the Team. In this case, the Author recommends the following:
    - In performing NBC reconnaissance, it is especially important to exercise extreme caution, as mines and booby-traps may be seeded throughout the area.
    - All Team Members don PPE, if available or provided by higher headquarters.
    - Always approach a suspicious site from upwind and from higher terrain, if possible.
    - Only two Team Members should enter the site; all other Team Members should be deployed to provide security.
    - Continuously monitor Team personnel for signs of chemical agent effects.
    - Avoid ditches or other low areas; these areas may have been used for dumping of hazardous materials. Look for dead vegetation. Avoid nearby streams that may contain runoff from the site.
    - Effluents from the agent manufacturing process may be drained into a nearby pond or pit. These effluents will not normally contain active agents, unless the area is contaminated; but these effluents may be hazardous. Note that if the material is radiological in nature, the pond/pit may have radioactive residue. If any samples are taken by the Team, use extreme caution;
    - Nearby grave sites, burn pits may be used to dispose of contaminated test subjects/animals and materials. Do not uncover anything buried. This should only be performed by a specially trained and equipped element. Be observant of animal remains (indicative of an agent release or experimentation) and the appearance of nearby vegetation (indicative of precursor chemicals or decontaminants). If these traces are detected, back off.
    - Also, be aware that if the site is not occupied, the occupants may be away for a brief period, perhaps for a meal, and may return. Or the site may only be periodically occupied for various reasons, but it may be routinely checked by security patrols. Information must therefore be gathered swiftly.

- ○ At a minimum, the Team should take notes and photographs of documents, facility and equipment layouts, clothing and equipment, refrigeration, precursor chemicals, petri dishes, container markings, warning labels, vacuum sealed doors, vaccine vials, test subjects/animals, etc. Take photos (digital) first, before disturbing anything; then use the photos to ensure disturbed items are restored to their correct positions.
  - ○ It may be best to leave everything as is, so that a specialized Team, NBC/EOD element can be inserted to exploit the site.
- Note that manufacture of small quantities of biological agent is possible with a minimal amount of low cost, simple equipment and minimal space (e.g. a garage). If the enemy has quality 'seed stock' or samples of a biological weapon pathogen, the chief obstacle to biological agent production is resolved.
- Lab personnel may be quartered in surrounding areas. The best circumstance would be to capture and evacuate an occupant or site worker (e.g. lab technician), especially if the person shows signs of elite status. Consider establishing a POW snatch ambush along the lab access road. These signs may include: security escort, a driver and expensive vehicle, distinctive clothing/uniform markings. Even though the persons absence would be a red flag that the enemy capability has been discovered, the POW may not only have technical knowledge, but he will have key information on his associates and – operational intent. If the facility is involved in biological agent development or production, the POW may have been vaccinated; his blood will reflect vaccine antibodies specific to the agent(s) with which the person has been working.
- IMPORTANT! Immediately ascertain if the site has been abandoned, or if the site has not been recently used, as evidenced by accumulated dust or other signs; but if it is still equipped or stocked – BEWARE! It may be the site of an accident/incident where an agent had been released. Contamination may be present.
  - ○ In this circumstance, do not touch anything and move cautiously and carefully away so as to minimize aerosolization of the agent.
  - ○ If Team Members have intruded into a biological agent laboratory or production site, they may already be contaminated. If this is possible, the Team should take soil samples and move to a <u>string</u> LZ for extraction (so as not to contaminate the helicopter interior or its crew);
  - ○ Higher headquarters should drop PPE to the Team Members, prior to exfiltration, so that the Team will not spread contamination;
  - ○ Aircraft crew members should be dressed out in PPE;
  - ○ Ideally, the Team should be extracted during/after a rainstorm. This will suppress aerosolization of an agent from prop-wash while the helicopter is hovering.
  - ○ Higher headquarters should establish a decontamination station and an isolation area that are located upwind from the Team drop-off point, for personnel and equipment decontamination processing and for medical care of personnel.
  - ○ Helicopters should be parked laterally to and downwind of the drop-off point and isolated if it is possible that they were contaminated. Uniforms and other items should be packaged for retention, so they may be analyzed for evidence of an agent.

Operating Environment-Specific TTPs:

- Such variables as extremes in temperature, humidity and elevation; aircraft type, condition, weight (including fuel), load balance and cargo weight may have significant and varying effects on the lift capability of transporting aircraft. Take the following steps during the planning process:

- ○ Weigh Team Members in full field configuration in preparation for a mission at high elevations.
- ○ Consult and train with pilots and/or crew as early as possible in the mission planning process.
  - ▪ Remember that aviation support elements are there to support the mission and the Team; do not hesitate to request additional support (e.g. additional aircraft, or aircraft of different capability) if necessary.
  - ▪ If an external aircraft load is required (e.g. to carry cross-country mobility equipment), ensure that aviation crewmembers prepare and rig the cargo prior to the day of departure. Remember that vehicular fuel tanks and containers must have airspace provided to allow for fuel expansion at flight altitudes/elevations.
  - ▪ Coordinate on insertion/extraction accessories, such as: ropes, ladders, tie-downs, cutaways, etc. Ensure these accessories accompany the Team to the Launch Site or are maintained on-hand at the launch site and are appropriately stored for immediate use. The T/L should double check to ensure that Launch Site personnel know how to assist the aircraft crew in the rigging of these accessories.

Jungle/Rainforest Operations TTPs:

- Ground, air observation and sensor/electronic surveillance capability is substantially impaired in dense jungle or rainforest. Intelligence requirements therefore are dependent on ground reconnaissance in these environments.
- Thin canopy may promote undergrowth that will substantially impede Team movement. Plan additional time for dismounted movement in this environment.
- A ray of sunshine will occasionally penetrate the jungle/rainforest canopy and if this occurs on a hill/mountainside, the break in the canopy may sometimes offer a view of surrounding terrain. Team Members should exercise caution and not expose themselves in illuminated areas. Light exposure pin-points Team Members' pupils that should normally be dilated to the otherwise 'eternal twilight' under canopy. Further, the Team Member presents himself as a superb target within an illuminated patch.
- Jungle or rainforest environments offer few and limited landing areas. Most LZs will be found alongside streams or in areas that are or had been prepared for agriculture. Enemy units may be well aware of the location of such LZ sites and may place these sites under continual observation. Therefore, helicopter insertion and extraction of Teams via ropes or ladders into less exposed locations may often be the most viable solution. Note that helicopter night-time insertions or extractions may be a hazardous proposition due to the proximity of the vegetation, uneven landform, impaired vision and depth perception.
- Dense canopy and dissected terrain will impair radio communications and GPS navigation capability.
- Waterways provide a means of surface/subsurface movement and

Figure 39. Heavy Undergrowth Abounds Beneath Light Canopy. (*Depositphotos.com*)

are an aid to navigation. Using waterways, which are danger areas, to accomplish insertion/extraction and movement has various limitations and risks, especially if the enemy is also using the waterway for movement. Much more intelligence and planning will be required, if waterways are to be exploited successfully.

Desert Operations TTPs:

- Factors affecting reconnaissance operations include scarcity of water, sparse vegetation, varying soil compositions and combinations (lava beds, sand, caliche and topsoil), extreme temperature variations, brilliant sunlight, and typically long-range observation.
- Lava beds may be found in most climates/environments, including desert. Team Members, navigating these areas, should wear protection to protect eyes and lungs from lava dust particles. Beware navigation errors in lava flow areas (see entry under navigation).
- Sparse vegetation and flatter terrain may promote faster movement and a T/L may be tempted to increase the Team's rate-of-march to cover distances in desert target areas. However, since the Team should navigate cross country while making the best use of folds in the earth, deviations en route will consume time. Additionally, the enemy may focus his observation and/or patrolling on earth folds (arroyos, wadis, etc.) as likely SR routes of approach. Subsequently, the T/L should factor-in additional time for desert movement.
- Desert movement is often restricted to darkness and/or to arroyos/wadis (especially during daylight). Thermal imaging equipment may enable enemy units to acquire the Team heat signature at considerable distances at night. Enemy use of thermo-optics is unlikely during daylight hours. Time permitting, the Team should navigate at night, while still taking advantage of all available folds in the earth (arroyos/wadis).
- Vehicle tracks in the desert (especially in caliche) may endure a long time. Example: WWII North Africa military campaigns established vehicle tracks that lasted for decades. Following these pre-existing traces, or those made more recently by enemy patrols, will help disguise the Team's track – especially if Team vehicles are equipped with enemy pattern tires.
- In the desert, an increase in the number of flies suggests the presence of water nearby.
- Large rocks and metal objects (e.g. destroyed armored vehicles, trucks, etc.) will be hotter and retain heat longer than the surrounding area due to solar loading. These objects can provide thermal screening for the Team.
- If elevated terrain is proximate to the objective or the Team route-of-march, beware of enemy sniper teams/OPs.
- Lightweight thermal-imaging devices are essential Team equipment during desert operations, enabling the Team to navigate, locate, observe or evade enemy positions/combatants and to enhance Team security. These lightweight optics, to include weapon-mounted optics, are important tools for locating, surveilling, and evading the enemy – and for examining the sufficiency of its own heat signature suppression measures.
- Animal or vehicular transport may be essential for over-land insertions, to reach an objective

Figure 40. An Example of Desert Caliche, aka "Desert Concrete".

over long distances and to carry extra burdens associated with the mission (e.g. camouflage nets) or Team survival (e.g. water). Vehicular transport should feature capabilities to suppress engine/exhaust noise and heat signature. Animal heat signature may be suppressed behind landform or by use of heat suppressive camouflage nets or blankets.

- Scarcity of water in the desert may create significant operational problems.
  ◦ Team route planning will often be governed by proximity to water sources.
  ◦ Water sources may dry up at certain times of the year.
  ◦ Non-combatant/civilian personnel are more likely to reside near the water supply. Dogs may alert to the presence of Team personnel.
  ◦ The enemy may concentrate near, or conduct surveillance of water sources (to interdict SR Teams).

Mountain Operations TTPs:

- Recruit master mountaineers to mentor and train Teams being deployed on operations in mountainous terrain. Master mountaineers can provide training prior to the operation and provide On-the-Job-Training (OJT) and mentoring during the operation itself.

German Army Group F, on the defensive during the opening months of 1944, in the face of Tito's Partisans in Yugoslavia, initiated offensive operations in May 1944 near Dvar,

> 'to destroy the Tito forces in their main stronghold. The operation was known as ROESSELSPRUNG, and it was planned to commit elements of the 1st Mountain Division, elements of the Division Brandenburg (designation of a special demolitions and sabotage unit), the 202d Tank Battalion, the 92d Motorized Infantry Regiment (Separate), an SS parachute battalion, and a number of Croatian units…. Though Tito himself managed to escape, the Partisan headquarters was captured, with its extensive communications system. The 1st and 6th Partisan Divisions were badly mauled in the fighting, suffering a total of 6,000 casualties, and an enormous stock of booty taken.'[23]

- Heavily dissected mountain terrain form, depending on season and foliage, normally provides good concealment and cover. Observation will vary according to season, vegetation, elevation/terrain form/dead zones and weather factors.
- Descending a ridge may be at least as physically challenging as an ascent. Different muscles come into play. This is especially true in shale/scree or uncertain footing, as leg muscles may be under constant stress. Teams should train in similar terrain prior to deployment.
- Enemy AA systems on elevated terrain may have superb fields of fire, so the T/L should collaborate with the flight leader and/or FAC on insertion/extraction planning, before final selection of LZs/DZs.
- Team insertion or extraction will be limited by varying wind and weather conditions, aircraft capability and availability of acceptable LZs/DZs. Insertion by parachute can be very risky in dissected terrain.
  ◦ Team Members of varying weights (e.g. US versus indigenous) and parachuting skill levels may find themselves broadly scattered. Lighter individuals will not descend at the same rate under parachute canopy as heavier individuals; shifting wind currents/drafts may separate Team Members to opposite sides of a ridge or even beyond adjacent ridges. Air density in cold environments, or occurring in a weather front, will slow descent and cause scattering. Equipment bundles may also be scattered and lost. Team Members may

---

23. Unnamed authors and contributors, 'German Anti-Guerilla Operations in the Balkans (1941–1944)', Department of the Army Pamphlet No. 20-243, August 1954, pp. 65-66.

be injured or killed during landing. Personnel possessing key mission equipment may be separated or lost. The time needed to reassemble the Team will consume mission timeline. Insertion by air landing or by rope/ladder descent from helicopters will avoid these vagaries, but admittedly will present other problems.

- ○ If feasible, insertion should occur at LZs/DZs that will facilitate route-of-march considerations and minimize mountaineering obstacles to Team travel.
- ○ Enemy, target dispositions or other operational features will limit Team insertion options, perhaps requiring a mission of longer duration over longer distances. In this case, pack animals or ground mobility equipment may be required. MSSs/caches may have to be established to support longer or multiple missions within a target area. (See MSS/cache information elsewhere in this book).

- Team Members must be in excellent physical condition, be acclimatized to environmental conditions, and skilled in mountaineering techniques. Ideally, Teams should have at least two Team Members who are skilled/experienced mountaineers, who will serve as master trainers to the remainder of the Team and lead climbers during missions. If mountaineer skilled/experienced US Team Members are not available, the Team should include mercenary/indigenous Team Members who are.
- Communications and GPS navigation from low areas may be made difficult or impossible by terrain form. Airborne communications capabilities or covert (automatic or manned) relay stations located on dominant terrain – providing line of sight – may be required. The Team may have to climb to elevated terrain to communicate with radio/SATCOM and may miss scheduled communications.
- An enemy may be able to employ thermal detection equipment in the shadow of dominating terrain under cold and overcast conditions.

Cold Regions/Winter Warfare Operations TTPs:

- Cold Regions procedures may encompass operations in mountains, in the sub-arctic and in temperate regions where extreme cold conditions may prevail during the winter season. Train for this.
- Teams are burdened by the additional weight of clothing and equipment needed for surviving and operating in extreme cold.
- In moderate to deep snow depth, Teams may have to travel on skis or snowshoes.
- Snow will develop a 'crust' due to cyclic freezing and melting. Melting may occur due to increased temperature and/or solar loading/radiation (sun exposure). To minimize the noise that attends traversing snow crust, consider moving through heavily forested (conifer) or shaded areas that have not been exposed to the sun.
- <u>Beware of 'Tree Wells'</u> especially when traversing deep snow and steep terrain. A Tree Well is formed beneath a conifer where the tree branches shelter the area beneath it from snowfall, permitting loose powder/pelletized snow (acting like quicksand) to form within the well. Team Members may accidently ski into or attempt to seek shelter in a tree well and can get trapped, sometimes upside down, and sometimes sliding down-slope beneath the snow surface, and die from suffocation. An estimated 20 per cent of winter mountain deaths are attributable to Tree Wells. If a void beneath a tree is to be used as a hide, cover or shelter, ensure the void is not on an incline or in deep snow. If caught in a Tree Well, the Team Member should avoid panic, 'hug' the tree, create a breathing space and await rescue. If rescue is not forthcoming, the Team Member must use slow, rocking motions to compact the snow and use the tree to climb out of the well.
- Snow Glare will impair Team Member vision. Use adaptive polarized lenses on snow goggles/glasses.

- The Team point man has the additional challenge of breaking trail; this is often an exhausting activity that will require frequent swapping of point duties with other Team Members.
- Time is always a factor in SR operations. The T/L must be conservative in calculating time-distance cross-country movement during his METT-TC planning and analysis processes and during mission execution to ensure the

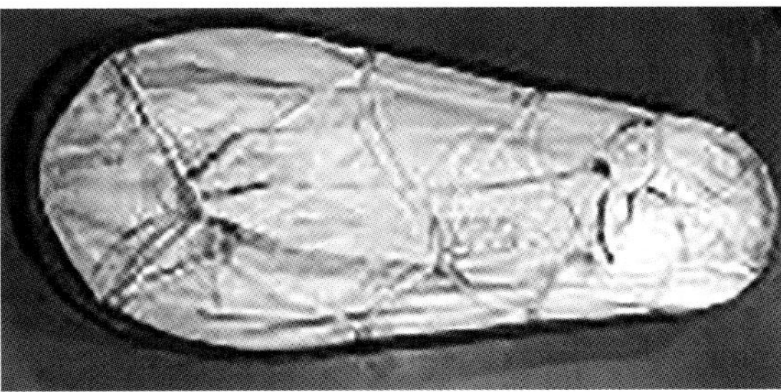

Figure 41. Ahkio. Loaded, with camouflage cover. Top View. (*Army Photo*)

Team reaches the objective and accomplishes the mission on schedule. Incorporation of a 'fudge-factor' to deal with time delays associated with unforeseen circumstances should be common practice in all SR operations, regardless of environment. But <u>everything</u>, including routine tasks, <u>takes more time in cold conditions</u>. And stealthy, covert movements will take yet even more time, as frequent deviations (e.g. to conceal the Team's trail) in navigation will be required. Always plan for extra time to accomplish the Team mission. If mission timeframe cannot accommodate additional movement time, mission risk will dramatically increase.
- If the Team must approach a target, establish an ambush, etc. in snow covered terrain; consider using an active (no ice cover) small stream to conceal Team tracks. If the Team is wearing footwear with the enemy tread pattern, tracks in snow or mud may not appear suspicious to enemy patrols.
- The T/L must seek ways to accelerate cross-country movement, especially in cold regions and/or where long distances and logistics burdens are involved – this may require mobility equipment.
- Use of dogsleds, caribou, All-Terrain Vehicles (ATVs)/Utility Task/Terrain Vehicle (UTVs) and/or snow-mobiles with an Ahkio, may be required to travel cross country with cargo.
  - A 'stealth' or suppressed Utility Task/Terrain Vehicle (UTV) may be equipped with snow tracks, providing excellent, and relatively rapid cross-country mobility; this vehicle can also tow an Ahkio or another type of sled with ease.
  - Once snow cover melts, an UTV is still cross-county mobile, when equipped with tracks or by swapping out the tracks for standard tires. Conversion kits also exist for snowmobiles, but are not as nearly cross-country capable as UTVs when used outside of snow/ice environments.
  - Utility snowmobiles are also available that are optimized for cargo and multiple passenger transport (including towed sled accessories) and rapid movement over snow and ice.
  - If captured enemy utility vehicles are available, consider using them, the enemy will confuse them as belonging to their own units. Wheels with enemy, or common-use civil, track patterns are recommended.
  - The utility of an Ahkio will be diminished once the snow cover is gone. An Ahkio may make significant noise during movement, unless the Ahkio is constructed of alternative materials – the Team may have to move more slowly to insure noise discipline. An Ahkio may be used to transport crew-served weapons. Ensure that training in battle drills encompasses putting these weapons into action.
  - When the Team moves into a NDP (night or day) the Ahkio should be placed on some kind of dunnage (e.g. small tree limbs), so that it won't be frozen to snow/ice/ground.
  - An improvised Ahkio may be more effective than a purpose-built version.
    - Consider using a Kayak of a durable, lightweight resin or alternative construction with a removable spray deck. Unloaded weight may be approximately 35lbs; modifications

can reduce this weight somewhat. This improvisation may provide cargo space similar to the Ahkio and also offers the utility of a watercraft. Kayaks designed for military application are in the US SpecOps inventory. Military Kayaks will typically be longer and with greater beam than those designed for civilian use; this greater size may make military Kayaks unsuitable as an improvised Ahkio. If this option is used, remember to bring a patch kit to preserve its waterborne capability.

- Consider using a 'slide sheet' of heavy gauge (≥1.5mm = 1/16in), flexible/pliable plastic, which can be fabricated by plastics specialty shops at a very reasonable price. The sheet should be sized for typical box dimensions and/or a human cargo of >6.5ft in length and should have sufficient length and width to allow wrapping of the plastic partly around the cargo to prevent intrusion of mud/snow. The sheet must also be modified with robust grommets for towing, belay rope attachment and for securing cargo. The sheet employs low-friction characteristics to perform as a cross-country multi-functional cargo sled, ground sheet, debris removal accessory, and litter/game or casualty transport conveyance. It can also be used to transport cargo cross country in conditions where snow and ice are not present. A slide sheet may impede movement across mud due to suction. It is lightweight, yet capable for cross-county movement of heavy items, even across hard-top/gravel surfaces; when not in use, it can be rolled up for carry/storage. Some of these should be stored in a MSS.

- Heavy snowfall may render enemy mines ineffective, especially if the Team is wearing snowshoes or skis – or if the Team has low-ground-pressure RV-type mobility equipment.

- During WWII, Soviet armored vehicles would create ruts in deep snow to provide paths for infantry. These tracks would be created in 'Go' terrain, conducive to armored vehicle passage, which should be identified in Team METT-TC and terrain/IPB analyses. This same technique can be used to support Team movement, if the Team has a snowmobile to break trail for Team Members who are foot/ski-mobile.

- As the Team trail is easily detected in snow-bound terrain, plan to occupy a surveillance position more distant from the target. Use remote observation equipment (e.g. trail cameras with data link) whenever possible to limit Team Member tracks/exposure.

- Accurate weather factors and forecasts are extremely important throughout the planning and execution phases of the operation. Ideally, a Team may deliberately plan for an insertion and subsequent movements just prior to a front moving in, while winds are still stable. Brisk winds and/or snow would subsequently help conceal the Team's tracks and would impede enemy movements/pursuit and attempts at aerial observation.

- In August 1942, the Russians were sustained with huge amounts of supplies by allied shipping into Murmansk. The German Twentieth Mountain Army was committed to the northern reaches of Finland, but military operations had reached a stalemate along the front. The Germans could air interdict the Murmansk-Leningrad rail line only sporadically, but these interruptions were short term. The Russians had developed military and civilian labor forces able 'to repair the lines and even damaged tunnels and bridges with surprising ease and speed. The Germans came to the conclusion that the only way to disrupt traffic along this important supply artery was to effect a thorough demolition of bridges and tunnels by trained sabotage units.'[24] This mission was entrusted to the 128-man Special Mission and Sabotage (SMS) Company which was activated as part of the Brandenburg Regiment which operated under the direct control of the Armed Forces High Command.[25]

---

24. Halder, *Small Unit Actions*, p. 179
25. A more fulsome account of this operation is found in the referenced 'Small Unit Actions during the German Campaign in Russia, DA Pam 20 – 269, Department of the Army, Washington DC, July 1953.

'Each squad had an able and experienced NCO and an expert Finnish interpreter who spoke fluent Russian. Two out of three of the men were former PoWs or deserters from the Red Army, originating from the Ukraine or other parts of Russia and opposing the Soviet regime. The others were so-called ethnical Germans who hailed from southern Tyrol, the Balkans, and the German settlements along the Volga. Three men in each squad had undergone engineer training and were especially proficient in handling and setting explosive charges.

The company was equipped with Russian or Finnish submachine-guns. Each platoon had one 80-mm. mortar and each squad one light machinegun. In addition, the company had two 75-mm. field guns that could be disassembled and transported in sections…. Attached to the company was a detachment of 18 bloodhounds and watchdogs with 6 Finnish handlers.'[26]

In mid-July, 'the SMS Company received orders to prepare for a long-range reconnaissance and sabotage operation against the Murmansk-Leningrad rail line.'[27] Mission duration was estimated at 2-3 weeks and depth of penetration, from release point to targets, was approximately 150 miles – primarily traveling by boat over an extensive network of lakes and streams.

- The company entered into a phase of intensive training and preparation for the mission, to include:
  - Training on a new, highly effective explosive charge developed by the Finns.
  - Extensive training of radio operators to include the adoption and use of the Russian 4-element code-group system, to avoid attracting the attention of Russian signal monitoring.
  - Extensive small boat training.
  - The company commander[28] took two VRs of the route and targets and availed himself of the advice of Finnish liaison officers who had experience in this AO.
  - Two fixed-wing aircraft were provided as dedicated support for resupply and medical evacuation.
  - Provision of Finnish reconnaissance/security patrols and motorized assault boats (to tow the SMS small boats) for the first leg of the mission.
  - Provision of special uniform items and supplies for use on the mission.
  - Establishment and manning of seven covert MSS/caches along the routes to and from the target.
  - Mortars were discarded (weight consideration) and were replaced with a rifle with grenade adaptor. Additional machineguns were provided.
- Comments and key points affecting the conduct of the mission:
  - Extended daylight during August advantaged Russian observation and required wise use of terrain to screen water navigation.
  - Local civilians fished the lakes and streams throughout the routes and prime fishing areas had to be avoided.
  - Three reconnaissance patrols were dispatched to observe the three planned targets. Each patrol established a cache site containing the explosive charges to be used on the targets. These caches, and their contents were later moved closer to their targets, based on observations by the reconnaissance patrols.
  - Dogs were used to detect Russian sentry positions allowing a team of Finns to silently kill the guards on the main target.

---

26. Halder, *Small Unit Actions*, p. 179
27. Ibid, p. 181
28. The commander of the SMS company was *Lauri Torni* aka Larry Thorne, a Special Forces legend who died on a SOG operation.

○ After execution of the attacks, the company was menaced by Russian reconnaissance and combat aviation assets and enemy motor patrol boats during its withdrawal along the water courses of the return route. The return journey was more hazardous than the advance.

○ The unit was intercepted and ambushed by a Russian company, as it entered a water course between two lakes. This was a natural choke point, which may have become obvious to the Russians from aerial reconnaissance of the withdrawing SMS unit. The unit lost all of its boats and many of its weapons during this engagement. The Finnish reconnaissance elements in support of the SMS were able to rapidly attack the enemy company from the rear, saving the SMS unit from further losses.

> 'Finnish use of long range reconnaissance patrols deep behind Soviet lines began during the Winter War, when selected soldiers were trained for clandestine operations. Three detachments of so-called 'Ski Guerrillas' were formed, and given the designations TO1 to TO3 – from Tiedustelusasto, 'reconnaissance detachment'. During the Continuation War, these troops were formed into long-range patrols to perform not only reconnaissance but also sabotage and other unconventional activities behind enemy lines. Patrols were small, most being only of section/squad or sometimes platoon size; but during the war a few company- and even battalion-sized groups, including specialist personnel such as engineers drawn from other units, were assembled for specific missions. The personnel of the LRPs were all expected to be at the peak of fitness, many patrols were made up of young athletes and top-class skiers. They had to endure great hardships during missions that might last for several weeks, and they were issued with 'pep pills' to keep them alert. During 1943, 50 patrols were set out, and in 1944 just under 100, with various missions including sabotaging the Murmansk Railway. One of the most famous of these raiders was Lauri Torni, who led a Jaeger company on so many spectacularly successful raids behind Soviet lines that the enemy put a reward of 3 million Finnish marks on his head (needless to say, this went unclaimed). Attrition rates in Torni's unit were high, and only three of his original men were still alive and uninjured at the end of the war.'[29, 30]

While examples of the larger Finnish operations may be considered well beyond the mission envelope of an SR Team, consider how such larger operations may be accomplished where the SR Team(s) operate in conjunction with an Exploitation Force, an allied Force, a large indigenous contingent or a Guerilla/Partisan unit. Examples of these larger patrols include:

Figure 42. Larry Thorne, aka Laurni Torni. U.S. Army Special Forces/SOG Officer. (*US Army SOCOM Photo*)

• Majewski raid on Mai-Guba along the Soviet Murmansk rail line during the Continuation War, conducted 100 kilometers behind enemy lines (January 1942). It was the largest long-range patrol and consisted of 1,600 men and 250 horses (for carrying supplies). Source: http://www.ww2incolor.com/finnish_forces/Major_TimoJohannesPuustinen_Knight%23117.html.

---

29. Philip Jowett and Brent Snodgrass, *Finland at War 1939-45*, Osprey Publishing, 5 July 2006, pp 31-2.

30. *Lauri Torni*, aka Larry Thorne, was a Mannerheim Cross recipient and later, a famous Special Forces soldier and member of MACV-SOG. He died in 1965 during one of SOG's first cross-border operations. In 2010 he was named as the first Honorary Member of the United States Army Special Forces Regiment. https://en.wikipedia.org/wiki/Lauri_T%C3%B6rni#Vietnam_War_and_death

Figure 43. Majewski Raid Departure, Mai Grubaan Roach Lake 1/14/42. (*Public Domain – Finland*)

- Puustinen raid (March 1943), consisting of a patrol of 600 men who raided the large supply depot along the Jeljärvi-Kuutsujärvi Railway (Russian Karelia), 50–60 kilometers behind enemy lines. The unit navigated along multiple waterways and large lakes and upon arrival in the target area, they,

  'attacked on skis in broad daylight, destroyed some 30 buildings including ammunition storages, stables, hay storages, petrol storage and vehicle repair shop. They also destroyed two bridges, railway station, railroad switches, two railway bridges, five trucks, tractor and twenty horses. Over 200 enemies were killed and possibly same amount were wounded. Thirty of his men were killed, most of them during return trip, and six were wounded.'

  Source: http://www.ww2incolor.com/finnish_forces/ Major_TimoJohannesPuustinen_Knight%23117.html.

- Some of the lessons to be learned from these operations include:
  - Avoid chokepoints during the approach and during evasion. These will be guarded.
  - If a 128-man company can penetrate 150 miles without detection, so can an SR Team.
  - The value of dogs is demonstrated in these operations.

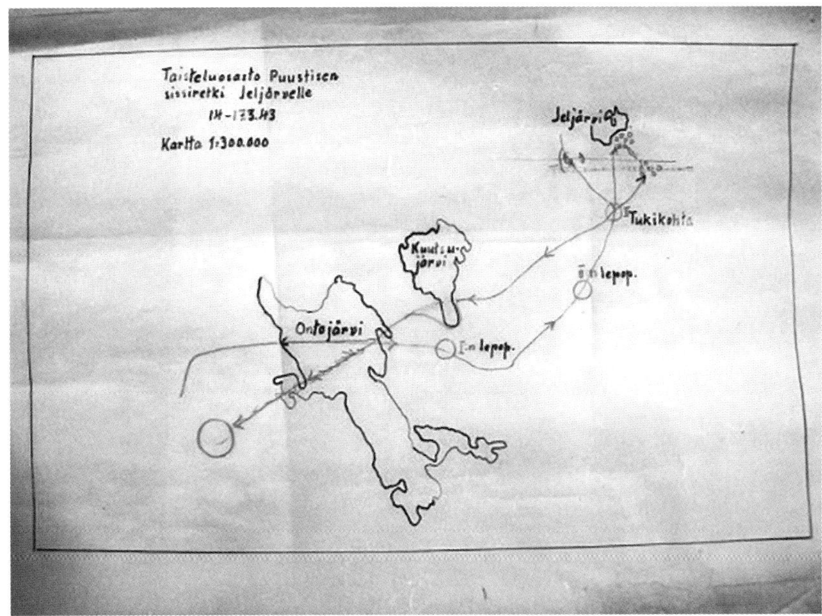

Figure 44. Strip Map used for Puustinen Raid (*Public Domain – Finland*)

- ○ Caches/MSSs are <u>fundamental to long-range</u> missions.
- ○ Stay-behind elements, supported by cache/MSS stocks employed. Or cache/MSS sites could be used for periodic or subsequent SR missions.
- ○ SR is instrumental in reconnaissance of routes to avoid detection during approach and withdrawal.
- ○ Use uninhabited islands to block enemy observation of SR movement over water or across ice.
- Prior to winter, booby-trap buildings that are likely to be occupied by enemy troops in cold-weather; better, booby-trap buildings that are likely to be occupied by an enemy headquarters. Use time-delayed explosives with incendiary capability; this munitions option will produce casualties (possibly high-ranking) and destroy headquarters' materiels and equipment. The Team should take at least one full set of acid/chemical delays (or equivalents) on missions, as application of these fuses will vary according to circumstances, targets and time requirements.
- If the enemy is tracking the Team in snow, they will likely use the Team's own ski or snowshoe tracks to avoid the effort of breaking their own trail and to speed closure with the Team.
  - ○ Place a mine/booby-trap in the ski ruts; especially where terrain and vegetation channelizes the enemy. Be aware that the pressure of snowshoes/skis on snow cover may be insufficient to detonate the mine/booby-trap, especially when additional snowfall occurs. An enemy will likely know of this difficulty and may not expect a mine or booby-trap as a result. Consider these techniques:
    - Cut a section of the ski rut from the path and then replace it once the mine is emplaced (camouflage).
    - Use a trip-wire fuse initiator, instead of pressure fusing. Use dental floss or clear plastic fishing line.
    - When planting a mine/booby-trap, ensure that the Team Member does not leave tell-tale impressions in the snow. This may best be done, by selecting a location on the trail where the Team has paused for a break or where one or more Team Members may have stumbled or fallen in areas of rough footing.
    - Use a bounding mine with a pressure fuse.
      - * A regular pressure mine/booby-trap may be initiated by the leading or trailing portion of the ski or snowshoe, rather than at the point where the combatant's foot is attached; a bounding mine will kill the combatant who initiates the munition, and it will kill or wound his companions unless the enemy point man has a long lead on his unit.
      - * Use the pronged fuse extension, painted white, and place an innocuous piece of bark or some pine needles over the fuse prongs, allowing the prongs to extend slightly above the compacted snow to ensure sufficiency of pressure.
  - ○ The Team should use its ski/snowshoe tracks to channelize the enemy trackers/pursuers into an ambush. Ambush considerations:
    - The 40mm grenade launcher is of limited utility in snowbound terrain. Unless the HE grenade is fired into vegetation, or at a vehicle, the round will bury itself into the snow, and either fail to detonate or substantially muffle the detonation, reducing its kill radius. 40mm CS and smoke rounds are generally useless in snowbound settings.
    - Rifle grenades are only moderately better, but are also of limited benefit.
    - Hand grenades are even more difficult to throw when the Team Member is wearing bulky winter clothing along with his equipment. The best use of hand grenades in a snowbound environment is employment as a booby-trap attached to vegetation above the level of snow and initiated by a trip-wire.

- The Claymore mine is the best option. The Claymore must be camouflaged to correspond with the background (partially painted white); this can be supplemented with green and brown local vegetation. The Claymore can be affixed to a tree trunk/limb, etc. above the snow line, using bungee cord (for instance).

- Deep snow may assist the Team in concealing its stationary observation posts; however, if the Team must move to conduct other operations, the Team may expose itself to detection by leaving trails/disturbed snow (at least until a subsequent snowfall or the formation of drifts). Route selection then becomes even more critical.
  - Teams should avoid crossing open areas and cling to areas populated with conifers/evergreens during movements, if possible. The counter-tracking techniques explained in this book can also be applied to snowbound environments with some modification.
  - If open areas must be crossed, use folds in the earth and terrain shadow to conceal tracks.
  - Remember that shadows cast by a low sun will make tracks detectable from the air, so keep to shadowed areas or bear in mind the sun's direction when crossing open areas.

- Team Members will compact snow as they travel; compacted snow will melt at a different rate than un-compacted snow. As the snow melts, Team tracks may be more easily detected; detectable even after subsequent snow-melt cycles. Bear in mind that snow under the protection of shadow will melt more slowly. Again, keep to the tree line and terrain shadows, and if possible, travel along terrain form that is sheltered from the sun.

- Travel in a file formation to avoid multiple trails, to minimize the risk of detection and to conceal the size of the Team, unless the Team is attempting to emulate enemy behavior. During movement, consider using existing animal trails; these may be difficult to detect in alpine areas where wildlife has moved to lower elevations and snowfall has been substantial.

- An exception applies to using the file formation in cross-country travel – when crossing frozen lakes, rivers, etc. the T/L may require an open traveling formation. These flat areas are often windswept, so signs of crossing may be eradicated within a short time. However, beware these danger areas, as they provide ideal kill zones with extended grazing fire. Note also that they provide an ideal setting for Team far ambush situations on enemy tracking

Figure 45. Claymore Mine Positioned in the Crotch of a Tree to Enhance Kill Zone.

Figure 46. Wrong! Use the Verge to Conceal Tracks. Notice How the Sun Casts its Shadow to Define the Tracks. (*Depositphotos.com*)

units. When using open formations, beware of losing Team Members in driving snow/low visibility conditions.

- Troops operating in cold weather environments require more rations or caloric intake than contained in normal rations. Special purpose, enhanced-calorie rations (Meal, Cold Weather (MCW) and Food Packet, Long Range Patrol (LRP) rations) have been developed for the US military. Consider adding water to freeze dried meal components early in the day and then tuck the rations into a warm inner pocket to facilitate meal hydration for later consumption.
- Use enemy equipment (e.g. skis, snow shoes, footwear, etc.) so that the enemy will confuse the trail as belonging to friendly patrols. Note: Make sure that ski poles are also of enemy design or the pole 'basket' is one that is in current use by the enemy.
- Thermal Signature Discipline. If the enemy is well equipped with thermal equipment this should be a significant Team concern. If the enemy's thermal detection equipment resources are limited, the equipment may only be found in the possession of combat or special units; rear echelon units are less likely to have this equipment unless it is to secure crucial/sensitive locations. The enemy will not operate thermal detection systems in sunlit areas/daylight periods, so again, travel along folds in the earth. Defeating Thermal Sensors:
  - Block IR signature using an ordinary 'space blanket' (Mylar foil sheet), which may also be used as a ground cloth. Test this during training.
    - If the space blanket is worn or used to conceal the heat signature of equipment (e.g. UTV), heat will build up and vent, the plume then becomes visible to nearby imagers.
    - Consider using the blanket as a screen or placing it out of intimate contact with the Team Member or equipment.
    - Note: In a hot climate/desert, use of a Mylar blanket may seem too cold to an IR sensor, blocking the normal background IR in warmer backgrounds.
  - Wearing several layers of clothing and/or a thick wool blanket will suppress thermal imaging.
    - If the Team Members are active, heat will build up and may vent through openings in the clothing, but the overall signature will still be suppressed.
    - If Team Members are inactive (e.g. in a surveillance hide position), less heat buildup and venting will occur.
    - Remember that much heat is irradiated from the head, neck, boots and hands; wear a hat, hood, scarf, camouflage netting and/or balaclava and wear gloves to suppress heat emitted from these 'hot-spots'. The shirt should always be worn with the sleeves down in all environments. Wear leggings or gaiters to partially cover boots and suppress heat emissions at the trouser cuffs.
  - Conceal Team Members under a blanket of leaves or pine boughs (and snow). A heavy/solid canopy of leaves/pine boughs will defeat an airborne IR sensor.
  - Blend in next to warm objects like large rocks, lava beds, etc. while they hold heat captured from the sun. This will last only a few hours after sunset.
  - Use an organic thermal/IR scope to view Team Members and/or their positions after dark to ensure that they blend in.
  - Thermal suppression covers must be easily and rapidly accessed.
  - If enemy rotary-wing aviation assets loiter in the area, remain in covered positions and take additional heat suppressive measures contained in this book. If Team Members are not in covered positions when the noise of enemy aviation is detected, take the heat suppressive measures contained in this section and lay face down in a depression, behind a tree, under leaves, etc. to further suppress heat signatures.
  - Put terrain, trees and/or brush between you and an enemy unit suspected of operating IR sensors.

- ○ Rain, falling snow, and fog will limit the effective range, or even defeat the detection capability of Thermal Imagers. Consider this in Team operational planning and during execution. These same weather conditions will impair NVDs.
- Radio communication is seriously affected by storms and atmospheric disturbances/effects. Proper frequencies selection and pre-launch radio checks are of extreme importance.
- Signs of enemy activity may be easier to detect in cold regions. Types of enemy equipment and their locations can be determined from the tracks they leave in snow and slush. To detect these signs, the Team may have to commit to reconnaissance through movement rather than operating from static positions. Again, use of enemy cross-country and mobility equipment may make Team tracks seem innocuous.
- The Team may resort to a warming area/hut/snow cave during extended operations. The Team must employ extraordinary measures to avoid thermal detection.
  - ○ Freezing environments may substantially deplete batteries for electronic equipment and other uses. Such equipment and battery spares should be periodically charged and/or kept in a warming hut/area until needed.
  - ○ Use ravines or depressions, well away from OPs/LPs, for the warming hut/area. The warming area/hut should be snow covered and/or concealed by conifers. The same would apply to conceal mobility equipment.
  - ○ Use of candles at night may generate sufficient heat for an enclosed warming area. No green needles or smoke-producing fuels should be used for heating in the warming hut, cave, etc.
  - ○ When building a warming hut/snow cave, bear these tips in mind:
    - Pick a spot with plenty of snow, and ideally with a natural protective barrier (e.g. earth mound, depression, fallen tree) where the shelter entrance will be located.
    - If the snow is not deep, collect it, using an Ahkio for transport, and pile it high at the designated spot.
    - Let the snow settle for at least an hour; preferably overnight for compaction.
    - The shelter dimensions should accommodate all Team Members and their personal equipment, at a minimum, half sitting and half recumbent. Cut long stakes and drive them into the snow mound to mark the interior wall, allowing a wall thickness of 2'.
    - Dig the entrance to allow entry of the largest, fully equipped Team Member and continue to dig out the interior. Stop digging when the stakes are encountered.
    - Ensure Team Members will sit/recline off the cold ground. Create snow platforms, which are warmer than the ground, within the living space. Improve the platforms with pine needles, etc. as time allows.
    - Weather-proof the entrance with a tarp and local materials, tree limbs, etc., ensuring that wind will not cause the tarp to flutter and cause noise.
    - Poke several holes in the roof to ensure ventilation. Ensure these are kept clear during periods of snowfall.
  - ○ Make a channel in the snow as a chimney/conduit for heat. Passage of the heat plume beneath a layer of snow will cool the exhaust.
  - ○ Warming areas/huts are best used during the day, during snowfall or during cloud cover/fog, when thermal devices are not active.
  - ○ Team Members should always use a thermal sensor/scope to examine Team concealment efforts and to detect any heat plume or heat trace. The T/L should direct immediate action to eliminate heat traces.
  - ○ Partial warming area/hut occupancy is recommended, unless temperatures are truly dire.
  - ○ Weapons should be kept outside (covered) or in an unheated area of the shelter to prevent condensation and subsequent mechanical freezing.

- ○ Choose weapons that are designed to function in extreme cold. This may require the Team Members to equip with foreign weapons.
- Team Members may have to occupy an OP/LP for extended periods of time, without the benefit of a warming area/hut.
  - ○ Ensure the OP/LP is prepared with overhead concealment (and cover if possible).
  - ○ If possible, position the OP/LP so that Team Member movement to/from the hide is screened by terrain fold, snow banks and vegetation.
  - ○ Observation ports/embrasures should be small enough to mitigate heat signature, yet offer adequate observation. Ports/embrasures that are not in use should be temporarily sealed with blocks of ice/snow.
  - ○ Multiple layers of clothes and/or a sleeping bag will further suppress body heat signature. Always wear a hat and/or full-face balaclava and a hood, as much heat is generated at the head.
  - ○ Author's Rule-of-Three: During surveillance, two men keep each other awake; one man sleeps.
- If feasible, use camouflaged remote cameras (e.g. game cameras) for observation of roads, facilities, etc. – to deter surveillance hide heat signature detection.
  - ○ Current technology, a Class-1 Bluetooth device has an approximate range of 100m, but consumes more power than shorter-range Class 2/3 devices.
  - ○ Be aware that use of such equipment may require periodic exposure of Team Members when they replace batteries, clear lenses of frost, etc. Be sure to conceal tracks when conducting maintenance.
  - ○ If the cameras transmit images via RF (rather than land line) in close proximity to a sensitive unit/installation, be aware that this energy may be detectable to a sophisticated enemy with scanners/RDF capability; but resorting to operation via hard wire implies an additional weight burden and a practical limit on wire length/transmission distance.
  - ○ Perhaps the best method of employing a game camera is to secure the camera to the top of an expedient pole. This allows easier camera positioning and reduced ToT to service the device.
- Use of foreign cross-country skis and bindings or show-shoes may require Team Members to accommodate to lightweight shoes with heavy socks and gaiters, such as are in use by Scandinavian military units, rather than issue boots. This enhances the ability of Team Members to rapidly shuck their skis or snowshoes to perform battle drills, and then rapidly don the skis/snowshoes. During movement, the feet of Team Members will not get cold, unless the footwear is immersed in water. If the Team Member will not be moving for a period of time, he should replace wet socks and don lightweight mukluks/over-boots.
- When the Team approaches danger areas, enemy troop locations, and ultimately, its objective, be aware that movement through snow crust and over ice requires exceptional noise discipline. Noise may travel further in frigid conditions.
- Travel over thick ice can be rapid and often more silent than through snow. However, the enemy's ability to observe over ice is enhanced; therefore, two Team Members should reconnoiter well in advance of the Team. If the Team is traveling through isolated areas, devoid of enemy and civilian populations, cross ice (e.g. rivers, lakes, etc.) rapidly. If the enemy's location is known, use intervening geographical features (e.g. islands, peninsulas, etc.) to screen the movement of the Team.
- Northern summer conditions are characterized by long periods of daylight, and numerous water obstacles and marshy areas. The use of boats to negotiate remote waterways, devoid of human activity, during periods of limited visibility, resulted in extremely successful operations conducted by the Finns during its Continuation War with the Soviet Union – as was done during the Puustinen raid (see above).

- Heating of vehicle oil may be required in very cold temperatures. Russian (and client state) armor are all equipped with electrical oil pre-heating elements for this very purpose. And automobile parking spots in Alaska and Canada and similar environs are often equipped with electrical connections for engine heating blankets. If Team vehicles are to be operated in very cold environments, and no heating element is provided, the Team may have to build a fire and warm the engine and crankcase over coals. Learn how to do this quickly, safely and with minimum signature. Additionally, Team Members may have to run engines periodically to keep the battery charged, especially if vehicular ancillary equipment (e.g. radio) is to be used. Carry some flexible tubing to divert the exhaust plume along the ground (preferably under snow) to minimize signature.

Operations in Swamp/Marsh TTPs:

- If the Team is operating in swamp or in lake and river/stream environments, OPs/LPs may have to be placed in trees. Team Members ascending to or descending from tree-top OPs/LPs should exercise care to minimize shaking of foliage. If a bird colony or primates are occupying the tree, select another tree.
- Caches:
  - If caches are established in trees, they should be made impenetrable to primates or other animals. If the location may be subject to violent storms, the cache should also be firmly lashed to the tree trunk or to its major limbs.
  - Caches may also be established on the ground or below water. If this is to be the case, the following steps are required.
    - A clearly identifiable terrain feature must be present close-by so that the cache may be found later.
    - If it is to be submerged, the cache must be thoroughly water-proofed and weighted.
    - Established on the ground or submerged, the cache must be firmly anchored, especially if the area is subject to flooding or tidal flow.

SR Operations Conducted in Built-Up Areas:

- SR Team Members may need to enter a building/room to take prisoners, to plant electronic devices, to plant booby-traps, to search for intelligence, to establish a surveillance post in a built-up area, etc. This is an inherently hazardous proposition as the structure/room may be occupied by civilians or enemy troops or may be booby-trapped.
- Team Members must be aware of the tactical implications of varying building construction practices/standards.
  - Older multi-level brick buildings may lack steel supports and may therefore have the thickest exterior walls at the base floor to provide sufficient strength to support successive floors; wall thickness will decrease at each level thereafter. Interior weight-bearing brick walls/columns may have the same construction. This will factor-in to wall breaching, munitions penetration, defensive position preparation and similar considerations.
  - Lightly constructed interior walls (non-weight-bearing) are easily penetrated by small arms ammunition. Be aware that such walls will not protect Team Members from enemy fires. Additionally, enemy combatants may be oblivious to these facts and can be effectively attacked (e.g. with a Claymore) through interior walls.
- If the facility is usable/habitable but appears to be abandoned, the T/L must ask himself why this is so.
  - The Team must observe the building from a distance during day and night to verify that it is abandoned. At night, the Team may scout in close to spot signs of occupancy and

to plant remote cameras closer to the structure. The Team should look for any sign of possible activity.

- Are lights observed at night in or around the structure? If the Team is equipped with thermal optics, are human heat sources detected?
- Upon approach, are recent human or tire tracks visible?
- Are windows intact or are they broken?
- Is the structure still receiving electrical power?
- Is there a night watchman/custodian/security element? Observe any shift changes and security element resupply.
- Are there any occupied guard posts or outbuildings close by? Is there secondary or recently installed fencing present?
- Is facility in use during weekend/holidays? Outside of curfew hours?

  - If the structure is truly abandoned, it may be a temporary condition. It may be earmarked for subsequent occupancy.

- If the structure, or any out-building, is occupied, verify if the occupants are civilian or enemy combatants, before attempting entry.
- If the T/L decides to enter the facility, his plan must include a secure approach and a secure and a detailed rapid exit/escape contingency, with overwatch. It is assumed that all SpecOps personnel receive training in building entry TTPs; experience in this type of operation is now commonplace. Nevertheless, a raid on a structure during deep penetration missions should be well planned.
- The T/L should consider clearing outbuildings or sentry locations before attempting the entry of the main structure.
- Initial Entry:
  - If the building is occupied, and raiding the structure is warranted, the T/L must decide whether to kill all occupants or take some captive. In either event, the raid must ensure that all sentries are taken out, that all exits are covered/blocked and that occupants are prevented from sounding an alarm. Silent kill weapons and techniques (discussed elsewhere in this book) are preferred so enemy troops in the vicinity do not respond to gunshots.
  - Do not use normal entrances, if possible. Enter the building from an upper level access point (e.g. window) or existing wall breach as regular access points may be booby-trapped or under observation.
  - If the building is moderately to heavily damaged, consider bringing rope for access support and safety purposes.
  - If the enemy is not immediately or subsequently encountered upon silent entry to the building, the Team must post security at key building locations/access points, while the remainder of the Team conducts a search.

    'When you enter a room full of enemy, kill the first one that moves. He is starting to think and is therefore dangerous.' Lieutenant Colonel Robert Blair 'Paddy' Mayne, co-founder of the SAS.

- Once the building has been secured, the T/L has several options.
  - If the main structure, or its outbuildings, had been occupied and subsequently cleared by the Team, the Team must conclude its business quickly and withdraw (with or without prisoners). TTPs found within this book should be used to foil pursuit or other enemy actions against the Team.
  - If the building is not occupied, the Team might occupy the main structure (or an outbuilding) to conduct surveillance of surrounding enemy activity. Alternatively, the Team may install remote cameras to conduct the surveillance from an offsite location.

- ○ While occupying the structure, the Team may execute a POW snatch or ambush of approaching civilians or enemy troops – preferably using silent weapons and appropriate TTPs.
- ○ The Team may plant booby-traps or remotely initiated devices, if the tactical situation requires operations outside the confines of the building/campus.
- ○ Time-delayed incendiaries may be the best approach if the structure is to be destroyed.

Subterranean Operations:

- An enemy will often conceal and protect its most sensitive and vital facilities below ground in tunnels, commercial mines or in caves. These facilities may protect C⁴ISR capabilities and high-level leadership and even vital logistics items, such as WMD warheads/materials, static launch sites, TEL hide locations, etc. These areas should be of special interest to SR operations. Ideally, SR or intelligence collection operations on these sites should occur prior to hostilities and/or while the sites are unoccupied.
- These facilities may be active sites, prepared well in advance of planned military operations; they may be abandoned from previous military operations or past conflicts with potential for reuse; or they may be natural geologic or private/commercial sites earmarked for military occupancy as need arises.
- Current and archived maps (friendly and/or enemy; of commercial and/or government/ military origin) should be examined for evidence (e.g. map symbols) of old mining operations, caves, likely terrain-form, etc. Aerial/satellite photography and signature collection (by intelligence support) should be taken of these sites, if at all possible, to determine recent activity/use. If evidence of these sites exists on archival maps, but are absent on current maps, this should be a 'red flag'.
- If not abandoned, the sites may be fully occupied or only under caretaker occupation. Special care must be taken in approaching any site that is occupied or guarded. If it can be determined that the site is unoccupied/abandoned, caution is still mandatory as the site may be alarmed, subject to security checks or booby-trapped. Once the entrance to the site has been discovered, the Team must determine the condition and utility of the site; if it is occupied or if trails/signs indicate recent use, utility may be assumed. The Team should then mount a surveillance of the facility to determine comings and goings, other entrances, communication wire/antennae, nearby defensive positions, etc. A well-positioned game camera would be recommended to remotely observe the entrance(s).
- ○ Once the facility status has been determined, other measures must be considered. If unoccupied:
  - ▪ The T/L might consider sending an element into the facility. The T/L establishes security and a command post at the entrance of the facility. The recon element should record distance and direction of travel, to map the tunnels/corridors and dimensions of the facility and to collect intelligence. The recon element must clearly mark (bread crumbs) its progress through the facility.
  - ▪ Augmented night-vision optics, to include thermal, are essential.
  - ▪ A breeze in the tunnel/corridor will indicate either an air circulation system or another access point.
  - ▪ Note that any incursion into a subsurface facility is risky, even if it is unoccupied. An enemy patrol/tracker team may come upon the Team after the recon element has entered the facility, or Team sign may have been reported by local hunters or inhabitants. Hazards within the facility might injure members of the recon element. A chance encounter with an unanticipated enemy within the facility offers substantially

      increased hazard, as room to maneuver may not be available. The recon element should consider the use of CS gas/powder if the tunnel breeze will carry the cloud toward the enemy; obviously, recon element members must carry protective masks.

- GPS will not work underground. If the soil or rock formations contain ore, a magnetic compass will not be reliable (tip-off: the needle spins and won't settle during movement) in determining direction.
- RF communications between the recon element and the main Team component will generally not be possible, except for a short distance past the entrance. Some other form of basic messaging (e.g. Return to CP; Enemy Approaching; etc.) between the elements is necessary.
- The recon element must be prepared to take extensive photos/videos within the facility. Automatic location/direction stamping will not be possible. The recon element may have to use a manual notepad to record/describe what is in the frame.
- If the facility is extensive or if its contents are of intelligence value, the Team should request special equipment and/or technical assistance, if this is possible. A non-GPS PLS and a laser measuring device may be necessary.

   ○ If the facility is in use/occupied/guarded.

- The Team should request covert Measurement And Signature Intelligence (MASINT) sensors from higher authority. The Team may have to relocate to communicate its requirement and will certainly have to relocate several kilometers to receive the devices and/or technicians to install them. The Team should covertly mark its back-trail to return to the site.
- If long term sensors are not available, the Team may have to continue its surveillance and area reconnaissance for as long as possible. If higher authority wants to continue the surveillance, the Team will either have to be resupplied or replaced with another Team.
- Higher authority may require that the Team conduct a Prisoner Snatch so that key information about the facility may be had through interrogation. If this is to be accomplished, then higher authority should consider supplying a recent corpse resembling the POW to forestall compromise. The POW should be stripped of his uniform and possessions, and these placed upon the corpse. The corpse should have died from an accidental event, if possible, and should be arranged in a scene that suggests accidental death. If higher authority is not capable of making these arrangements rapidly, then they must be content with the result. In the absence of a corpse, the enemy will make every effort to locate its missing soldier to determine if the site has been compromised.

   ○ In all events, the Team must take every measure to thoroughly obliterate evidence of its presence in the area.

Offensive Operations TTPs:

    'Certainly, there is no hunting like the hunting of man, and those who have hunted armed men long enough and liked it, never care for anything else thereafter.' Ernest Hemingway, April 1936.

- A lax adversary may sometimes be found in rear areas, especially in the early stages of a conflict. In this setting, the security threat to the Team is mitigated, offering the T/L a lot of operational latitude. A lax security environment presents an opportunity to do some real damage to enemy units, infrastructure and capabilities with reduced risk to the Team. In these circumstances, SR Teams should first focus on identified/verified high-priority targets and set aside lesser targets until later.

- Once an SR Team has inflicted substantial damage on the enemy, the T/L and FOB may feel inclined to enjoy mission success and conduct an exfiltration of the Team. This may be exactly the wrong decision! Unless the Team has casualties, is being hotly pursued, is plagued with equipment failures or beset by other core problems, the T/L should exploit the disrupted enemy/chaotic situation, rather than give the enemy time to recover and reestablish operations.
- Friendly guerillas/partisans may possess mortars or other heavy weapons. Where SR Teams are operating in UW environments, guerilla/partisan units might be tasked to provide support for SR operations. See definition of guerrillas and partisans later in the book.
- Where friendly guerrilla units are active, attacks on trains or convoys may not be the optimum use of the Team – unless the Team spots a high priority 'World Series' target.
- When attacking an enemy SOF unit, use overkill to inflict maximum casualties or destroy the unit. This can have a devastating effect on enemy morale and psychology. If the unit reconstitutes, replacements will often be green, degrading the operational effectiveness of the unit. If the Team is not effective in its attack, the enemy unit may later reply in kind as a point of honor.
- When severing an enemy communications landline, cut out/remove a section, or more than one section. If the enemy has not provided excess wire loops along the standing line, the repair crew may have to splice in a wire segment, which is more time-consuming and problematic especially for coaxial or fiber-optic cable. Cutting an enemy landline will result in the arrival of a communications repair crew, which may be accompanied by a security element. It is best to make the cut during severe weather; the enemy may attribute the interruption to the storm, falling limbs, etc. This would normally present the Team an opportunity to collect POWs, especially if the repair crew is unaccompanied by security. Additionally, if the landline is fiber-optic, the crew may respond with a vehicle transporting hard-to-replace specialized tools and equipment. Destroying the vehicle and its equipment could cause a local cascade of problems, until the arrival of replacement equipment. In the absence of landline, the enemy may resort to radio transmissions (that could reveal the location of enemy headquarters to friendly Signal Intelligence (SIGINT) assets) or to couriers which would slow the flow of communications.
- When operating at strategic depths and in the vicinity of inhabited areas, consider leaving false trails and evidence to cast blame for Team operations on locals (e.g. bandits). This may have the following results:
  ○ Enemy suspicions of a Team presence in the area may be shelved.
  ○ The enemy may suspect locals of being members or sympathizers of a bandit gang or of a guerrilla/partisan band and may react by punishing the inhabitants by taking hostages, conducting spot executions of locals, relocating inhabitants, seizing or destroying property, establishing curfews and other restrictions on the local population.
  ○ Some punitive measures (for example: curfew, population relocation, restrictions on movements) taken by the enemy against the locals may alienate civilians and serve the security interests and tactical flexibility of the Team. Locals may blame and become openly hostile to the Team if the ruse is not convincing.

Counter-Reconnaissance Operations TTPs:

- Successful enemy reconnaissance and guerrilla units survive and succeed partly because they employ relevant tactics, techniques and trade/field-craft; superior knowledge of the terrain and a near obsessive fixation on security and stealth. Therefore, one of the most appropriate resources to deploy against enemy reconnaissance/guerrilla units would be US-led SR Teams, which would possess the equivalent or superior knowledge, skills and traits.

- To become adept at counter-reconnaissance operations, the SR unit should seek additional training to complement reconnaissance training, and to expand the inventory of trade/field-craft skills. This training might include combat tracker courses such as the Malaysian Man Tracker course (Jungle/Rainforest), and perhaps other military tracking courses that are specific to terrain types (desert, cold weather environments, etc.). Note that tracking skills have a shelf life; they must be frequently exercised.
- More than likely, the SR Team Members will never reach the field-craft skill level of a primitive indigenous tribesman, who has had to depend on concealment/tracking/hunting skills throughout his life for his very survival. These indigenous tribal personnel can be instrumental to successful counter-reconnaissance operations, can substantially improve the overall tracking skills of the Team, and can be essential to Team survival. These trackers can often be hired as mercenaries for trivial wages. If you incorporate an indigenous tracker into the Team, he must be thoroughly trained in Team TTPs, or he will become a tactical liability.

> **True Account:** A SOG Reconnaissance Team was operating within a critical North Vietnamese Army Base Area located adjacent to a major artery of the Ho Chi Minh Trail. The Team approached Laotian Route 110, fish-hooked to observe its back-trail and established an OP/LP to conduct a road-watch. The following day, an NVA tracker team, led by a Laotian tribesman in a loincloth, was spotted by one of the Team's indigenous commandos as they approached the Team's 'hide' location. The tribesman halted when he spotted a Claymore mine three meters before him, which had been deployed along the route of approach by a SOG indigenous commando. Apparently, the tribesman had never seen such a device before, because he summoned forward the leader of the tracker unit and pointed it out to him. The expression on the NVA team leader's face, when he recognized the device, was immensely hilarious to the commando as he detonated the Claymore.

- Be aware! An experienced, elite enemy reconnaissance team may employ some of the same tactics, techniques and tradecraft cited in this book.
  - FM 31-20-5 has been translated into Russian, and probably other languages as well.
  - Additionally, some elite foreign SpecOps units have exceptional experience, specific TTPs, and superior field-craft skills. While we do not perceive friendly foreign SpecOps units to pose a threat in foreseeable conflicts, they may provide training and military assistance to nations that do.
- The counter-reconnaissance Tracker Team should be less encumbered than the enemy RT that they are pursuing. The Tracker Team should therefore be able to move more swiftly and more stealthily than a heavily burdened enemy reconnaissance team. Clear communication, via hand and arm signals or other means, between the point-man and the Tracker T/L is essential.
- If the Team closes on the enemy tail-gunner and the Team point-man suspects that he <u>might have been</u> spotted, <u>he must immediately engage without hesitation</u>. The tip-off as to detection may not be clear cut; if the enemy combatant believes that he does not have a decided advantage in a chance encounter, he may pretend that he has not spotted the Team point-man (or other Team Members) so that he can then move to a firing position of advantage or spread an alarm. If the enemy has his weapon at 'high port' or has assumed a cheek or stock weld, even if he is not looking in your direction, <u>open fire</u> immediately.
- The classic 'Hammer and Anvil' tactic remains among the best methods of destroying a RT.
  - The counter-reconnaissance Tracker Team (the Hammer) reports its location to the blocking force (the Anvil) and any changes in the enemy team's azimuth of travel. The Hammer also monitors the location(s) of the blocking force.

- ○ When the enemy RT is estimated to have closed within approximately 200 meters of the Anvil's position(s), the Hammer coordinates with the Anvil more closely.
- ○ If the enemy RT navigates toward its presumed target without much variation in its azimuth, the Anvil may establish a deliberate ambush with hasty defensive positions. Once the enemy RT is engaged by the Anvil, the counter-reconnaissance Hammer should then establish an ambush formation, as the enemy RT will likely attempt to break contact along its back-trail.
- ○ The counter-reconnaissance Hammer may try to channelize the enemy RT toward restrictive terrain (e.g. a river or cliff) or flush the enemy toward the Anvil, by firing signal shots. But an experienced RT leader will not fall for this tactic and will change direction multiple times and use his own TTPs to evade or ambush the Hammer.
- ○ If the enemy RT changes azimuths frequently, the Anvil should establish hasty area ambush positions (see ambush techniques later in this book), occupying available cover, preferably on elevated terrain and with observation of likely routes available to the enemy RT, e.g. defilade areas such as ravines, along streams, etc.
- ○ If the Anvil is formed in an 'L', 'Y' or 'V' concave configuration, all-the-better, as the arms of the formation will serve to block the enemy's movement alternatives. Once set in its ambush position, the Anvil notifies the Hammer; the Hammer may then rapidly close on the enemy RT, initiating a firefight, and spreading out in an assault formation to press the enemy RT toward the Anvil. The enemy RT will likely perform an immediate action drill to break contact with the Hammer, unknowingly breaking toward, and into, the kill zone of the Anvil. The Hammer's objective is to flush the enemy into the Anvil's kill zone – while taking care to avoid fratricidal fires with the Anvil.
- If no suitable blocking force (Anvil) is available, the Tracker Team must assume the role of Hunter-Killer Team.
  - ○ The best time to hit the enemy RT would then be when the enemy RT is preparing to move out from its noon mealtime break or from its NDP and <u>after the enemy RT has recovered any anti-personnel devices (e.g. directional mines)</u>. In mountainous/hilly terrain, the enemy RT will typically flee using the military crest or will move downhill (e.g. towards an LZ/safe zone). The SR tracker team can travel along ridge tops and move much faster to outpace the enemy RT; this provides the SR Team the opportunity to establish an ambush in front of the enemy RT.
  - ○ If the enemy RT is well trained and its leader is skilled, beware the enemy's use of fish-hooking to observe its back-trail during communication/meal breaks or at the NDP. The enemy may establish a pattern as to when it makes scheduled communications; if these communication windows are known/become evident, the Tracker Team should maneuver to take up ambush positions with maximum stealth. In any event, the Tracker Team should be especially alert around meal times and as evening approaches, so as not to blunder into a disadvantageous encounter when the enemy may be in a defensive configuration. Should the Team's point-man come upon a suspicious azimuth deviation, especially a ninety degree turn unwarranted by terrain or vegetation, he should immediately suspect that the enemy team has executed a fish-hook maneuver. He should immediately signal the team down or to take cover and then should confer with the T/L.
  - ○ Then the Tracker Team's challenge is to identify the proximate location of the enemy perimeter and to establish an ambush (preferably an 'L' formation) that anticipates the enemy's movement as he later departs the defensive perimeter. This will require considerable stealth, as the RT will be very observant once it is in its perimeter.
  - ○ The enemy RT's likely direction from its defensive position may be the azimuth upon which the enemy had been traveling prior to making its fish-hook maneuver.

- ○ The short leg of the 'L' ambush (ambush formations are covered later in this book) should be oriented across this expected route-of-march and the hunter-killer T/L should ensure that a crew-served weapon is allocated to the short leg. If the enemy team departs the perimeter along the expected direction, the short leg should initiate the ambush. The long leg of the 'L' will be oriented in proximity to the perimeter and back along the back-trail.
- ○ This plan is more effective if a barrier terrain feature/obstacle or other danger area limits enemy route choices. Accordingly, the hunter-killer T/L should examine his map to properly position his Team to pin the enemy against the danger area.
- ○ Once contact is initiated, the Tracker T/L may opt to assault/sweep from the long arm of the 'L' to exploit the assault by fire. He must use a clear signal to lift fires from the short leg of the 'L' to avoid fratricide. If the enemy RT flees into a ravine, the Tracker Team members should quickly move parallel and rain grenades down upon them.
- ○ The enemy RT is unlikely to assault the 'L' formation in this scenario; RT immediate action/Battle Drill would typically require breaking contact and withdrawal in such a disadvantaged situation. Caught in an 'L' ambush, the enemy would then have a choice of two general directions from the point of contact – away from both legs of the ambush.
  - If the enemy attempts to withdraw along its back-trail, where the enemy leader may have designated an RP, they will do so while under fire or assault from the long leg. The Tracker Team may make the enemy RT pay dearly for this by firing a command-detonated mine/booby-trap along the back-trail.
  - The enemy RT leader may have established a RP or withdrawal direction leading toward an LZ, a linear terrain feature (e.g. stream) or defensible terrain. If the Tracker Team can anticipate this, the T/L may dispatch a 2–3 man element along the anticipated route of withdrawal.
  - If the enemy's perimeter was located on high ground/defensible terrain, the RT may return to reestablish the perimeter to make a last stand. The RT leader may be more inclined to select this option if he is burdened with WIAs. The Tracker T/L should transition to an assault/sweep of the enemy <u>before</u> it can return to its perimeter positions.
- ○ Once the fire fight is over, the Tracker Team should care for friendly casualties (if any) and secure enemy prisoners and captured materials. This done while maintaining security over enemy prisoners and against RT counter-attack. Evacuation of friendly casualties and enemy prisoners will require Tracker Team assets. Depending on his remaining resources, the Tracker T/L may continue the pursuit of the enemy RT.
- Upon any engagement, the enemy RT will attempt to break contact using a Battle Drill designed for withdrawal. Remember, if the Tracker Team is able to wound even one enemy, the enemy RT's firepower will be substantially reduced; drastically so, if the enemy RT carries the wounded man.

SR in the Counter-Insurgency (COIN)/Counter-Guerilla (CG)/Counter-Terrorism (CT) Role:
<u>Guerrilla</u>: A person who engages in irregular warfare especially as a member of an independent unit carrying out harassment and sabotage. Definition also applies to a member of a Resistance or insurgent unit.

<u>Partisan</u>: A member of a body of detached light troops engaged in making forays and harassing an enemy. Partisans may consist of conventional units/troops that were cut off from their main body of forces, but continue independent operations in the enemy rear areas. Partisan forces will be resupplied by the main body of forces, but will also use captured enemy stocks.

- As has been pointed out elsewhere in this book, some of the most successful operations mounted in Iraq and Afghanistan against the al Qaida and Taliban were when SpecOps operators and CIA analysts/targeting personnel worked in concert. This arrangement integrated the Operator's tactical knowledge, combat experience, and common sense/pragmatism to the all-source intelligence analysis and targeting processes, and it enabled SpecOps personnel to rapidly react to breaking intelligence.

- This collaborative enterprise was seen as a brilliant insight by senior leadership within the SpecOps and CIA rather than an application of Lessons-Learned from the Vietnam experience. In fact, the Vietnam-era CIA PRU Program used this model to devastating effect in dismantling the Viet Cong infrastructure in many areas of Vietnam.

'Success in antipartisan warfare … is contingent upon carefully gathering all facts for evaluating the partisan's command structure, intelligence system, mobility in occupied territories, and relationship to the civilian population. The more examples from practical experience that are available for analyzing these factors, the better prepared will be those who might be called upon to lead the fight against the partisans.'[31]

- Guerilla/partisan forces may oftentimes be within a day's journey of a local support structure (e.g. auxiliary, intelligence apparatus) within the rural civilian community.
- A competently established enemy base camp will have three security zones:
  - Observers near Counter-Insurgency force installations and along routes toward the enemy base camp/AO to provide early warning of enemy operations.
  - OPs/LPs along high-speed routes of approach to the base camp.
  - Field fortifications within the base camp.
- A typical base camp configuration will likely have the following features:
  - Located in a draw (in hilly/mountainous terrain) with dense foliage and access to water.
  - Ridge fingers of the draw will have fortified positions and fields of fire.
  - The kitchen/mess area will be centrally located.
  - Latrines will likely be outside the bunker line and downstream or away from the camp's water source.
  - There will always be a primary entrance/exit (high-speed) and at least one concealed emergency exit opposite the high-speed route of approach.
  - Automatic weapons positions will dominate the high-speed approaches.
  - Concealed bunkers/foxholes will be near to the sleeping platforms/huts for the command group. The bunkers will have interconnecting trenches/tunnels.
  - Locations of mines/booby-traps may be marked with white cloth on trees or bushes that face the enemy fighting positions.
  - Camps in swamps will be located on dry ground.
  - Camps with no stream-fed draws will either have a well or it may be located within 150 to 200 meters from a water source. If the camp has been long established, water may even be conveyed to it via a pipeline. Note that the Team might interrupt the pipeline to draw an enemy repair crew into a trap.
  - Village buildings/huts in remote areas of the countryside may be a covert base camp.
  - Insurgents will usually regulate woodcutting. For example, areas for woodcutting may be as far as one hour (e.g. 3 miles) from the base camp. Evidence of extensive cutting may indicate a large base camp. Cut bamboo/wood will be used for sleeping platforms, huts, fires and overhead cover for bunkers.
  - Access (1 to 3 kilometers) to roads, major trails or the jungle periphery. If the base camp is closer than 200 to 500 meters, this indicates a lazy or incompetent cadre.

---

31. Halder, *Small Unit Actions*, p. 262

- In the vicinity of an enemy insurgent base camp, the T/L should establish a well concealed RP where the Team can drop rucksacks and excess equipment to enhance stealth.
- If the SR Team is deployed from vehicles to conduct counter-guerilla/antipartisan operations in the vicinity of an infiltrated community, the Team should expect that the guerilla/partisan force will be notified immediately. If friendly RDF/signal intercept capabilities are available, they could be covertly deployed in anticipation of such warnings to the enemy force; the communicator(s) could then be captured by host nation police/security forces. Covert deployment of the RDF/intercept capability should be tightly held information.
- The SR Team may also expect that they will be followed or tracked by members of the enemy guerilla auxiliary. The Team may anticipate this by executing an ambush or using a sniper element.
- Deceive the enemy guerilla/partisan forces by leaking a false report or false indicators as to the location of pending operations. This can be accomplished in various ways, such as an 'inadvertent slip' to a prostitute or bar girl; a VR using host nation aviation assets; providing a false notification to local host nation authorities; conducting an obvious ground reconnoiter of a prospective (false) staging area; etc. If the Team is deploying by air, the aircraft should depart the helipad in the direction of the false AO/Target Area; if deploying via ground transportation, the convoy should follow a circuitous route, initially toward the false AO/Target Area, then in a roundabout road-march, preferably at night after curfew, to the actual dismount point. To prevent the enemy from using a tail on the convoy, host nation security forces could establish highway checkpoints.
- If the enemy realizes that his base camp has been compromised, he will flee to an alternative base camp location. However, the enemy may likely return to the original base camp to recover <u>concealed</u> documents, and/or precious supplies/equipment.
- If the SR Team is operating in the target area, undiscovered by enemy guerilla/partisan forces, they may be positioned to intercept the withdrawal of the enemy force to the camp after it conducts an operation. The SR unit headquarters must monitor appropriate communications traffic, so that the SR Team may be notified of the Guerilla attack and plan its actions in a timely manner.
- While surveilling the enemy, the Team should take its meals at the same time as the enemy. Cooking aromas will obscure any food aromas coming from the Team. However, if enemy guerilla/partisan troops have been starved of rations, the scent of Team rations will clearly give away Team presence.
- Guerillas/Partisans must have food, medical supplies, personnel replacements/recruits and intelligence. These items are either provided voluntarily by local sympathizers/auxiliaries, are seized from the population, and/or are provided from foreign sources. Friendly local intelligence sources may know the names and addresses of local sympathizers/auxiliaries and may have other information that would be crucial to a COIN effort.
  - Monitoring or surveilling the comings and goings of known sympathizers/auxiliaries may establish patterns where routine contacts are made with insurgents. This task is best accomplished by law enforcement sources and COIN agents.
  - Once patterns are established, the SR Team can be deployed to track, observe sympathizers/auxiliaries when they attempt to make contact with the insurgents.
  - Enemy insurgents rely on food supplies furnished by locals (willingly/unwillingly). The key window of opportunity (for both the insurgents and SR units) is when crops are ready for harvest.
  - If enemy forces have been starving, they may become desperate enough to raid a food source. The SR T/L should anticipate this.

- Guerrillas/partisans may assist villages in harvesting and planting crops which in turn will be used to supply enemy combatants, some insurgents may be members of the community. Harvest/planting time is optimum to observe insurgent participation in these activities. If the insurgents do not participate, but merely take delivery of foodstuffs, this too may be observed by the SR Team. Enemy guerilla/partisan leaders may visit villages just prior to harvest/planting time to cultivate good will or menace the population. Additionally, the local farmers may furnish pack animals to transport quantities of food; tracking of burdened animals and carts is easily accomplished if done in a timely manner. In these situations, the Team may then track the movement of personnel and food supplies to insurgent destinations.
- Foodstuffs and other supplies must be transported to guerilla/partisan caches and field locations. At harvest time, SR Team may plant a covert beacon on the farmer vehicle/cart, or the Team may plant a covert/disguised transponder among supplies that are destined for the insurgents. The British Police Special Branch used homing devices taped to radio receivers of the type known to be used by the CTs. The Special Branch ensured some radio sets bugged in this way were made available at attractively cheap prices to Chinese shops identified as covertly supplying goods to the CTs. When they operated the radio, it transmitted a signal, allowing spotter aircraft flying overhead with a receiver to fix the location of the CT camp.'[32]
- Insurgents will seek additional medical supplies in anticipation of an operation (expecting casualties) or immediately afterward to replenish stocks consumed in the treatment of wounded. After engagements, the insurgents will recruit replacements and may also seek to temporarily supplement their medical staff with local doctors/nurses; all local medical personnel should be placed under surveillance. If the insurgents maintain pack animals, surveillance of the local veterinarian by police intelligence may be productive. A veterinarian (local or visiting) may also be a source of medical supplies, to include animal medications that are also common to human treatment.
- Fabrication or stockpiling of coffins in a village may presage pending insurgent operations and insurgent affiliations.
- If the guerilla/partisans are under pressure, they may decide to move their base camp, some of their caches and personnel under medical treatment to alternate sites. To do this in a timely manner, the insurgents will require transport (pack animals, carts, etc.). Some of these assets may be furnished by local supporters, if insurgent organic assets are inadequate.
- Guerilla/partisan main force units may seek to supplement their food supplies by growing their own crops in open areas. These areas may be detected by aerial assets if the insurgents are amateurs. Note that insurgents will normally not grow crops in uniform rows and may mix crops to blend in with existing foliage that will not be readily apparent from the air. The enemy will also plant on a schedule so that their crops will ripen and be available for harvesting in a staggered timetable. If these crops can be detected, the SR Team should easily find trails leading to caches and to base areas.
- Insurgents may occupy villages and intermingle with local civilians, using them as human shields. This may be done for both propaganda/PSYOPS purposes and for tactical reasons. When friendly forces arrive on the scene, the insurgents may ambush infantry/security elements, but they will then withdraw before they can be encircled. The SR Team should be waiting on insurgent routes of withdrawal.

---

32. Leon Comber, *Malaya's Secret Police 1945–60: The Role of the Malayan Special Branch in the Malayan Emergency* (Institute of South East Asian Studies: Singapore, 2008), 86.

'Today American military officers frequently study CORDS as a model for effective coordination of military and non-military counterinsurgency programs. There is some sentiment for implementing a similar reorganization in Iraq and Afghanistan, where interagency operations have been fraught with troubles.… As of the writing of this chapter [2007], the White House has not attempted to implement a major reorganization to increase collaboration among the US agencies that operate in Iraq and Afghanistan.'[33]

'On both the American and South Vietnamese sides, higher headquarters often insisted on taking custody of prisoners captured in the field, in order to extract strategic intelligence. In almost all cases where the prisoners provided tactically useful information to upper level interrogators the information was never passed down until its utility had expired.… The Phoenix program was, first and foremost, an attempt to achieve what in the twenty-first century is one of the most desired and most difficult objectives for the US government: systematic sharing of intelligence information among intelligence agencies.'[34]

'Counter-Terror Teams [later renamed Provincial Reconnaissance Units], which the CIA controlled completely, … copied Viet Cong methods. These small, elite groups of North and South Vietnamese men collected intelligence on the VCI [Viet Cong Infrastructure] and other VC, and then captured or killed them, usually at night.… Another program, known as the Kit Carson Scouts and run largely by the US military, organized VC defectors into small, elite units. Most often, these units performed reconnaissance missions or guided US military units.…'[35]

'The most effective Allied forces in the village war were the Provincial Reconnaissance Units. Some of the misconceptions about the Phoenix program have arisen from the mistaking of PRUs for Phoenix.… the PRUs were a highly secret paramilitary organization [of up to 6,000 personnel] that operated in dangerous areas and at night more often than most other South Vietnamese forces. Most PRU members served in their native areas and thus had familiarity with and contacts in their operational regions. Because of their tactical prowess, and their success in amassing intelligence, they typically dealt heavy losses on the enemy.… They … did inflict remarkable damage, capturing or killing between eight thousand and fifteen thousand Communists nationwide per year.…

'Although nominally under the authority of South Vietnamese officials, the PRUs were in fact completely run by the CIA.

'The South Vietnamese men selected for PRUs generally had a clear hatred of the Communists: many had relatives who had been killed by Viet Cong or the North Vietnamese Army. Some were former Viet Cong – Communist defectors were some of the toughest anti-communists in the conflict.… The most important reason for the superb performance of the PRUs, … was not the background of the rank and file but the quality of the leaders [who were selected by CIA cadre].'[36]

'The PRUs were among the few Allied forces that regularly operated at night and in VC-controlled territory.… The PRUs infiltrated into hamlets in small numbers

---

33. Mark Moyer, *Phoenix and the Birds of Prey*, University of Nebraska Press, Lincoln & London, 1997, p. 370.
34. Ibid, p. 374-8.
35. Moyer, *Phoenix and the Birds of Prey*, p. 38
36. Ibid, p. 384-90.

and captured or killed VC, set up small ambushes across the countryside, and swept through hamlets to find hidden VC. A wealth of accurate intelligence allowed them to surprise the enemy again and again. Operating in areas known to be clear of other friendly forces, the PRUs sometimes dressed like the enemy and carried the enemy's weapon, the AK-47.'[37]

- The PRU kill-to-capture ratio was 2:1
- PRU generated actionable tactical intelligence. PRU members had family connections who also provided actionable intelligence. PRU did not share intelligence with Phoenix centers because they knew the centers were infiltrated with enemy agents. Direct coordination between intelligence (CIA) and action units (PRU), shortened reaction time to actionable intelligence. PRU also acted unilaterally on self-generated intelligence.
- One key to PRU success was the recruitment of indigenous personnel who 'hated' the enemy. Another key was the recruitment of former insurgents (ralliers), who knew enemy TTPs well. Use of former enemy troops as an indigenous team member recruitment source should not be ignored. Recruitment inducements:
  - Higher pay
  - Better living conditions (quarters, food)
  - US medical treatment
  - Hatred of the enemy
  - Bonuses and bounties
  - Better leadership
  - Esprit
  - Bonding
  - Preferred weapons/equipment
  - Level of support
  - Operational independence
- During COIN operations, the Team may encounter a solitary rural civilian, farmer, shepherd, fisherman, hunter, trapper, etc. who may also be an enemy insurgent, guerilla, partisan or sympathizer – or he may not be aligned with the enemy at all. The civilian must decide to run, hide or fight when he encounters a Team. The T/L must decide whether to kill, contact or capture the individual. Some guidelines: upon the encounter, observe his behavior/actions, to include where he glances initially; this may be a tip-off as to an intention to run or hide, or the presence of a weapon or a companion. Capture, search and question this individual immediately for essential elements of information, to include:
  - Personal information.
  - Location of mines/booby-traps. Such devices are present for a reason; find out why.
  - Location and activities of enemy units and enemy cache locations.
  - Location and identity of enemy auxiliaries/sympathizers and/or agents.
- Once spotted by a local, the T/L has no choice but to assume compromise of the Team, and possibly its mission. The situation may be salvaged if the T/L can employ deception to mislead the enemy or other measures to achieve an operational advantage.
- If the local cooperates and provides the information sought, persuade the individual to act as a guide. He may not be able to read a map or use a compass, and may not know much about military equipment, so ask questions accordingly. Example: 'This trail crosses a stream within the next kilometer, are there any intersecting or hidden trails that lead to an enemy

---

37. Ibid, p. 170

encampment, prior to the stream crossing? After the stream crossing?' Employ him as a guide, but watch him closely for deceptive behavior.

- If he declines to act as a guide, he may have lied in response to the questions and may even be trying to place the Team in jeopardy. Consider taking this individual a prisoner and evacuate him to higher headquarters.
- If more than one individual is captured, the prisoners will not be persuaded to cooperate in front of fellow prisoners. Consider treating them as POWs and evacuate them to higher headquarters.
- If the individual is cooperative, he may have potential as a friendly source/agent; if so, request disposition instructions of higher headquarters. If his activities or behavior are suspicious in any way, or if he is uncooperative/ambivalent, evacuate him to higher headquarters, covertly if possible. Regardless of his allegiance, he may well possess valuable intelligence. Alternate courses of action may include:
  - Drop off a sniper/observer team to hide and observe the behavior of locals who may have detected the Team. Once the Team is out of sight, these 'civilians' may recover hidden weapons to undertake operations against the Team, may try to track and observe the Team, may use a communications device or make a dash to spread the alarm and report the Team location. Assuming that RoEs are permissive, kill these individuals – silently if possible. If Team Members are using enemy weapons and munitions, blame for the killings may be assigned to enemy troops.
  - If individual(s) are captured but are subsequently to be released, make a show of [false] Team intentions. Covertly observe his/their behavior and/or activities (e.g. using a sniper/ observer team) as the Team sets out on a false route. If his/their behavior denotes hostile intentions (e.g. running to report the Team), kill him/them (RoE permitting).
- Of the five primary missions of US Army Special Forces, all five relate to some aspect of Counter-Guerrilla Warfare. Much of the US military Unconventional Warfare expertise resides within the Special Forces domain; this special knowledge of UW operations also implies expertise in countering enemy Unconventional Warfare operations and insurgencies, as the US Special Forces Foreign Internal Defense/Counter Insurgency (FID/COIN) Special Forces mission area reflects. As terrorism is a primary tool of Guerrilla and Underground operations, the USSF Counter-Terrorism (CT) mission again has a UW linkage. Further, Special Forces Direct Action operations (independent or in coordination with other SF or conventional unit operations) may be targeted against enemy UW or insurgent forces. All of these mission areas are supported by or executed in conjunction with SR operations. In Counter-Insurgency operations, reconnaissance patrols are tasked to 'Find', 'Fix' and 'Destroy' guerrilla units.

> 'There are two types of offensive operations employed against insurgent forces. The first is at the local level where US forces (SOF or trainers) work with local authorities to find, fix, and destroy local insurgents who seek to exert control in the communities, cities, and regions. These forces are normally small but well armed. Examples of this type of insurgent force include the Viet Cong in South Vietnam, the FMLN in El Salvador, and al Qaeda in Afghanistan and Chechnya. They move freely within the population and use raids, ambushes, and small hit-and-run attacks intended to drive out occupation forces or destabilize established authorities. The second type of offensive operation is conducted by regular army formations....'[38]

---

38. Field Manual 3-07.22 'Counter Insurgency Operations', Headquarters, Department of the Army. Washington, DC, 1 October 2004, Chapter 3, Section III, Paragraph 3-38.

'The greatest promise of success lies in carrying the fight against partisans beyond the immediate vicinity of threatened supply lines and right up to the enemy's strongholds and rallying points. Careful reconnaissance is a paramount requirement for such operations.'[39]

German Anti-Guerrilla Operations in the Balkans (1941–1944)

'Abwehr and OKH ... shared the services of a number of special purpose units of highly diverse organization, equipment, and function. Most notable was the Brandenburg Division, specially established for long-range penetration, sabotage, and antipartisan warfare. In the latter role, the Brandenburgers formed cadres for the Jagdkommandos (ranger detachments) which, after 1943, operated against the partisans in the Balkans and Russia.'[40]

'A highly effective offensive weapon was found in the Jagdkommando (ranger detachment), designed to seek out and destroy guerrilla bands.... Physically hearty and trained to live in the open for extended periods of time, they depended little on supply columns and could pursue the guerrillas, often burdened down with wounded, families and impediments, into the most inaccessible areas. When the situation required, the Rangers would put on civilian clothing, disguising themselves as Chetniks or Partisans, to work their way closer to their wary enemy. In the event they came upon major guerrilla forces, the Ranger detachments ... would keep them under observation and inform Battalion or other higher headquarters. While awaiting reinforcements, they would attempt to gather additional information on the guerrilla strength and dispositions.'[41]

Counter-Guerilla Operations During the US Civil War

'Major General John M. Palmer, Union commander in Kentucky, baffled by the activities of Confederate sympathizers in keeping the guerrillas informed of his troop movements, organized a guerrilla command of his own. He commissioned one Edwin Terrill, leader of a small Union guerrilla band in Spencer County, to undertake the pursuit and capture or destruction of Quantrill. Terrill, a deserter from the Confederate army, acted promptly and made contact with Quantrill on April 13, 1865. He never lost contact, pursuing and harassing the Missouri gang without cessation.

Finally, on the morning of May 10, Terrill caught up with Quantrill resting on a Spencer county farm. The usually wary Missourians were taken completely by surprise and several were killed, while Quantrill was shot in the spine and partially paralyzed. He died in a military prison at Louisville on June 6.'[42]

'Of all the counterguerrilla devices attempted during the war, only those proved effective which met the guerrilla bands with well-trained, disciplined, and hardy troops. These succeeded only when they maintained unrelenting pressure on the guerrillas'[43]

---

39. Unnamed authors and contributors, Department of the Army Pamphlet No. 20 – 240, 'Historical Study, Rear Area Security in Russia, The Soviet Second Front behind the German Lines', Department of the Army 25, DC, 31 July 1951, p. 35.
40. Gunter E. Rothenberg, Isolating the Guerrilla, 'The German Experience in World War II'. Historical Evaluation and Research Organization, 2233 Wisconsin Avenue, N.W., Washington, D. c., 20007, 1 February 1966, p. 172
41. Department of the Army Pamphlet No. 20 – 243, 'Historical Study, German Anti-Guerrilla', p. 48
42. Marshall Andrews, Isolating the Guerrilla, 'The American Civil War', Historical Evaluation and Research Organization, 2233 Wisconsin Avenue, N.W., Washington, DC, 20007, 1 February 1966, p. 41
43. Ibid., p 49

The Emergence of Modern Era US Counter Guerilla Doctrine

'Prerequisites for Successful Guerrilla Operations

- Civilian Support
- External Support
- Favorable Terrain
- Effective Leadership
- Unity of Effort
- Discipline
- Use of Propaganda
- Intelligence Effort
- The Will to Resist

The subversion, destruction, or denial to a guerrilla force of any one or more of these prerequisites will hit at the very core of the organization and make the eventual destruction of the force an easier or perhaps unnecessary task.'[44]

- Destruction of a guerrilla insurgency requires a multifaceted approach and an in-depth treatment of such subject matter goes mostly beyond the scope of this book. However, SR, employed in the COIN mission area, is essential to destroying a guerrilla insurgency.

'Points of greatest guerrilla force vulnerability are –

- Support of the civilian population
- Food and medical supply
- Command structure
- Morale
- Arms and ammunition supply'[45]

'Harassment of the guerrilla force, primarily by ground patrols, aerial 'hunter-killer' teams, and aerial surveillance is initiated against the guerrilla force to locate it and keep it under pressure. When an enemy element is located, an offensive reaction with adequate combat power is initiated without delay to destroy it.'[46]

'Once the area under the control of the guerrilla force has been definitely determined, harassment operations are restricted to this area. They are conducted primarily by the use of –

1. Reconnaissance patrols … to locate guerrilla units and bases.
2. Combat patrols … and raids … against known and suspected enemy bases, installations, patrols, and outposts.
3. Aerial 'hunter-killer' teams….
4. Ambushes….
5. Marking targets.
6. Mining probable guerrilla routes of communication.
7. Continuous aerial surveillance.'[47]

---

44. FM 31-16, 'Counterguerilla Operations', HQ Department of the Army, Washington 25, DC 19 February 1963, pp 9-11.
45. Ibid. p. 21
46. Ibid. p. 31
47. Ibid. p. 49

- Note: Tasks 1 – 2 and 4 – 6 above are well suited to the capabilities of a properly trained and equipped SR Team.

> 'These harassing operations are conducted day and night. Operations at night are directed to the movement of the guerrillas moving about on tactical and administrative missions. Operations during the day are directed primarily at guerrillas in their encampments while resting, regrouping, or training.
>   'When a guerrilla element is located during harassing operations, the friendly force making contact with it engages the enemy and destroys it if it has sufficient combat power…. Often immediate reaction to hastily discovered guerrilla forces will consist primarily of a pursuit. In such cases, efforts are made to envelop the enemy force and cut it off from the rear. Once the escape of the guerrilla force has been blocked, it may be destroyed by the pursuing force.'[48]

United States Special Forces Mobile Guerrilla Force (MGF) Operations in South Vietnam:

- Organized in 1966 and led by Special Forces personnel, four experimental Mobile Guerrilla Force units were organized, trained, and equipped to operate independently in remote areas of Vietnam considered to be Viet Cong base areas. The MGF was essentially a company-sized force that operated against an enemy guerrilla force (Viet Cong unit) deep in enemy controlled territory. The MGF concept was adapted from Lessons-Learned derived from French 'Groupement de Commandos Mixtes Aéroportés (GCMA)' operations during the French-Indochina War:[49] with the GCMA as the French 'equivalent of the US Special Forces in Vietnam with a similar mission running SOG type operations…',[50] and corresponded to WWII Jedburgh operations; and British Special Air Service (SAS) operations during the Malaysian insurgency of the 1950s. The MGF would deploy into an enemy base area for a period of 30–60 days to collect intelligence, to seek out and destroy enemy units, base camps and logistics storage sites and to interdict enemy Lines of Communication (LoCs). Each company-sized MGF consisted of a 12-man Special Forces 'A' Detachment and 149 indigenous troops organized into an headquarters element, three small infantry platoons and an organic reconnaissance platoon (in lieu of a weapons platoon).
- The core of the operational concept, and the key to the tactical success of the MGF, was to leverage the more sophisticated training, patrolling skills and capabilities of the SF-led reconnaissance platoon to recon and secure each successive MGF operations base camp and resupply DZ, to patrol in advance of the MGF Line-of-March, and to conduct reconnaissance of the enemy. The reconnaissance patrol would locate and conduct surveillance of enemy infrastructure or personnel for rapid maneuver and tactical engagement by the follow-on MGF infantry platoons, supported by tactical air assets.

MGF Operations TTPs:
The best force mix, or combination of combat units, to harass and destroy a guerrilla force, may include:

- A Mobile Guerrilla Force (MGF) similar to the Vietnam-era SF MGF (see above), comprised of a SF-lead company of native light infantry and at least two SF-lead Recon Teams (see note below), incorporating or supplemented with (if available).

---

48. FM 31-16, 'Counterguerilla Operations', p. 51
49. 'French Indochina, The First Vietnam War', http://www.gia-vuc.com/gcma%20history.htm
50. 'French Indochina'. http://www.gia-vuc.com/frenchindochina.htm

- ○ Trained, experienced trackers. Enables the Recon Team to better interpret signs of the enemy and to track enemy elements over varying terrain.
- ○ Dog teams. Enhanced dog hearing and smell will alert the Team to the presence of enemy combatant base camps, patrols, ambushes.
- ○ Guerrilla defectors as indigenous guides or scout elements. Especially valuable if the defector was formerly a member of an elite enemy unit or if the defector is locally recruited and familiar with the area. These assets can be outfitted with the uniform and equipment of enemy combatants and used as 'road-runners' – to openly travel on enemy trails to pinpoint enemy encampments, positions, etc.
- ○ Sniper teams, if terrain, vegetation and other conditions permit.
- ○ Interpreters.

Note: Two Recon Teams are recommended, as it is optimum to rotate RTs due to mental and physical exhaustion associated with the multiplicity of assigned tasks and the stress of continual concentration/focus. If a single RT is employed and is mauled in an engagement, the effectiveness of the entire MGF may be impaired.

- It is optimum to have a second MGF, as supplemented, in reserve, to rotate with the initial MGF, to maintain pressure on the enemy. The MGF is expected to pursue the enemy as rapidly as is prudent, to exhaust the enemy. In this process, the initial MGF itself will become exhausted.
- The MGF should have:
  - ○ Priority or dedicated close air support.
  - ○ Air transport and aerial reconnaissance.
  - ○ Provision of vehicles and/or pack animals, if possible. The MGF, and especially its RTs, must have mobility that is equal or superior to that of the enemy. This means that RTs, in particular, must be minimally burdened and that part of their field load may have to be transported by supplementary means.
  - ○ On-call airmobile light infantry reaction/response force. Used to (primarily) establish blocking forces (e.g. Hammer and Anvil operations) and as an assault force if the enemy units are compelled to defend from static positions.
- The MGF RTs have many duties (see above), all of them vital to the MGF, but the primary focus is to find an enemy force and to attain and maintain contact with the enemy force and to 'fix' the enemy unit/element for the remaining MGF elements to engage and destroy.
  - ○ 'Maintaining contact' may be achieved by following, but not engaging, enemy elements; or it may be achieved by engaging the enemy in running gunfights.
  - ○ 'Fixing' the enemy involves forcing the enemy to stand and fight; this may be achieved if the enemy force is exhausted, burdened with WIA or if restricted by obstacles.
  - ○ At some point, an enemy unit will split into smaller elements to throw off a pursuit, or to mitigate the threat to the main force. Teams may split to follow these enemy elements. Note: The enemy unit may split into separate elements, so that one element may lead a pursuing Recon Team into an ambush by another element.
  - ○ Because RT personnel are 'stripped down' to lighten their field burden (to include rations), rotation of Recon Teams, or operating in split Team elements may be necessary.
- The RT exploits the weaknesses of the enemy guerrilla force by:
  - ○ Applying unrelenting pressure on the guerrilla force and destroying the morale of the enemy combatants.
  - ○ Pursuing the enemy to drive them away from support of civil populations and depriving them of their access to local food, logistics and intelligence support.
  - ○ Finding and destroying base area infrastructure and logistics stores of medical, armament and food supplies. During an active pursuit, the Team should not pause to find/

destroy infrastructure or stores; this can be left to MGF infantry elements or follow-on exploitation forces.

- ○ Inflicting casualties on the enemy to harass him to exhaustion or to 'fix' the enemy for elimination in Hammer and Anvil operations or by encirclement.
- ○ Rapidly exploit information provided in the field interrogation of captured, surrendered or deserter combatants who plausibly offer voluntary compliance or compliance in return for food, medical treatment, etc. – prior to evacuation to the MGF or higher HQ. This may lead to discovery of logistics caches, base areas, booby-trapped/mined areas, the presence of other enemy units, location of ambushes, etc. of immediate value to the Team.
- If the SR Team is not operating within the context of a MGF, many of the points above remain valid, but situational adaptation will generally be required.

> 'General Bell felt that the only way he could terminate the insurrection in the region under his command was by cutting off the income and the supplies of the insurgents and at the same time pursuing them with sufficient persistence and vigor to wear them out…. Most importantly, it was absolutely essential to make it impossible for the insurgents to procure food by forced contributions….
>
> 'Bell continued to pursue them persistently, not waiting for them to come out of hiding, penetrating into every mountain range, and searching every ravine and every mountain top. The American forces continually found their barracks and hidden food in the most unexpected and remote hiding places. They burned hundreds of small barracks and shelters as fast as the insurgents would build them. They destroyed their clothing and supplies. Finally, the guerrillas ceased to stay in one spot for longer than 24 hours. They were on the run.
>
> 'Bell maintained as many as 4,000 troops in the field at one time, keeping them supplied in the mountains even where roads did not exist. They camped by companies at strategic points on trails, each sending three or four detachments with five or six-days' rations to bivouac at points radiating several miles from the company base. The detachments would leave their rations in charge of one or two men and search and scour the mountains both day and night. In this manner, it was rendered unsafe for the insurgents to travel at any time, and, no longer having any retreat in which to hide themselves, they became so scattered and demoralized that they were constantly being captured and surrendering in large numbers.'[51]

General Ambush TTPs:

- Definition: An ambush is a surprise attack from a concealed position on a moving or temporarily halted target to harass, interdict or destroy the target or to temporarily seize the target in order to secure information, confuse an adversary, capture personnel or equipment, or to destroy a capability, culminating with a planned withdrawal.
- Team Members will seldom know what targets will present themselves for a deliberate or opportunistic (hasty) ambush. Targets passing on a road or high-speed trails could range from a dismounted single soldier to large infantry units; from a relatively soft logistics element to a mounted combat unit (infantry/armor) or to a nuclear weapons convoy with heavy security.

---

51 Linnea P. Raine, Isolating the Guerrilla, 'The Philippine Insurrection, November 1899 – July 1902', Historical Evaluation and Research Organization, 2233 Wisconsin Avenue, N.W., Washington, DC, 20007, 1 February 1966, pp 103 – 107. Regarding the operations conducted by Brig. Gen. J. Franklin Bell in 1902.

- ◦ If the Team is on a deep penetration mission with scant, non-existent or non-responsive fire support, the Team should favor the 'Far/Remote Ambush' techniques if its aim is to harass or interdict the enemy – or if the Team wants to enhance its survivability.
- ◦ If its aim is to collect intelligence (e.g. capture documents, materiel, prisoners) then it must favor the 'Near Ambush'. In this context, patience in selecting the proper target is essential.
- ◦ In all cases, the Team must plan, and take precautions, in the event of discovery/compromise while in the ambush site. These steps include evasion planning/withdrawal route establishment; use of decoys/deception; flank protection; use of terrain/obstacles, mines/booby-traps; and denial of high-speed routes of approach. In denying routes of approach/flanks, a single mine/booby-trap may not be enough to deter a combat element that is willing to endure casualties – unless these devices are supplemented by obstacles/terrain form.
- • Easy Target: Kill the driver of a moving vehicle (single/convoy), firing (e.g. a Claymore) from his left flank (assumes he is operating a left-side driven vehicle); this will cause him to fall away to his right, his grip on the steering wheel will then turn the vehicle sharply to the right. This will cause the vehicle to flip or to run off the road; even better, if the road has a drop-off on the right – resulting in a catastrophic accident for crew, passengers and cargo. The same effect can be achieved by shooting the driver of a tracked vehicle (when he is not buttoned up); but he may veer in an unpredictable direction from the direction of fire, depending on the steering mechanism used in the vehicle.
- • Convoy Ambush:
  - ◦ Once the US began using helicopter gunships to provide rapid response to NVA ambushes on convoys during the Vietnam conflict, the enemy learned to conduct their daylight ambushes during overcast/bad weather and/or to limit ambushes to 10 minutes duration. An SR Team might follow suit in future conflicts.
  - ◦ As convoys will normally have security escort vehicles, the Team should normally employ far ambush TTPs during <u>interdiction</u> operations to avoid Team casualties.
  - ◦ If the enemy uses quick reaction gunships with night-vision/thermal optics, the Team should use far ambush TTPs using remotely fired weapons/ordnance. The Team should be located in field fortifications with overhead cover and take all measures to suppress IR/thermal signature. If the Team is to conduct ambushes during daylight in sparsely vegetated terrain, the same measures should apply to prevent aerial detection of the Team. If the enemy gunships do not employ night-vision/thermal optics, execute the ambush at dusk. Onset of darkness will cover Team withdrawal.
  - ◦ A convoy or armored escort driver, and vehicle occupants might not hear brief duration small-arms fire over the sound of diesel engines, especially if the shots occur well behind or advance of the vehicle. Of course, shots passing close by will be heard, as the rounds crack the sound barrier; but the origin of the shots may not be evident, unless the shooter makes an error (poor camouflage, dust, muzzle-flash).
  - ◦ The best location to initiate a convoy ambush is on a grade at a hairpin turn. Heavily laden trucks may reduce speed to as low as 4mph at the turn going up/down hill. If the road is heavily pot-holed, heavily laden cargo trucks will slow significantly. And slow, heavily laden trucks on steep inclines may actually have to be pushed or towed by more powerful vehicles.
  - ◦ Convoy trucks may attempt to drive through the kill zone, while escorts engage. If enemy doctrine/practice follows this pattern, consider using two engagements. (1) engage with remote devices (decoy) to distract the escort; (2) second ambush further up the road, to engage the cargo vehicles. The enemy may not dismount to perform counter-ambush TTPs unless the vehicle is disabled or the convoy is stopped.

- ○ Enemy units in convoy, during the early stages of a conflict, or deep in secured areas may not have proper security measures or counter-ambush TTPs in place. A skilled Team will be very effective in this circumstance. This was demonstrated on 23 March 2003 near An Nasariyah, Iraq where the 507th Maintenance Company was caught in a hasty ambush resulting in the loss of eleven US soldiers killed, nine were wounded and seven were captured. Vietnam-era convoy operations Lessons-Learned were clearly forgotten.
- ○ Convoy commanders are sometimes placed well back in the convoy so that they may coordinate the response to an ambush. If the convoy commander can be identified, he should be a priority target, as he will control the convoy's communication link to aviation quick response. Convoy cargo vehicles are unlikely to have radios, but security vehicles probably will.
- ○ The Team should avoid using the same ambush locations repeatedly. The enemy could mine/booby-trap the area or he may plan artillery RPs on previous ambush sites.
- ○ Dress in enemy uniforms/equipment for near ambushes. On several occasions during the Vietnam conflict, NVA donned Vietnamese army uniforms and walked along the sides of roads, attacking in close combat when convoy vehicles passed.
- ○ When an enemy convoy approaches its destination, soldiers gain a sense of security and become lax – a good time to hit the convoy with a remote device.
- ○ Disabled vehicles left unattended should be booby-trapped.
- ○ During night ambush operations, limit using individual weapons as the muzzle flashes will attract enemy counter fire. Use remotely detonated devices, mortar fire, 40mm instead.
- ○ Smoke and/or CS will obscure the Team from enemy counter fire and should be employed during withdrawal.
- ○ Along with security vehicles, fuel trucks are to be considered a high priority convoy target, as leaking and ignited fuel will often envelope other vehicles and the flames will block the road.
- Use embankments, paddies, dikes, hedgerows to trap or channelize an enemy unit.
- The enemy will, at some point, converge on an attack/ambush site to rescue wounded and retrieve KIAs, to launch a pursuit, to retrieve materiel, to clear a choke point, to recover transport, to repair transportation infrastructure, to extinguish fires, etc. The T/L should take opportunities to attack such secondary targets, unless the Team is under pressure to rapidly leave the site of the initial/primary attack.
- ○ The Team/element must have observation of the primary attack site.
- ○ Ensure the Team always has remotely detonated devices (e.g. demolition charges, mines, booby-traps) that are rigged for rapid deployment. Each device should have a SD capability.
- ○ The Team can attack secondary targets using remotely detonated devices, using far ambush; by direct fire (line-of-site permitting); by sniper fire; by indirect fire (e.g. mortar) and by CAS.
- ○ After the initial ambush, the Team might relocate to a secondary ambush position to ambush rescue/recovery efforts.
- ○ Always assume that the enemy has tracker dogs, which can be deployed in pursuit of the Team. The dogs and/or handlers should be priority targets, if they arrive on the scene. The Team must always use a scent obscurant/deterrent (e.g. CS/capsicum powder, animal lures, etc.) not only at the Team's departure point, but also at other points along the route of withdrawal – as the enemy may attempt to cut across the route of withdrawal with tracker dogs if they have 'wised up' to scent obscurant/deterrent practices.
- An SR Team may conduct an ambush to:
- ○ Interdict high-value and/or opportunistic targets.
- ○ Execute a battle drill (hasty ambush) in reaction to a chance contact.

- ○ Take prisoners or obtain materials of intelligence value.
  - ○ Eliminate trackers or pursuing enemy forces.
  - ○ Provide security while other Team Members perform other tasks.
  - ○ Obtain supplies to continue deep operations
- Deliberate vs Hasty Ambushes
  - ○ Deliberate ambushes are planned. Hasty ambushes are executed as a battle drill against opportunistic targets or chance encounters.
  - ○ If the Team has sufficient time, it can prepare dummy positions that are not too well concealed. The dummy position should also have remotely controlled remotely initiated devices (i.e. grenades), that will attract the attention and the fires of an enemy, causing battle drill deployment. The actual ambush may occur on the enemy flank as the enemy turns and maneuvers toward the perceived threat.

> In July 1942, the 19th Panzer Division attacked Russian positions in the vicinity of Nikitskoye. The Russians established a line of anti-tank guns, 'emplaced in pairs for mutual support, dug in so that the muzzles were just above the surface of the ground. Between each pair of guns was an additional antitank gun mounted on a two-wheeled farm cart. The cart-mounted guns were camouflaged, but no effort has been made to conceal them.
>
> As the German tanks advanced, the dug-in guns fired a volley, then ceased…. the Germans noticed the guns mounted on carts, and moved toward the newly discovered targets. As soon as a German tank turned to bring the cart-mounted guns under fire, it was hit from the side by Russian antitank fire from the concealed positions.'[52]

  - ○ Deliberate ambush planning considerations include:
    - ▪ Reconnoiter/surveillance of the ambush site.
    - ▪ Rapid movement to an RP that is outside an enemy sweep zone. If the enemy does not conduct sweeps near the ambush site, the Team may move directly to the ambush position.
    - ▪ Security or not? Security can be provided by security elements or by use of sensors/cameras or by use of mines/booby-traps. Security can also be provided by an element of the ambush formation (e.g. 'Z' ambush).
    - ▪ Assault or not? Assault will be necessary if the purpose of the ambush is to take prisoners or enemy materials of intelligence value. If the purpose of the ambush is destruction (e.g. WMD convoy) or harassment, then assault by an SR Team wouldn't be necessary. If the Team establishes the ambush to take on opportunistic targets (e.g. convoy, lone courier, etc.), the Team should have a designated assault element that can launch at the discretion of the T/L. Note that an assault bears increased risk to Team Members.
    - ▪ Time-on-Target (ToT). The Team may not know if other enemy elements are nearby. The ambush (and assault, if employed), must be executed rapidly and the withdrawal executed promptly.
    - ▪ Withdrawal routes (primary and alternate) to an RP or an extraction LZ.
    - ▪ Coordination with air assets, including CAS and exfiltration capabilities.
    - Tasks:
      - * Site selection
      - * Mining and booby-trap prior to execution and afterwards to deter pursuit
      - * Formation selection

---

52. Halder, *Small Unit Actions*, p. 111

  * Assault assignments
  * POW capture and handling
  * Search of enemy WIA/KIAs, vehicles, cargo, etc
  * Demolition/Destruction
- Ambushes are classified by category, type and formation.
  ◦ Categories: deliberate or hasty
  ◦ Type: Near, far, point or area
  ◦ Ambush Formations: Point, Linear, 'L', 'V', 'Y', 'Z'

Deliberate Near Ambush TTPs:

- A Near Ambush is often defined as occurring within grenade throwing range of an enemy (FM 7-92), but may be further (governed by terrain and vegetation).
- The Team may expect that an enemy will react to a near ambush in much the same manner as US forces do. Some foreign doctrine may prescribe immediate flanking by soldiers not in the kill zone. Team Members should seek out enemy doctrine via the FOB S-2 to understand how they will react to a Team Ambush, so that the Team can train and execute to counter such reaction.
- The surprise and shock of a well executed ambush will often cause an enemy unit to panic, to hesitate in executing its battle drill and to defer/deter pursuit. This outcome is more likely if the Team presence in the area is not known by the enemy, if the enemy unit is comprised of inexperienced or non-combat troops, if the ambush occurs in areas deemed safe zones by the enemy (rear areas), and/or if the ambush takes out the leadership of the enemy unit.

True Account: An experienced SOG T/L established a deliberate ambush site on an enemy MSR in southeastern Laos, in an area well concealed from aerial detection by interwoven canopy. For reasons of site selection (terrain/vegetation), the ambush site was positioned on the opposite side of the road from the Team line of withdrawal to an extraction LZ. The T/L was hoping to ambush a small element of enemy troops or even perhaps a solitary courier/combatant. After only about an hour's wait, a column of enemy infantry troops appeared on the road, traveling southeasterly in platoon strength. This unit was followed by a stream of other elements. The enemy column, now estimated as a battalion of infantry, stopped to rest and eat its noon meal. By stroke of fortune, when the column halted, the battalion headquarters stopped directly in the Team's kill zone. Enemy troops moved to the verge of the road to establish cooking fires and began gathering wood, some troops approached very close to the Team position; so close, in fact, that one of the Team indigenous commandos, convinced that he had been spotted, detonated his Claymore. The Team immediately engaged with small arms and detonated its remaining deployed Claymores. Two Claymores cut down enemy troops in the kill zone; flank security Claymores cut down enemy troops to the right and left of the Team ambush position. As the Team quickly crossed the road to head toward its extraction LZ, Team Members could see nothing but bodies on the road and along the verge; the rest of the enemy battalion had disappeared. It wasn't for another 15 minutes until the Team heard enemy weapons fire. The Team was extracted with no casualties.

- The Deliberate Near Ambush uses any ambush formation, except perhaps the 'V' formation, at the discretion of the T/L.

Hasty Near/Far Ambush:

- The Hasty Ambush is opportunistic and unplanned. Hasty ambush planning considerations include:

- ○ Execution as a Battle Drill normally performed during movement. Typically, the Hasty Ambush is a simple formation closely associated with a meeting engagement battle drill.
- ○ Seldom involves deliberate site selection.
- ○ Opportunistic upon sudden discovery of enemy infrastructure, roads, trails, enemy activity or upon a chance meeting. If time and circumstances allow, the Hasty Ambush may evolve to a Deliberate Ambush.
- ○ Team has not been detected. Team may or may not have detected an approaching enemy.
- ○ Mining and booby-traps may be incorporated to deter pursuit.
- ○ When the Team encounters a road or high-speed trail during movement, the T/L should consider deploying the Team into a hasty ambush as a precaution and/or in consideration of an incidental opportunity. This provides the T/L better options and places the Team in a better tactical posture in proximity to a linear danger area. The Team must be prepared for enemy combatants to appear suddenly from either direction on a road or high-speed trail and be primed to defeat the target or to capture a prisoner; deploying into a hasty ambush is also recommended if the T/L wants to move forward to reconnoiter/photograph the road/trail. If the hasty ambush site is acceptable, the T/L may then take the necessary steps to plan and prepare a deliberate ambush; if the site is suboptimal, the T/L may withdraw the Team to a position that over-watches the back-trail (fish-hook) and then he may conduct a Leaders Reconnaissance (e.g. to find an ambush or surveillance site nearby), or the T/L may, after consulting his map, lead the Team to move by bounds along the road/trail to find a more optimal ambush site, to collect intelligence or to find enemy unit or installation locations adjacent to the thoroughfare. A high-speed trail may be defined as a well-groomed, well-maintained trail broad enough for two-way or more-than-one combatant abreast and capable of supporting cart/bicycle traffic.

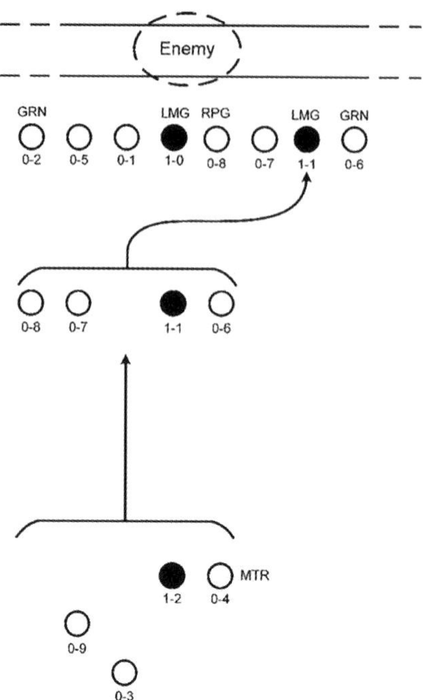

Figure 47. Deploying from Immediate Action Battle Drill Formation to Line/ Hasty Ambush Formation.

Figure 48. High Speed Trail. Not well groomed/maintained. Note bamboo husks.

True Account: A SOG RT with an experienced T/L was assigned a mission into Northeastern Cambodia on the Laotian border. The Team picked up trackers immediately after insertion and spent the next two days trying to lose them in very rough terrain – and was finally able to ambush the enemy squad. But it subsequently became clear that the enemy had mounted an all-out search for the Team, deploying at least a company of troops to sweep the terrain.

The Team evaded the sweeps for the remainder of that day and for much of the following day, but had reached its point of exhaustion. At midday, the T/L established a hasty perimeter in dense brush on a ridge to briefly rest his Team. The Team could hear enemy elements moving down-slope and on adjacent ridges. Within a few minutes, enemy personnel approached the Team position along the ridge. This enemy element was led by an officer/advisor, tall in stature with Chinese features, wearing an unusual uniform – and he was accompanied by a radio man. While the Team observed, the enemy element took a pause, while the officer communicated on the radio. One of the enemy soldiers undid his fly and urinated into a bush where one of the Team's indigenous grenadiers was hiding. The Team Member cocked the hammer on his .45 pistol and the click caught the attention of the enemy soldier. At the point of the pistol discharge, 'all hell broke loose' and the Team cut down the remaining enemy in the immediate proximity. The Team moved rapidly toward an extraction LZ with the enemy in hot pursuit. Extensive CAS was necessary to extract the Team; the Team was fortunate to sustain only two casualties.

- A Hasty Ambush may be either a near or far ambush, depending on terrain, vegetation and other circumstances, that is typically employed as a Battle Drill. The Team will generally deploy into the ambush formation from the march and/or at the discretion of the T/L.
- The Hasty Ambush is usually deployed into a basic line or 'L' formation; but deployment can assume other formations if the Team is trained to do so. Security elements are not automatically/immediately deployed.
- The 'Point' formation may be a Hasty Ambush (or a Far Ambush) that is generally deployed astride or to the near flank of a high-speed route of approach. It would typically be deployed to blunt enemy hot pursuit of a Team withdrawal (or the withdrawal of a guerilla unit from its base). Sniper teams can accomplish this function.
- Deployment into the Hasty Ambush formation should be rapid, silent and flexible enough to address a threat or target approaching from the front or flank of the Team's line of movement. As in all drills, Hasty Ambush deployment must be practiced. Ideally, this formation should be a variant of the Team immediate action/Battle Drill for the sake of simplicity and speed of execution.
- The T/L should consider deploying the Team into a Hasty Ambush formation in the following circumstances:
  - During movement when the Team detects an enemy presence or encounters enemy troops in a chance encounter.
  - During movement when the Team encounters a road or trail, or another danger area (e.g. stream, commo wire, enemy facility). This will set up the Team to respond to happenstance and opportunistic encounters, especially where an enemy target (possible prisoner, courier, command vehicle, etc.) or threat (e.g. patrol) may suddenly appear.
  - During an enemy pursuit/hot pursuit.
  - At the discretion of the T/L, and if the Team is appropriately trained and equipped, the Team may execute a silent or POW ambush.
  - The T/L may transform the Hasty Ambush into a Deliberate Ambush at his discretion and if time and circumstances allow.
- From the Target (Objective) RP, the T/L moves forward to conduct a reconnaissance of the prospective ambush site. An ideal ambush site will be a choke point that is bounded by obstacles/impediments that will keep the enemy combatants in the kill zone and/or restrict his ability to maneuver and conduct his counter-ambush battle drill.
- After the recon, the T/L may deploy a surveillance element or OP forward to monitor enemy activities and routines in the vicinity of the ambush site. If target surveillance is required

for a long period the element personnel might have to be rotated for food and hygiene. The T/L may subsequently conduct a Leaders Reconnaissance, taking element leaders to their planned ambush positions. Note that this movement bears some additional risk of discovery.

Author's Solution:
Instead of establishing a loitering surveillance element or OPs/LPs, consider using electronic surveillance means that can be monitored from the main Team hide location. Micro-cameras are cheap, easily concealed, are available with built-in night-vision capability and can transmit images wirelessly to a tablet IT device.

- Prior to deploying the surveillance element or the ambush elements to their positions, rucksacks are dropped at the release point (which then becomes the primary Rally Point).
- Prior to moving into the ambush position, the surveillance element is recovered and debriefed; the T/L develops and briefs his plan.
- Ambush Roles and Responsibilities:
  - Flank Security: METT-TC, and/or SOP, may dictate whether flank security is deployed; if it is deployed, a grenadier should be included. A Claymore mine should also be deployed to each flank.
  - Anti-armor: If armor is present in the AO, deploy off-route mines, such as Explosively Formed Penetrators (EFPs), AT mines or anti-armor weapons to cover the kill zone, and to the flanks if warranted.
  - Main/Assault Element: If the ambush is to capture a POW or seize intelligence materials, the main element will include an assault element with a minimum of three men.
  - Rear Element: Always anticipate trackers or the discovery of the Team trail.

Point Near/Far Ambush TTPs:

- A Point Ambush may be either a Near or Far Ambush, depending on terrain, vegetation and other circumstances. The ambush element, typically one to three Team Members, should normally position together on the same side of the enemy route of approach to facilitate withdrawal.
- Positioning should also optimize weapons capabilities so that the enemy can be effectively fired upon from as far away as possible; this would suggest that Team Members use a slight rise along the enemy route, maximizing the linear (long axis) aspects of the enemy route as a kill zone. If the Team Members are positioned on a rise along the trail/road, the reverse slope would screen Team Member movements, and allow Team Members to be positioned on both sides of the trail/road.
- The ambush element should briefly engage the enemy pursuit with maximum fires for no more than 30 seconds, and then should speedily withdraw, to either lay a false trail for the enemy to follow, to rejoin the main body, or to set up its next ambush position.
- If time and terrain allows, the element may deploy mines/booby-traps (with SD) in the path of enemy pursuit. Remotely initiated off-route/directional mines (e.g. Claymore) can also constitute a point ambush against the flank of an approaching enemy. In these circumstances (e.g. during a hot pursuit), mines and booby-traps must be pre-rigged and ready for instant deployment.
- If terrain and vegetation accommodates, a sniper team can serve as the ambushing element. Snipers may also be positioned to the flank of an enemy route of approach.

Simply put, a Far Ambush is typically executed outside hand grenade throwing range between combatants, and is often initiated via command detonated mine/charge; or crew-served weapons fire.

Ambush Site Selection TTPs.

Site selection and positioning of personnel will be governed by METT-TC, and the ambush formation selected. Formations may include: Point, Linear, 'L', 'V', 'Z', 'Y'.

- If the direction of enemy travel is known, consider aligning the linear formation or the long axis of the 'L' or 'Z' ambush along the right flank of the enemy. Seventy to ninety per cent of the world population is right-handed and nearly all military weapons are designed for right-handed use. Attacking an enemy from his right flank will provide Team Members a momentary advantage stemming from:
  - Any left-handed enemy combatants who are not well trained may fumble at a right-handed safety under the stress of a surprise attack.
  - The right-handed enemy must simultaneously release his safety, hit the ground/seek cover and turn 90 degrees to the right, in order to bring his weapon to bear. It is more difficult to bear a weapon right, than to bear it left. Coupled with the element of surprise, this buys the Team valuable seconds, where even an elite enemy element may not maximize return fire. During this critical period, the Team must mass their fires, attaining and maintaining fire superiority – even over a numerically superior unit.
- The terrain for the ambush should meet certain specific criteria, but these criteria are for guidance only and can be supplemented or abridged as the situation may warrant:
  - Provide concealed positions to prevent detection of the Team from the ground and air.
  - Enable the Team to deploy, flank and/or divide the enemy.
  - Provide fields of fire from Team positions.
  - Permit emplacement of key or crew-served weapons to provide accurate, sustained fire.
  - Permit the Team to set up observation/surveillance posts for early enemy detection.
  - Permits covered movement of Team Members to ambush positions and from these positions to routes of withdrawal.

Ambush Formations TTPs:

Point Ambush Formation TTPs:

- The point ambush is used by a Team element to harass the enemy, to interdict/delay enemy pursuit or to lure an enemy into a trap. It can be executed by a single Team Member/sniper. The ambush position is best located behind cover at the bend of a road, along a path, back-trail (e.g. during a pursuit) or in the vicinity of the anticipated line of advance of an enemy element.

    'A point ambush involves patrol elements deployed to support the attack of a single killing zone.' – FM 90-8

- The ambush is initiated by weapons fire or by the use of a command-detonated mine (Claymore) or other remotely operated weapon/munition.
- Additionally, mines/booby-traps (e.g. the Pursuit Deterrent Munition) may be seeded along sides of the kill zone, so that further casualties are inflicted as the enemy seeks to take cover or maneuver. Note that mines/booby-traps used in an ambush will generally not be recovered by the Team, so they should always be equipped with a SD capability.
- The Team Member(s) may withdraw and repeat the tactic as frequently as needed. The <u>Area Ambush</u> is a series of Point ambushes. See Point/Area Ambush illustration below.

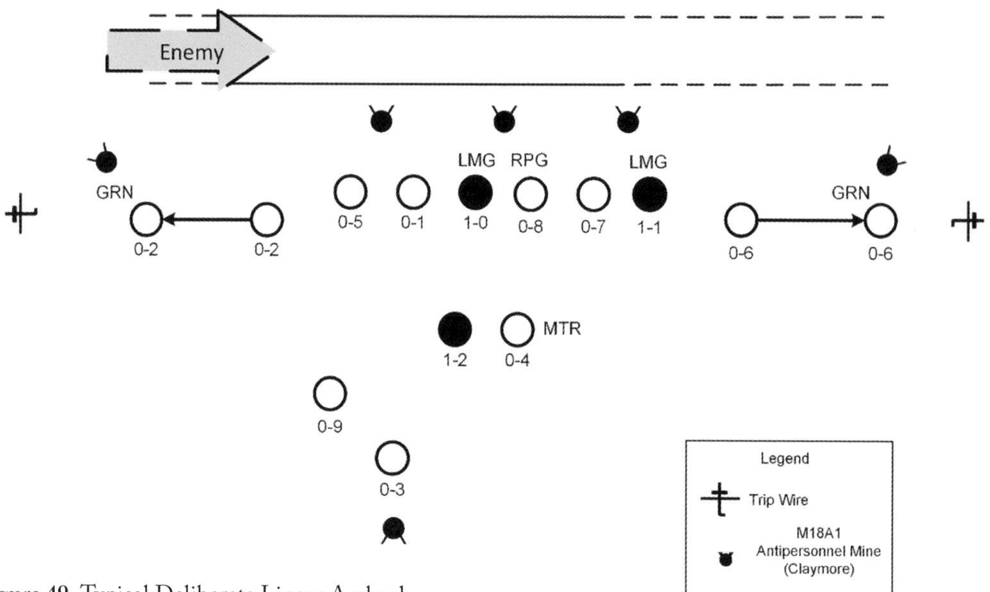

Figure 49. Typical Deliberate Linear Ambush.

Linear Ambush Formation TTPs:

- Deployed as a <u>deliberate</u>/<u>hasty</u> category ambush formation. As a hasty formation, security may not be immediately deployed. The hasty ambush may be later converted to a deliberate formation.
- This ambush is situated parallel along a road, path, back-trail or to the anticipated line of advance of an enemy element, and it allows a maximum amount of weapon fires to be brought to bear on the flank of a linear target.
- Mines/booby-traps may be seeded along both sides of the kill zone, and to the flanks of the ambush position so that once the ambush is initiated, further casualties are inflicted on a fleeing or maneuvering enemy. The ambush is initiated by signal of the T/L, by weapons fire or by the use of a command detonated mine (Claymore) or remotely operated weapons/munitions.
- Mines/booby-traps may be an alternative to discrete security positions.

'L' Ambush Formation TTPs:

- Sometimes considered to be a combination of a Point and Linear Ambush, the formation consists of a long leg situated parallel along a road, path, back-trail or to the flank of the anticipated line of advance of an enemy element, and a short leg that lies perpendicular to the anticipated line of advance of an enemy element. It provides crossing fires along the flank of the target and down the long axis of the enemy force.
- Again, mines/booby-traps may be seeded along both sides of the kill zone, and to the flanks of the ambush position so that once the ambush is initiated, further casualties are inflicted on a fleeing or maneuvering enemy. Initiation of the ambush is the same as in the Linear Ambush formation above.

In 217 BC, the Carthaginian General Hannibal lured the army of Roman General Flaminius into pursuing him around the northern periphery of Lake Trasimene. Using deception, Hannibal then seduced the entire Roman main body into a massive 'L' shaped ambush kill zone formed by the Carthaginian army; the lead Roman elements were blocked by a strong arm of infantry, the Roman left flank was then assaulted by light infantry launched from ambush positions it held on a ridge that paralleled the Roman left; simultaneously, Hannibal's forces closed off the road to the Roman rear. Hannibal effectively used deception to channel the Roman forces into an ideal ambush site where (1) the Roman right was blocked by a water obstacle, (2) where the van and rear of the Roman army were blocked and (3) where the Carthaginians used high ground and constricted terrain (chokepoint) to rush the Roman flank before they could form up in battle array.

'Of the initial Roman force of about 30,000, about 15,000 were either killed in battle or drowned while trying to escape into the lake…. Another 10,000 are reported to have made their way back to Rome by various means, and the rest were captured.'[53] This battle was one of the largest, most successful ambushes throughout military history. While this engagement involved large bodies of troops, the formation is applicable down to even small combat elements (e.g. split Team).

'V' Ambush Formation TTPs:

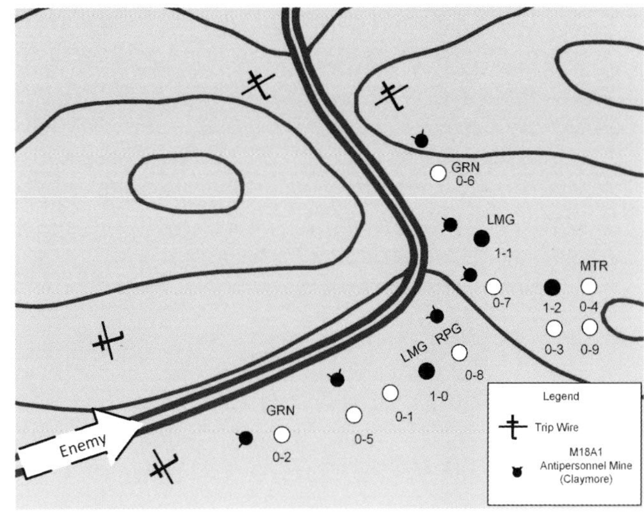

Figure 50. Typical "L" Ambush.

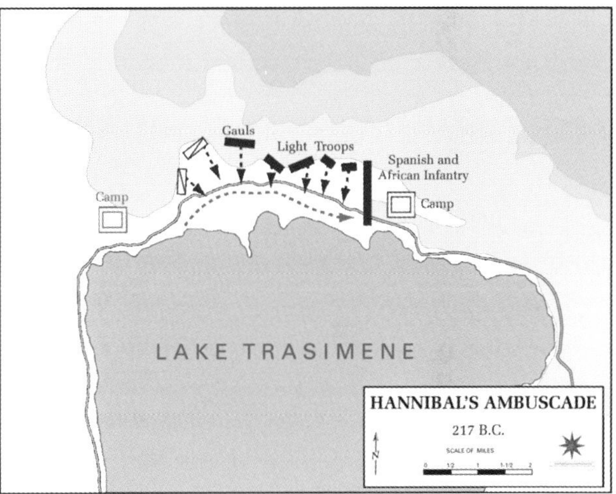

Figure 51. Lake Trasimene Ambush. Romans in Red; Carthaginians in Black. (*Public Domain. Frank Martini. Cartographer, Department of History, U.S. Army Military Academy*)

- The 'V' Ambush is positioned with its mouth open toward the enemy advance. A favorite of the Viet Minh, Viet Cong (VC) and NVA, it was used in elevated terrain as well as jungle. The ambushers, either in good concealment along the legs of the 'V', or lying in wait until the enemy point had passed and then creeping closer to the enemy (when flank security is used). The 'V' ambush was virtually undetectable by enemy point or flank security until at least a portion of the enemy force was in the kill zone. Fire is directed down the enemy axis of advance, with plunging, interlocking fire from each leg across the 'V'. The 'V' ambush also lent itself to the use of controlled mines/booby-traps.
- This complex formation is best used where the two legs are situated on elevated terrain, above a valley, a mountain pass or a large ravine, so that multi-level, intersecting fires may be brought to bear on an enemy force.

---

53. 'Ancient History Encyclopedia', http://www.ancient.eu.com/image/166/

- The apex or intersection of the legs of the 'V' is typically the most elevated position of the formation and should be positioned to fire down the long axis.
- Positions on either leg are located on sequentially lower elevations from the apex so that their fires are directed downward at the enemy and not across toward friendly opposite positions. This formation will take longer to establish, but it can be extremely effective against a numerically superior force.

This formation was used by the Viet Minh on several occasions against French motorized columns in Indo-China. Example: The Battle of Mang Yang Pass (aka Battle of An Khe) where a force of 700 Viet Minh troops established several 'V' ambushes to destroy the well-armed/equipped French Groupement Mobile No. 100 (strength of 2,500), during its convoy to withdraw its forces from An Khe in 1954. During five days of fighting, the French force lost 500 KIA, 600 WIA and 800 captured combatants (a casualty rate of 76 per cent), along with '85 per cent of vehicles, 100 per cent of artillery, 68 per cent of signal equipment and 50 per cent of crew-served weapons'.[54] This was one

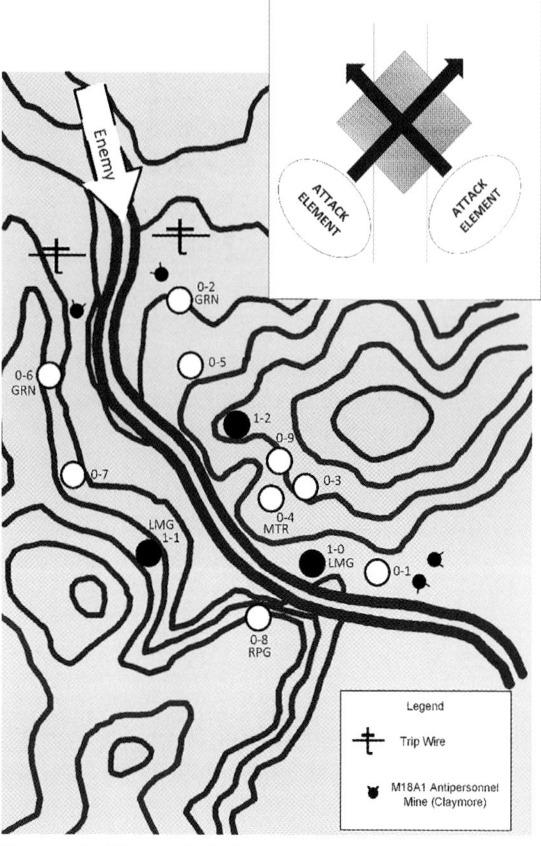

Figure 52. Typical "V" Ambush with Plunging Crossfire.

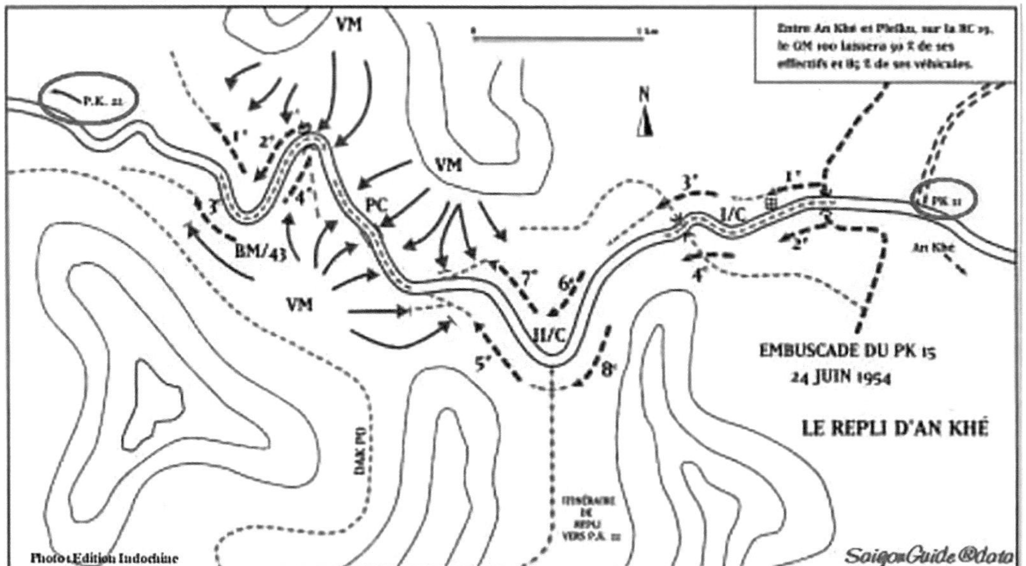

Figure 53. Battle of Yang Mang Pass in 1954. Note Viet Minh (VM) positions on both sides of the highway and the multiple linear and "V" ambushes. (*Public Domain*)

---

54. 'Battle of Mang Yang Pass', https://en.wikipedia.org/wiki/Battle_of_Mang_Yang_Pass

Figure 54. The Mang Yang Pass as it appeared in 1969. (*U.S. Army*)

of the worst French defeats during the French Indochina War. In exchange, the Viet Minh lost approximately 147 KIA and 200 WIA.

This ambush formation was used repeatedly from 1967 through 1969 by the NVA, at the same location.

'Z' Ambush Formation TTPs:

- The 'Z' Ambush is essentially a variant of the 'L' ambush. The design of the 'Z' anticipates that the enemy will attempt to flank the long leg of the formation; so a second short (blocking) leg is used to create a blocking position/secondary ambush opportunity. The 'Z' was often used during the Vietnam War, and was even employed by large NVA formations (e.g. battalions, regiments) against large US/allied units.
- The 'Z' ambush can be effectively used against numerically superior enemy units. The blocking leg/element, comprised of two to three Team Members, can deploy anti-personnel mines/booby-traps (with SD features), to include an array of Claymore mines, anti-vehicular mines and/or EFPs, to inflict substantial casualties on a flanking enemy unit. If the terrain on the flank will allow (restricted) maneuver (e.g. via a logging road) by an enemy armored vehicle, an anti-tank mine/weapon may remove the threat. The SR Team must ensure a route of withdrawal.

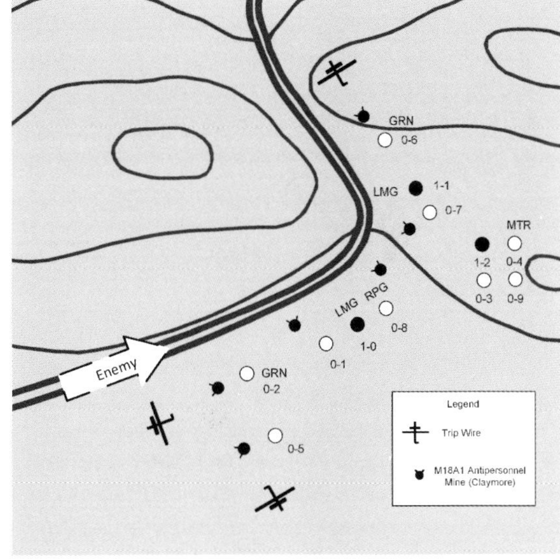

Figure 55. "Z" Ambush is used to counter a flanking maneuver.

'Y' Ambush Formation TTPs:

- The 'Y' Ambush was a specialized ambush formation devised by the NVA to be used specifically against SOG SR Teams. The base of the formation was a dug-in position (or positions) established laterally across a ridge. The formation normally possessed two wings, each consisting of several positions established on the slopes to each side from the base position and descending rearward and diagonally from the base.

- Since the enemy unit could not predict whether the Team would approach from along the ridge top, along either military crest or down a ravine on either side of the ridge, the 'Y' formation was created to be flexible enough to any of these routes of approach. If the Team approached along the ridge top, the entrenched positions would then engage the Team along its long axis and would likely inflict casualties on the unit; the wings would maneuver, approaching along the military crest

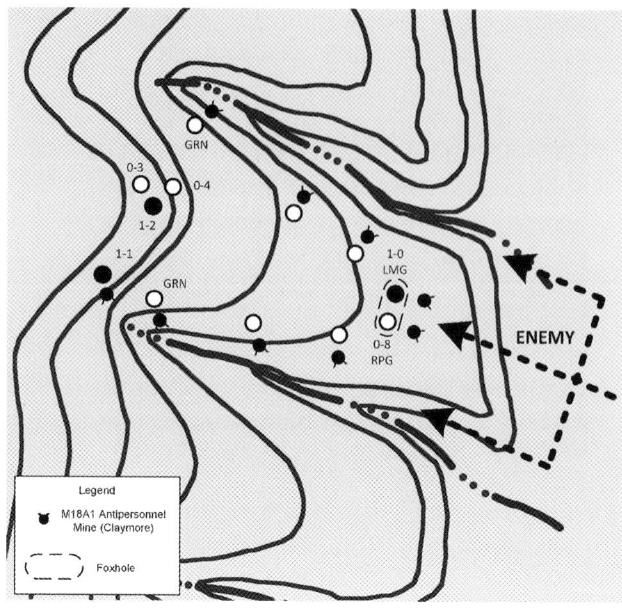

Figure 56. Typical "Y" Ambush. Enemy Line of Approach is Unknown.

and up-slope on both sides, flanking or enveloping the Team. If the Team approached along a military crest or along a ravine, individual enemy combatants on the wing would engage with intersecting fires and enemy combatants from atop a ridge would emerge from their entrenched positions and from the opposite wing to flank the Team. The formation can be employed by a single squad or can be employed by a larger unit to cover a broader area/series of ridges.

<u>True Account:</u> The most experienced T/L at SOG's FOB-2 received a mission to conduct area reconnaissance of a target area that had long been ignored, but radio intercepts and other intelligence, indicating an increase of enemy activity, revived interest in the target area. Within a day of insertion, the T/L found little evidence of activity, and had nearly concluded that the target area was a dry hole, when the Team encountered a high-speed trail on a ridge line with some signs of recent use. Rather than employing stealthy reconnaissance patrolling techniques to explore the rest of the target area, the T/L decided to move quickly along the trail with the prospect of finding something of intelligence value within the remaining mission window. The Team followed the trail, as it traversed ridgeline-to-ridgeline, for three days, with no result; but near the end of the third day on the trail, the Team was traveling along the top of a ridge, when a camouflaged top to an enemy foxhole burst upward and an NVA soldier shot the Montagnard point-man in the back of the head as he passed the concealed position. In an effort to recover the body of the indigenous Team Member, the heavily armed Team engaged in an intense firefight at extremely close quarters with the entrenched enemy, until a grenade could be thrown into his position. It was at this point that other enemy combatants began firing on the Team's flanks, compelling the Team's withdrawal and subsequent extraction. <u>What went wrong</u>: Team Members believed that they were in a dry hole and lost tactical focus. With no trackers and seeing scant sign of enemy activity, they parted from recon SOP/protocol by remaining on the same trail for three days, inviting a cunning and resourceful enemy to ambush the Team. Three days was far too long an interval to follow the trail. The Team could have resorted to other methods to trace the trail to likely enemy

locations. For instance, the T/L could have estimated the trace of the trail and where the trail would likely converge on other terrain features in the target area. In association with this information, knowledge of enemy tendencies might then point to base camps, logistics depots, trail-road junctions, etc., toward which the Team could then navigate. Upon his return, the T/L traveled to SOG headquarters in Saigon to exchange information on the new enemy tactic, only to discover that the tactic had been known to SOG's intelligence branch; sadly, this valuable intel was never circulated by SOG HQ to the FOBs.

Area Ambush Formation TTPs:
The Area Ambush 'involves patrol elements deployed as multiple, related, point ambushes) – FM 90-8.

- The Area Ambush was extensively used by the SAS on enemy trail systems during the Malaysian Counter-Insurgency campaign. It can also be used in conjunction with a raid.

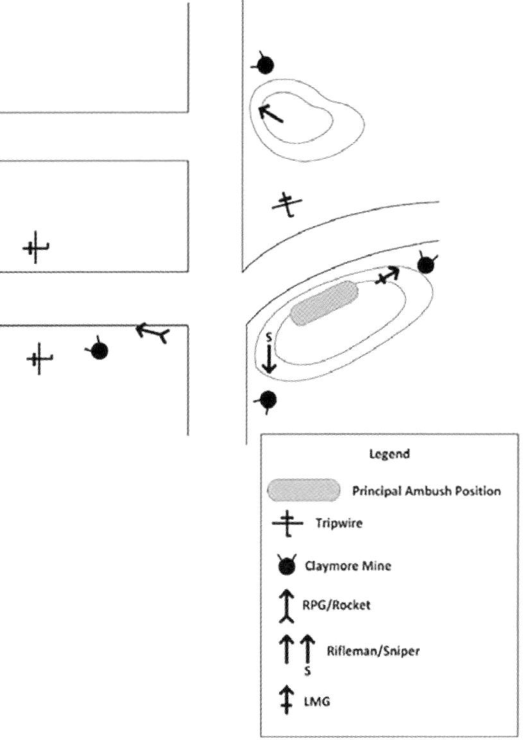

Figure 57. Example of a Team Area Ambush on a road/trail network.

- The Team may attack (by fire) from various angles/directions.
- If the enemy attempts to assault (Battle Drill) the ambush position(s), the ambush element(s) can withdraw and other ambush elements can attack by fire.
- A Team may establish a series of point ambushes around an insurgent base or enemy rear area installation egress points as part of a raid operation, where an enemy is attempting to escape from expected aerial attack/bombardment or to counter a raid/attack by fire.
- Each element must plan a safe route of withdrawal to the Team RP, ensuring that it does not cross other Team point ambush sectors of fire.

Advanced Ambush TTPs:

- Use smoke to obscure withdrawal/movement. Use WP or another immediate acting smoke, coupled with standard smoke grenades (or CS grenades) to provide a more lingering obscuration effect.
- When the Team reaches the vicinity of the planned target or discovers an opportunistic target, it should move to a rally/release point that is far enough away so that Team presence will not be detected by routine sweeps/patrols. Fish-hook into this position to observe your back-trail.
- Use large shaped charges, combined with cratering charges to create a road barrier.
  ○ The substantial crater(s) formed by shaped/cratering charges, especially when employed at a chokepoint and in conjunction with a deliberate ambush, can be used to trap an enemy unit in a kill zone.
    ▪ Best Practice: The crater should be emplaced at a bend in the road in advance of an expected convoy. This will block forward movement of the convoy. The demolition element must have a covered approach/withdrawal.

- Two charges must be detonated: (1) a shaped charge will bore a deep, relatively narrow hole through the road surface. This allows (2) a cratering charge to be placed within the bore-hole resulting in an optimum crater that is both wide and deep. Both charges can be placed and each detonated in less than five minutes by a trained element.
- The enemy may use a scout vehicle that precedes the main convoy. It may be prudent to allow the scout to proceed through the ambush site unscathed. The security element may engage the scout if it tries to join the fray. It may also be necessary to engage the lead vehicle to delay the convoy long enough to create the crater.
- A second crater may be emplaced some distance to the rear of the convoy, if the terrain and road contour accommodates and if the convoy is small enough. This too is best accomplished at a bend in the road. If the convoy is fairly large, there are a variety of means to interrupt the convoy between serials, allowing time to create the crater.
- The pair of craters will trap the convoy, front and rear, and will limit access by other enemy forces converging on the ambush site. As noted elsewhere in this book, these sizeable charges would have to be stockpiled in caches/MSSs in advance of such use.

○ Cratering Operations during BDAs. Attempting to cut a road with B-52/B-1 strikes with hard bombs (as opposed to precision weapons) is almost always ineffective. And attacking roads with precision/guided bombs, while accurate (unless under solid canopy), generally provides only abbreviated interdiction of such targets, as the enemy can rapidly, reroute convoys, use detours, repair bomb craters, etc. As SR Teams may be dispatched to conduct BDAs or exploit bombed-out areas for intelligence value, the Team should consider taking a 40lb shaped charge and a 40lb cratering charge to opportunistically interdict the road and to supplement the effects of an airstrike.

- If possible, the placement of the charges should occur as rapidly after insertion as possible. The Team then can proceed on its primary BDA mission.
- Primers for the charges should be prepared in advance and be immediately available to dual prime the charges after they are placed. Each charge can then be successively and rapidly emplaced and

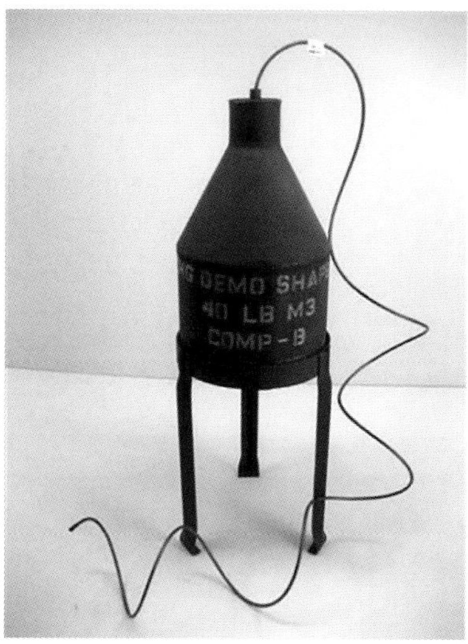

Figure 58. 40lb Shaped Charge on its stand-off legs. (*Army Photo*)

Figure 59. 40lb Cratering Charge w/ "overkill" quantity of C-4 "booster" demolition blocks. (*Army Photo*)

detonated in minutes. Note that it is advisable to use <u>at least </u>one block of C-4 as a booster for the cratering charge; the C-4 block would be the priming point for the demolition charge.

- Shaped charge stand-off, using the legs provided, or using a field expedient, is <u>essential</u> for the explosively formed jet of super-heated gases to form, which is necessary to create an adequate bore hole for the cratering charge.
- Bring sandbags and entrenching tools in the event that the shaped charge stand-off legs must be stabilized on uneven ground.
- The completed succession of detonations will create a substantial crater.
- The Team can place the charges with precision to maximize interdiction effectiveness. This could include placement at a chokepoint, at a crossroad, at a junction with a rail crossing, at an over/underpass or on a mountain road bounded by a precipice.

- Consider using an off-route mine to attack wheeled or tracked armored targets in road ambushes. An off-route mine can be used effectively in nearly all ambush formations.
  - While some obsolete and/or field expedient off-route mines use a High-Explosive, Anti-Tank (HEAT) rocket as the kill device, an EFP device is much preferred; it is cheaper, can be locally fabricated as a field expedient, can kill its target at a stand-off distance, can effectively be fired through intervening light debris or vegetation to hit its target and is easily concealed.
  - A HEAT rocket device is more easily detectible; because it must have both ends of its launcher clear of earth cover and it must generally have an open line-of-flight (unimpaired by vegetation) to the target.
  - Note that an HE artillery projectile can be used as an off-route mine. The fragmentation of the projectile would obviously be effective against personnel, light armor and light-skinned vehicles; but the base of the projectile, when pointed at an armored vehicle will have sufficient mass to defeat heavy armor protection. The base of an HE projectile may sometimes remain largely intact upon projectile detonation, forming an EFP.
  - Command-detonation options are preferred for initiation, consisting of either electronic or electrical (e.g. firing wire) means.
  - Find a way to cause the enemy vehicle to slow, pause or stop during travel. Example: Where the vehicle must make a sharp turn.

Author's Solution:

- *An electronic Remote Firing Device, such as the MK 152 Remote Activation Munitions System (RAMS), or a civilian equivalent, is preferred to electrical firing wire initiation. The RAMS offers much more rapid deployment, has a much longer range (5 kilometers), is faster to emplace than a firing wire and can be fired from a remote location to address the target perpendicular to or along the long axis of the kill zone or even from above the target on a slant.*
- *If the off-route mine is offset from the observer-target line-of-sight, its detonation will deceive the enemy as to*

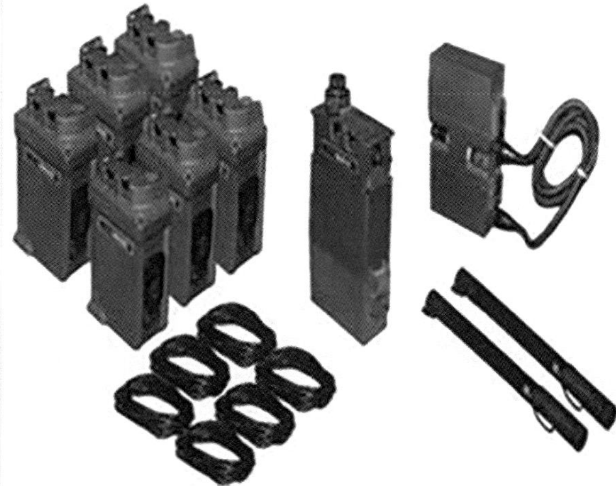

Figure 60. Mark 152 Remote Activation Munition System (RAMS) components. (*Public Domain – Army Photo*)

the location of Team personnel. This can be especially productive if the detonation occurs on the opposite side of the road to the Team's location. A successful kill may block the road; enemy personnel will then likely dismount from convoy vehicles and take up positions on the opposite side of their vehicles to the detonation (but closer to the Team's ambush kill zone), making them easy prey for the Team's subsequent attack by fire. If the enemy uses a Reaction-to-Near-Ambush Battle Drill (immediate action drill), and mistakenly assaults the seat of the detonation (vice the actual Team location), the Team may remotely initiate Claymore mines when the enemy enters a pre-planned kill zone; alternatively, the area near the seat of the detonation may be seeded with trip-wired booby-traps.

- *Use an innocuous roadside visual marker to gauge when the target is in the kill zone.*
- *Note that even a small diameter EFP, if properly aimed, is capable of knocking out an armored vehicle.*

POW Ambush TTPs:

- The POW Ambush is a specialized form of Near Ambush. A POW snatch can also be executed as a Raid.
- Time on Target (ToT): Minimizing ToT, through planning, training and execution, is even more <u>essential</u> in a POW snatch than in other ambushes that incorporate an assault (e.g. to gather information or material of intelligence value). The POW(s) will typically be armed, may be wounded and can be expected to struggle – all of this operates against efforts of the Team to reduce ToT and Team Member exposure.
- Pick targets carefully. Be patient. Small convoys, detachments or single vehicles/foot soldiers are much less risky. Avoid convoys with armored escorts in POW ambushes.

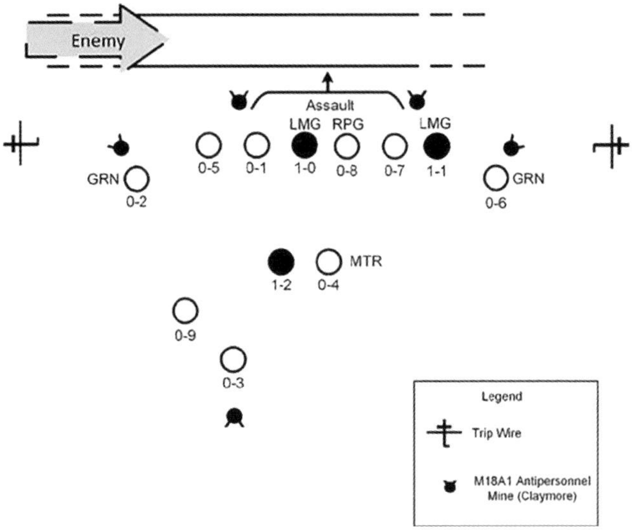

Figure 61. Deliberate Linear/POW Ambush Formation.

- Take out the lone or lead convoy vehicle when it is beginning to ascend a grade and shifts to a lower gear. Even better, if the vehicle is climbing a grade on a curve or turning a corner.
- Use an EFP, Claymore or other command detonated device to knock out wheels/tires, transmission, axles, engine.

At least two SOG SR T/Ls theorized that a simultaneous detonation of two demolition charges or two Claymore mines (positioned at angles to leave a small gap in the kill zone) would have a concussive effect sufficient to knock out a solitary NVA soldier. On each occasion, this was found in practice not to have any other effect than to cause the enemy combatant to sprint down the trail/road at a record breaking pace.

- Use security and/or obstacles (mines, abatis), preferably at a turn in the road, or a constricted point or curve, to block enemy response.
- Beware of infantry passengers. Use concentrated fires or a mass-casualty producing device (e.g. Claymore) to defeat this threat immediately. An aimable device, such as a Claymore (see elsewhere in this book), may be ideal.

- The driver of a soft-skinned vehicle will have a grip on the steering wheel; if he is shot from the driver's side, he will fall toward the opposite side of the cab, retaining for an instant, his steering wheel grip. This will cause the vehicle to veer sharply toward the passenger's side of the road. Remember this when you select an ambush position/location and formation.
- Use 3-man elements to take the driver and the other occupant of the cab. The other cab occupant may be more valuable than the driver. Neither the driver nor the other cab occupant will be able to wield their individual weapons effectively, unless they are able to dismount. The element should therefore take them quickly.
- POWs, wounded or not, should be dragged off the road/trail immediately upon capture to limit exposure of the Team Members.

Deliberate vs Hasty POW Ambushes TTPs:
A POW <u>snatch</u> <u>mission</u> infers a deliberate ambush. The T/L considers the variables of METT-TC to plan the deliberate operation. The Team must practice this type of ambush prior to departing on the mission. In addition all Teams should periodically rehearse their actions for <u>hasty</u> POW ambushes as a variant of the hasty ambush. Planned and hasty POW ambush tips follow:

Silent Ambush/Raid:

Use of Incapacitants	
Advantages	**Disadvantages**
Quickly incapacitates the individual(s), preventing him/them from using weapons accurately.	Not all incapacitants will be effective on all targets. The target may panic and run upon exposure to the gas and may have to be chased down.
Incapacitants include CS powder, Pepper Spray, Mace, CS grenades, vomiting agents.	Slow discharge of CS grenades permits enemy reaction; suggesting that they should be used on static sites, cantonments.
May prevent the use of dogs being used in the pursuit of the SR Team.	CS powder will cling to the PoW (and friendly) clothing which may affect helicopter crew members (if they do not have masks).
In warm/hot weather, a heavy-duty, pressurized Mace or Pepper-Spray dispenser can be used on vehicle drivers proceeding at a modest speed (windows open). The spray can have an immediate effect if it strikes the enemy in the face; the driver will stop or crash his vehicle in these conditions. Larger dispensers may have an effective range of over 10 meters.	Team Members wearing protective masks will have obscured vision. An enemy PoW may pull off the protective mask during a struggle.
Enemy personnel may have to put on protective masks to enter the area, which will hamper their vision and therefore their effectiveness.	Use of CS powder, Pepper Spray, Mace, or vomiting agents in a silent snatch must be done at close range. Use should be supplemented by suppressed weapons employment, if the silent snatch is unsuccessful, or by standard (non-suppressed) engagement in last resort.
Can alone facilitate silent capture if enemy consists of only 1–3 individuals.	When target(s) are part of a larger body of troops, a well-trained enemy, familiar with the vicinity, can avoid the incapacitants and pursue the Team parallel to its line-of-march.

Advantages	Disadvantages
Can be supplemented by suppressed weapons or Claymore firings, e.g. if enemy consists of more than 1–3 individuals.	Somewhat less effective against mounted targets, as vehicles will pass through the cloud; though the driver may be impaired, the vehicle may continue through the 'kill zone'. Note that troop carrying cargo trucks create a draft that will suck the agent into the cargo compartment.
Tips:	
• The Team may prepare to execute a gas ambush by Team Members mounting their protective masks on top of their heads. It will then only take a couple of seconds to don the masks just prior to initiating the ambush. • Use caution in moving up on a wounded, but armed enemy. Use electro-shock weapons to further subdue him. • Use an interpreter or Team Member linguist to demand the surrender of the enemy combatant(s). If the Team does not have linguistics skills, Team Member should be trained in necessary basic phrases. • Pay attention to wind direction and speed.	

'Five Russians belonging to a reconnaissance patrol jumped [a] soldier at [a] machinegun, threw ground pepper into his face, pulled a bag over his face, and disappeared into the night.'[55]

Use of Silent Weapons	
Advantages	**Disadvantages**
The noise from a silenced/suppressed firearm or electroshock weapon is minimal. Silenced firearms are found in either pistols, sub-machineguns (using pistol cartridges) or rifles using subsonic rounds.	Suppressed firearms (e.g. M-4), using standard ammunition, will deceive an enemy as to the origin of the shot(s), but the crack of a normal (supersonic) round will still be heard as it passes enemy ears. The movement of the action is audible to the rear/flanks of the shooter.
A well-executed and well-placed shot will stop an enemy and prevent him from returning fire using his individual weapon.	Small caliber (9mm or less) silenced firearms may be ineffective in POW snatch operations, requiring multiple shots or very good marksmanship. While a .45 caliber has the knockdown power to incapacitate an enemy, its wounding effect may be mortal. A wounded POW may die of shock or loss of blood before proper treatment can be given. Regardless of caliber, a disabling firearm shot, used in POW ambush, will normally be directed at a leg to hobble the subject. • If the shot ruptures the femoral artery, the target will likely bleed out before he can be exfiltrated. • A wounded POW will have to be assisted and possibly carried, thus slowing down the Team during its withdrawal.
An electroshock weapon disrupts muscle functions to incapacitate. Electroshock weapons include: • Hand-held stun weapon or electric baton/prod. Range: 0in to 2ft	• Hand-held stun or electric baton/prod weapons have a very short range. This type of device is not recommended as it may require continuous contact with the subject for as much as five seconds to incapacitate. During this interim, the subject may still retain sufficient motor functions to resist and to shout.

---

55. Halder, *Small Unit Actions*, p. 260

Advantages	Disadvantages
• Projected probe weapon (e.g. Taser®): Effective range of 25ft. Military grade weapons can mount on the Picatinny rail of the primary weapons. When the stun weapon is fired the report is minimal. A three-shot Taser® is available. • 12 GA shotgun cartridge (wireless). Range 100–300ft. This option permits repeated shots. • Prototype weapons are not discussed in this book.	• Electroshock weapons (hand-held stun, electric batons/prods, or Taser-type projected probe weapons) are often ineffective on targets wearing heavy winter clothing and/or LBE. • The 12GA with wireless cartridges has a loud discharge and is therefore unsuitable for a silent ambush.
Pepper Spray can incapacitate a subject for up to 40 minutes. • Is silent when a pressurized spray can is used. • A spray canister is capable of repeated shots. • Range is up to 18ft; 25ft is possible with a pepper spray gun. • Must be directed at a subject's face; and is particularly effective if the subject's eyes and/or mouth are open. It has a nearly instant effect on the subject's airway, causing gasping, coughing and difficulty in breathing • Subject's ability to shout is typically suppressed; especially if the subject's mouth is open when dosed.	• A pepper spray 'gun' has a loud report. • The subject may still be capable of resisting after being sprayed. • It would not be wise to spray upwind into a stiff wind. • If subject's eyes and/or mouth are closed, effects will be delayed. • Generally short range.
Tips:	
• Use caution in moving up on a wounded, but armed enemy. Use electro-shock weapons to further subdue him. • The Team should only capture a single enemy combatant, if the prisoner is not mobile (due to wounds). More enemy prisoners can be captured if they are fully mobile. If several enemy combatants are wounded, the T/L or the Team Medic should select the one that is most mobile and most likely to survive. • Use an interpreter or Team Member linguist to demand the surrender of the enemy combatant(s). If the Team does not have linguistics skills, Team Member should be trained in necessary basic phrases. • Pay attention to wind direction and speed when using pepper spray.	

No Fire/Silent Capture	
Advantages	Disadvantages
If properly done, there is little noise to give the Team's presence or location away.	Requires the Team Member(s) to surprise the target and to: Convince the combatant to surrender quietly, or use physical blows (hands, bludgeon, sap) and/or choke holds, to subdue the target. Excessive blows may kill the subject.
A live and healthy POW is the best kind.	Target may have the opportunity to struggle and shout before being subdued. Subduing the POW may require two Team Members.
May be appropriate for disoriented enemy combatants subsequent to an airstrike.	Generally only useful on a single individual and when the target is isolated (e.g. when the target is answering the call of nature).

Advantages	Disadvantages
Tips:	
• Use caution in moving up on an armed enemy. Use electro-shock weapons to further subdue him. • Use an interpreter or Team Member linguist to demand the surrender of the enemy combatant(s). If the Team does not have linguistics skills, Team Member should be trained in necessary basic phrases.	

<u>True Account</u>: **A SOG SR Team encountered a high-speed trail in southern Laos at midday. The veteran SR T/L and his point man left the main body of the Team to reconnoiter along the trail and very quickly came upon a single NVA soldier sitting by the trail preparing his noon meal. His weapon was leaning against a tree three paces away. The T/L (via his point man) commanded the soldier to surrender. Instead, the soldier screamed and lunged for his weapon, requiring the Team Members to kill him. <u>What went wrong</u>: It was nearly predictable that the T/L might encounter enemy personnel along a high-speed trail, but the Team was not trained/ prepared to use physical force to capture an enemy soldier. Had the Team element jumped upon the lone soldier, the element might have easily overcome the individual and come away with a POW. Note that the Team was not equipped with a silent capture weapon/device.**

• Ambush with full weapons discharge (including deliberate, chance/meeting engagements, hasty ambush). Team weapons, Claymores and other explosive devices will inflict casualties among enemy troops who are present within the kill zone.

Ambush with Full Weapons Discharge	
Advantages	**Disadvantages**
Team may have a selection of POWs from among the WIA.	Wounded POW may have to be carried from the ambush site.
Offensive (stun) grenades are effective in incapacitating a target, when used in confined spaces. Effects of offensive grenades or demolition charges are marginal against personnel in the open.	The noise of the explosive going off will alert enemy forces in the Team's vicinity.
The Team can engage multiple combatants with a combined effects attack.	Disturbed soil/vegetation, blood trail and other signs will mark the Team's ambush location. This will aid the enemy in his attempts to pick up the Team's trail.
Proper engagement techniques will ensure that the enemy is stopped within the kill zone.	A demolition ambush, using the concussive effects of explosives, is ineffective in the open and should not be used. To be effective, substantial amounts of explosives would be required and are not feasible in SR operations – especially where other courses of action are available.
A violent, aggressive SR engagement may thoroughly disorganize and confuse enemy combatants, preventing them from reacting effectively.	
Tips:	
• Use caution in moving up on a wounded, but armed enemy. Use incapacitants/electro-shock weapons to further subdue him. • The Team should only capture a single enemy combatant, if the prisoner is not mobile (due to wounds). More enemy prisoners can be captured if they are fully mobile. If several enemy combatants are wounded, the T/L or the Team Medic should select the one that is most mobile and most likely to survive. • Use an interpreter or Team Member linguist to demand the surrender of the enemy combatant(s). If the Team does not have linguistics skills, Team Member should be trained in necessary basic phrases.	

◦ <u>Chance Contacts/Meeting Engagements</u>: Both friendly and enemy forces may see each other at the same time.
- A Team must always be prepared to take a POW during a chance encounter, especially with a small enemy element. The Team should use a variant of its hasty ambush drill (discussed elsewhere in this book) to effect the capture. Train for this.
- If the enemy combatant carries his weapon in a nonchalant manner, the chances of taking a prisoner are improved. The SR Team has the advantage, as Team Members always move with weapons ready. The Team will undoubtedly achieve/attain fire superiority and inflict casualties on the enemy before the enemy unit can even move weapon safeties to fire.
- If the enemy element is armed with inferior weapons, the Team may be able to quickly overwhelm the enemy element.
- Use caution in moving up on a wounded, but armed enemy. Use incapacitants or electro-shock weapons to further subdue him.
- Use an interpreter or Team Member linguist to demand the surrender of the enemy combatant(s). If the Team does not have linguistics skills, Team Member should be trained in necessary basic phrases.

POW (Ambush/Raid) Snatch TTPs:
General: A POW <u>snatch</u> is a specialized <u>deliberate</u> ambush or raid. If the SR Team has been selected for a deliberate POW-snatch mission and been given a target area, the following actions should be taken in addition to those normally taken for a reconnaissance mission. Note that a POW snatch is often a priority supplemental mission to all other assigned SR missions.

- Study the map to find an appropriate location to conduct a deliberate POW snatch within your target area. The T/L should conduct a visual reconnaissance to familiarize with target area terrain, select LZs, evasion routes, and rally points. Record any new trails discovered in the target proximity and pick tentative POW snatch positions.
- Finalize plans for primary and alternate LZs and routes of approach to and from the tentative POW snatch location.
- Rehearse POW snatch procedures and put as much realism into training as possible. Train Team Members on the use of handcuffs, gags, blindfolds and transport of a wounded prisoner. Consider the distance from the tentative POW snatch site to the planned extraction LZ and practice porting a wounded prisoner an equivalent distance. Other actions that need to be explained, practiced and rehearsed are:
  ◦ Hasty POW Ambush Battle Drill/SOP.
  ◦ Team movement into position.
  ◦ Signals to be used.
  ◦ Concealment of personnel and equipment.
  ◦ Actions to be taken by each Team Member, if discovered by the enemy, while occupying the POW snatch position.
  ◦ Employment of Claymores or other devices.
  ◦ Use of decoys and/or field traps.
- Considerations for selecting the location of the POW Snatch Site. Generally speaking there are several promising sites to stage a POW snatch:
  ◦ Best to find a location where the enemy soldier feels secure, and/or a time when his guard is down – but at a time and place where he is accessible.

<u>True Account</u>: A SOG Reconnaissance Team was assigned a priority road–watch mission in Laos. The second day into the operation, while the Team was nearing its

intended observation position, it encountered a large, well-concealed, unoccupied vehicle park that appeared recently constructed. The truck park was 'very well-groomed' and was probably awaiting its first occupancy. At one end of the park, there was a new, sizeable, unoccupied bamboo hut built on 'stilts' that offered any occupants an elevated view of the park. Most intriguing to the T/L, was a large, newly dug, open-pit latrine located 15 meters from the hut at the fringe of the jungle. The Team lingered overnight, in position at the verge of the vehicle park in the hope that a convoy would arrive and that an opportunity to capture an enemy combatant at the latrine would present ... before striking the occupied park with TACAIR. However, no vehicles arrived and the pressure of the assigned mission coupled with anticipation of an enemy tracker team required that the Team leave the nearly ideal POW snatch site the following morning.**

- Optimum times might include: periods of peak heat, especially after a meal; after a security patrol has swept the area; at dawn, upon awakening etc.
- Locations might include isolated outposts. Some isolated positions (such as bridges, railway tunnels, dams, canal locks, power substations, etc.) may be guarded by alert personnel making a snatch much more difficult. Surveillance is necessary to gauge guard shifts and to determine guard discipline and level of alertness. AA or radio relay sites may sometimes be optimum; and they must be provisioned with food, water and expendables, usually by a vulnerable single vehicle. Other areas would include swimming/bathing/watering holes, vehicle parks, latrines.
- Isolated or infrequently used roads/trails used by individuals, small parties of soldiers or single vehicles, especially roads/trails located in relatively secure areas between logistics installations, base camps and other key points, particularly if they are not well patrolled or secured, etc.
- Other considerations:
  - Establish a release point at a secure distance (depending on terrain and vegetation, etc.) from the ambush or raid site. Dispatch site recon/surveillance elements from this location.
  - Always search the vicinity of the prospective ambush site, e.g. flanks, for trail networks, facilities, outposts, or any signs of the enemy which might affect the task. If the prospective site is unsuitable, clover-leaf to identify another prospective site. Be patient and be picky. It is critical to select a good ambush site.
  - Designate at least two extraction LZs and two RPs and ensure that each man knows the direction and approximate distance to each before moving into position.
  - Brief the Team.
  - Prior to moving into the snatch position, provide a SITREP to notify HQ and alert the FAC of Team intentions.
  - Cache rucksacks prior to moving into ambush/raid positions. Orient the carrying straps in the 'up' position for faster and easier recovery and donning during withdrawal.
  - Double check all weapons and essential equipment prior to departing release point.
  - Have a pre-cut/pre-assembled stretcher positioned at the release point for movement of a casualty or a WIA POW.
  - If POW capture opportunity does not appear within two–three days, relocate.
- Designate Team Members to accomplish the following tasks.
  - Handcuff, gag, blindfold, and search the prisoner, in that order.
  - Treat prisoner wounds.
  - Prepare to carry or assist a wounded prisoner. This may include precutting a stretcher/travois and positioning the item(s) at the objective RP or ambush site.

- ○ Carry prisoner equipment and weapon.
- ○ Establish false signs at the ambush site, if the situation and time permits.
- If the time and situation permits, examine captured maps and other documents, at an RP, for opportunistic and/or immediately actionable information.
- If the Team has taken casualties, rapidly determine the tactical variables and act decisively. The Team SOP and Team immediate action/battle drills may have to be modified on the spot.
- Navigate rapidly to the nearest LZ. If the POW is disabled and cannot be ported due to the tactical situation, or if the Team has casualties, find a location with sparse vegetation and request extraction by string or ladder. Or summon a QRF/Bright Light Team.
- Make a security check of the extraction LZ.
- Guard prisoner(s) at the LZ.
- If required, rig the POW for string/ladder extraction.
- How best to evacuate a POW during a string/ladder extraction.
  - ○ Regardless of the condition of the POW(s), the evacuation aircraft should land thereafter so that the POW(s) can be transferred to the aircraft interior. A reception party should be deployed at the launch site as the strings/ladders approach the ground.
  - ○ WIA POWs would be best extracted by the chase aircraft, and should be secured by at least one US Team Member throughout the extraction; once inside the aircraft, the POW should be secured to the aircraft floor/bulkhead or be under positive control through the evacuation, ensuring he does not attack friendly personnel, or attempt to throw himself off the aircraft. If additional personnel are provided, from the Launch Site/Bright Light Team/QRF, to secure the POW during the extraction, the Team may continue operations in the target area and even exploit opportunistic information acquired from the POW. Whether on a string, ladder, or carried internally, ensure positive control of the POW(s) at all times and ensure that weapons/munitions of friendly troops are secured.
  - ○ Wounded POW:
    - ▪ Insofar as possible, the POW(s) should be paired with a medic on the string/ladder. If a Team Member is also wounded, the medic's first priority, in accordance with triage considerations, should be to the Team Member.
    - ▪ If medically feasible, handcuff the POW(s) to the ladder/string, aircraft interior stanchion and/or otherwise ensure that the POW(s) cannot struggle
    - ▪ If the POW is wounded, the prisoner should be exfiltrated in the chase aircraft (if the LZ is large enough for a landing), which should have a medic on-board; the same rules of control apply. If no chase aircraft/medic is provided, a Team medic, if available, should accompany the POW. This aircraft may fly directly to a field hospital; Team Members/US security should remain with the prisoner until properly relieved (e.g. by headquarters or intelligence personnel).
- Civilian prisoners (friendly/unfriendly to the US or its allies) may be more valuable than enemy military personnel. Locals have intimate knowledge of the vicinity and may know exactly where enemy units, facilities and capabilities are located. Alternatively, enemy soldiers below the rank of sergeant may have very limited information. Enemy troops, depending on rank/position, military specialty, etc., may be generally or entirely ignorant of their own location, the location of enemy units, order of battle, pending operations, and so on.

Far Ambush TTPs:

- Establish a far ambush to destroy targets in open terrain with sparse vegetation. Use a combination of remotely initiated munitions and mines/booby-traps (preferred) or long-range fires (ranging from sniper weapons to anti-armor missiles to CAS, if available). Best to use long-range fires if the location of the Team will not be revealed, if restricted terrain will deter enemy pursuit, if an escape route is proximate and/or if the Team is mounted. Near ambush options may be undertaken if cover and concealment for the Team can be established, e.g. for POW capture – where the Team can pick-off a lone combatant or a small detachment. Much effort may be required to camouflage Team Members in this circumstance.

- Select the kill zone where the enemy is channelized, where his maneuver is restricted, and/or where cover and concealment is limited.
- Use mountain roads or other back-county roads that are opposite to, and run parallel to, a gorge, river or other obstacle to prevent enemy maneuver on the Team position and making enemy counter-fire more difficult.
- Deliberately immobilize an enemy WIA on a road, a well used trail or other opportunistic site. Best if he is conscious and vocal and leave him out as bait. This is a variation of a TTP used by snipers since the US Civil War, if not earlier. Move to a far ambush location or establish a sniper position to attack other enemy personnel as they arrive to assist the wounded man. Moving to a more distant position will enhance the Team's withdrawal/evasion options in the face of a numerically superior force and/or air support and will minimize the effectiveness of enemy counter-fire. Inflicting an abdominal wound may be the best option to immobilize the combatant; a thigh would is too likely to sever the femoral artery – a mortal wound that will rapidly render the man unconscious from blood loss and cause him to bleed out shortly thereafter. Additionally, if the combatant is wounded during close combat or a near ambush, there may be a brief opportunity to place a mine or booby-trap along the rescuers' expected path of approach – use of a mine/booby-trap requires that the device be previously rigged and available for immediate deployment.
- Kill as many enemy as possible before they take cover; but, if possible, mark the following targets in priority order:
  - <u>Radio operator (or the radio itself)</u>: This combatant can communicate with other units in the area (including blocking forces) and can summon air/fire support and/or rapid response forces.
  - <u>Officer or NCO in charge</u>. In several military organizations, individual initiative is discouraged and lower ranks are not taught basic skills that are common among western military personnel. In Unconventional Warfare or COIN settings, discipline and morale may disintegrate without officer or NCO leadership.
  - <u>Crew-served weapon teams or armored vehicles</u>. These pose a substantial threat to the Team. Destroy lead and trailing security vehicles to trap a convoy in restricted terrain to destroy other vehicles in the convoy (with their cargos) and to block pursuit by combat vehicles. Shoot the armored vehicle driver in the first burst of fire.
  - <u>Tracker/Dog Team</u>. Killing or wounding the tracker or dog will destroy or substantially reduce the enemy's capability to pursue the Team.
- Enemy Counter-Ambush Battle Drills emphasize assaulting through an ambush, much like US doctrine.
- If possible, use the sun (at sunrise/sunset) to blind the enemy at an opportune tactical moment (e.g. a deliberate ambush).
- Use a decoy/explosive device to suspend debris (e.g. snow, dead leaves, disturb green leaves, dust, etc.) in the air to attract enemy attention or to draw enemy fires. This may cause the enemy unit to deploy and expose its flank.
- Ambush and raid formations may be governed by Team training and experience, Team composition (US vs indigenous) and size, state of communications, Team weapons and munitions, available cover and concealment, weather and light conditions, enemy units

and patrolling activity, the nature of the target and the control requirements of the T/L. All of these variables must be factored into the T/L's decisions on deployment and formation selection.

- <u>Time on Target (ToT)</u>: Minimize the exposure of Team Members to enemy countermeasures. Training and rehearsals are essential to minimizing ToT.

Opportunistic Ambush of Disabled Enemy Equipment (to include armor) TTPs:

- Vehicles will often break down during administrative and tactical movements. It is common to attempt roadside repairs, rather than towing the vehicle to the owning/maintenance unit. A recovery operation will cause a concentration of targets at the site of the breakdown. This may be a superb opportunity to take prisoners or to destroy opportunistic 'prey'. The Team may wait for a concentration of support to arrive on the scene, presenting multiple targets.
  - ○ Some breakdown sites may almost be predictable (e.g. swampy or muddy areas, stream crossings, steep ascents).
  - ○ The Team may have to wait for a convoy to pass before attempting an ambush or POW snatch on a solitary broken-down vehicle. It may also be prudent to use mines or obstacles to prevent oncoming convoys from interfering with the Team's attack.
  - ○ The site of the breakdown/accident may have the following targets within the Team's Kill Zone:
    - ▪ <u>Personnel</u>: Vehicle crew and passengers; maintenance personnel; wrecker/recovery vehicle crews; security personnel; unit supervisory/leadership personnel; Material Handling Equipment (MHE) operator and additional personnel for logistics vehicles (in the event that cargo has to be transferred to other vehicles). Several of these personnel may have valuable, even vital skills (e.g. Heavy-Wheeled Vehicle and Armored Vehicle Mechanics, etc.); others may have valuable information (supervisory/leadership personnel).
    - ▪ <u>Equipment</u>: The broken-down/damaged vehicle, wrecker/recovery vehicle, security personnel vehicle, supervisory/leadership vehicle, maintenance personnel vehicle, MHE (to transload palletized cargo) and MHE transport vehicle, alternative transportation vehicle(s). Some of these vehicles (e.g. wreckers, transporters, and recovery vehicles) are low-density among the owning units and would be difficult to replace in the short term.
    - ▪ <u>Cargo</u>: Some cargos may include vital end items, components and repair parts that may cause a cascading effect on unit readiness. If the broken down vehicle is carrying POL or ammunition, it can easily and rapidly be booby-trapped or remotely detonated to destroy other vehicles in subsequent convoys.
  - ○ Maintenance personnel will likely have to use lights if repairs must be performed at night, especially during administrative moves in rear, presumably secure areas.

<u>True Account</u>: In December 1970, a broken down tractor-trailer was stranded after dark on Highway 19 near the An Khe Pass. Two armored gun trucks remained to provide security. Subsequently, a replacement tractor with gun truck escort, a V-100 armored car, and a wrecker converged on the breakdown site – attended by approximately twenty-five security, transportation and maintenance personnel. The NVA mounted a hasty far ambush against this target of opportunity.

- Any truck that is set afire during an ambush will create a screen of smoke that may benefit the Team – or the enemy. Beware ammunition trucks set on fire.

- Downed Aircraft. Regardless of whether the aircraft is friendly or non-friendly, the enemy may be expected to send a search party to the crash site.
  - Enemy Aircraft: Locating a downed enemy aircraft provides the SR Team the prospects of prisoners, intelligence materials/documents and setting a trap for an enemy search/rescue party.
  - Friendly Aircraft: Locating a downed friendly aircraft provides the SR Team the prospects of rescuing/recovering air crew personnel, salvaging or destroying materials/documents of intelligence value, destroying the aircraft itself and again setting a trap for an enemy search party.
  - The ability of the SR Team to pinpoint the location of the crash site will be enhanced if friendly aviation assets, to include drones/UAVs can rapidly be made available. In open terrain, this task is obviously more achievable.
  - The SR Team may employ a variety of ambush TTPs in anticipation of an enemy search party. The Team must be able to locate the crash site before the enemy is able to do so.

Ambush Considerations for Priority Targets TTPs.

- Ambushing/destroying Priority Targets can have cascading effects on a convoy and can have outsize effects on enemy operations and capabilities. Priority targets in no special order include:
  - <u>Officers, Noncommissioned Officers (NCOs) and Key Personnel</u>. In many armed forces in the world, only officers or mid-senior ranking NCOs in certain critical specialties (e.g. fire direction center personnel; command-and-control staff specialists) are taught land navigation or are permitted to possess maps. Furthermore, operations plans/orders, warning orders and similar directives are tightly controlled and restricted to these select personnel. Subsequently, producing casualties among these select personnel will have a cascading effect on operations.
  - <u>Political Officers</u>, where these personnel are used, and where they can be identified, they should be equal to enemy commanders as a target priority. They are often accorded co-commander status and are often vested with summary execution authority to compel troops to fight in high mortality operations. In some other countries, Military Police are vested with similar authority, to compel compliance with the policies and orders of the high command; and in yet other countries, certain elite, politically reliable, military units may also be vested with arrest and summary execution authority. Killing these personnel may substantially reduce the combat effectiveness of combatant units.
  - <u>Communications specialists/units</u> may operate and maintain complex communication systems. High-level/key communication systems and/or signal units in convoy can easily be detected by the specialized vehicles that they operate. For instance, satellite communications vehicles will mount dish antenna (in transport configuration). Mobile communication systems for C2 organizations may also easily be detected by specialized antennas. Killing or wounding communications technical personnel or destroying/disabling key communication systems could have game-changing consequences on the battlefield.
  - Capture POWs, especially Couriers, Officers, CP Personnel and Military Police. These categories of enemy combatants will possess substantially more information than lower enlisted ranks and specialties.
  - <u>Canal Locks, Sluices and/or Canal Traffic</u>.
    - These targets are vital if canals are being used in the provisioning of enemy units.
    - Locks/sluices are controlled by engines/machines and fixtures that will be difficult to replace if damaged or destroyed. Thermite grenades are preferred for destroying this equipment.

- Locks/sluices may be guarded, especially if they are located near concentrations of enemy troops. The bulk of the guard force will likely not be top-tier or experienced combat troops; however, some of the troops may be combat veterans temporarily allocated to security duties (and who may be recovering from wounds or injuries) pending frontline reassignment.
- A canal barge, sunk in a narrow canal or at the gates of a lock, will block canal traffic. A thermite grenade would ensure that the vessel cannot be raised and repaired.

◦ Command, Control and Communications (C^3) Units/Facilities.
- If possible, identify the type of unit to which the C^3 unit/facility belongs by observing the type of equipment used, the presence of security and similar tip-offs.
- A sophisticated enemy will position its communications antenna and other emitters at some reasonable distance away from a major HQ.
- Be on the lookout for communications landline while conducting operations; the landline will normally be suspended from trees or poles and will often have a minor footpath beneath it that would be used to lay and maintain the landline. Depending on enemy sophistication, the type of landline used may indicate its importance, its purpose and the type of unit that it supports. Two-wire (e.g. WD-1) landline may be used at the Battalion level or lower, unless the unit (e.g. air defense) requires a landline that will bear significant traffic. Coaxial cable may be used to support several units, a higher level headquarters, or to communicate large amounts of data. Optical fiber will carry substantial data and voice traffic and would likely be used between major headquarters and to major subordinate units; due to its low signature and resistance to Electronic Countermeasures (ECM) and Electro Magnetic Pulse (EMP), it may be used by sophisticated air defense systems. Follow the landline to a terminal; some of these nodes may be isolated and vulnerable, but beware of landline maintenance personnel who may trace and service the communications landline on a periodic basis.

◦ <u>Electrical Transformers</u>. An easy target, which can be destroyed with small arms fire and which is especially vital if rail transport is electrically powered. Electrical transformers may not be stockpiled and may often be manufactured individually; replacement of transformers will not be timely. Where the destroyed transformer blacks out an area where enemy units/facilities are located, the enemy must resort to generators to supply his power requirements. This places an increased burden on POL resources. It may also knock out some telecommunications capabilities.

◦ <u>Tank/Heavy Equipment Transporters</u>. Transporters are typically soft-skinned (vulnerable) vehicles that are used to move heavy equipment and armored vehicles swiftly, over moderate to long distances, to marshaling areas or other locations in preparation for operations – and to evacuate damaged equipment to maintenance locations for repair. Tanks are not driven long distances cross country, because they are subject to mechanical failures and because they consume enormous amounts of fuel; without transporters, serviceable tanks will sustain substantial road wear and degradation to tracks, suspensions, engines and transmissions, if they have to be road-marched to their destinations; if tanks cannot be evacuated to maintenance locations, maintenance units must expend substantial resources in roadside repair of tanks in vulnerable locations. Destroying a tank transporter that is laden with a tank, may destroy both the transporter and the tank. Destroying a number of tank transporters will have a cascading effect operationally and logistically on armored units across the battlefield, as the enemy would have little choice but to road-march his equipment to participate in operations; this will affect enemy OPTEMPO and/or degrade his equipment availability as a result.

○ <u>Fuel Trucks</u> are vital to convey petroleum from railheads, pipelines and other sources to mobility equipment, combat vehicles, missile Transporter, Elevator, Launch (TEL) vehicles, aircraft and power generation equipment (to include generators that power communications equipment, radars, and command and control systems). Destroying/ damaging fuel trucks at a critical moment can have a cascading effect on enemy tactical operations. Fuel trucks are soft skinned vehicles that are easily destroyed. Armor units, in particular, consume huge amounts of fuel.

○ <u>Material Handling Equipment (MHE)</u>. Destruction of Rough Terrain Forklifts (RTFLs) and other MHE will substantially impair the enemy's capability to arm, maintain and sustain its units in the field. Ammunition logistics units, in particular, will be severely impacted, especially in intermodal transfers and handling of containerized and palletized shipments. Other units substantially affected would include heavy field maintenance units, engineer units and other units handling commodity shipments in bulk (palletized or containerized). RTFLs are rarely road-marched during unit movements, and then, only short distances. Road-marching MHE, even moderate distances, will render the equipment unserviceable in due course. Consequently, this equipment will be routinely displaced using organic transporters (tractors with low bed trailers); disabling or destroying the transporters (especially if they conveying MHE) will have a broad effect on MHE availability. To replace this equipment, the enemy will resort to impounding civilian construction equipment with similar characteristics; but there will be a finite number of Rough Terrain MHE available in the civilian sector. If Rough Terrain MHE is not available, the enemy would be constrained to using warehouse-type MHE that can only be used on hard surfaces, making Ammunition Storage Points (for instance) more easily detectable and vulnerable. RTFLs use large dimension, specialized tires that are easily disabled by small arms fire or caltrops and that are difficult to repair or replace.

  ▪ The RTFL, Rough-Terrain Crane and other tactical-MHE, use special, wide, all-terrain tires that are easily deflated by weapons fire or by the use of caltrops. Caltrops come in a variety of configurations and can be employed against most wheeled vehicles. Expended 40mm casings can make very effective field expedient caltrops, as can ¾-inch wide (or wider) strips of discarded steel banding material, bent into a 2-inch curlicue; banding is used to palletize supplies/equipment (e.g. especially ammunition) or to secure pallets to cargo beds of long haul trucks/tractor-trailers – curlicues will act as cookie-cutters and leave large holes in the tires that are not repairable by conventional means. Caltrops are best used at night or in periods of limited visibility, but whenever used on a road they are best placed on the opposite (down-slope) military crest from the vehicles direction of travel and at hairpin turns; the driver will not see the caltrops until it is too late to stop. Caltrops will often ruin a tire to the extent it is not repairable in the field. Camouflage the banding so that the caltrop cannot be detected in the headlamps beams. Use of the 40mm casing expedient will reveal the presence of the Team, but the banding expedient may be viewed as accidental.

Figure 62. OSS Caltrop. Hollow spikes allow air to bleed out even if caltrop remains embedded in the tire. (*CIA Artifact - Public Domain*)

  ▪ Large RTFLs (e.g. in the 6,000 to 10,000lb range) are critical in the transfer of containerized shipments from railcars or long haul trucks/semi-trailers to Ammunition Supply Points (ASPs), or to other vehicles especially at unimproved transfer points/rail sidings. They are also critical in retrograde of empty containers to transportation assets. Further, they are used to transfer large missile/rocket pods. Disabling or destroying these MHE would significantly slow ammunition provisioning to key enemy units and during enemy operations. Replacement of ruined large tires is difficult in the field and is an enemy supply chain burden.

- Rough terrain cranes are also used to transfer shipping containers, missiles and rocket pods from rail cars and long-haul road transport; however, they are not nearly as effective or as rapid in transfer operations as the RTFL. Cranes are essential in transfer to/from watercraft. Rough terrain cranes also use large, difficult to replace tires that are vulnerable to weapons fire and caltrops.

- Smaller RTFL (e.g. 4,000lb), especially shooting-boom RTFLs, are critical to rapid stuffing/un-stuffing of shipping containers and are universally used in hard surface and unimproved surface supply points for pallet handling and loading of unit transport assets. These smaller RTFLs use vulnerable, hard to get rough terrain tires.

- Supply points (e.g. ASPs) are frequently moved according to the ebb and flow of operational need. Rough terrain MHE can only self-deploy short distances; organic or supporting trailer transport is generally required for movement of MHE; without trailer transport, rough terrain MHE would soon become disabled if road-marched, even over moderate distances. So, destruction of prime movers would have a cascading effect on the capability of supply units/points to relocate with potentially substantial effect on enemy operations. And without a full complement of MHE, transloading of supplies would then be more dependent on manual labor.

- <u>Anti-aircraft Artillery/Missiles Systems and Radars</u>. Self-Propelled Anti-aircraft Systems, units and Resupply Vehicles and stationary AA systems pose a threat to aviation assets that are in support of the Team; destroy these systems before they can engage critical aviation assets. AA missile systems often require road access (especially for large missiles and radars), and must be sited on terrain that facilitates weapon line-of-sight and the use of radar, IR or optical target acquisition and warning systems; the weapons must be positioned to provide broad, unobstructed firing zones to protect/over-watch critical enemy capabilities. The same criteria generally apply to AA gun systems and their target acquisition sensors. Some large AA missiles, in transport configuration, may resemble WMD capable missiles/ rockets in convoy, but they will not have the same level of security. When these systems are displacing on the battlefield, it is generally a predictor of planned enemy offensive or defensive operations.

- <u>Weapons of Mass Destruction (WMD) logistics convoys and WMD capable systems (missile and large rocket) units</u>. WMD logistics convoys, which resupply firing units with warheads or missiles, motors and other essential components, have a distinctive signature; this signature should be made known to US Team Members. The signature of WMD firing units is made obvious by the presence of massive Transporter, Elevation and Launch (TEL) systems. Teams should abandon nearly all other missions to engage these targets, regardless of risk, for obvious reasons. If these units can be engaged at choke-points, creating a convoy delay, US/allied air assets could have the time to penetrate enemy airspace and destroy WMD columns, but in deep penetration operations, timely CAS or fire support is unlikely. In technologically advanced militaries (e.g. Russia, China, etc.) missile and large rocket units are considered elite organizations and their C^2 and firing unit personnel are highly trained. TELs and their resupply vehicles will be accompanied by strong security escorts that are organic to the missile unit. A large security presence may be a tip-off to WMD presence.

Be aware that the destruction of a TEL, or a resupply vehicle, may precipitate a substantial secondary explosion of the motor propellant and may involve the dissemination of WMD material. Destroy the systems from upwind, using remotely fired weapons/munitions to mitigate Team risk. As WMD systems and logistics convoys will normally have a strong security escort, remote attack is wise. All Surface to Surface Missile/Rocket Systems have vulnerabilities, but some are more vulnerable than others.

- All TELs run on very large truck tires that can be deflated by weapons fire or the use of large caltrops. This assumes that run-flat tires are not used.

- <u>Some Keys to Detection</u>: TELs are large multi-wheeled vehicles that are constrained to roads that can accommodate the substantial weight, width and height of the vehicle, so IPB analysis may be productive in forecasting deployed locations. Enemy decoys and deception will be used. TELs will likely occupy terrain that screens their location (e.g. 'dark' side of a ridge) and will seek heavily vegetated positions. The TELS will have the capability of moving through light vegetation, so mines/IEDs should be used at choke points along escape routes (once the system is flushed out) that the vehicles <u>must</u> use. Expect that prepared positions will be constructed to protect and conceal these systems at initial and subsequent field locations.

- The coalition experience during the Iraq War, in attempting to interdict SCUD TELs and their support vehicles (noted in Chapter 1), indicates that locating these systems in large geographic areas (even with satellite observation platforms) and then attacking in a timely manner, is quite difficult. Subsequently, SR Teams may have to take direct action when the opportunity presents. Rocket/Missile Forces Security elements will take extreme camouflage measures to include taking steps to conceal traces of movement. The firing units and their support vehicles will go dark and communicate only in the most secure manner. Security elements will be heavily armed and professional.

- Large missile systems will move during darkness and during conditions that will deny aerial/satellite observation (unless they are flushed out) and they will also use terrain features, whenever possible, to screen movements. The T/L should be aware of friendly programmed aerial and satellite reconnaissance over-flights of the target area, as lapses in coverage will be optimum times for enemy movements – especially of high-priority units/ equipment.

- A remotely fired EFP can be very effective in attacking a TEL (and its missile) or support vehicles. Ideally, the EFP might be oriented as a mine to fire upwards through the TEL, resulting in a very large detonation (missile motor/fuel and warhead contribution) with the possibility of CBRN contamination (WMD warheads).

- Enemy hide positions may be prepared well in advance of a pending movement. Early telltales may be the presence of bulldozer tracks, fresh evidence of earth moving activity, proximity to roads and road junctions, natural camouflage and protective terrain.

- Many TELs are both conventional and WMD capable firing units; always assume that the TELs and resupply vehicles are carrying WMD warheads. Once a WMD/large missile system is in its hide position, it will only move subsequent to firing, to address new targets or upon discovery of a threat.

- Enemy WMD firing units are unlikely to co-locate with other units in base areas for collective protection.

- Warhead and missile motor resupply vehicles will be located separate to the TELs. They will collocate only to conduct weapon transfer tasks. If a TEL relocates to another hide location, its resupply vehicles may move with it to take advantage of protection furnished by the security elements. Alternatively, the resupply vehicles may be pre-positioned at the subsequent/alternative firing position where the TEL will join it after firing. If the

Figure 63. Russian Topol-M on the MZKT-79221 TEL. (*Public Domain*)

TEL must exchange warheads (e.g. conventional to WMD), the resupply vehicle will likely move to the TEL location. TELs may be located well away from other enemy units, and behind terrain features, so that other enemy units are not caught in a counter-strike envelope. After firing a WMD missile, the TEL and its resupply vehicles may move upwind of the firing position to avoid weapons effects and contamination inflicted by a WMD counter-strike. Ideally, the WMD-equipped TEL and WMD warhead resupply vehicles, should be positioned so that neither are downwind of the other – and be separated by terrain features and distance to survive a counter-strike.

- The WMD/large missile system will have at least two different long-range field communication systems. One type, the primary fire command system, may be a satellite van with dish antenna, which will be connected by land line to the CP and not too close either to the CP or the firing units. Other (both AM and/or FM) communication systems will also be deployed as back-up to the SATCOM and for the conduct of more mundane tactical and administrative functions. If reliable telephone and/or fibreoptic cables can be tapped into, this would be a preferred communications method in enemy rear areas where WMD or large missile system are likely to be located.
- Generally speaking, it will be difficult for the SR Team to anticipate TEL movements, or movements of other high-priority targets. The T/L should consider ways that these systems can be forced or flushed from their hide locations. If the Team plans to flush these targets from their hide locations, the Team should plan an ambush or plant mines on anticipated enemy escape routes. Note that TELs are very large vehicles; their dimensions will restrict what thoroughfares and terrain can be used during relocations.
- Nuclear weapons have redundant safety features to prevent a nuclear detonation if the warhead is damaged or impacted by ordnance (except perhaps by another nuclear warhead); the chief threats to the Team from a WMD missile are: (1) being caught within the explosive blast envelope of the warhead and/or the missile propellant, or multiples of the same if several warheads (nuclear and conventional) and missile motors are caught in the same detonation, (2) being caught in a counter-strike area or (3) being caught within the field of contamination from WMD materials. Therefore, given the explosive nature of the missile propellant and warheads and contamination potential of WMD warheads, the Team's (far) ambush site should be as remote as possible and upwind as well.

- ◦ All TELs have various control panels and generator boxes evident on the side of the vehicle. If these can be disabled, the entire system can be disabled (until repaired).
- ◦ TELs consume a lot of fuel. Track a fuel truck that has a security escort. Attack a TEL while it is refueling, if possible.
- ◦ Some TELs mount a shroud for the missile. When the missile is ready to fire, the shroud will be opened and the missile will then be more vulnerable to small arms attack. Other TELs fire the missile directly from its cocoon; the missile is vulnerable to a Man Portable Air Defense (MANPAD) system (e.g. Stinger). Yet other TELs (e.g. SCUD) do not shroud the missile at all.
- ◦ Operation of a TEL and preparation and firing of the missile requires a skilled crew. Killing the crew, when they dismount from the TEL, will disable the system.
- ◦ Several missile systems use liquid fuel motors; the missile must be fueled with a volatile (and vulnerable) mixture of fuel and oxidizer, before the missile can be launched.
  - The missile fueling operation is hazardous; the fuel and oxidizer are provided by separate logistics vehicles. The SCUD family, for instance, uses kerosene as the fuel and Inhibited Red Fuming Nitric Acid (IRFNA) as the oxidizer – which is highly exothermic in the presence of water vapor or in the presence of the fuel and a spark. Some liquid fueled missiles use fuels that can spontaneously explode in the presence of an oxidizer.
  - Attacking the fuel/oxidizer trucks will stop pre-launch preparation. If attacked while fueling is being conducted, especially as the oxidizer is being added, the TEL, missile and the fuel/oxidizer vehicles and any crew present, may all be destroyed.
  - Mobile missile systems that use liquid fuel include: the SCUD Family of tactical missiles; Chinese DF-3A medium-range ballistic missiles; the North Korea KN-08 and the BM25 Musudan intermediate-range ballistic missiles; the Russian RS-24 Yars and RS-26 Rubezh ICBMs. Some of these systems may have proliferated to other countries. As of 2017, North Korea possessed two liquid-fueled long-range ballistic missiles and was conducting aggressive R&D to field a solid-fueled long-range ballistic missile capable of launch from a mobile platform.
  - Small arms attack on solid-fuel missile/rocket propellant is also effective, but the effects may not be immediate or dramatic as in the case of liquid fuel. The strike of a small arms projectile will create a void and fracture in solid propellant that will likely cause the propellant to detonate after launch. The closer the strike of the bullet to the rocket/missile engine, the sooner the detonation. Use of a shaped-charge or EFP munition may produce an immediate effect.
- Maintenance Units/Personnel. Frontline and key combat support units rely heavily on maintenance units to keep their systems repaired and available. Maintenance units use specialized tools and equipment, and trained maintenance personnel, that are difficult to replace. The more sophisticated the end item, the more specialized and valuable the maintenance tools, equipment and personnel become. Direct and General Support maintenance units are staffed with technicians and have minimum combat power. These technicians are not easily replaced, especially in the higher skill level disciplines (e.g. fire control; sensor systems; C^{4}ISR systems; large rocket/missile systems and ancillary equipment (e.g. TELs); heavy equipment mechanic; etc.). Killing numbers of key specialty technicians can have a substantial effect on the unit readiness/equipment availability of key combat and combat support organizations.
- POL Pipelines. Pipelines will be found in rear areas; the pipelines will be patrolled, but typically by rear area security personnel. Lengths of repair pipe, valves, etc. may be positioned at intervals along the pipeline. If a length of pipeline is breached, the enemy will

deploy a repair crew with security to the location. If possible, attack the pipeline at a shutoff valve or at a pipeline branch; leave booby-traps behind to create casualties of security and repair personnel. Consider the use of Thermite grenades rather than explosive devices.

- <u>Aviators, Aviation Maintenance Personnel and Aircraft</u>.
  - ◦ An airfield is a large area that is difficult to secure.
  - ◦ Aviation personnel require substantial training to become pilots or maintainers. Killing or wounding them has the potential to ground numerous aircraft. Replacement pilots may be inexperienced and therefore more vulnerable during the conduct of their sorties.
  - ◦ Once a combat aircraft has left the ground, it becomes a serious menace to the Team – especially ground-attack helicopters (e.g. Mi-24 Hind or the Mi-28 Havoc). Before attacking an airstrip/airfield, observe the flight path for departing aircraft. If the Team is equipped with MANPADs, attack the aircraft before it reaches maneuver speed or altitude. More on this elsewhere in this book.
- <u>Bridging Units</u> are essential to mobility operations. Destruction of bridging equipment and the trained engineers responsible for bridging operations can constrain enemy maneuver on the battlefield and impair operations at critical points.
- <u>Railroad Locomotives, Tanker Cars and Ammunition Rail Car Cargo</u>. Locomotives, if destroyed, are difficult to replace. If the enemy has enough rolling stock, they may use spacer cars to limit the damage caused by the detonation of an ammunition car. Destruction of a moving locomotive will likely cause derailment of most or all attached railcars. Destruction of a locomotive will cause the other cars in the train to be immobilized until a replacement locomotive can arrive – making the entire train vulnerable to further attack. Tanker cars supply fuel trucks, which are essential to enemy mobility. An off-route mine HEAT or EFP munition strike could detonate an ammunition-laden rail car, which could destroy a good portion of the train and damage the rail bed. Effective harassment and interdiction of an enemy rail capability will require the enemy to transload equipment and supplies to vehicular carriage. This will have the following results:
  - ◦ Transloading sites make lucrative targets, where rail carriage, vehicular carriage, logistics units/personnel/equipment and troops are concentrated.
  - ◦ Vehicular carriage must be reallocated from other priorities to support emergency transloading operations, affecting not only logistics functions but also priority enemy combat operations.
  - ◦ Use of vehicular carriage, vice train carriage, will cause the consumption of fuels to escalate and increase the wear and tear on cargo vehicles.
  - ◦ The enemy will likely have to reroute approaching trains, impairing provisioning.
  - ◦ Vehicular carriage must carry transloaded cargo via road convoy to the required destinations over heavily trafficked road networks, offering lucrative targets to friendly forces, air attack/long-range fires or to further Team interdiction.
  - ◦ Other Rail Targets.
    - ▪ Destruction of railroad repair cars and recovery equipment will significantly delay retrieval/removal of damaged locomotives and rail cars and the repair of damaged roadbeds, ties and rails. Machinery, equipment and supplies critical to war production are often transported by rail to industrial facilities.

'In the fall of 1943 four supply trains were destroyed simultaneously at the Osipovichi railroad station, and all traffic on that line had to be suspended for a long time. Investigations revealed that a magnetic mine had been attached … to one of the tank cars of a gasoline train. When the mine went off it set the car on fire, and the spreading blaze soon enveloped the entire train. An ammunition train standing nearby was ignited and blew sky high, setting fire to an adjacent forage train. Finally, a fourth train loaded with 'Tiger' tanks suffered the same fate and also burned out completely.… Moreover, the explosion of the ammunition train had caused considerable damage to many of the switches, so that the line itself was no longer in operating condition.'[56]

Interdiction of Enemy Rail Transportation:

- Ammunition and Fuel Material Handling Equipment (MHE) and vehicles used in transloading, may be specialized and difficult for an enemy to replace. Attacking these assets may substantially increase enemy labor commitments, slow down logistic operations and therefore impair enemy plans and capabilities. For instance, large tires on Rough-Terrain Fork Lifts (RTFLs)/MHE are difficult to repair or replace and may not be stockpiled forward and immediately accessible. In the absence of military RTFLs, the enemy may seize commercial equivalents (typically used in the construction industry). Stripping the industry of this equipment creates its own problems for the enemy
- Attack mobile construction units (rail, bridging, etc.) or mine their access roads. This will impair the enemy's ability to repair damaged track, bridging or rolling stock.
- Security concerns may result in reduced train speeds at night.
- The enemy will give first priority to defending choke/critical points along its rail system. The Team should generally focus on lower risk, less defended/patrolled corridors and infrastructure.
- Dummy railcars may be used as decoys. An effective practice is to stage destroyed railcars in an open area rail siding, nearby to an area with overhead cover that is the true location of the siding.
- Flat cars loaded with rocks with the locomotive in the center of the train. The flat cars are designed to initiate mines or demolitions while avoiding damage to the remaining rolling stock (especially the locomotive).
- A surge in rail traffic may imply preparation for an offensive, or could be deception operations. SR observation of rail (including the use of remote sensors/cameras) may be able to identify cars by type and even by cargo.
- In areas with mostly unimproved roads, rail lines may be of increased importance, particularly during the winter or wet seasons due to roadway impassability.
- The SR Team may plant long delay, nonmetallic fused mines or demolition charges, with SD features, to demolish rail sections or other rail assets weeks or months after the Team has departed the area.
- Block rail lines in isolated, channelized areas and damage rail switches to trap and fix a train for destruction by friendly supporting fires.
- Robust and redundant road/trail/rail networks (forward and lateral), that support unit movements and logistics distribution, are of vital importance to military units in the field. Where these are not sufficiently available, water routes may be necessary.
  - In a general conflict, these networks provide a target-rich environment for SR Teams. It is a primary purpose of SR Teams to identify/detect and interdict such networks and capabilities, and the units they support – and more specifically, those that represent capabilities most critical to enemy operations. The TTPs contained in this book are necessary to achieving the primary purpose of SR Teams.
  - These networks are not quite as vital to partisan/guerilla/insurgent organizations of modest size, as these organizations can subsist off of local supply sources, supplemented

---

56. Unnamed former German Panzer Army G4, 'Historical Study, Rear Area Security in Russia', pp. 25-6.

by limited covert deliveries from a state/non-state sponsor. The SR unit purpose remains the same, but the TTPs employed will vary. Once these organizations reach a certain size or level of activity, or as they attempt to transition to conventional operations, SR Teams must employ other TTPs.

- Attacking the enemy in his rear support areas will cause him to divert resources to rear area security; effectively supplementing or conflating the SR mission purpose with an 'Economy of Force' mission purpose, compelling the enemy to the commitment of additional resources for security. In response, the enemy may concentrate Combat, CS and CSS units and functions for mutual protection. This approach seemingly leaves clustered units more easily detectable and vulnerable to attack; which notionally, may assist the SR Teams in their operations. However, enemy units in Laos and Cambodia were clustered into large Base/Sanctuary Areas; the combination of heavily dissected terrain (natural cover); heavy and continuous canopy (natural concealment); co-locating of front line combat units in Base Area sanctuary locations; a robust, dedicated security presence (including SpecOps), a huge road/trail network (route redundancy) and aggressive patrolling was successful in mitigating the vulnerabilities of base clustering.

> 'In any event, the results of guerrilla warfare and of partisan attacks employing guerrilla tactics were serious enough in themselves to need no embroidery. One historian of Civil War guerrilla activities, Virgil Carrington Jones, has estimated that Confederate guerrillas held back as many as 200,000 Union troops from the active armies, Col. John Singleton Mosby, himself a most active and successful Confederate partisan leader, wrote after the war that, with no more than 200 men, he was able at one time to force detachment of 30,000 troops from the Union Army of the Potomac.'[57]

> 'Areas in close proximity to the front are always the scene of strong concentrations of forces which have firm control over the local rail and road net and are in position to keep the local population under close surveillance…. An entirely different situation prevails in the rear areas where the vastness of the country, sparsely covered by German troops, presented a constant problem…. The protection of … supply depots involved a variety of problems…. Since these installations were to include warehouses for all classes of supply, as for instance rations, clothing, ammunition, fuel, medical and veterinary equipment, as well as motor vehicles and spare parts, the need for security forces grew considerably as operations progressed…. The mere fact that some of the largest supply installations might well assume the proportions of a medium-sized city may offer an indication as to the number of security troops that would become necessary.'

Ammunition and fuel supplies were heavily revetted/protected by existing terrain features or were specially constructed in rear support areas, where the Russia air attack became a theat.[58]

- <u>Other Priority Targets</u>, in no special order, are:
  - Armored vehicles.
    - Track-mounted AA gun systems (e.g. 12.7mm to 40mm), sometimes used in convoy security, represent a grave threat to the Team. If any of these weapon systems are spotted, avoid engagement and wait for a softer target. However, the presence of such weapon systems in convoy security may signify a high priority target traveling in the convoy. If this is the case, consider engaging the escort first before engaging other high priority targets in the convoy. Strip this target from the convoy using remotely-

---

57. Andrews, Isolating the Guerrilla, 'The American Civil War', p. 27
58. Unnamed former German Panzer Army G4, 'Historical Study, Rear Area Security in Russia', pp. 4-9.

fired anti-armor weapons that are initiated from firing positions away from the Team's location.

- Killing an armored vehicle that is employed in the convoy protection role will create a road obstacle to other vehicles, if the ambush site is properly located (e.g. at a choke point), making other convoy vehicles vulnerable to further attack by the Team or air interdiction. Armored vehicles bristle with protected weapons (main gun and machine guns) that represent a significant threat to the Team. This protection-level may signify a high priority target elsewhere in the convoy. If a lone tank is spotted in convoy duty, engage it with anti-tank weapon or munitions (especially if remotely-fired). Note that Armored Fighting Vehicles (AFVs), especially wheeled AFVs, are more likely to be convoy escorts than tanks. If tanks are part of an armored column, use remotely fired anti-tank weapons, rather than direct-fire anti-tank weapon systems.
  - Infantry Fighting Vehicles (IFVs) and Armored Personnel Carriers (APCs), both tracked and wheeled, combine infantry combatants with vehicle organic weapon systems; as such, they carry many more anti-personnel weapons than a tank, and therefore represent a greater threat to the Team. Armor protection for these vehicles is inferior to that of tanks and they are more easily defeated by the organic weapons and munitions carried by a Team.
  - Armored Vehicle Recovery Vehicles. Incapacitated or battle-damaged armored vehicles can be restored to operational condition – if they can be recovered by specially designed armored recovery vehicles. Destruction of an armored recovery vehicle will keep several armored combat vehicles from repair and subsequent use in operations.
  - Mounted (e.g. on soft-skinned vehicles) heavy/crew-served weapons. This may include medium to heavy machineguns, automatic grenade launchers, mortars or recoilless rifles mounted on or towed by vehicles and may also include weapons in the hands of troops and vehicle crewmembers carried on soft-skinned vehicles carrying medium machineguns, flame weapons, sniper weapons, grenade launchers/rifle propelled grenades in the hands of mounted troops or vehicle crewmen.
  - Dismounted personnel may be equipped with key or crew-served weapons, especially snipers, machineguns, flame weapons, grenade launchers/rifle propelled grenades or Anti-Tank rockets/missiles.
- Observe villages and structures for the presence of enemy personnel and operational use.
  - Unoccupied buildings behind enemy lines, or in the path of an advancing enemy, may subsequently become occupied by enemy units. If such structures are likely to be used, the Team might place sensors to monitor occupancy remotely.
  - An abandoned/derelict industrial facility may be ideal to accommodate an enemy mobile headquarters within building interiors. Other structures that are located near road junctions or that have access to power and/or communications lines will also be desirable to headquarters units.
- If weather, terrain and vegetation permits, force the enemy out of an assembly/marshalling or built up area; or destroy critical equipment and stocks by using fire. Forcing them out into the open will leave the enemy vulnerable to air attack or long range fires.

Raid TTPs:

Raid General TTPs:
- <u>Definition</u>: A raid is a surprise attack from a concealed position on a static target to temporarily seize the target in order to secure information, confuse an adversary, capture personnel or equipment, or to destroy a capability, culminating with a planned withdrawal.

- In rear areas, especially among CSS units, the enemy may be lax, ill-disciplined, poorly led or trained. If the enemy is unaware of the SR threat, sloppy behavior may be more prevalent. Behavior indicators would be:
  - Smoking
  - Cooking fire or fires for warmth
  - Loud talking and camp noise
  - Weapons not ready for action
  - Poor security
- Most classes of supply (except munitions or POL) are difficult to destroy or sabotage, especially if the stocks are revetted or separated by distance. If the Team encounters enemy logistics stockpiles, the Team may not have the capability to destroy the supplies. Example: How would a Team destroy several tons of rice – or spare parts?
  - Have a plan on how to quickly, effectively and efficiently destroy or sabotage various classes of enemy supplies!
  - Fire (flame) is the universal solution to destroying masses of supplies. If conditions of weather, terrain and vegetation permit, this may be the best approach to destroying enemy logistics facilities/stockpiles. In deep penetration operations, beyond the reach of friendly air support, this may be the only option. Use of napalm or other thermal weapons may be ideal.
  - The best time to initiate the destruction of enemy stocks is when supply transfers are being conducted. For instance, if a convoy of cargo vehicles arrives to load or unload ammunition, that would be the preferred time to attack, because trucks, rail cars, MHE, drivers and supply crews might also be destroyed.
  - Mark the ends or the center of mass for the logistics facilities/stockpiles with beacons, IR markers, time-delayed or remotely controlled marking devices. This will facilitate air attack, at a later time, by aircraft equipped with the optimum ordnance (e.g. incendiaries). This will also allow the Team to move off a safe distance until the air attack is over, and then return to exploit the chaos and enemy response.
- If it is necessary to assault an enemy machinegun emplacement during a raid, attack from the enemy's right, as the machine-gunner will have more difficulty in traversing to his right, and the left-feed ammunition belt may crimp in the process; to get the Team Member(s) in his sight, he may have to displace the machinegun or remove it from its mount. Even if the enemy is firing with a bipod, he will have similar problems in shooting to his right.
- To destroy wooden trestle bridges use Thermite grenades instead of explosives. By weight, the incendiary approach is much more efficient and effective against flammable targets. A fire, begun with a few Thermite grenades may nearly consume the entire structure, while substantial amounts of explosives may be required to knock down a few supports or drop a section of bridge.
- Minimizing ToT is essential to Team survival. DO NOT overreach in the attack; limit the raid objectives rather than risk Team annihilation. Keys to rapidly executed raids include:
  - The plan must be simple.
  - Locate where enemy guards, gun emplacements and reaction forces are located. Consider integrating mortar (mix of HE and smoke) and overwatch machinegun or Mark 19 40mm (mobility systems) fire to suppress these enemy threats. If smoke or flame weapons are used, these targets will be obscured, use range cards, stakes or other devices to align Team weapon aim points.
  - Raiding of airfields and other key installations that are located in flat, featureless terrain such as desert/arid land, savannah, etc. will require Team ground mobility equipment.

- Conduct reconnaissance and surveillance of enemy security measures, to include installation access points, sensors, ground and aerial patrolling, barriers/minefields, etc. Satellite/UAV imagery will not provide this granularity. Deployment of remote digital cameras within the enemy's security zone may be essential for surveillance of vital installations.
- The raid should be carried out at night and should be planned for <u>periods of inclement weather</u> and/or when moon illumination is minimal.
  - The earlier the attack occurs during hours of darkness, the more time the Team will have to withdraw under its cover.
  - Relatively few airfield or logistics facility security personnel and none of the aircrews or maintenance personnel are likely to have individual night-vision equipment.
  - Low ceiling and/or inclement weather may mean that most aircraft will be grounded and vulnerable to attack; these conditions will suppress an aerial response to the Team during its withdrawal.
- As in all tactical operations, use terrain to cover and conceal the approach to the target and to screen Team elements from enemy observation and defensive fires.
- Thermite grenades may often be the most effective munition to destroy aircraft, logistics stores and ground equipment; but this technique generally requires intimate contact of the munition with the target.
  - A Thermite grenade is a fairly heavy object and cannot be thrown very far. It may actually be overkill in its standard configuration. It is not necessary to melt entirely through the hood and the engine block to disable a truck engine; or to disable a tank turret. Half the quantity of thermite would suffice in many circumstances.
  - The US Thermite grenade has a near-instantaneous fuse. This fuse should be replaced with a longer delay or use a remote initiation to allow the Team Member to take cover/ concealment. A Thermite grenade (e.g. of foreign manufacture) equipped with a longer delay, would allow a sling to be used to launch the Thermite grenade much further than it can be thrown by hand (explained later in Chapter 4).
  - A Thermite grenade will ignite the aluminum skin of an aircraft, which will burn at a very high temperature. If the aircraft is in a hanger or if it is in close proximity to other aircraft when it is attacked it may take other equipment/material with it when it burns and/or detonates.
  - To disable/destroy an armored vehicle, place the thermite device over the engine, fuel or ammunition compartments; inside the barrel of the main gun; over the gun breech (requires interior access); at the main gun mantlet or between the turret and body (to create a weld). Vehicles with aluminum composite armor may ignite and consume the vehicle entirely.
- Where vehicles/aircraft are dispersed, or where the Team Member cannot ensure intimate device-to-target contact, use explosive devices that incorporate a multi-purpose warhead (e.g. Shaped/EFP, fragmentation and incendiary effects). The devices can be placed a short distance from the target and would be effective against light armor, light-skinned vehicles and aircraft and a broad spectrum of logistics items in storage. The devices will provide penetration, and the incendiary component should ignite POL, munitions, aluminum aircraft skin and other combustible materials.
- A mortar round mix of HE and WP can be used to penetrate the skin of aircraft or refueling vehicles, etc.; the WP can ignite the fuel.
- The use of small, armed drones may be an excellent approach to attacking an enemy airfield or aviation assets.

- ▪ Pros: Low detectability, especially at night and during noisy enemy operations. Provides a precision, stand-off capability that mitigates risk to the Team and enhancing Team withdrawal; capable of penetrating open hangars, revetments and other structures; capable of reconnaissance/surveillance, target identification/selection, repeated sorties and engagements per sortie.
- ▪ Cons: Limited cargo, limited range/endurance. Cumbersome to transport. Drones and ordnance, support items, etc. must be cached. Team Members require training. Small ordnance size results in reduced ordnance effects on targets, especially armored systems.
- ▪ Armament: Ordnance should be dual/multi-purpose (e.g. armor-piercing, fragmentation, incendiary) to allow penetration of aircraft skin and/or to destroy the aircraft or its critical components. The drone itself may be operated as a glide-bomb or ordnance may be launched, dropped (bomblets, grenades/mini-grenades) or placed (incendiary, specialty munitions) by the drone. Fusing may be point-detonating/immediate, time-delayed, or remotely initiated. Given small ordnance size, the best ordnance to destroy an armored helicopter would be an incendiary.
- ○ Some attack helicopters (e.g. Russian Hind and Havoc helicopters) have armor incorporated in their design; they represent a substantial threat to a Team during and subsequent to its attack on an airfield; they must be a priority target during the operation. When attacking these aircraft on airfields or in revetments, concentrated fire of a .50-cal machinegun (vehicular mounted preferred) firing suitable armor-piercing (AP), armor-piercing incendiary (API), Saboted Light Armor Penetrator (SLAP), Saboted Light Armor Penetrator-Tracer (SLAP-T) or the High Explosive Armor Piercing Incendiary (HEIAP) will destroy these aircraft. Ammunition belts incorporating an incendiary munition with the AP effect would be preferred, as it will ignite fuel, hydraulic fluid, and other inflammable stores. As effective, if not more so, to include use against personnel and ammunition stocks would be the vehicle-mounted Mark 19 40mm grenade launcher (or foreign equivalent) firing HEDP munitions (incorporates a shaped-charge). Acquire intelligence on aircraft vulnerable points.
- ○ Use captured enemy uniforms, equipment and vehicles, if possible, for covert entry/egress and to confuse enemy pursuit.
- ○ Revetments with overhead cover must be attacked with drones or from ground level (e.g. using armed mobility equipment). If revetted positions contain attack helicopters or other ground attack aircraft, the Team's must make a determined effort to destroy these protected aircraft first.
- ○ Pilots and maintenance crew require extensive training and are not easily replaced. They should be a priority target. Pilots and maintenance personnel will attempt to rescue aircraft not destroyed during and immediately following an attack. This may be an optimum time to kill these personnel, while they are out in the open, with time-delayed devices or with mortar fire. Pilots will attempt to get flyable aircraft airborne and those flying ground attack aircraft (fixed/rotary) will immediately commence a hunt to find and kill the Team. The Team should do its best to ensure no attack aircraft leave the ground. An attack on the aircraft and aviation personnel may be accompanied by cratering the runway, ToT permitting.
- ○ Beware of combat aircraft that are already airborne which have just departed the airfield on a sortie, or are returning to base, as these aircraft may be immediately employed against the Team, especially during its withdrawal. A mobile air-guard equipped with MANPADs may be prudent. Another reason to attack during bad weather.
- ○ Snipers may also be effective in overwatch. They should target crew-served weapons, security force personnel and pilots. A .50 caliber sniper rifle will be effective in destroying enemy aircraft and ground equipment.

- ○ One Team Member on each vehicle should carry a silenced weapon for initial entry to the airfield.
- ○ Obscurant smoke, WP and Thermite grenades should be carried on each Team vehicle.
- ○ Consider a remote raid by bombardment if the target is deep in enemy controlled territory, where the enemy possesses air superiority, where the enemy security zone is heavily patrolled/observed, if Team exfiltration is not immediately possible, or where withdrawal and evasion is especially problematic.
  - Raid by bombardment uses numerous rockets to saturate an enemy concentration/ facility.
  - It is most suitable where supplies are in open storage and where mobility or aviation systems are concentrated and parked in the open. It may also be suitable against structures, revetted equipment or stores/equipment under canopy, especially if the bombardment is heavy enough.
  - Raid by bombardment planning is simple, but preparation is time-consuming. Steps include:
    * Selection of munitions, warhead types and fusing.
    * Fabrication of expedient lightweight launchers and aiming and firing devices.
    * Training and testing to create firing/range tables, if necessary.
    * Team insertion, including mobility equipment.
    * Reconnaissance/identification of approaches and firing positions.
    * Selection, establishment and camouflage of cache/MSS sites.
    * Delivery of munitions, firing components and launchers to guerilla/partisan unit LZs/DZs or to LZs/DZs near Team cache/MSS sites. Moving items from LZs/ DZs to cache/MSS sites. Movement may be performed by guerilla/partisan porters, pack/draught animals (with carts, ahkios, etc), ground mobility equipment or a combination of means.
    * Advanced preparation of munitions.
    * Movement support by guerilla/partisan personnel (e.g. porters) requires enhanced security measures. Betrayal is possible. Quarantine porters (if used) in advance of the operation and thereafter until the mission is complete. Ensure porter accountability throughout the execution of the mission.
    * Movement of munitions, firing components and launchers forward to firing positions.
    * Launcher loading, set-up and aiming.
    * Extensive camouflage of launchers and firing devices.
  - Electrical initiation of the missiles may be accomplished using a delay device set to initiate well after the Team has withdrawn and made its escape – or the initiation could be executed using remote/RF signals.
  - An assortment of US/foreign air-to-ground or surface-to-surface rockets may be feasible in this application, such as 122mm 'Katyusha-style' rockets or even 2.75-inch Hydra 70 air-to-ground rockets. For instance,
    * Hydra 70 has an effective range of approximately 8,000 meters and a maximum range of approximately 10,500 meters (fired in the aerial configuration with standard launchers). Used in a surface-to-surface role, the range would be somewhat less, but still outside an enemy airfield or installation security zone; range testing and establishment of firing/range tables would have to be conducted.
    * Warheads for the Hydra 70 would include White Phosphorus (WP), HE, HE high-capacity, HEDP, APERS and sub-munition dispensing. Fusing for these warheads may include Point Detonating or Air Burst/Proximity. Attacking with a mix of warheads and fusing would have a devastating effect on a target. Different rocket

configurations will have different flight characteristics and ranges, testing would be necessary for each configuration.
- Lay a false departure trail, if feasible, prior to the attack. Escape at night during inclement weather and by separate routes, if feasible. Leave behind time-delayed/remotely-initiated decoys (e.g. false firefight devices) on the opposite side of the perimeter to deceive the enemy as to the point/route of withdrawal. And leave behind anti-vehicular, anti-armor and off route mines (with SD) to discourage pursuit. In open terrain concealment of tracks may be prudent, if done at night; drag brush or chain link fence behind the vehicles. Deceive enemy air by masquerading in enemy vehicles. Before dawn, move vehicles into prepared, camouflaged hides. Disperse personnel away from the vehicle hides and into prepared, camouflaged defensive positions with overhead protection.

The SAS conducted a raid at *Sidi Haneish* airfield during the night of 26 July 1942, destroying a significant number of aircraft; it is an operation worth studying. While this was a Direct Action operation, a similar operation could be conducted by an SR Team – if the target of opportunity is worth the risk and reduced in scope.

- Note that an airfield occupies a lot of real estate and is therefore difficult for ground troops to secure. So, a raid by a Team equipped with mobility equipment, on an airfield (or another thinly defended enemy facility) may be feasible. A good example of a vulnerable facility would be a rail-siding where supply (especially munitions, POL) transfer operations are being conducted. The conduct of the raid would require a thorough reconnaissance of the target and its surrounding area, to include routes of approach and withdrawal.
- Munitions and weapons must be appropriate and effective in terms of the target(s). For items susceptible to fire, the Author would recommend a fragmentation device combined with a secondary incendiary effect.
- Approach, task execution, ToT and withdrawal must be very rapid.
- The withdrawal and evasion phase is almost as important as the attack. If a successful withdrawal is improbable, the raid should be shelved. A desert raid poses special risks, requiring specific route selection precautions, as the enemy is likely to use air assets to pursue the Team (enemy will have air superiority).
- Planning/preparation should encompass daylight hide locations during the approach, withdrawal and evasion phases. Vehicle hide considerations include:
  - Hide vehicles behind vegetation, under camouflage netting and within shadows.
  - One vehicle should eradicate tracks of other vehicles before seeking its own hide position.
  - Team Members should occupy defensive/hide positions away from vehicles.
  - Pay attention to shadow drift as the sun continues its path. Move vehicles/camouflage accordingly.
  - Run vehicle motors periodically during the day to charge batteries. Designate one or two vehicles to carry flexible tubing to attach to a vehicle exhaust to minimize heat signatures at night.
- In deep penetrations, villages may have bordellos that attract officers or NCOs. This could be a ripe target for a raid or POW snatch.

Rail Network Raids TTPs

- When the Team attacks or destroys rail facilities/infrastructure, it is considered a raid or sabotage; when the attack is against rolling stock in motion, it may be considered an ambush.
- Mine/booby-trap enemy railcars, vehicles (and their cargos); especially if the devices can be initiated remotely or by an anti-disturbance feature, at critical chokepoints, or when they

are being unloaded at their destinations. For instance, a booby-trapped pallet of artillery projectiles unloaded and stored at a munitions depot or ammunition supply point could have a catastrophic effect through inter-round/inter-stack conflagration. If the pallet is delivered directly to the using unit, this too can be devastating. Alternatively, tag an enemy vehicle with a tracking device to determine its destination for subsequent destruction by supporting fires.

• If possible, delay-fused mines, demolitions, incendiaries, booby-traps, affixed to a POL car should preferably be attached near the air space above the fuel line – because fire requires oxygen; but these would easily be detected. Alternatively, mines or demolitions, attached below the air space, can be used to breach the compartment, in combination with an incendiary to cause the conflagration. Further, the Team might use an EFP, combined with an incendiary capability, to attack the fuel car without having to physically mount a device on the fuel-car; the EFP can be fired upwards from the ground through the tank car. Note that the Team should not rely on magnets to affix a device to a POL car, as the tank-car may have an anti-magnetic coating.

• Perhaps the ideal opportunity to destroy ammunition or POL supplies is during trans-modal transfers. During these operations, a detonation or fire may destroy rolling stock; vehicles; MHE; transportation, security and ammunition personnel and rail infrastructure.

> 'The French Resistance, supported by the British SOE, sabotaged vital machinery at a factory in France that was essential to the manufacture of German tank turrets. The amount of explosive used in this sabotage was miniscule as compared to what would have been required in a bombing attack (the total amount of explosives used by the Resistance in the sabotage operations from 1940-44 was a mere 3000lbs). When the replacement machinery arrived from Germany months later, SOE operatives and Resistance fighters attacked the rail car and destroyed the equipment in the rail yard before it could even be transloaded. The factory would not produce the vitally needed tank turrets through to the end of the war.'[59]

• Some rail networks (e.g. in Europe) are electrical, using overhead electrical wires or third-rails to power electric locomotives. Shooting electrical transformers that supply power to the rail network is an easy and effective way to stop a segment of local rail traffic. Transformers are typically not stockpiled and manufacturing replacement transformers is time-consuming.

• A sabotaged rail switch can divert a train onto a siding, spur or another track. A thermite grenade can freeze the switch/switch-stand into one position, if this is desired (best in daylight so thermite flare cannot be detected). The Team should also sabotage the switch by disguising the point indicator with a false placard (for instance) pointing in the wrong direction. Additionally, some rail signals can easily be sabotaged to reflect safe track and train speeds for areas where the Team may have sabotaged the rail or a switch. Note that some switches are electrically linked to a dispatcher control system that would automatically alert a dispatcher to a misaligned switch. Note also that rail signals are often not universally standard. The SR Team should discover what signals prevail in a given AO if they intend to devise a sabotage method. Imagine a train moving at full speed, suddenly diverted to a siding or spur occupied by railcars engaged in ammunition transfer operations; or suddenly transferring at a rail junction into the path of an oncoming train.

• One of the most lucrative rail targets will be specialized maintenance cars used in Maintenance of Way, Track-Laying, Signal Repair and Recovery. Maintenance of Way equipment can be expensive, complex, automated systems that are difficult to replace. Normally, two Recovery Cranes are required to lift/recover a locomotive and are vital to clearing a wrecked train.

---

59. M.R.D. Foote, *SOE in France*, (digital rendering without pagination).

Figure 64. Italian Maintenance-of-Way Equipment. Notice electrical lines above. (*Public Domain*)

Figure 65. Example of a Breakdown (Wrecker) Crane Used in Accident Recovery. (*Public Domain*)

- It may be unnecessary to destroy rails with munitions; other, more silent, mechanical means are available that are capable of unseating rails from their ties. Thermite grenades can supplement the mechanical means by super-heating the rails, allowing them to bend more easily. Test the use of modified jacks, or block and tackle for this task. Note that some countries use concrete, rather than wood ties.

  'Retreating Russian forces often buried mines with long delay fuzes under the tracks where they might blow up as much as three months later.'[60]

- Key locations along the railroad network will be guarded; these include: trestles, railroad passenger and freight stations, fueling stations, military logistics rail sidings, rail yards, tunnels and turntables, etc. Other than the key locations, rail lines, overall, will not be guarded, but they will be patrolled periodically. Personnel assigned to guard/patrol duty will likely be second or third tier troops, reservists or home guard personnel; however, be aware that veteran combat troops, who may be recovering from wounds, may temporarily be assigned to rear security duty. Also, combat units located in nearby marshalling/assembly areas may be called upon to temporarily provide rear area security support – so do not underestimate rear security troops.
  - Damaging rail signals can cause the collision of trains, prospectively doubling the destructive potential offered by other types of attack. Or damage of signals can cause a train to halt so that it can be ambushed or made vulnerable to attack. If the Team has appliqués (opaque, transparent, luminous) that can quickly be pasted over the signal light (to change its color or block the light), the same ends may be achieved. Rail switches can be frozen or destroyed with a thermite grenade; this too may cause a train to halt or to collide with another train.
  - Attacking the cab of a locomotive with long range or remotely detonated weapons may kill train engineers activating the dead-man's switch and causing the train to halt.
  - A train can be attacked from below and from above with relatively small demolition devices/charges; for instance: mount a camouflaged, remotely-detonated EFP or off-route mine below an overpass to attack a locomotive or ammunition and/or fuel cars as they pass beneath. An EFP can also be mounted beneath a train trestle so that it will fire upwards (e.g. directed at a tanker car); if the EFP also has a secondary incendiary feature, several

---

60. Unnamed former German Panzer Army G4. 'Rear Area Security in Russia', p. 25

other railcars and much of the bridge may go up in flames. The Team could employ some trickery to slow the train down, such as a tree-fall across the track.

- ○ Use mines and booby-traps to kill enemy patrols and repair/recovery personnel.

- It may not be necessary to use demolitions to destroy a concrete railway overpass, or for that matter a concrete road overpass/bridge, where the rail traffic passes beneath. A very hot fire will severely weaken concrete to the extent that it will subsequently collapse under enemy traffic.

- ○ Local materials ignited beneath the overpass may do the trick, but that would be labor and time intensive.
- ○ Ambush cargo laden enemy vehicles or halted rail cars as they pass beneath the overpass.
- ○ Steal a fuel truck, park it on top of the overpass, open the valves and ignite the fuel.

Defensive Operations TTPs:

- There are occasions when a Team must operate from a static position. These occasions include NDP, surveillance locations, ambush and defensive positions. NDP and surveillance procedures are discussed elsewhere in this book. If the Team becomes trapped or surrounded by an enemy force, or if the Team is on-the-run from a more mobile enemy force, the Team may have to assume a temporary defensive posture pending an extraction or a breakout operation.

- Given US Air Superiority over the Target Area, the enemy can be expected to use the following approach to overwhelm the Team in its defensive position: (1) if sufficient forces are immediately available, the enemy may attempt an immediate assault to overrun the Team; (2) if sufficient forces are not immediately available, the enemy may attempt to closely surround the Team, while more enemy forces are rushed to the scene; otherwise the enemy will employ blocking forces at likely avenues of egress; (3) the enemy may employ reconnaissance-by-fire to provoke Team defensive fires in order to pinpoint Team defensive and crew-served weapons positions; (4) the enemy may reposition some of his anti-aircraft assets to over-watch the vicinity of the Team perimeter and likely LZs (a trap for subsequent US extraction/rescue operations), using the Team as bait; (5) the enemy may position his assault elements prior to dark, and using cover and concealment, may move these elements forward as close as possible to the Team (e.g. outside grenade-throwing range) to inhibit the use of close air support; (6) the enemy may attempt to rush the Team at dusk; intermingling of US and enemy forces will preclude the use of US CAS; the enemy may use artillery/mortar bombardment, if these assets are available, to inflict casualties and to force the Team out of its defensive positions and into a break-out attempt.

Attacks from Enemy Air/Fire Support.

- During deep penetration operations, Teams might not receive enemy air/fire support attacks unless they have been detected and located. Detection/location would most likely happen when the Team must cross a danger area; but it may also happen if the Team has been sloppy (e.g. exposing its location to thermal optic sensing, etc.).
- Enemy rear area units and activities will typically not possess artillery or mortar weapons; and rear area security forces may seldom have support of such weapons. Engagement by these weapons typically indicates the presence of combat arms units, which may be co-located with combat support or CSS units for common defense, located in sanctuary locations or in temporary marshalling areas, deployed in nearby counter-insurgency operations – or have been allocated to dedicated counter-reconnaissance operations in a sensitive location.

- If the Team is attacked by air, or by artillery/mortar bombardment, it may indicate that the Team has been observed or is under observation by a counter-reconnaissance/special operations element, an aerial platform, a Forward Observer (FO) or an enemy OP.
- Indicators that an enemy air attack is imminent includes:
  - Observer aircraft making passes or loitering in the vicinity
  - UAV activity
  - Orbiting attack helicopters
  - Marking round(s) fired by observer aircraft or signals (smoke, flares, etc.) from a ground observer
  - High performance aircraft making a pass over/near the Team location
- Indicators that an enemy artillery/mortar attack is imminent includes:
  - Observer aircraft making passes or loitering in the vicinity
  - UAV activity
  - Marking round(s) or signals (smoke, flares, etc.) from ground observer
  - Flashes (at night) from artillery tubes in the distance
- As mortars are used for high-angle fire, the Team may be able to hear mortars firing from a kilometer or more, before the impact of the rounds. The type of mortar can be determined by the sound made during firing. A deep timbre will signify a larger caliber tube. The size of the tube will reveal the size or type of unit firing on the Team; for instance, a 60mm mortar may be a company weapon, a 82mm mortar may be a battalion weapon, etc.
- The best policy is avoidance; adherence to the TTPs in this book should help in this ambition, but once observed and engaged, the immediate task of the Team is to rapidly escape from the observer's view.
  - If the T/L believes that the attack is imminent, the Team should 'run like hell', using terrain form and vegetation to screen its movement.
  - Once the ordnance begins to impact, smoke/dust from detonations may obscure the observer's view of Team movement. The Team can also use smoke grenades to screen movement; but further use of smoke (during movement) may only inform the observer of the Team route of withdrawal and location.
  - The T/L should consider changing direction as soon as possible and as often as necessary as the observer may continue the fire mission along the expected direction of Team withdrawal.
  - If a tactical UAV is used to track the Team movement, the Team must disable it.
- Perhaps the worst circumstance may be when enemy attack helicopters are used against the Team. They may possess enhanced survivability and day-night operability and carry a broad array of ordnance, to include rockets, missiles and guns guided by target acquisition systems. Attack helicopters launched from nearby may have substantial endurance over the target area. If the target area has areas of dense vegetation (especially solid canopy), the Team must move to that area to evade gunship target acquisition. If the Team is operating in a target area devoid of concealment, and is without countervailing air support, an enemy attack helicopter may be difficult to evade. The T/L will have to dig deep into his bag of tricks to mitigate this threat. Some thoughts:
  - The Team has few armor-defeating weapons with sufficient range and effectiveness to be of any value. If the gunship comes to a hover, a .50 cal sniper rifle may be the only weapon with utility at an extended range. A Rocket Propelled Grenade (RPG) launcher has dismal results beyond a couple hundred meters on a moving target.
  - Ideally, the Team will conduct its operations in conditions that favor cover and concealment. If weather, terrain and vegetation are not in the Team's favor, fire and smoke plumes generated by a deliberate fire may save the Team.

- Smoke grenades, especially if the smoke can defeat thermal optics, will be helpful, until the Team exhausts its inventory.
- Use ravines, if available. This would generally require the gunships to attack along a single axis, allowing the Team to concentrate its fires. Quick thinking by the T/L may allow the Team to ambush a gunship. This would entail leaving a sniper behind to occupy a dominating terrain feature, while the rest of the Team occupies a ravine, thereby influencing the gunship to fly a vulnerable flight path. However, firing on the gunships will likely cause continuation of attacks, until fuel supply is low. So, alternatively, the Team may take a passive posture, hoping that the gunships will believe mission success and return to base. If the Team occupies the ravine close to nightfall, Team Members may dig laterally into the sides of the ravine to evade aviation night-vision optics.
- If the gunships are expected to have night-fighting capabilities, the Team may be better able to evade during daylight. It may be possible to outlast the gunship as it may be near its fuel reserve limits, so digging in at night may help the Team survive until the gunship departs.
- Infantry or security forces may converge on the Team location while gunships pin down the Team. The T/L should avoid this circumstance if at all possible. Otherwise, when the security forces close on the Team, the Team might resort to a breakout that causes the aviation assets to disengage for fear of fratricide.
- Some gunships may simultaneously carry a small detachment of infantry or SpecOps personnel. These troops can be inserted to verify the Team location, condition, etc. Ambushing these dismounted troops may serve to the Team's advantage, as the gunship may withhold fires if fratricide is possible.

SR Team in the Defense TTPs:

- Note the locations of bomb craters and fallen trees during mission preparation (e.g. VRs) and after insertion. Be attentive for defensible locations during movement.
- If the Team is forced into a defensive posture, the Team should occupy the best ground possible – a position that limits enemy covered and concealed approaches as much as possible.
- In a defensive posture, it is standard practice (time permitting) to construct both primary and alternate fighting positions. Rationale:
  - If primary positions are lost to enemy assault, Team Members must resort to alternative positions.
  - Friendly casualties may require the Team to reduce perimeter size to eliminate gaps in defensive fires.
- The T/L must consider excavation of defensive positions based on tactical conditions as they develop on the ground. If the Team must occupy a defensive location, the Team must waste no time in developing defensive positions.
  - Save valuable time and effort by occupying ground that is more defendable and that requires less improvement. If a bomb crater (or a hole created by a large uprooted tree) is available, it is far easier to improve by burrowing into the lip of the crater/hole than to build fresh foxholes. Also use folds in the earth to provide cover and minimize digging.
  - The T/L should position elements to guard primary routes of approach, but especially those that offer covered/concealed approaches. Team elements then should prepare hasty defensive positions, improving on natural folds in the earth, standing and fallen trees, bomb craters, etc. with hasty excavations.
  - Defensive Position Construction, time permitting.
    - The defensive positions should be on slightly elevated ground. If using an existing bomb crater/large hole or fold in the earth as the core of the defensive position, burrow

into the side(s) of the crater/hole/swale at an angle so that every Team Member has some protection from mortar fire, air-burst munitions or ordnance fired from an aerial platform.

- In conditions of friendly air superiority, call an airstrike, using hard bombs, to create craters that can be used as defensive positions. Using and improving the craters to form defensive positions will save time and energy for a Team being pursued and/or in extreme jeopardy.

- Avoid, if possible, preparing a defensive position on ground that is heavily laced with large (e.g. deciduous) tree roots.

- Define the dimensions of each defensive position and begin with excising any sod or vegetation using an entrenching tools or a heavy bladed knife/machete. Save sod/vegetation/overhead cover for use as camouflage once excavation is complete.

- In extreme cold regions, even a heavy bladed knife or pickaxe will have difficulty hacking through the soil frost line to dig a position. The Team must rely heavily on terrain form in this situation.

- Use the side edge of the entrenching tool blade to rapidly scrape or shave away the preliminary layer of earth. This hasty technique creates a shallow hole that can be occupied in the prone position to offer some defilade to enemy fire. Save topsoil for camouflage of the berm; lower layers of soil may be of different color to the top soil.

- As digging continues, use a folding saw and/or a heavy bladed knife to cut through roots. Earth spoil should be established around the side of the hole in a berm, leaving a firing lip/ledge. Packed soil protects better than loose soil, so pause periodically to pack the berm.

- Cease digging (temporarily) once the position is crotch high and then collect materials from outside the perimeter for overhead cover and camouflage and to create firing lanes.

- Recommence digging until the lip of the position is at sternum height. Begin placing overhead cover, in at least three layers, using the berm to support the lattice of overhead layers of cover. Pack earth between layers. If bamboo is used, stalks of 4in diameter can be cut open to pack dirt into the voids. The overhead protection should slant backwards to shed rainfall.

- Top overhead protection with sod collected from elsewhere within the perimeter and other vegetation for camouflage. Consider transferring entire plants (with roots) around the position, especially if the position is to be used as a long term hide/OP.

- Carve out firing ports/embrasures into the berm, aligned with the firing lanes cleared earlier. Retain sod or other material to seal embrasures at night from thermal optic detection. Embrasures should be large enough to acquire and fire upon targets, but small enough to deter enemy assault measures.

- Improve the position with grenade sumps, recessed shelves, and a platform to keep Team Member's feet from emersion in water. Fabricate a device to periodically bail water from the position.

  ◦ Keep Team Member 'dead time' minimal. The Team leadership must ensure that improvements to the defensive posture are organized and continuous.

- Beware of seeking cover behind rocks or on rocky ground. Establishing defensive positions in rocky ground is difficult, even impossible; this may result in Team Members seeking refuge among rocks/boulders. If the enemy then attacks with RPG or rocket rounds, tank projectiles, artillery/mortars and even hand grenades, rock splinters will combine with ordnance fragmentation to create an even more lethal environment. Find soft ground to establish defensive positions.

- Open ground, with ground cover, but without canopy, is preferred for defensive positions where friendly aviation superiority exists. FAC/CAS support will more readily spot the Team perimeter without concealing canopy. Enemy fires, in a forested position, may cause wood/bamboo splinters to combine with fragmentation, creating a more dangerous environment.
- Time permitting, the Team should then dig alternate defensive positions, ideally just outside hand grenade throwing range and up-slope from the primary positions. Again, take advantage of natural protection. Mark firing lanes/FPF sectors at these positions.
- Both primary and alternate positions should have grenade lanes cleared or marked for night defense so that thrown grenades do not rebound off vegetation toward friendly troops.
  - If vegetation might inhibit the trajectory of a thrown grenade, cut down the vegetation. Throwing lanes (and markers) should include paths of flight <u>above</u> underbrush/lowest layer of canopy.
  - At primary positions, grenade lanes should be oriented toward covered routes of approach. If the Team has indigenous Team Members, redistribute some of their grenades to US Team Members; indigenous Team Members may not have the arm strength/throwing range of Americans.
  - Remember to conserve most of the grenades for night combat.
  - Author's Note: The grenade throwing range can be greatly extended by using field expedients described elsewhere in this book.
- If time and the situation permits, plant mines and booby-traps (with SD capability) on primary routes of approach and/or at likely enemy crew-served weapons positions. All Team Members should know where these items are placed.
- Deploy Claymore mines so that they can easily be recovered or reoriented towards a maneuvering foe.
  - Attach Claymores to the end of poles (bamboo, saplings, etc.) for easy reorientation/recovery to thwart an assault and to secure the mine from damage of enemy fires. Remember that firing the Claymore will shatter the pole, creating flying splinters, so take cover when firing the Claymore(s) in this configuration.
  - Claymores may be strapped to trees, positioned in front of a tree (with the operator taking cover behind the tree), or may be positioned close-by in front of defensive positions.
  - Ensure some Claymores are reserved for use from alternate defensive positions and for FPF.
- If possible, send out a couple of Team Members to cause a distraction (e.g. attaching lines to remotely rattle bamboo/brush, setting time-delayed or remotely initiated demolition charges, firing a weapon from a false position, etc.) several meters away from primary positions – to deceive the enemy as to the true location of Team positions.
  - This may expose the enemy flank when he attacks in the direction of the disturbance. It would therefore be best to create this disturbance within a firing lane or where the Team can mass its fires.
  - Hang a long-duration light stick on brush or on a low limb of a sapling near dusk; plant booby-traps/mines close by. Use a long line to cause the movement of the vegetation and the light stick to attract enemy fire or cause the enemy to assault into a trap. If the light stick is suspended at or slightly above head height, the enemy may fire too high and enemy mortar fire may be misdirected as well.
- If the Team is under several layers of canopy, the canopy may block visible marking of the Team location to friendly air assets (and essential night-time strobe light marking). The marker may have to be elevated through the canopy to be visible to air assets. Techniques for this are noted elsewhere in this book.

- Thwart the enemy from the primary defensive positions for as long as possible; the T/L must use his judgment on when to withdraw his men to alternate positions. If he believes that an assault on primary positions is imminent, he should consider a covert, preemptive withdrawal to the alternate position to deceive the enemy. As the enemy assaults the empty positions, the enemy troops will be exposed to concentrated fires from alternate positions.
- Expect the enemy to 'prep' Team defensive positions with artillery, mortar and even air bombardment. When this occurs, immediately relocate Team Members to protected positions to mitigate the bombardment; reoccupy the primary positions once the fires are lifted, as the enemy will then mount his assault.
- Use CS grenades to break up an enemy assault and impair enemy vision. Be careful not to set a fire that may consume Team defensive positions.
- It may be essential to survival for the Team to conduct a breakout from its defensive perimeter. Breakout TTPs are noted elsewhere in this book.
- If conditions (terrain, wind direction and speed, fuel available) are right, consider lighting a fire to force enemy withdrawal.
- The enemy may be expected to concentrate fires on positions they have identified. Team Members may counter this by:
  - Temporarily withdrawing to alternate positions, allowing fires to fall primarily on vacant positions, and then reoccupying the primary positions prior to an enemy assault. It is important to be able to detect when the enemy is about to commence his assault.
  - Preparation of dummy positions (e.g. in front of the primary positions) to cause the enemy to fire on and assault deceptive positions. Team Members can increase the believability of dummy positions by using 550-cord to move vegetation remotely or make noise.
  - Note: If a Team OP/LP is detected, the occupants must move to another position.
- If the enemy has air superiority and is likely to resort to a 'napalm'/thermobaric attack, the Team <u>must</u> abandon its positions and execute a rapid breakout using evasive movement in doing so. A single napalm canister can cover a significant linear area. Closing with the enemy during breakout will deter enemy use of flame weapons.
- If enemy aircraft possesses night-vision or thermal-observation equipment, Team defensive positions must have overhead cover and camouflage. Advanced camouflage nets may not be carried by dismounted Teams, but should absolutely be carried on Team ground mobility equipment.
- If the Team must defend a position, consider erecting abatis along vehicular high-speed approaches for vehicles (e.g. armor). If the barrier causes the vehicle to veer off, it may expose a vulnerability to Team fires (e.g. side armor). To accomplish this, the Team must possess the means to rapidly cut down medium sized trees.[61]
- Spoiling Attack:
  - Identify enemy assault jump off positions (map analysis or movement detected).
  - Plant noisemakers, mines/booby-traps or command detonation/remote munitions.
  - Attack these positions, when appropriate, using CAS, indirect fire or command detonation/remote munitions attack.
- A major threat to a Team in a defensive posture is an infantry attack supported by armor and artillery fires.
  - The Team will normally have few weapons capable of defeating armor, unless the Team is carrying AT rockets. Other than AT rockets, a well aimed .50 sniper rifle is capable of defeating light armor at a distance; the 40mm HEDP round may be effective at close distances.

---

61. Note: The Author has intellectual property for this purpose.

- ○ If an armored threat is likely, the T/L should consider a breakout before the attack can be launched. Tip offs to an armored attack include:
    - ▪ An approaching dust plume
    - ▪ Engine/track noise
    - ▪ Falling trees, movement of vegetation along lines of approach
  - ○ Create the conditions under which a breakout may be possible. This would require the use of CAS or area suppression via long-range fires (missiles).
- A noise maker delay can be used to deceive the enemy as to the Team location and its direction of movement.[62]
- Sniper in Tree Top: use smoke to obscure movement.

Counter-Ambush TTPs:

- Where friendly forces enjoy air and fire support superiority, it is standard practice for enemy combatants to press an assault and close with a Team, to prevent the employment of TACAIR/artillery in support of the Team – for fear of fratricide.
- <u>Battle Drill:</u> Team Members caught inside the kill zone should:
  - ○ Seek cover and return fire. Priority of fire should be against crew-served weapons.
  - ○ Throw/project smoke and/or CS grenades, and fragmentation grenades if within close combat range.
  - ○ After smoke forms/fragmentation grenades detonate, increase rate of fire and assault (using Fire & Movement, and the 3-second rush) to breech the enemy ambush formation. <u>Use terrain form</u> during the assault to avoid enemy line of sight weapon engagement.
  - ○ If it is necessary to assault an enemy machinegun position, attack from the enemy's right, as the machine-gunner will have more difficulty in traversing to his right, and would have to reorient the machinegun.
  - ○ Once Team Members have penetrated the enemy line, they should move to covered positions and fire laterally along the long axis of the enemy ambush formation. Team Members may use trees, rocks and terrain form for cover as they fire along the axis, using the obstacles to shield their backs from enemy fires to their rear and/or use Fire & Movement to clear enemy positions laterally.
  - ○ Team Members outside the kill zone should attempt to flank enemy positions. As Team Members roll up the enemy long axis, they should:
    - ▪ Keep their heads on a swivel and stay alert for enemy movement and camouflaged enemy positions.
    - ▪ Avoid shooting other Team Members as they assault through the enemy ambush.
    - ▪ Use grenades judiciously clearing positions down-slope.
    - ▪ Firing 40mm grenades through vegetation may initiate detonations in proximity to other Team Members. Beware of the arming distance for 40mm grenades.

Breaking Out From Encirclement TTPs:

- There were occasions where SOG Teams had not planned for, or practiced methods to, 'breakout' from encirclement and were subsequently overrun by enemy forces. The consequences of a breakout attempt should be discussed openly within the Team before a deployment. Breakout techniques must be practiced.
- It is difficult to evacuate litter casualties during a breakout operation, unless the Team is equipped with vehicles.

---

62. The Author has intellectual property for a device that can provide this capability.

- If a breakout is successful for any Team Members they should transition immediately to SERE.
- The T/L must apply tactical wisdom to decide if the Team should take up a defensive position, use TACAIR on an enemy encirclement and await rescue by a RF – or if a breakout is appropriate. The chief issues are:
  - What are the METT-TC factors that weigh on available courses of action?
  - When will the enemy have enough combat power to overrun the Team?
  - What conditions warrant leaving behind a wounded Team Member?
- Once encircled, the sooner the Team attempts to break out, the better the chance of success with the least number of casualties. Delaying the decision allows the enemy to mass and deploy his forces and take up defensive positions.
- Training and Preparation for Breakout. Breakout requires planning, some of which will be adapted on the fly.
  - Rucksacks and equipment, especially sensitive items, to be left behind must be destroyed by someone.
  - KIA must be left behind. Someone must remove any classified documents (SOI, notebooks, maps, etc.).
  - The last two Team Members in a formation have the responsibility of rear security, tactical deception and assisting those who may be wounded during movement. Additionally, they should recover classified documents, time/situation permitting, from casualties during breakout.
  - Indigenous Troops often have a code of loyalty/honor that requires them to stand and die with their American Team Members, rather than leave an American behind. The senior Team Member must countermand this tendency, as the firepower and mutual assistance of all Team Members must be directed to the breakout effort.
  - Bear in mind that successful completion of your mission depends on getting information back to headquarters. SR professionals will understand this.
- <u>Breakout Execution TTPs</u>. The following actions should take place:
  - Breakout is best accomplished during periods of low visibility (night, fog, snowfall, rain). If the enemy fires on the Team during breakout in these conditions, they will tend to shoot high and miss. The enemy may have difficulty spotting the Team's trail or pursuing the Team in periods of limited visibility.
  - Team formation should be governed by conditions on the ground and by the immediate situation and METT-TC considerations.
  - Team route should also be based on the situation. If the Team is carrying a wounded Team Member, select a route to ease the burden of litter bearers.
  - If the Team is in possession of a POW, the enemy soldier should be bound, gagged, and abandoned in a protected position. The POW must not be harmed by Team Members during Breakout.
  - The Team forms up in a concealed RP.
  - The Team dons protective masks.
  - Set time-delayed demolitions at the RP just prior to movement. Detonation times of the devices should be staggered and should be set so that initiation occurs after the Team has started movement. Detonations will draw enemy fire away from the Team.
  - If possible, reconnoiter the initial meters of the path from the Team RP and remove dead vegetation, sticks, etc. (assumes weapons engagement has not commenced).
  - The Team covertly moves forward, crawling if necessary, using as much cover and concealment as possible, until detected or enemy activity or positions are spotted. Ideally, the Team movement will be with the wind.

- ○ <u>On command</u>, launch 40mm CS rounds, and/or CS rifle-grenades from covered locations, accounting for wind conditions and cloud/plume drift, such that the cloud passes through actual and suspected enemy positions and across or in the direction of the breakout. Firing done away from the Team main body may cause misdirection of enemy fires. Initial 40mm and rifle grenade smoke rounds should be placed in front of enemy crew-served weapons.
- ○ If the Team is compelled to breakout in daylight/illuminated conditions, launch smoke rifle-grenades/40mm to conceal Team location and movement – again accounting for wind drift. Launching of rifle-grenades should be done away from the main body of the Team, so that enemy fires are misdirected.
- ○ Move swiftly along the selected route, making as little noise and disturbance as possible. The Team should move forward in bounds, from terrain fold to terrain feature, so that the Team may rest momentarily and take the best cover possible in the face of fires/pursuit.
- ○ The Team should not fire on the enemy during movement unless the Team is receiving accurate, effective fire, as this will confirm the Team position, will cause the enemy to increase his rates of fire and may cause the enemy to maneuver on the Team.
  - ▪ If the Team must return fire, Team Members should shoot with enemy weapons, if so equipped, to deceive the enemy into thinking that they may be firing on their own troops. Additionally, Team Members may attempt to deceive the enemy by calling out to them in their native tongue, pretending to be comrades.
  - ▪ Team fires should first be concentrated on enemy crew-served weapons. The Team should cease fire as soon as it has suppressed the enemy weapons and continue swift movement.
- ○ Tail-gunner(s) should deploy decoys/grenades along the back-trail, especially near where the last firefight occurred. Indiscriminate use of short time-delayed munitions will alert the enemy to Team direction.
- • <u>Breakout Supporting Fires TTPs</u>: Artillery, helicopter gunships and TACAIR, if available, should be used to assist the break out attempt.
  - ○ Artillery: SR Teams will generally have artillery support only when operating in a COIN environment. Artillery fire, if available, can be effective in breakout operations and it may be the only fire support available during inclement weather. During breakout execution, direct artillery (HE, ICM and smoke) fires to be 'walked' in front of the Team in the direction of the breakout. The Team Member directing the artillery fires must know his craft and be competent to direct fires on the move. A GPS capability may be extremely useful in this 'danger close' engagement, at least to designate the jump-off location. As artillery is an area weapon, it lacks pinpoint accuracy (except for laser guided munitions); the longer the gun-target distance, the less precise the fires. Ensure that your breakout direction is not along the gun-target line (to reduce the threat of fratricide). Improved Conventional Munitions (ICM) dispense submunitions – very effective in open canopy. See 'Supporting Fires TTPs' below.
  - ○ Helicopter Gunships: Gunships are preferred over artillery and can provide precision fires. They can provide almost continuous, timely, accurate and flexible close fire support (even at night), firing directly to the Team front, rear and flanks during the break out attempt. Gunships must be able to either see the Team and/or the enemy, or be able to spot Team marking, and know Team direction of travel to be effective and to eliminate fratricide. The Team must often direct gunship fire by adjusting from the strike of the rounds and rockets.
  - ○ CAS: Tactical airstrikes can also assist the Team in breakout operations. CAS assets will have a more extensive capacity and variety of ordnance than helicopter gunships. Note that dropped ordnance (bombs, bomblets, napalm) from most high-performance aircraft

represents an enhanced 'danger close' hazard. It is common practice for an enemy to close with a Team to avoid airstrikes. Guns and rockets, as employed by CAS aircraft like the A-10, can fire in close proximity to the Team. AC-130 type gunships can provide superb CAS, but these large, relatively slow-moving aircraft are especially vulnerable to AA fires and will generally not be used where the enemy has air superiority. Breakout is best supported by 'guns' versus dropped ordnance. As in the case of helicopter gunship support, CAS guns can clear a path for the Team during the breakout effort.

What to do if Captured:

- Team Members should not expect POW protection under international laws/treaties when engaged in COIN/counter-terrorism operations, operations against a country not signatory to such laws/treaties, operations in a country where the US is not in a state of declared hostilities or where Team Members are not themselves fully compliant with international laws/treaties. These circumstances place captured Team Members at the mercy of their captors; and likely, mercy would not be forthcoming. Capture in these circumstances may result in the very worst of outcomes.
- SR Team Members should be very attentive to SERE training. Attempt escape at first opportunity, especially while in transit from the capturing unit to the next station in the POW evacuation chain.
- Hide small escape aids in various points of your uniform. Use these aids as soon as possible, before they are discovered and removed by your captors.
  - Examples of Escape Aids: surgical blades; surgical/piano wire, fishing/Kevlar® line; incendiary matches; and a shim (to disengage the ratchet-pawl on plastic handcuffs/cable ties).
  - Note that the Team Member will be bound through much of the initial POW and transfer process; secrete the aids in locations where access to them can be achieved while bound.
  - Expect the uniform to be searched at each transfer stage of the evacuation. The least thorough search will be conducted by the capturing unit.
  - At late stages of the evacuation, captors may replace the uniform with prisoner attire.
- Do what is necessary to delay or impede evacuation to the next station in the POW evacuation chain. For instance, slow down the pace by feigning injury.
  - Slowing down the evacuation will give higher headquarters the time it needs to take compensating security measures.
  - The higher up in the POW evacuation chain, the more professional (and brutal) the interrogation or treatment.
  - The enemy may try to compel a Team Member to communicate with their parent unit. This may be done by captors to induce a rescue attempt – a trap, with the Team Member as bait. Once the Team Member has been missing for a specific period of time (by SOP), the Team's parent unit will ask a question during communication that will require of the Team Member a set response which will indicate whether or not he is communicating under duress. It may be in the Team Member's interest to prolong the message traffic to facilitate friendly Radio Direction Finding, if available, to locate the Team Member's position. If RDF is not available, HQ may tell the Team Member to stand-by, to switch to another frequency, or to employ other delaying tactics to provide the time necessary to arrange for RDF or other measures. Other message traffic may be used to further ascertain the Team Member's status; based on replies, the HQ may also communicate a code word that will indicate intended action by HQ.
- Some captors have a reputation for exceedingly brutal treatment and will not hesitate to use extreme interrogation measures or use horrific methods of execution for propaganda

purposes. If this fate is likely upon capture, the use of an 'L-pill', secreted on your person, may be a preferred option. This is not a fictional or theoretical scenario; 'L' pills were commonly available to SOE/OSS operatives; while no SR personnel were known to carry them, they were available to SOG personnel upon request.

<u>True Account</u>: In 1967, a CCN SR Team was overrun, three days into their mission in Southeastern Laos. One rescued survivor (the One-Two) reported what happened. A NVA counter-reconnaissance (SpecOps) team captured the SR Team intact, except for the One-One (Assistant T/L) and a Montagnard, who managed to escape. Led by an English-speaking officer, the enemy tied the T/L to a tree and slit open his abdomen, spilling his intestines. The enemy unit then used a flame-thrower on the T/L and burned him alive. The One-Two was allowed to live, so that he might later report the event to intimidate other SR personnel. The story was confirmed the following day, when Sergeant First Class Fred Zabitosky[63] leading a Bright Light mission, recovered the charred remains of the One-Zero. The remains of several Montagnards discovered at the site, were not recovered due to enemy pressure; they were also burned to death.[64]

- Some enemy Lessons-Learned regarding POWs:
  - An enemy will be fully aware of the US military's informal code of 'no one left behind' and the expressed intent to do everything possible to rescue/recover US MIA, KIA and POWs and those trapped/surrounded by an enemy force.

<u>True Account</u>: In June 1967, a CCN SOG Hatchet Force of approximately 100 men were inserted into Target Area Oscar-8, to exploit an Arc Light of nine B-52s dropping bombs onto the presumed location of the 559th Transportation Group, the control center for the Ho Chi Minh Trail. Upon landing, the Hatchet Force found itself immediately surrounded by numerically superior NVA forces and cut off by belts of anti-aircraft artillery. During the battle, lasting four days, attempts to support and ultimately rescue the trapped Hatchet Force resulted in the following losses: twenty-three American KIA/MIA (Special Forces, and aircrews); approximately forty-six Nungs (indigenous commandos); one A-1 Skyraider; one F-4 Phantom; two helicopter gunships; one CH-34; and one CH-46. Of the Special Forces personnel, one Sergeant First Class Wilklow (WIA) was laid out, exposed in a clearing with a signal panel, and deliberately used as bait by the NVA AA gunners, who waited for three days for rescue aircraft to appear in their sights. On the evening of the third day, SFC Wilklow was able to crawl away down a hillside and through heavy underbrush, and was subsequently rescued the following day. The NVA severed the heads of KIA SF and mounted them on stakes.[65]

  - The enemy is likely to move POWs frequently (e.g. weekly), especially during overcast conditions that would obscure aerial/satellite detection or when satellites are not in a position for surveillance.
  - The enemy may be expected to use POWs as bait and may set up a trap for Bright Light/ RF/SERE operations, which may include use of MANPADS. Note that the enemy can use decoys and deception too.

---

63. SFC Zabitosky was awarded the Medal of Honor for another action that took place in approximately the same time frame.

64. Account extract from John L. Plaster, *SOG: The Secret Wars of America's Commandos in Vietnam* (Simon & Schuster, New York, 1997), pp 87-90.

65. Ibid, pp 90-94.

- If the enemy is obsessive about OPSEC, US rescue/recovery forces should always be suspicious of enemy security lapses. If something is too good to be true, it probably is.
- The sooner a rescue/recovery operation is mounted, the more likely it is to succeed. This requires the Bright Light/RF to be on stand-by, with dedicated air assets, for immediate use/timely response.

Supporting Fires TTPs:

- In COIN operations, or in other rare circumstances when effective artillery is available to Teams, plan artillery Registration Points (RPs) to use:
  - In adjust-fire missions
  - To disengage from an enemy
  - To conceal or distract the enemy from Team movement
  - To orient the Team on its location
- Send selected Team Members (18B/C) and the T/L and Assistant T/L to train with artillery forward observers or receive training from them.
- Without immediate, responsive fire support to engage fleeting or high-priority targets, the enemy may escape/evade engagement. SR efforts to spot these targets may then be entirely wasted. Response to targeting intel must be timely. SpecOps engaged in deep penetration operations may have no fire support within range of their target area or on call (e.g. cruise missiles) in a timely manner. The ability of SpecOps units to identify 'World Series' targets deep inside enemy territory makes it essential that they receive priority of fires (and targeting resources). This might include:
  - Direct support fires
  - Pre-positioning of fire support assets
  - On-call aviation/ability to divert
  - Preplanned targets/RPs
  - Hand off of targets to target acquisition elements (drones/satellites)

  In 1970, MACV-SOG T/Ls were given automatic priority of fires/air support, by declaring a 'World Series' target. The threshold definition for a World Series target was a minimum of a battalion of enemy troops or a significant enemy convoy. This designation and the related Spot Report information would immediately be passed to the Airborne Command and Control Center (ABC³), which would then rapidly coordinate necessary air assets to engage the enemy target. Through this process, the ABC³ could divert a B-52 sortie from its preplanned target to support the Team. The World Series designation would normally be invoked if the target was massed, fixed or temporarily stationary sufficiently long for a reasonable attack window.

- All Team Members <u>must</u> be able to Call for Fire <u>while on the move</u>.
  - Maintaining an abbreviated Call for Fire procedure on a 'Quarterback's Wristband' may be helpful.
  - Practice Call for Fire during all field exercises and other field training when appropriate.
  - Conduct training with artillery FOs to acquire 'tips of the trade' and proficiency.
  - Apply the WeRM formula described elsewhere in this book for fire adjustment purposes.
  - SR Teams should establish pre-planned RPs for fire support that should be located within the mission area of interest and/or near the expected target location, on suspected enemy locations, and in the vicinity of prospective exfiltration LZs. The T/L may update/add other RPs after insertion. Note that the FAC/relay and supporting artillery unit should have the coordinates of the RPs.

- ○ Once the Team has found its target, the fire support provider (artillery, FAC/CAS) should be alerted and placed on standby. The Call for Fire/Request for Air Support may have to be passed through relays which will cause a delay in execution.
- ○ Teams on the move (e.g. while being pursued) should use a modified Shift-From-A-Known-Point Call for Fire procedure. Note that this entails a deviation from normal procedure, so the modification must be pre-arranged with the supporting unit. The Team or higher Headquarters may be able to coordinate exclusion of blanket 'Danger Close' notifications. Also, the Team should not Call for Fire from the vicinity of the gun-target line. Call for Fire format (TTPs) are found at Appendix D. Note: The supporting unit normally requires notification of 'Danger Close' conditions, when they exist. The T/L may elect not to make the notification or may inform the supporting unit of risk acceptance.
- Request for Air Support format/TTPs are found at Appendix D.
- Assuming that the Team is within range of friendly supporting bombardment, delay fused projectiles/bombs can be used to create craters for hasty defensive positions. This may be done while the Team is being pursued and must resort to a defensive posture. These craters can also be done surreptitiously, if other barrages/bombings are occurring in the same time frame.
- To minimize the possibility of fratricide, the Team should not be positioned along the Gun-Target line (either before or behind the target).
- The AOB/FOB should consider establishing a covert, <u>temporary</u> mountain-top fire support base (or several of them) furnished with a 120mm mortar, a competent crew and a security element.
  - ○ The crew should train on firing the mortar from elevated terrain to lower elevations.
  - ○ The crew and security force would likely have to rappel onto the mountaintop (perhaps at dusk) and the supplies and equipment delivered to the cache by sling. The personnel would be then be withdrawn (e.g. by stings) and then only deployed for support upon commencement of local operations.
  - ○ A mortar and its ammunition could be stocked at a well camouflaged MSS. But the weapon and ammunition must then be moved to its firing position.
  - ○ This mortar position could provide general fire support to SR missions within range of its tube (over 7,200 meters for the HE projectile).
  - ○ Equipped with the PGM (Precision Guided Munition) kit the mortar would be capable of striking point targets with accuracy, without laser designation.
  - ○ Mortar support from friendly UW forces may not be wise, as mortar crews will not normally have sufficient training/practice.
  - ○ The mortar position should only be operational temporarily and then evacuated before the enemy can locate the position. The position can be re-established as a MSS for later use.

Aviation Support TTPs:

General Aviation Support TTPs:

- Employment of air assets in support of SR operations, is much more available and responsive in permissive environments; and far less available and responsive in deep penetration operations.
- The terminal velocity of a falling projectile casing from an aviation system may approach 200mph. Depending on munition caliber, this could be a fatal surprise to a Team Member.
- 'Clear Below' when firing weapons from helicopters after leaving an LZ; other aircraft (e.g. gunships) may be passing below your extraction aircraft.
- A C-130 gunship (e.g. Spectre) is a relatively slow-moving aviation asset that is vulnerable to anti-aircraft fires and typically employed at night. If the gunship comes under enemy anti-

aircraft attack, the aircraft will normally be withdrawn, unless the anti-aircraft systems are rapidly destroyed by other aviation assets that are night-combat capable. It may be possible to initially direct the approach of the gunship to the Team location by the sound of the aircraft engines. The aircraft commander may also be willing to drop a flare to assist the Team in directing the aircraft closer to the Team from the flare location. During engagement, make adjustments from gunship tracer impact area to ensure full target coverage. This may be difficult or impossible in dense vegetation/canopy.

- In SR AOs, Teams generally operate far beyond friendly artillery range; and air support is generally distant, if available at all. If the Team is entering a high-risk mission envelope, request the presence of a FAC (or an airborne radio relay) to linger in orbit nearby. Then, if contact is imminent, air assets can be staged in advance of a firefight to reduce reaction time.

Forward Air Controller (FAC)/UAV and other Air Support TTPs:

- The FAC may sometimes be considered second in importance only to the Team's firepower during a mission, since the FAC controls aviation support. Team Members should learn all they can about FAC procedures. Proper use of FAC support could mean the difference between whether or not the Team returns intact from a mission.
- A FAC, with an experienced observer, may be the best method for Close Air Support (CAS) coordination. Direct coordination (Team to strike aircraft) is generally not the best method, unless the Team has a USAF coordinator attached or if the Team is using a laser designator.
- In situations where an SR Team turns off its radio (e.g. to conserve battery power, to 'go black', or in extremely close proximity to enemy combatants, etc.), a FAC can alert the Team to urgent communications; 'the FAC … would fly in the vicinity of the recon team and rev the engine. That was the signal to turn the radio on.'[66]
- Project Delta (and SOG) used FACs to coordinate SR Team/platoon/company insertions and extractions, direct CAS and perform radio relay functions. FAC pilots and ground support personnel were actually attached to Project Delta (aka B-52); this was not the case with SOG FACs. SOG FACs had a much larger portfolio, specifically to interdict the Ho Chi Minh Trail, but their priority was to support SOG ground operations. The Delta model, where FACs 'lived, trained and operated solely with the recon teams … made them well versed regarding recon team tactics, techniques and procedures … and resulted in a close professional and personal relationship between the FACs and the Project Delta recon men.'[67] This collocation is much preferred as it enables close collaboration in mission planning, preparation and execution.
- The FAC is effective in marking targets and may have sufficient loitering time, allowing the FAC to address targets using multiple sorties of different aircraft with different capabilities and ordnance combinations.
- A high-performance aircraft, dropping ordnance at high-speed, can represent a threat to the Team unless precision weapons are used.
- Note: Given the area and distances that a FAC in General Support (GS) may be required to cover, pilots may navigate and operate using large scale charts, not smaller scale topographical maps. If the FAC is in a Direct Support (DS) role, the pilot should have both aviation charts and large scale topographic maps. If a FAC has a 'rider' on board, the rider will certainly

---

66. COL Al Greenup (Ret), 'No Glamour No Glory – A Ground Pounder's Perspective of Project Delta Forward Air Controllers in the South East Asia War', 'The Drop' Magazine, Winter 2017 Edition, Special Forces Association, Fayetteville, NC. p. 106
67. Ibid, p. 105

have appropriate topographical maps. Where CAS is available, pilots of high-performance aircraft are very unlikely to have smaller scale topographical maps.

- The sensor package aboard a UAV/drone can be very capable in identifying and locating friendly and enemy forces on the battlefield, and is also capable of limited CAS. But these capabilities may not be available or effective in heavy/continuous canopy or in non-permissive air support environments.
- A FAC or strike leader pilot can pose a genuine threat (e.g. a misdirected strike) to the Team if he is not aware of SR best practices. Erroneous targeting is even more likely in circumstances where laser target acquisition is not possible and/or where high-performance aircraft are used. These problems can be alleviated if the FAC is accompanied by a 'rider', an experienced SpecOps soldier who can assist the FAC in air-ground coordination and who provides the FAC pilot with tactical wisdom. The rider will also be capable of directing artillery support.
- The FAC aircraft is substantially limited in bad weather (e.g. during rainy seasons). This means that CAS and message relay will also be limited. Have a plan to mitigate these problems.
- The Team must survive enemy contact/engagements until the FAC and aviation assets can respond to your location; plan and train for this as well.
- Supporting aircraft may have mixed ordnance (e.g. hard bombs, napalm, dispensed munitions, missiles, a variety of rockets, etc.) some of which may be relatively unsuited to the target or for close combat situations; if it's a diverted aircraft, the ordnance may be tailored/programmed for a pre-planned target. The FAC or strike leader will normally advise the Team of the ordnance on-board; if he doesn't, the T/L should make it a point to ask – especially if the support is provided by allied/coalition aircraft. Team Members should be familiar with the capabilities and shortcomings of the various aviation assets and the capabilities and effects of the ordnance they carry.
- The FAC/strike leader generally must know Team location before he will clear a strike on the target. There may be exceptions to this rule when:
  - Geographic bounds are established for the strike.
  - Strike is to be delivered on a RP or on a target directed from a RP (e.g. from RP #3, left 200, up 100).
  - T/L takes responsibility for 'Danger Close' conditions.
- Signaling a FAC:
  - A signal mirror is the best way of covertly signaling the FAC to mark Team position – if in the open during daylight. It will be somewhat less effective in signaling a high-performance aircraft. The mirror is directional and should not give away Team position.
    - If the sun is obscured by clouds one can still signal an aircraft by placing a strobe or high-intensity flashlight against the signal mirror; this requires some practice.
    - The Author's recommended method of using the signal mirror: (1) acquire the sunshine onto the mirror face and shine the sun spot onto the palm of the hand, held upright in the direction of the aircraft, (2) acquire the location and direction of the aircraft at the tip of the index finger of the upright hand, while moving the hand to track aircraft movement, (3) use a slashing motion of the sun dot, across the palm and past the tip of the index finger and thereby at the aircraft.
  - Use smoke, flares, pen guns, and tracers, all of which produce easily detectable signatures, as a last resort for marking Team position. The Author recommends carrying only the canopy-penetrating version of the pen gun; other versions will often ricochet off limbs and never clear the canopy. Notify the aircraft before firing a flare/pen-gun since it may be mistaken for an enemy tracer. Never fire the flare directly at the aircraft.

- A strobe light (especially an IR strobe) is an excellent marker for signaling aircraft at night, but it could disclose Team location to the enemy; remember that the strobe cannot be seen under solid canopy.

  Author's Solution:
  Bend tall bamboo or a tall, slender secondary tree to the ground and attach the strobe to the top; initiate the strobe and allow the bamboo/tree to rise back up so that the strobe can be elevated at least one canopy layer up.

- Pilots should identify the color of smoke used by the Team/element on the ground only after it has been thrown and once the plume can be seen from aloft. The Team should not identify the smoke color to the pilot, but rather the pilot describes the color first and then obtains verification from the Team. Otherwise, if the Team frequency is monitored by the enemy, and he hears your color disclosure, he may duplicate the color of the smoke to spoof the pilot as to Team or target location.
- Violet and red smokes are the best colors to use in jungle/rainforest; yellow and green smoke may merge and blend in with background vegetation colors and may not be as easily detected; this will be a consistent problem under inversion conditions, as the smoke will often not plume above the canopy, but will cling close to the earth and even drift down into ravines and other low-lying areas.
- When using a transponder or beacon do not point the tip of the antenna at the aircraft; the tip has a weaker signal than laterally from the long axis of the antenna.
- If contact is made with the enemy while the Team is in a dense jungle or rainforest, use WP grenades to mark the Team location for the FAC. The heat of the white phosphorus will cause the smoke plume to quickly rise above the canopy, regardless of inversion conditions. Warning: Use of WP during dry conditions in temperate region forests (deciduous or conifer) may cause a fire.

Figure 66. White Phosphorus Plume Penetrating Rain Forest Canopy. (*Justice*)

- ○ If the Team is in the open at night, a flashlight may be placed inside of a 40mm grenade launcher barrel and aimed directly at the aircraft (if aircraft location can be determined). This shields the light from lateral observation by the enemy.
- ○ When directing a FAC to the Team position, use the clock system. Note: The nose of the aircraft is the twelve o'clock position. This system can also be used to generally direct the FAC when the aircraft can only be detected by sound (e.g. Team is under canopy or in a ravine).
- ○ Use cardinal readings to direct aircraft until the FAC has position located. Once the FAC has the Team located, use azimuth readings and distance in meters to the target.

Shift From a Known Point/Reference Point (Air or Artillery Fire Support) TTPs:

Author's Tip: The T/L can direct a strike from a known point (e.g. a RP or clearly identifiable feature), even while on the move, in the following manner:

- The FAC ought to have the same map and RPs plotted as the T/L; the FAC pilot must be trained in the procedure.
- T/L to FAC:
  - ○ <u>Target Description and Activity</u>: 'FAC Oscar-seven, this is Paradigm. Fire Mission; immediate; danger close. Vehicle Park under canopy with five 152mm tracked artillery vehicles, five trucks and one command vehicle preparing for movement.' [This notifies the FAC that he has a fleeting target opportunity and allows him to select air assets and ordnance type.]
  - ○ <u>Target Location</u>: 'From RP#1, Right 300, Up 200' (in meters).
  - ○ <u>Team Location</u> (while this step is standard doctrine, it is <u>not always necessary</u>): 'From RP#1, Right 300, Down 100. Under cover.' [The enemy cannot determine Team location without knowing the location of the RP].
  - ○ <u>Team Intentions</u>: Observe/Adjust Fire; Relocating to another position, Team moving from North to South, etc.
  - ○ <u>Additional</u>: Advise of the type of ordnance to be used. (This will allow the Team to take protective measures.) Recommend approach from NE to SW (to avoid fratricide).
- Other Considerations:
  - ○ Make adjustments or provide feedback for the FAC after each aircraft delivers its ordnance, if the strike or its effects can be observed. Aside from the tactical relevance of such report, the feedback provides a morale boost to FAC and CAS personnel.
  - ○ Ultimately, the FAC and/or strike aircraft pilots must decide their flight path/approach to avoid obstacles and anti-aircraft fire. The Team should recommend a strike aircraft flight path to the target to avoid the airstrike approach from passing directly over the Team position. Note: A strike that occurs fractionally too soon or too late could deliver ordnance directly on the Team position. This has happened before! The chances of this occurring may increase if the team is moving or if the ordnance is not designed for precision attack (e.g. cluster bombs, napalm).

<u>True Account</u>: A SOG Team was being hotly pursued along a ridge line in the southeastern Laos rainforest by a numerically superior NVA unit. The moderately experienced T/L paused briefly to call in an airstrike, using a WP hand grenade to mark the Team location, and then the Team continued its withdrawal along the ridge. The strike resulted in three Team Members, including the T/L, being struck by flaming napalm. <u>What Went Wrong</u>: Normally, USAF CAS was provided to SOG Teams by A-1 fighter-bombers, however, the T/L had declared a Prairie Fire Emergency, while the A-1s were over half-an-hour away.

Subsequently, the FAC diverted an immediately available high-performance aircraft armed with napalm for the strike. The T/L did not inform the FAC of his direction of travel, he did not inquire about the ordnance to be employed and he did not request that the airstrike be delivered along a specific flight path (e.g. behind and across the direction of travel).

- ◦ A FAC can effectively direct a Team in contact to a LZ or to defensible terrain and can also direct airstrikes to 'prep' an exfiltration LZ with ordnance prior to Team arrival.
- ◦ A FAC can assist the Team in breaking enemy contact before the arrival of strike aircraft. A low pass over canopy or firing marking rounds may make the enemy think they are being attacked, thus causing them to seek cover or withdraw.
- ◦ The FAC can create bomb craters (delivered by CAS) for an imperiled Team that they can use for expedient defensive positions or a string LZ.
- When FACs may be inappropriate:
  - ◦ When US or allied air assets do not possess air superiority
  - ◦ Highly lethal, guided anti-aircraft systems are extensive
  - ◦ Deep penetrations
  - ◦ When their presence impacts political/operational deniability or betrays the presence of Team or guerilla/partisan operations
- The armed UAV has substantial limitations in the CAS role.
  - ◦ If the enemy has air superiority in the AO, UAVs may not long survive
  - ◦ Armed UAVs have a limited payload
  - ◦ Armed UAVs may be severely limited over jungle, rainforest, dense forest or other environments that hinder the use of laser designators or target identification.

Chapter 4

Sustainment

Logistics

- Non-Traditional Acquisition: US Special Operations Command (USSOCOM) has its own acquisition activity that under certain procedures may acquire experimental/developmental, specialized military and off-the-shelf materials and equipment. USSOCOM may also obtain specialized equipment/items through intelligence organizations.
- Team clothing and equipment must often be 'sterile' or of such widespread use that plausible deniability may be attached to its wear in denied areas.

Team Resupply:

- Maintain an updated/verified listing of equipment (e.g. weapons, LBE, specialty items) and sized uniform items (e.g. boots, trousers, blouse, hat), for Team Members to facilitate resupply, if required. This is particularly important for deep penetration/extended missions. Add to this list any essential medical or personal items (e.g. eyeglasses) required by Team Members. Keep this listing in a binder at the SR unit headquarters.
- The SR unit headquarters should maintain a master catalog of supply items, segregated by supply class. Each supply class section should contain an item listing identified by a simple code designation. For instance: the code '5M1' could indicate Class V (ammunition) item ('5'); mine ('M'); Claymore ('1'). A simple system, like this, could simplify and abbreviate Team resupply ordering via voice/text communications in the field. Certain Team Members (T/L, Assistant T/L, and Communications Sgt.) would carry an extract of the catalog, tailored to the Team needs.
- Prepared resupply bundles should be secured in locked, serviceable containers (e.g. shipping containers) to prevent pilferage and losses due to rodent/insect infestation and weather. These containers must also be placed on dunnage to provide drainage. Some bundles may require secondary barrier protection to insure that they are sealed against vermin infestation – especially if food is stored in the bundles. Vermin will also burrow into clothing and other equipment to nest, so multiple layers of packing may be required. Any bundles packed with food, which are vulnerable to vermin infestation, should be kept elevated off the floor (e.g. on poles or on hooks suspended from the container ceiling. The containers should be inspected on a scheduled basis to ensure content serviceability and to ensure infestation is detected/prevented. Packing lists, with inspection and expiration dates (e.g. food), should be kept with the containers.
- If the Team is to be deployed with All-Terrain Vehicles (ATVs)/Utility Terrain Vehicles (UTVs), consider the following factors:
  - Do not fill fuel containers/tanks to capacity if they are to be delivered by aircraft at high altitude.
  - Be aware of thermal signature from ATVs/UTVs. This may require equipment modification or navigating/sheltering behind screening terrain/camouflage.
  - Consider using small enemy 'Jeep-like' light utility vehicles to confuse enemy surveillance. If possible, use enemy tactical vehicle marking protocols. Use of enemy vehicles may

enable the Team to move cross country during daytime or to join enemy road convoys at night. If enemy convoys operate predominantly at night, the noise of convoy vehicles will screen Team vehicle noise.
  ○ Purpose-built ATVs/UTVs should be equipped with:
    ▪ Quiet engines/exhausts
    ▪ Reliable/low-maintenance engines, with essential maintenance/repair tools and spare parts
    ▪ Cargo compartment/bed
    ▪ Camouflaged net
    ▪ Pioneer tools
    ▪ Fuel can(s), extra POL and siphons
    ▪ Run-flat tires; tire repair kit and/or replacement tire(s)
    ▪ IR beams (secondary)
    ▪ Possible Accessories: Winch, compass, etc.

Water TTPs:

- Train to conserve water and exercise consumption discipline. The more water carried, the heavier the load. Remember that water may be needed to reconstitute dehydrated rations. Also, soldiers are always seized by an overpowering thirst resulting from the adrenaline rush after a firefight.
- During dry seasons or in arid conditions, Teams may have to conform its line-of-march in general proximity to water sources. Once acclimatized (through training), personnel can more effectively moderate water intake. Note: Again after an adrenaline rush, personnel will crave water.
- During deployments to arid regions, US troops are, as a rule, strongly encouraged to 'hydrate'; but SR personnel rarely have the luxury of an abundant and reliable water resupply during missions. Note that Bedouin tribesmen drink water in the early morning and in the evening. Liquid intake is supplemented with small amounts of tea, loaded with sugar. And Apache youths in the Old West, would participate in games to test their endurance by racing before dawn for ten miles over rough terrain carrying a mouthful of water, without swallowing any. Similarly, SOG SR Team Members would often drink water only during meal/communication breaks and NDP, with an occasional sip during movement breaks.
- Intermittent streams flow strongly during the wet season and dry up in rainforest/jungle environments during the dry season; if Team Members empty their canteens/bladders during mission cross-country movement, they will be compelled to move to (and perhaps linger near) water sources to replenish; the enemy may anticipate this and beef up surveillance of water sources during the dry season. Teams employed in COIN operations should bear this tendency in mind in anticipating similar enemy behavior.

Food and Ration Discipline TTPs:

General Food/Ration TTPs:

- Eat well prior to movement to the launch site. Bring additional rations to the launch site to be consumed prior to launch, as a launch is often delayed. Team Members may need the energy after insertion, especially if the Team is pursued.
- Consider not eating for the first day following insertion.
  ○ The Team Member will be more alert and sleep more lightly. The meal(s) consumed before launch will supply adequate energy.
  ○ Eat twice a day after the first day. It is important to eat something at each mealtime.

- Dehydrated rations require water. If the ration is not rehydrated prior to consumption, it will self-hydrate as it passes through the digestive tract – dehydrating the body incrementally in the process. There is a trade-off between the light weight of dehydrated rations and the additional demand for water, which may tether the SR Team to limited water sources. Where water sources are abundant, this is not of much concern; in other climates/conditions it's a significant vulnerability.
- Enhanced rations are typically required in cold regions during winter; cache 'Rations, Cold Weather (RCW)' in a MSS during fair weather, or prior to onset of winter/bad weather. Monsoons and environments where heavy and/or continuous cloud cover is likely infer the possibility that the Team may be stranded in its target area beyond its mission window – the Team should carry additional/emergency rations for these situations. The Team leadership may have to monitor consumption.
- Consider preparing the meal well in advance of consuming it, by rehydrating the rations where water sources are present, and stowing the rations in the bellows pocket of the field trousers for consumption at designated mealtimes. The meal should be sufficiently wrapped to reduce food odors/leakage.

<u>True Account</u>: An experienced SOG T/L would carry a mix of indigenous and US dehydrated meals (scavenging only those components he desired and discarding the rest) and occasionally a few cans of C-rations (to include a couple of tins of fruit). He would hydrate one ration per day and place the ration in the bellows pocket of his trousers. This single ration would supply him two daily meals – essentially a half-ration regimen. It was normally enough to satisfy his hunger, provide sufficient daily energy, reduce his weight burden, reduce his water requirement, and keep him alert throughout the day and lighten his sleep. When he felt his energy flag or he felt hunger pangs, he would open up a C-ration can at meal times to supplement his caloric/sugar intake. For a seven day mission, he would carry the equivalent of 3-4 days of rations and he would further reduce his load by carrying fewer canteens than other Team Members, opting instead to carry more ammunition.

Mealtime TTPs:

- Ideally, meals should only be taken twice a day: (1) near midday, simultaneous with scheduled communications break, and (2) during evening communications break, prior to moving into the NDP. A morning meal should be avoided, as this time of day, prior to movement from the NDP, is peak danger time for the Team.
- Food, including sweets, should not be consumed during movement, unless the Team takes a break in a secure location and has deployed security. Never eat food at a surveillance hide due to aroma and hearing issues.
- Remember that the mastication of food will reverberate through the sinuses to your ears, impairing hearing; pause mastication during meals frequently to listen.
- Always deploy Claymore mines prior to meal breaks. Food odors will suggest to enemy trackers that you are taking a meal break – an optimum time for them to attack the Team.
- Establish a perimeter prior to dusk and consume the evening meal (and conduct communications) there – do not perform these functions at the NDP.
- Establish the evening meal/commo perimeter early enough to allow time for food consumption, to compose and transmit messages and move to and establish an NDP by dusk.
- Ideally, no more than half the Team Members should eat at any one time. The rest of the Team should be on security and alert.

Maintenance TTPs

General Maintenance TTPs:

- All Team Members should carry the tools (cleaning rod, brush, wrench, etc.) and materials (lube) necessary to maintain their individual and assigned crew-served weapons. Tools/materials excess to immediate needs, such as may be required in deep penetration/long duration operations, should be cached.
- All Team Members should be cross-trained on weapons assigned to other Team Members, to include maintenance training.

Primary and Secondary Individual Weapons Maintenance TTPs:

- Some weapons require more care than others. Weapons with reliability problems or those that require constant maintenance should not be used in SR.
- After training, prior to and after an operation, thoroughly clean and lubricate individual weapons and magazines and ensure magazines are free of debris. Always carry proper cleaning equipment on operations and carry a small vial/tube of lubricating oil for your weapon(s). A primary cause of weapon (M-16 family) malfunctions is lack of lubrication; second cause is carbon fouling. Only one Team Member at a time should clean his weapon and disassembled parts must be secured.
- Lubricate <u>daily</u> during an operation:
  - Oil the weapon bolt, bolt carrier and selector switch <u>daily</u> and <u>quietly</u> work the components back and forth, especially during rainy season, or key moving parts may freeze from rust/corrosion.
  - Especially in moist environments or when time permits after a stream/river crossing.
  - Lubricate as soon as time permits after a firefight. Firing the weapon at its cyclic rate will heat the bolt and bolt carrier and burn off earlier lubrication; if not attended to, this becomes a cause of malfunctions.
  - WD-40 displaces moisture, prevents rust and corrosion, lubricates parts and magazines and cleans weapon bores.
  - In cold regions/freezing temperatures, use very light lubricant or WD-40 (vegetable oil may suffice)
- Field strip and detail cleaning of weapons <u>during</u> operations is normally only necessary when on long duration missions, after a malfunction or possibly after a heavy firefight – and then, only when the Team is in a relatively secure location (e.g. an MSS).
- Ensure ejection ports are kept closed; check this after a firefight or maintenance.
- Place a cap on the muzzle (or suppressor) to keep dirt, water and dust from the barrel.
  - Ensure an air gap is permitted to allow drainage of condensation/moisture/water.
  - Use tape to cover the muzzle if a cap is not available.
  - Quietly 'break the seal' at the chamber to allow drainage from the barrel daily and/or after immersion of the weapon in water, etc., by partially retracting the bolt. Re-seat the round using the bolt-assist.

Crew Served Weapons Maintenance TTPs:

- Crew-served automatic weapons are treated/maintained the same as individual weapons (above).
- External trigger/firing mechanisms for mortars, rocket launchers are treated the same as individual weapons (above).

Mobility Equipment Sustainability/Maintenance TTPs:

- Mobility equipment should be selected, aside from its operational capabilities, with an emphasis on high Reliability, Availability and Maintainability (RAM). Mobility equipment should only require basic operator level and emergency care (e.g. fuel, filters, wiper blades and lubricants/fluids) – without needing specialized tools.
- If the Team deploys with captured and/or acquired enemy mobility equipment, high RAM characteristics may not be inherent and parts may be scarce. Equipment maintenance will be dependent on captured enemy parts and supplies or those acquired from former enemy client states.
- Foreign mobility equipment may have to be modified for SR use. This would include the installation of a liquid-filled vehicle compass that is adjusted for vehicle metal mass.
- Tools/materials necessary for operational/emergency maintenance should be carried on/in the vehicle. All tools/materials excess to immediate needs should be cached.
- Vehicles should be equipped with very robust tires (treads and sidewalls) with a flat-run capability, if possible. Replacement tires should have the same capabilities.

Communication Equipment Sustainability/Maintenance TTPs:

- Tools/materials for communication equipment maintenance should be carried on/in the vehicle where such equipment is installed.
- Communication equipment consumables include batteries, microphones and ear-sets. Carry only those that may be required for replacement during the operation. All tools/materials excess to immediate needs should be cached.

Transportation TTPs:

Helicopter resupply will reveal Team location to the enemy, unless the helicopter crew uses deception to mitigate this problem.
- A helicopter can make one or more false landings/drops before or after dropping its actual resupply bundle(s) … deceiving the enemy as to which LZ was the actual.
- Alternatively, the helicopter could simply fly over two or more Drop Zones (one actual, the other(s) false) at speed, and kick the bundle over the actual DZ as it passes.
- Additionally, a booby-trapped bundle can be left behind on a false LZ/DZ, for the enemy to recover. This will cure enemy attempts to capture bundles.
- Helicopters, employing advanced technology noise suppression, are scarce, guarded assets and might not be used in resupply operations.
- Some Target Areas may be so completely covered by continuous canopy that suitable LZs/DZs will be scarce. Alternatively, those LZs/DZs, which may be present within a Target Area, may be so few in number that the enemy may place them under surveillance or station enemy units or anti-aircraft weapons to engage insertion or resupply aircraft.

The related photograph depicts a UH-1D helicopter approaching a Laotian hillside LZ. Notice the layers of canopy in this photo; this area might be considered fairly open by rainforest standards, but the LZ would have been considered 'marginal' in SOG operations, as the slope of the hillside would not permit helicopter landing (helicopter main rotor blades could strike the slope or trees); subsequently, this LZ would have required ladder or ropes for insertion or extraction and tossed bundles for resupply.

- Where an enemy has airspace dominance, parachute canopies suspended from tree limbs, may be easily detected.

Figure 67. UH-1D Approach to a Landing Zone – Center of Photo. (Buckland)

- Airdrop resupply to a Team under heavy canopy is feasible if the drop can be made with adequate precision and the package is configured to penetrate high canopy. Additionally, aerial detection of resupply LZs/DZs that are located under continuous canopy requires either (1) marking with WP munitions, which may inform enemy units of the Team's location; (2) use of a radio beacon (detectible to RDF); or (3) use of a GPS/PLS, many of which have some limitations in heavily dissected terrain and multi-level canopy.

  During the Vietnam conflict, the 'experimental' US Special Forces Mobile Guerilla Forces (MGF) frequently used modified, 500-pound napalm containers converted to contain resupply bundles delivered by A-1 strike aircraft, sometimes using the cover of local airstrikes to further conceal the resupply operation. Similar means were employed by the British SOE and American OSS during WWII, where this technique was pioneered.

- If the T/L believes the Team might be weathered-in during its mission, e.g. by cloud cover or rain that would obscure the DZ from resupply aircraft, the T/L must plan for the use of (and train with) a beacon or PLS device for blind delivery. Air Force cargo aircraft may boast of computerized airdrop capabilities that can calculate spatial (e.g. GPS) and weather (wind) factors to deliver bundles with precision, but in heavily dissected/vertical terrain, claims of high accuracy are doubtful.
- Remember that a beacon/PLS device may generate a detectable signature, and that this, coupled with use of these devices from elevated features in Target Areas with dissected terrain, brings additional OPSEC concerns. During weathered-in conditions, and in view of the effects of updrafts and cross-drafts that always occur in mountainous terrain, parachuted bundles may scatter across several ridges, losing at least part of the resupply. Multiple canopies may further attenuate beacon/PLS signals, complicating bundle delivery.
- The Team must develop supply lists for pre-planned or contingency resupply operations that are tailored to the mission and the T/L's concept of operation. Some lists can be incorporated into the unit SOP. The Team should work with 'Riggers' to configure and pack its pre-planned resupply bundles. The Team can have planned or contingency resupply packages delivered on-call by designating and using 'spare' codes from its SOI.

- SpecOps headquarters may acquire foreign/enemy equipment to use in SR missions or other (e.g. DA) missions. This equipment may include cargo and observation aircraft, a variety of vehicles and an assortment of enemy armaments. Some cargo aircraft may be covertly armed. If the Team intends to use a captured or acquired enemy vehicle, determine any enemy IR markings which may exist; ensure that these markings are not obsolete, if so, ensure current enemy IR markings are replicated on Team vehicles. These markings are used by sophisticated militaries to ID friend-foe to prevent fratricide.
- Horse transport requires forage; whereas reindeer and other draft/pack animals (e.g. Llama) can forage for themselves in austere environments. Some draft animals require special handling and may require the recruitment of native handlers, at least until Team Members acquire the requisite skills.
- If draft/pack animals are to be used on a long-duration mission, consider these issues:
  - Pack animals are usually best maintained and provided from a friendly guerilla or partisan base to support SR operations.
  - Braying of a pack animal is prevented by a surgical procedure performed on its vocal cords.
  - Pack animals should be trained to tolerate the sudden loud noises that accompany combat engagements.
  - Use pack animals to transport supplies from an LZ/DZ to a distant MSS.
  - In the absence of pack animal support from friendly guerilla or partisan units, a covert AOB or MSS may be necessary to pen and maintain the animals. In this situation, some personnel must remain with the animals to secure and care for them. Animal pens must be located where enemy aerial and/or thermal detection will be deterred. The grazing area must accommodate the number of animals.
  - Handlers must be trained in animal care, transport/portage rigging, cross-country animal movement techniques and tactical patrolling skills. The handler's first obligation in a tactical engagement is to control and protect the animal and its cargo; this may be best accomplished by rapidly moving the animal to terrain-form that provides cover from enemy fires.

'Each division received a reindeer transport column with fifty reindeer for the primary purpose of facilitating the supply of raiding detachments and reconnaissance patrols.'[1]

Supply and Equipment TTPs

Footwear and Foot Care TTPs:

- Boots with a zipper closure or a lace-up with zipper option are generally unsuitable for SR.
- Boots are worn <u>continuously</u> during an operation, except for sock exchange or foot inspection. Seek advice of veteran SpecOps personnel regarding footwear, and/or tread design, that they may have used in the AO where the Team will deploy. Combat boots used in Iraq and Afghanistan are produced by several vendors; while they appear much the same, their quality and construction may vary. Ensure that new boots are well-fitted, suitable to the AO and mission, and that they are properly broken-in before wearing them on a mission or you will risk debilitating blisters and lesions.

Author's Solution:
To accelerate the breaking-in of new boots, soak new boots and wear them until they dry, doing this repeatedly over the course of two-three days while in a garrison environment. In the evening, insert

---

1. Unnamed German Generals and General Staff, 'Military Improvisations', p. 63

shoe-trees and/or wadded paper into the boot to prevent shrinkage. This process allows the boots to better conform to individual feet. This worked well with new jungle boots which were notoriously stiff when issued and may be similarly necessary if boots of foreign manufacture (e.g. with enemy/'sterile' tread design) are to be used.

- Jungle boots issued during the Vietnam conflict possessed vents to evacuate water and had mud-shedding treads. These boots are still considered by many as the best footwear for jungle environments. Veterans would break in a new set of jungle boots by soaking them in water overnight and then wear them during training the next day while they dried out.
- Consider using boots modified with a standard enemy tread design to fool trackers/enemy patrols.
- Consider using boot inserts/insoles. The 'SOF Sole', among other vendors, produces a variety of quality inserts. Before using them on an operation, test them in garrison and during field training.
- To change socks, especially in the rainy season, wait until the Team has occupied its evening commo break position. No more than two Team Members should change socks at one time. Team Members should never take off <u>both</u> boots at the same time. Remove one boot, then one sock; don fresh sock and re-don boot; then repeat with the other foot. If the Team is attacked during this process, only a maximum of two Team Members may be forced to abandon a single item of footwear.
- Use socks that wick moisture away from the foot and that offer a thick cushion. While issue boot socks are quite adequate, consider purchasing high-end socks that employ a dense, durable weave of Merino wool that are comfortable in most environments.
- Each Team Member should keep fresh socks where he can get to them easily during commo breaks; not buried deep within his rucksack. Used wet socks should be wrung out before storing in a small cloth bag that can wick away moisture (e.g. Goretex®) while shielding the socks from rain or water immersion.
- Socks may be used as short-term field expedient boot covers to suppress tread impressions to confound trackers. This approach has limitations: material may fray and leave sign; material may collect debris, mud/snow/ice; traction of the boots will be reduced
- Consider the use of 550-cord as boot laces. In survival situations, interior strands of the cord can be used for snares, tripwires, fishing line and other applications. Alternatively, consider braided Kevlar string.
- <u>Author's Recommendation:</u> Avoid 'compression' socks, often hyped by outdoor wear outlets. They become uncomfortable during extended wear and may inhibit circulation, especially as feet swell during extended time in movement. Inhibiting circulation during operations in wet weather environments would contribute to and increase the severity of 'Immersion Foot' (See Medical, below). If worn at all, compression socks should not be worn at night. Note: It is difficult to don/change out compression socks, especially over swollen feet.
- At least one Team Member should carry a small Foot Care Kit consisting of Moleskin, Folding Scissors; Foot Powder; Nail Clippers; Pin/Needle and a Lighter. Although SpecOps personnel are professionals, they still may neglect the care of their feet; T/Ls, Team Sergeants and/or Team medics should spot-check Team Members' feet periodically during missions (and training).

Common Individual Equipment and Supplies TTPs (US Personal Wear/Carry):

Carry on Person:

• Map and durable protractor	• SOI
• CS Powder/capsaicin (Squeeze Bottle)	• Claymore Mine (Neck Bag) or alternative
• Survival Items: 　◦ Para-cord (550-cord) & tripwire/fishing line (w/ hooks) 　◦ Fire-making materials 　◦ Bouillon Cubes	• Compass • Triangular Bandages (worn)
• Cable/Folding Saw	• Notebook and writing instruments
• Camera (Select Personnel)	• Signal Mirror and Panel
• Multi-tool/Folding Knife	• Toilet Paper
• Insect Repellant (w/ high per cent DEET)	• Water Purification Tablets
• Skin Camouflage	• Pen Flare (canopy penetrating)
• Small flashlight with UV and red/blue filters	• Hat
• Whistle or other audible signal device (T/L)	• Gloves (personal and rappelling)
• Goggles/ballistic eyeglasses (clear lenses)	• Appropriate Medical Items: 　◦ Specific antidotes/treatments 　◦ Morphine Syrettes 　◦ Personal use medications 　◦ Anti-Malarial Tablets
• 'Quarterback's' wrist card	• Field (Rigger's) belt
• Fatigues/utility uniform or similar alternative field clothing	• Boots and socks
• Knee pads*	• Gaiters
• Primary and Secondary Weapons	• Seasonal Uniform Items
• Ear Plugs	• One Day's Ration

* Note that pockets can be tailored into trousers at the knee (interior) to allow the insertion of removable high-density foam knee pads.

- As SR Teams are likely to conduct cross-functional missions (SR, DA, COIN, etc.) in multiple environments, the T/L should consider choosing equipment that will serve multiple roles/purposes as well.
- Most lightweight protective masks are only sufficient to protect against riot agents. These masks may be sufficient in AOs where the use or presence of lethal agents have not been detected.
- A map protractor should have indelible markings that will withstand insect repellant/solvents.
- Notebooks that use waterproof paper are preferred, as are 'tactical' pencils/pens capable of writing on wet paper. In cold regions, the damp paper and writing instruments can freeze if not carried in interior pockets.
- Suppression of Heat Signature: Graphite is used in IR suppressive smoke to degrade IR optics capability. Graphite can also be found in commonly available commercial paints and other products. Consider using spray paints with graphite or blend graphite granules in with dyes to enhance the camouflage capabilities of Team uniforms – but <u>test before adopting</u>. Other additives that can be used for the same purpose: Titanium Oxide, Carbon Fibers, Glass particles can be used in clothing dyes and on camouflage weapon wraps.

- Wearing underpants on missions in high-temperature/high-humidity environments will result in chafing and a raging case of 'jock itch'. Military-issue foot powder, or the commercial equivalent, has a fungicidal component. If you get 'jock itch', use the foot powder; if the rash is fungal, the powder will cure the problem. If it is bacterial, it will not. Note: the powder may sting somewhat.
- During mission preparation, soak uniform items with insect repellent, especially around the tops of socks and boots, along the belt-line, at the fly (if button closure) and at sleeve cuffs to stop leeches and other insects. Keep sleeves rolled down, regardless of discomfort; sleeves protect the arms from insects, sun-burn and cuts from vegetation; they also provide camouflage.
- Ghillie Suit Pros/Cons:
  - Pros:
    - Most suitable for use in a static position for purposes of sniping, observing/surveilling
    - Has utility in open areas with limited natural camouflage
    - Breaks up the outline of the user
  - Cons:
    - While made of lightweight materials, the suit is bulky and takes up a lot of space in carriage
    - Will snag on vegetation. Snagging will cause movement of vegetation, creating more noise during movement and will leave more trail sign
    - Suitable only for movement over short distances in warm to hot environments or moderate distances in cool to cold environments. In high temperature and high humidity, potential for fatigue and heat injury are real – especially if any exertion is required. Temperature inside the suit can register 120 degrees even in moderate climates.
    - If water-soaked, the suit will increase substantially in weight, depending on fabric. In crossing a swift stream or river current, of if the Team Member loses footing or drops into a hole, risk of drowning is markedly increased.
    - Will retain moisture and will impair wicking from the individual uniform.
  - Comment: Should be stored in a MSS/cache. The suit may be worthwhile for several tactical and environmental conditions.
- Gloves will protect hands from thorns, insect bites, plants with stinging hairs, etc., and will eliminate the need for hand camouflage. Ensure the gloves are high quality, non-shrinking material, that they offer manual dexterity and a good grip and provide for ventilation and moisture wicking. As fingers may swell, ensure your personal gloves are a size larger than otherwise required. To increase manual dexterity, SOG SR personnel would cut off the thumb and forefinger portions of the gloves.
- If the nature of the mission requires deniability, consider wearing sterile uniforms, clothing of a third country (neutral) manufacture or commercial equivalent field clothing. If the mission is a 'black operation', consider wearing enemy uniforms.
- Consider the use of glow-in-the-dark 550-cord to mark firing sectors, grenade throwing lanes; routes of withdrawal; location of key items, etc. in the dark. Multipurpose string can be used to measure the approximate circumference of trees.
- Have the field uniform altered so that fishing line or Kevlar string is sewn in between the double-seams of the trousers and blouse/jacket and so that the line/string may be easily retrieved in survival/E&E situations without cutting into the seams.
- Each Team Member should have one wide-mouth water bottle/container capable of withstanding direct flame; to boil/sterilize water or to cook food in survival situations. Ensure the container is padded so that it will not make noise during movement.

- An SR Team deployed on a deep penetration, long-duration mission may have to live off of captured rations. At least two Team Members should carry a 'P-38' type can opener to open ration tins. The P-38 is cheap, very small/lightweight and available commercially. It's a superior can-opener to those found on utility knives.
- Use a branch of leafy vegetation to make a spray paint camouflage pattern for uniform and equipment items.
- A Team engineer or T/L might find a compass with an inclinometer useful for general terrain reconnaissance; it can be used to determine slope and tree height (if the top of the tree/feature is visible from the ground). To estimate tree or terrain feature height, use the inclinometer and two stakes; walk away from the tree until the ground to treetop angle is 45 degrees, forming an isosceles triangle; the stakes can be used as 'sights'. The distance walked will equal the tree height. The 45 degree angle can also be estimated by a Team Member without reference to an inclinometer using the same procedure. Angle estimation will be problematic if the ground is uneven or falls away; so the Team Member must select a direction from the base of the tree that offers the most level contour.

CS or Capsicum Powder TTPs:

- Several Team Members should carry CS/Capsicum powder in plastic squeeze bottles. Sprinkle this powder on empty ration containers (before covering the containers with earth) and around refuse burial pits to prevent animals from digging them up once you have buried them.
- The Team should use CS/Capsicum powder on the back-trail (during dry weather) to confound a tracker dog's sense of smell.
- Additionally, CS/Capsicum powder can be sprayed into the air and downwind toward habitations or domestic herds where dogs may be present. Capsicum powder, nearly undetectable to humans in trace amounts, may impair a dog's sense of smell such that the dog may not be able to alert to the presence of the Team for an extended period. Also consider using a booby-trap/mine (with SD) to kill the dog and/or its handler. Obscure Team Member scent when planting such devices.

Author's Solution:
Concentrated animal scents and lures, made from urine or glandular secretions, are available through commercial purchase. The use of these concentrations on the Team's back-trail may be useful in throwing off tracker dogs by (1) overloading the dog's olfactory glands, by (2) attracting game animals to the immediate vicinity and by (3) distracting the dogs to pursue game versus their human objective. Additionally, dogs may become carried away and alert to the animal scent by barking, which would disclose the presence of the enemy dog team to the SR Team.

- A primary camera (preferably waterproof) should either be carried in the cargo pocket of the T/L's fatigue trousers or it should be carried (secured) inside his fatigue shirt. The camera should be as easily retrievable as the T/L's notebook. In monsoon conditions, some cameras might be enclosed in, and used with, an underwater camera case. The camera and used rolls of film or memory chips will contain intelligence information, ensure that these items are retrieved if the T/L is wounded or killed. Other Team Members may also carry cameras.

Knives TTPs:

- Always carry two knifes (minimum); an all-purpose sheath knife carried on LBE; a folding knife (ideally a multi-tool) on your person. Author's Comment: A stiletto-type fighting knife is not recommended as its general utility is limited. The sheath knife should be both a multi-functional tool and weapon (e.g. a combat-utility/survival knife) that should have the following preferred characteristics:

- Blade:
  - A 6-7in blade with partial blade serrations or a serrated/saw-back spine. Serrations could be a plus, as sawing is less noisy than chopping.
  - Pointed blade. As a weapon or survival tool, the knife should be capable of delivering thrusting wounds.
  - Robust blade thickness.
  - Subdued blade with rust preventative treatment.
- Grip:
  - Full-tang with a robust non-slip grip. Grip should be firmly wedded to the tang.
  - Robust butt cap that can be used as a hammer that can deliver a stunning blow.
  - Lanyard hole(s) allowing the knife to be used as a spear point.
- Full or partial guard to prevent fingers slipping up onto the blade.

- Alternatively, a large, broad-bladed combat knife (e.g. Smatchet, Kukri, etc.) may be a good choice instead of the all-purpose sheath knife; and a good compromise alternative to carrying both a utility/survival knife and a dedicated machete. The large-bladed knife may be useful in cold region environments due to its utility in breaking through the soil frost line to establish an OP, hide location, cache/MSS, etc. Author's recommendation: the broad-bladed combat knife should be a multi-functional tool and weapon (e.g. a machete-like utility capability and a weapon) that should have similar preferred characteristics to the sheath knife but larger and heavier (1-2lbs.).

- A machete may also be a good addition for one or two Team Members, if brush or light-timber cutting is likely. This may be required to cut trail for pack animals. Many of the preferred characteristics of the large knife (above) apply to the machete. The machete must be robust and capable of holding an edge, with a non-slip grip/handle (preferably with a guard).

- A small non-metallic blade, handcuff key and flexible picks can be secreted within the Team Member's belt and uniform. Picks/shims should be sufficient to release the lock on zip-lock handcuffs. The enemy may eventually strip a captured Team Member of his clothing and replace them with prison uniforms; this will not occur before the Team Member is evacuated to a POW holding area or compound. A sophisticated search of the Team Member will not normally be performed by field troops; therefore, it is essential for the Team Member to attempt his escape while he still retains his secret tool set.

- Lock picks to enter a locked structure or to steal a locked vehicle. Picks should be appropriate to locks used in the region; different locking mechanisms may be used in certain parts of Asia.

- Use a 'Quarterback's' wrist card to maintain daily call-signs, report formats and other quick reference information for quick retrieval, or to take hasty notes. Remember that this item should also be retrieved from US wounded/KIA, along with notebooks and SOIs, as it may contain cryptographic/COMSEC and OPSEC information.

- Belt TTPs: SOG SR personnel almost universally discarded the issue subdued web belt and buckle for the field expedient 1.5-inch web strap and buckle used for lashing down aerial delivery packages and pallets (A7A straps). This was later to be called a Rigger's Belt and is available commercially or can be locally fabricated. Wear of the narrower issue belt would cut into the waist/hips of SR personnel and would sometimes cause abrasions during extended wear in the hot, moist Southeast Asian climate conditions; the abrasions would sometimes become infected; this was especially the case when the trousers' cargo pockets were weighted with gear. Further, the Rigger's Belt was considered robust enough (approximate breaking strength of 4,000+lbs; working load of approximately 1,400lbs) to be used in emergency rappelling/extraction contingencies in the absence of a purpose-built extraction harness. Author's Recommendation: Acquire a Rigger's Belt, but ensure that it has at least an additional foot of length for the following purposes:

- Knot the additional length at the buckle to ensure that the running end of the belt does not slip through the buckle under mechanical stress (example: during a string extraction).
- In the field, prior to defecating, the Rigger's Belt should not be fully unthreaded from the buckle when trousers are dropped; feed enough of the additional length through the buckle to allow the trousers to slip down the legs. If an enemy comes upon the Team while the Team Member has his trousers around his ankles, all he need do is pull up the trousers and jerk the running end of the belt to tighten it at the waist. If a belt does not have additional length or is fully unthreaded from the buckle, the Team Member will have to engage one hand to hold up his trousers while he is scrambling to rejoin the Team, recover his gear and engage the enemy.

- Flashlight TTP: All American Team Members should carry a small flashlight with a UV, red and/or blue lens filters. Consider using the same color lens filter that is commonly used by the enemy. Also, the flashlight should also have a strobe function; this can be used in position marking/evasion signaling, as a decoy or to 'blind' a night pursuit.
- Documents TTPs:
  - Always carry maps, SOI and notebooks (if the notebook is not all-weather) in waterproof containers/plastic bags. Maps can be water-proofed, but rubbing of the map against other items or uniform cloth material may remove or smear map markings.
  - An all-weather notebook and all-weather pen(s) are preferred.
  - All Team Members should carry these items in the same pockets/locations (by SOP).

Carry on Load Bearing Equipment (LBE):

• Ammunition	• Emergency Radio
• Strobe Light/Flashlight with Strobe Function	• Intra-squad Radio (if used)
• Sheath Knife	• Grenades (Several defensive and/or offensive, 1 Smoke/CS)
• Swiss Seat (rope/web strap) or alternative	• Primary Extraction Harness
• Karabiners (2-3)	• Plastic Hand Cuffs/plastic straps
• Medical/First Aid Pouch (other than medic)	• Gas/Protective Mask (Lightweight)
• Rifle Grenades (if used)	• Magazine pouches/Canteen covers
• Canteens/Water Containers	• Fuses & Time Delays; booby-trap devices for rapid deployment
• Small Monocular	• Tourniquet and Battle Dressing(s)
• Survival Items	• Secondary (or even tertiary) Weapon, including silenced and/or non-lethal weapons
• Elastic Bands (various sizes)	

LBE TTPs:

- Tactical Vest LBE: If the Team Member is to use a tactical vest, the Author recommends:
  - The Modular Lightweight Load Carrying Equipment (MOLLE)-type system. This allows the user to select the highest quality vest accessories and/or accessories that can be placed on the vest for optimum accessibility. MOLLE accessories can also be tailored for MOS/mission specialization and variability.

- ○ The vest should have no VELCRO® closures. VELCRO® closures make far too much noise when being unfastened. Use another closure method.
- ○ The vest must have quick release features to disconnect the vest from the user in emergencies.
- ○ The back of the vest should not have permanently attached accessories that would prevent the carry of a backpack/rucksack.
- ○ A water bladder, if carried, may be carried in a vest/LBE compartment or in compartment within a rucksack. The bladder should have a wide fill opening.
- ○ MOLLE vest accessories:
  - ▪ Magazine pouches must be optimally placed so that the magazines may be reached regardless of body position/orientation (e.g. face-down, prone). Magazines containing ammunition that is most often used should be placed for speedy magazine exchange; specialized ammunition should be carried in magazines that are less accessible.
  - ▪ If a haversack/pouch (e.g. carrying a Claymore mine) is to be worn suspended from the neck, the haversack/pouch should be positioned high up on the chest and magazine pouches positioned so that the haversack does not impair magazine retrieval.
  - ▪ Access to the magazines should not be impaired by magazine pouch flaps/covers/ fasteners. The Author recommends that the magazines be retained in the pouch by friction, rather than by straps, to permit rapid magazine exchange. The friction should be sufficient to retain the magazines throughout cross-country movement and Battle Drills. However, during rope insertions, rappelling or similar activities, ensure that the magazines are firmly secured in the pouches (e.g. using bungee cord/fasteners).
  - ▪ Magazines should be carried inverted in the magazine pouch, with the butt-plate uppermost. This will shield the magazine interior from debris, etc.
  - ▪ Other pouches should be positioned by priority of tactical necessity and frequency of use.
  - ▪ Accessories that would make a noise when struck by a branch (for instance) should be avoided/muffled.
  - ▪ Pouches that fit closely to the face of the vest are preferred to those that protrude, as protruding items will snag on vegetation.
- Ensure all LBE have quick releases. Water crossings (including quicksand/quick mud) present a drowning hazard; the Team Member must have the ability to rapidly shed his equipment to avoid drowning.
- Seldom remove the LBE, day or night (exceptions: to defecate or to don/remove clothing layers). When donning/removing clothing, ensure no more than two Team Members do so at a time. This process follows the rationale associated with sock changing (see Footwear above).
- If the nature of the mission requires deniability, consider wearing sterile/enemy LBE, LBE of third country (neutral) manufacture, or commercial equivalent LBE. If the mission is a 'black operation', consider wearing enemy or third country LBE.
- Canteen covers, rather than standard ammo pouches, can be used to carry 20-round magazines (e.g. for use by indigenous troops). They hold plenty of magazines, are easier to open, and the Team Member won't need so many (less capacious) ammunition pouches on his web gear. Canteen covers are also useful for carrying offensive/defensive hand grenades, rifle grenades and similar bulky items. Note that 'VELCRO'/hook-and-loop closures make a distinctive sound that, when opened, can be heard meters away, so modify closures if necessary. If using a zipper closure, use wax or soap to lubricate the teeth.
- Secure your emergency radio in its carry pouch to your harness/LBE.

- Fuses, time delays and/or booby-trap devices/items carried on your LBE are for rapid deployment, e.g. when the Team is being pursued or during a Battle Drill. When not driven by haste, carry extra items in/on rucksacks.
- Subdued karabiners should be connected to Team Members' LBE and rucksacks. They should be attached to robust, weight-bearing straps or fittings that will not easily break or rip off when its weight is suspended from a rope, an Extraction Harness Rig or ladder.
- For survival, Team Members should carry a tube of bouillon cubes, a bottle of purification tablets and perhaps a small pack of caffeine tablets/candy. One bouillon tube dissolved in one canteen of water, will provide energy in emergency situations.
- Plastic strap fasteners should be carried on both the web gear and on the rucksack exterior. These fasteners can be used in many ways: rapid attachment of snares and booby-trap devices to trees and shrubs; handcuffs; fabrication of field expedients; construction of sleeping platforms, etc. These fasteners come in varying lengths, widths and locking configurations.
- The number of Team Member water containers should be based upon weather conditions and availability of water in the Target Area.

Extraction Harness TTPs:

- Teams operating in mountainous terrain may require that Team Members carry a karabiner and length of rope to be used as a Swiss Seat (expedient) for rappelling and safety purposes. Some LBE may incorporate extraction rig features (e.g. Vietnam-era STABO rig), but some current (issued) LBE do not. Subsequently, a Swiss seat, or other rig, may be essential to string or ladder extractions, especially for wounded Team Members. The rope may also be helpful for lashing a wounded comrade or POW on a Team Member's back or litter or for assisting Team Members up steep slopes or similar tasks.

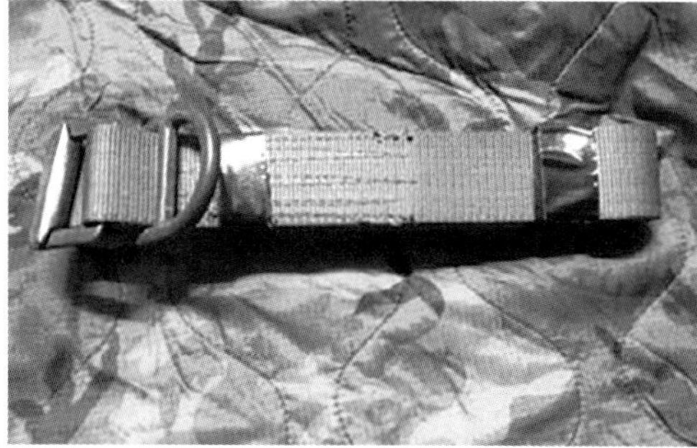

Figure 68. Hanson Rig in folded configuration. Note that the strap is not laced through the buckle and is therefore not fitted for a Team Member or rigged for immediate use.

- Note that a top heavy Team Member, especially if wounded, while using a Swiss Seat during extraction may invert (flip upside-down) while in flight and fall out of the seat. <u>The Author's Recommendation</u> is to don a Hanson Rig for extraction purposes; the Hanson Rig is a shortened A7A nylon cargo strap (with buckle), or equivalent, supplemented with one or two karabiners as being much preferable to a Swiss Seat, especially if wounded or KIA Team Members, or a POW are to be extracted on 'strings'. As is the case with a Rigger's Belt, the webbing in a Hanson Rig will have an approximate breaking strength of 4,000+lbs and a working load of approximately 1,400lbs. The Hanson Rig may be donned over Team Member LBE

Figure 69. A ladder extraction at altitude. SOG Team Members may either ascend to the helicopter cargo bay, or may secure themselves and their equipment to the ladder with Karabiners.

and the Team Member then connects to the 'string' using a karabiner; the rucksack would be hooked separately onto the 'string' using a separate karabiner. The process for donning the Hanson Rig is illustrated nearby. Worn in this manner, the rig forms a modified Swiss Seat (with a shoulder strap); it is nearly impossible to inadvertently fall out of a properly worn Hanson Rig.

- ○ Tip #1: The length of the Hanson Rig is typically measured/tailored to the height and girth of a specific, fully-equipped Team Member; but the Author recommends that additional length should be needed in the event that the rig must be used for someone else of a larger size (e.g. a POW or a wounded comrade). Excess length can also be used to better secure the Team Member in the harness.
- ○ Tip #2: Strap running-end/excess should be tied off at the buckle to ensure the strap does not escape the buckle under mechanical stress/load.
- ○ Tip #3: The rig must be folded and taped and attached with a karabiner to the LBE, with the running end already threaded into the buckle. The strap might also be threaded through a shoulder pad, as the Hanson Rig can also be used for a travois harness or litter strap/sling.
- ○ Tip #4: The Team Member's weapon can be fastened to the Hanson Rig D-ring using a karabiner during extraction or to a karabiner on the LBE.
- ○ Tip #5: Helicopter crew members in support of SR Teams should also be equipped with Hanson Rigs and should be properly trained in its use.

Figure 70. SOG Team Leader "Squirrel" Sprouse Demonstrating the Donning of the Hanson Rig (*U.S. Army*)

Hanson Rig Donning Process:

Hanson Rig Donning Process Description	
Illustration	**Description**
Figure 71 Number 1	The strap is draped from the shoulder with the buckle level with the collarbone. Free end (excess) should be tied off at the buckle to ensure buckle does not slip.
Figure 71 Number 2	The rear section of the draped shoulder strap is pulled diagonally across the back and then brought forward. Buckle remains in position.
Figure 71 Number 3	The strap section brought forward from the back is pulled across the abdomen and the soldier grips the shoulder and diagonal strap sections together. Feet are placed more than shoulder width apart. Free hand pulls rear section of strap forward from between his legs and is then merged/gripped together with the shoulder and diagonal straps to form a seat/cradle.

Illustration	Description
Figure 72 Number 4	Strap sections from the shoulder, the diagonal and crotch are joined together with a karabiner (gate upwards).
Figure 72 Number 5	Excess strap is brought around the back and crosses the chest/abdomen to be tied off at the front shoulder strap section. This is an optional safety measure and is always wise to employ; it is especially recommended for evacuation/extraction of WIA/KIA and prisoners. If not used, the strap should be tucked away/tied off.
Figure 72 Number 6	Rear view of the completed Hanson Rig
Figure 73 Number 7	View of the soldier with Hanson Rig donned over his LBE. The D-ring at the buckle can be used to secure the soldier's individual weapons (karabiner typically required).
Figure 73 Number 8	Front diagonal view of donned Hanson Rig. The soldier will hook his seat/cradle karabiner to an existing loop/karabiner on the rope 'string'. The rucksack or other equipment may also be connected to this attachment point using another karabiner. When the soldier is lifted off the ground, the strap will shift and stretch upwards so that the karabiner is upper-chest high; the soldier will therefore not invert during flight.

Figure 71. From Left to Right: #1 Drape; #2 Pull strap section from rear to front, #3 Form the seat. Note free end excess at left of demonstrator.

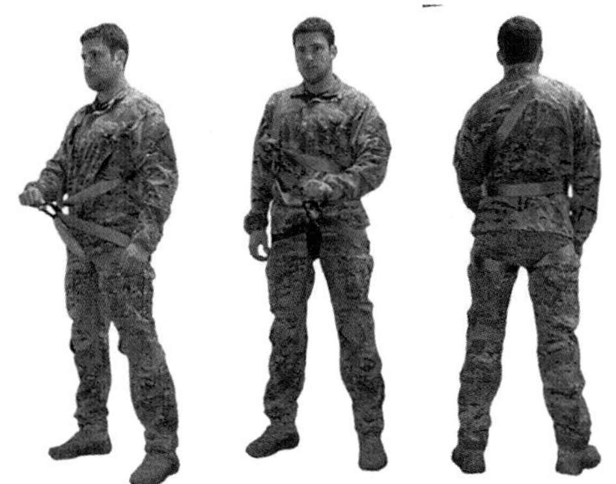

Figure 72. From Left to Right. #4 Seat fully secured w/ Karabiner; #5 Excess strap brought around chest and tied off; #6 Rear view.

Figure 73. From Left to Right. #7 Demonstrator with Rig donned over LBE; #8 View of demonstrator's left oblique.

Carry in/on Rucksack/Haversack:

• Waterproof Bags	• Extra Ammo
• Poncho or ground cloth	• Karabiners
• Sleeping Gear	• Food
• CS/Smoke/Thermite Grenades	• Canteens/Water Bladder
• Camera lenses and other accessories	• Mines, Demolition Charges, IEDs, Fuses & Time Delays; booby-trap devices
• Socks	• Entrenching Tool
• Specialized Equipment, as required (e.g. wiretap devices, beacons/transponders, sensors, propaganda materials, etc.)	• Unitary Anti-tank munitions (if enemy armor presents a threat)
• Spare Batteries	• Night-Vision Goggle (NVG), Laser Designator, Binocular/Spotting Scope
• Lightweight Body Bag	• Weapon cleaning equipment, lube, etc.
• Hygiene and Personal care kit: moleskin, small scissors, foot powder, toenail clippers, needle and thread, butane lighter, antiseptic, etc.	• Coil of rope/equivalent, as required
• Hundred-Mile-An-Hour tape	• Bungee cord of varying lengths
• Machete or saw	• Camouflage Cover/Outerwear
• Water filter/purifier (2/Team)	• Seasonal Items
• Wire Cutter (at least 1/Team)	• Haversack/Utility Bag

- With few exceptions, that which is carried on the Team Member's person and that which is carried on the Team Member's LBE should constitute what is most essential for immediate utility, fighting, navigation, survival and mission accomplishment. However, some mission essential gear must be carried in a rucksack and/or a utility bag/haversack, due to weight, bulk or frequency of use.

Rucksack TTPs:

- The rucksack will act as an anchor during water crossings whenever water enters the pack. This could prove fatal if the Team Member enters deep/swift water. Mitigate this problem by:
  - Lining the main compartment of the rucksack with a dark colored water-proof bag, or a large, heavy-gauge (e.g. 6-mil, reinforced) plastic bag, tied off to provide buoyancy.
  - Separating cargo by function (e.g. clothing items, PSYOPS materiel; survival items; etc.) and place them into sealable, smaller (1-2 gallon) heavy-gauge plastic bags for additional buoyancy and to mitigate the possibility of a leak in the main compartment bag. Food rations need not be segregated in this manner. Mark (night/day) the exterior of the bags to identify contents.
  - External rucksack pockets and separate, detachable bags used to carry gear should be similarly lined. Reserve this space for items that may require immediate/rapid access such as booby-traps, mines, medical items, ammunition, demolitions, etc.
  - Water containers can be emptied to produce additional buoyancy during water crossings.
  - A 6-mil, reinforced plastic bag (e.g. 4 gallon) encased in a sandbag, which can be used to gather water, can also be used as a flotation device.
- Use haversacks/detachable bags for use in Battle Drills/rapid response. Items contained in these bags might include: Claymore mines, booby-traps, demolitions and special devices.

- For short duration missions, a tactical vest with a haversack/lumbar pack may be sufficient.
- The profile of the Team rucksacks should resemble those used by the enemy; this profile, viewed at a distance by an enemy observer, may alone deceive the enemy into identifying the Team as friendly (to them). This may best be achieved by using actual enemy rucksacks. If the nature of the mission requires deniability, consider using sterile rucksacks, rucksacks of third country (neutral) manufacture or commercial equivalent rucksacks. If the mission is a 'black operation', consider using <u>enemy</u> rucksacks.
- Team Members should shake/test their rucksacks, haversacks, and other separate/detachable bags to identify and correct rattling of equipment/cargo before departure to the Launch Site.
- Cold Weather Outerwear: In temperate to cold environments, with heavy vegetation/underbrush, Team Members will become soaked from morning dew. 'Broken-in' (where rustling of the garments are suppressed) Goretex®, or uniform items with similar properties may be appropriate for wear in NDP and during subsequent movement during inclement weather.
- Personal Hygiene Items: Especially for long duration operations consider the following items that may be carried or stored in a MSS. These items should only be used where the Team is in a secure location.
  - Additional 'field use'/moisture resistant toilet paper. Kept in a plastic bag.
  - Toothbrush, toothpaste and floss.
  - Scentless soap, a non-scratch dishwashing scouring pad and scentless deodorant can moderate but not eliminate body scents/odors. Skin pores will exude oily or waxy matter that accumulates (especially in hot-humid environments), called sebum, which, when broken down by bacteria produces a strong odor, leaving an enduring scent trace (for dogs) that can withstand rainfall. Sebum may accumulate in an oily paste on the skin that can be transferred by the hands to gear and vegetation.
  - Anti-bacterial wipes (odorless). Helps treat lesions, sores, rashes; reduces body odor.
  - Shower/Body wipe (odorless). Body odor acquired during 3-4 days in hot-humid environments becomes rank enough for an enemy combatant, passing nearby, to detect. Use body wipes to reduce body odor.
- Water Filter/Purifier. At least two water filtration/purification devices should be carried per Team. These devices should be capable of filtering out pathogens as small as viruses (sub-micron in size) and be capable of rapid function.
- Sleeping Gear:
  - Do not use sleeping bags insulated with feathers/down. Feathers will inevitably escape to reveal the presence of the Team. Further, when down bags get wet, the feathers lose their insulation properties and the bag becomes very heavy.
  - Swamps are found in nearly all regions of the world. If the Team is to operate in swamp, or in even in hot-wet climates during monsoon season, Team Members may have to sleep in hammocks. If the swamp is located in temperate or winter/cold regions, the hammock will have to be supplemented with a sleeping pad (or a field expedient) between the hammock and occupant to provide insulation.
- Rainwear is typically a bad idea on SR missions.
  - The patter of rain on the rainwear material will impair Team Member hearing, especially if the hood is worn up. Additionally, the hood will obstruct Team Member vision.
  - Rainwear makes far too much noise during movement.
  - Wearing of raingear during or subsequent to exertion, or in hot-wet environs will soak the wearer in a stew of sweat and trapped condensation, defeating the purpose of the rainwear.
  - Goretex®, or equivalent, that is 'broken in' (to reduce fabric rustling), may be more acceptable, as this material 'breathes'.

- ○ Ponchos are acceptable for general SR use as they serve multiple purposes: ground cloth, rain fly, use as a litter, floatation device and other field expedient applications. However, a subdued/camouflaged plastic composite tarp would be better and have more flexibility.
- Team Members should carry a small roll of 'hundred-mile-an-hour' tape or subdued elastic material to silence all snaps and buckles. Ensure that bare metal is subdued with flat (not gloss) spray paint. Check these measures during the pre-mission inspection and ensure that camouflage discipline is maintained throughout the operation.
- Bungee cord may serve many purposes: expedient fastening of items to LBE or rucksacks; attaching demolition items, Claymore mines, sensors, etc. to tree trunks/limbs, etc. Ensure that the bungee cords are subdued in color, and come with subdued/sheathed end-hooks. Note that some bungee cords are adjustable to varying lengths.
- A lightweight (500lbs+ capacity) pulley may come in handy to raise heavy objects up a cliff face, to erect a deadfall, to establish a tree cache, etc. Consider such an item for a cache/MSS or for mounted operations.
- Very high strength rope, manufactured in a tubular webbing configuration, is commercially available, and is lighter and more compact than conventional climbing rope. Whether you carry a rope coil or its tubular equivalent, ensure it is carried in a rapidly deployable configuration that will not result in knots or kinks. The Author can recommend three alternatives for rapid, dependable deployment:

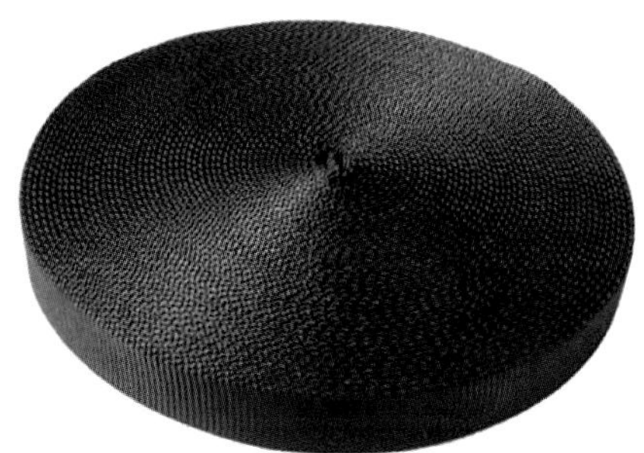

Figure 74. A 150' roll of Tubular Nylon Webbing. Tensile strength of 4,000=lbs.

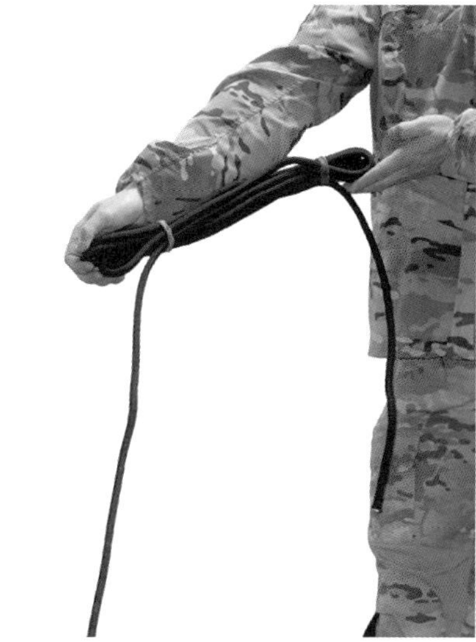

Figure 75. Example of a "Rigger Roll" on a length of rope. A sequence of these would be used along the remaining rope. These rolls can be folded into a canvas cover/deployment bag.

Figure 76. A "Rigger Rolled" STABO Extraction Line Contained in a Deployment Bag. (*Hardy*)

- An elongated 'rigger roll' canvas cover weighted with a small bag of sand. The weight will allow the roll to be thrown down a cliff face or dropped from a helicopter; and the cover can either be secured to the end of the rope/webbing or it can be allowed to automatically detach once the rope is fully deployed. The rope/webbing is looped back and forth, within the canvas wrap/cover, through heavy duty rubber bands that are secured to two series of loops sewn into the interior side of the wrap (referred to as a 'rigger roll'). Once this is accomplished, the roll is folded over/rolled up and secured in a closed position. The detached configuration will allow the Team Member to 'unthread' the rope from a hot karabiner after rappelling, without having to manipulate the karabiner gate. The dimensions of the bag, for dismounted use, should be such that it can be conveniently carried on the rucksack. The roll can be fabricated locally or by a supporting Rigger Detachment. The rubber bands will weaken/rot over time; so they must be replaced periodically. Longer rope coils (100ft+) can be rigger rolled, for helicopter deployment, using a modified duffel bag.
  - Preferred Alternative: An elongated canvas/nylon 'arborist rope bag' weighted at the bottom with a 'packet' of sand (preferred) and having a drawstring closure at the anchor end (if deployed from a helicopter). The anchor end should have a robust loop-type knot, with a heavy-duty karabiner attached. The bottom of the bag can have a sewn-in loop through which the deploying rope/webbing end may be threaded or the rope/webbing end can be attached to the loop using a heavy duty rubber band so that the bag can automatically detach, as above. The rope/webbing is then coiled into the bag. Again, the dimensions of the bag, for dismounted use, should be such that it can be conveniently carried on the rucksack. The bag can be commercially purchased or fabricated locally.
  - Tubular webbing or rope can be carried in a 'rigger roll' or rope bag configuration or it can be carried in a coil. If the webbing is in a coil configuration, it will be time-consuming to reconstitute the coil after the webbing is deployed.
- A throw weight with a lightweight line (e.g. 550-cord) can assist in deploying rope laterally or above. The standing end of the line would be attached to the deploying end of the rope.
- Tree climbing gear may be useful in heavily forested areas. Consider that caches may be secured/hidden among the branches of large trees, rather than buried. This may be an option if the soil is rocky or if the Team is operating in swamp or areas subject to monsoon flooding (e.g. the Amazon basin). If this type of cache is to be considered, some or all Team Members should be taught tree climbing techniques. Also, if the enemy is using river crossings/ infiltration routes or supplying their forces by water, a well positioned treetop position can provide good surveillance vistas.
- Carry <u>at least</u> two entrenching tools per Team. Entrenching tools are needed to bury trash and body waste, to bury mines, to create cache sites or to dig fighting positions. The US issue entrenching tool is compact in its folded configuration, but

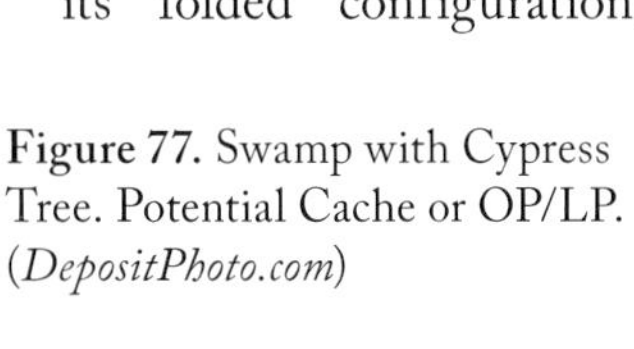

Figure 77. Swamp with Cypress Tree. Potential Cache or OP/LP. (*DepositPhoto.com*)

the folded sections make noise when they contact one another; this item must be muffled. The entrenching tool should be carried for easy access. The carrier for the US issue entrenching tool is designed for carry externally on a rucksack, but the carrier will undoubtedly hook on vegetation in this configuration; a hard plastic carrier also makes noise when it comes in contact with trees, brush, etc., so consider fabricating a soft carrier instead of using a hard plastic. Author's recommendation: a single-section, strongly-built entrenching tool, such as the Russian design 'saperka' with sharpened and serrated edges – therefore, suitable for digging, cutting and close quarters combat.

Figure 78. A Typical Stand of Bamboo. (*Public Domain*)

- The Team should carry at least two lightweight <u>body-bags or lightweight hammocks</u>. The Lightweight Body Bag, aside from its obvious intended purpose, can serve as a ground cloth, a bivy, or a litter (with reinforcing straps). It may be wise to use spray water-proofing, especially at the seams, as some versions of the bag are permeable. Waterproofing the bag will enhance its use as a ground cloth/bivy; it should also mitigate a blood trail.
- It may be necessary for the Team to cut bamboo or small-diameter trees to build a litter, sleeping platforms (e.g. in swamp), field fortification overhead cover or to clear an LZ.
  - Green bamboo or small diameter tree sections can be cut for litters prior to moving into an RP/ambush location to transport Team WIA or captured enemy items. Remember to cut lashings in advance as well. Practice this in training.
  - Cutting of vegetation to be used for a litter, or for other purposes, should be done away from the Team NDP/defensive position, back-trail, kill zone, etc. Woodchips, sawdust, cuts, stumps and disturbed ground are red flags to a tracker team or enemy patrol. Exception: when the telltales are deliberate for deception purposes.

Author's Solution:

All Team Members should carry a cable saw/wire-saw/folding saw; a relatively silent and rapid way to cut bamboo and small trees. Note that a 'wire' saw can easily break if not used carefully; specifically, keep hands a foot or more apart while sawing and do not place excessive pressure on the wire during the cut. A cable (chain) saw is more robust and is preferable; further, a Team Member can use long lines at either handle (one weighted) so that the saw can be used on overhead limbs. A rust resistant folding saw is effective, compact and lightweight, though it may be noisier in use and may not cut as quickly. If a folding saw is to be used, note that fine teeth should be used to cut bamboo or hardwood; medium teeth are best for soft wood and large teeth are best for green wood.

- Each Team should carry a small tube of luminescent paint or a small roll of luminescent tape/string to mark equipment/personnel for night operations and to mark firing sector pegs.
- Carry one extra pair of socks, plus fungicidal foot powder, especially during the rainy season. The foot powder is also a good remedy for jock itch.
- Each Team should consider carrying 1–2 thermite grenades for destruction of friendly or enemy equipment. If the Team is mounted, each vehicle should also carry a thermite grenade.
- A Claymore bag/haversack, attached to the rucksack, is extremely useful to carry night-vision optics, extra handsets, mines, booby-traps, or other special items that must be immediately

at hand for rapid deployment. This gives easy access to such items while on patrol or if the rucksack must be ditched or cached.

- All rucksacks and LBE <u>must</u> have quick releases, to allow Team Members to rapidly shuck the pack, if necessary, to facilitate rapid movement, to avoid drowning or in other emergency situations.
- The NVG and its harness should be stowed during the day. It will be worn when occupying LPs/surveillance positions, in night ambush positions and securing an NDP.
- Use waterproof bags within your rucksack to protect items while on patrol.
  - Remember that a false step during a stream crossing can send a Team Member into deep and/or swiftly flowing water. Also remember that indigenous Team Members may not be swimmers.
  - Waterproof/plastic bags can improve the flotation characteristics of the rucksack. If the rucksack does not have adequate flotation characteristics, it will fill with water during a deep stream or river crossing, and become incredibly heavy when the Team Member attempts to climb out of the water.
  - If crossing rapidly flowing or deep water, even with the assistance of a rope line, a non-buoyant rucksack will act as an anchor, quickly submerging the Team Member. The Team Member will then have to rapidly shed the rucksack (quick releases are absolutely essential) or he will likely drown.
  - If the rucksack is buoyant, take care that the Team Member does not invert – with the rucksack on top and the Team Member underwater. Check rucksack buoyancy before using it for flotation. The Author recommends that the rucksack be made buoyant and attached to a rope line to be floated across the stream separate to the Team Member, or that the Team Member use it as a flotation device. Stream crossing must be practiced.

Mission Support Site (MSS)/Cache TTPs:

- The T/L, and AOB/FOB operations section should consider establishing MSSs and/or cache sites for long duration/deep penetration missions and for possible use in evasion or ground exfiltration.
  - A cache is 'a source of subsistence and supplies, typically containing items such as food, water, medical items, and/or communications equipment, packaged to prevent damage from exposure and hidden in isolated locations by such methods as burial, concealment, and/or submersion, to support evaders or isolated personnel in current or future operations.' [JP 3-50 Personnel Recovery and the DoD Dictionary of Military and Associated Terms].
  - A MSS is used to support a specific mission or sequence of missions (to include deep penetration/long duration operations) in a target area; the MSS may be used as an assembly area and will contain a cache.
  - A patrol base bears some similarity to the MSS, but it is typically occupied for only 24 hours and may contain a minimum of supplies necessary to the patrol mission. A patrol base is used to launch reconnaissance and/or combat patrols within the mission duration. A patrol base does not typically contain a cache.
- Establishment of MSSs/caches, in anticipation or in advance of future operations (to include pre-hostility periods), will limit the exposure of friendly air assets to enemy counter-air measures; will enhance OPSEC; reduces repeated use of/reliance on limited numbers of suitable LZs/DZs; facilitates deep penetration/long endurance operations and cross-country mobility; facilitates SERE operations and may enable operations requiring the commitment of more powerful weapons and munitions or specialized equipment.
- Ideally, the MSS will be in the vicinity of a secure water source. See the Chapter 2 paragraph on Terrain Analysis to identify secluded water sources.

- Upon approach to a prospective MSS, search the area for signs of human and animal activity. An area with much animal activity and no/very limited human activity is probably a good location for an MSS.
- A wood-burning camp stove that uses wood-gas and/or 'volcano' operation burns small twigs, leaves or other combustible cellulose scraps in a very efficient manner producing a very hot, smokeless gas that will boil water in 3–5 minutes (depending on fuel and altitude) with similar results for cooking. Functionally like the Dakota Fire Pit (see description elsewhere in this book), but more efficient.
- Use airborne (UAV mounted) thermal sensors during the planning and preparation phase to assist in the selection and placement of a MSS/cache(s). Thermal imagery can screen for humans and cattle in certain environmental conditions. During/after MSS emplacement, the Team should use a thermal sight to check the MSS for inadvertent heat signature (e.g. while occupied).
- MSSs/caches may be established independently by higher headquarters at various locations throughout the AO. The best time to establish these locations is well prior to the start of hostilities, while travel is less restricted. After hostilities have commenced, higher headquarters may dispatch a SF AOB, separately or in concert with SR Teams, which would be instrumental in assisting the SR Team in establishing MSSs/caches.
- Considerations in the establishment of an MSS/cache:
  - The distance from the FOB, AOB or the launch site
  - The terrain of the target area/area of operations. This includes soil type and ease of excavation, drainage and accessibility
  - Areas of enemy or civilian activity
  - Local resources/sources
  - Type of operation
  - Area to be covered/supported
  - Access to water
  - Seasonal weather conditions
- If a Team has mobility equipment, it can use trailers as mobile MSSs/caches. Once the team identifies a suitable site, it can park and thoroughly camouflage trailers. If the mobile MSS/cache is going to be established for a long period of time (e.g. months), the trailers should be placed up on blocks. As the Team continues its operations, it can return to the site, retrieve the trailer(s), and move it/them to another location.
- MSSs can be operated in the following manner:
  - The MSS can be occupied by the SR Team, especially when Team Members require medical care; or when the Team must enter into survival mode during periods of extreme weather.
  - The MSS can be temporarily operated by AOB personnel, on a safari basis, while the AOB is being established.
  - The MSS can be briefly accessed for replenishment so that the Team can quickly return to its mission activities.
- Once the MSS or cache has been established, a cache report must be prepared containing the following data:
  - Type of cache
  - Method of caching
  - Contents
  - Containers used
  - General area description
  - Immediate area description, to include clear instructions on intermediate and final RPs

- ○ Cache location. Precise and detailed
- ○ Emplacement details
- ○ Operational data/remarks, to include access routes. Approach routes should offer concealed approaches in all seasons
- ○ Date of emplacement and estimated cache duration
- ○ Sketches, diagrams, photos
- ○ Other
- If water is abundant near the MSS, consider including a sprayer in the MSS. Not only can this item be used for NBC decontamination, but it can be used for personnel showers during long duration missions.
- Submerged Caches:
  - ○ Water condition should be cloudy year-round to avoid detection of the container.
  - ○ Cache containers can be placed in saltwater on a temporary basis pending near-term collection.
  - ○ Cache containers should be recoverable regardless of season. Unless the cache is established on a short-term, temporary basis during moderate weather, submerged caches are not recommended in cold regions where thick ice might deny access.
  - ○ All items being placed in a cache container must be thoroughly dried or mold/mildew will erupt. Items contained in cache containers must be protected with multiple water-resistant/waterproof wrappings.
  - ○ Communication gear/materials should be divided into separate cache containers.
  - ○ Containers must be sufficiently weighted and anchored to prevent the item from surfacing, drifting with current or displacing with flooding. Weights/anchors can be attached directly to the containers or to a net type mooring (used to secure multiple containers). Weight sufficiency for the containers should be tested in advance. Mooring cables/lines should be robust. A block and tackle may be necessary to retrieve cache containers.
- Items of different metals should not be placed in close contact to avoid corrosion.
- Consider caching a hunting blind and/or deer stand (with climbing stick). These items may be useful in the AO and at a hide location, depending on seasonal vegetation, target area elevation and observation restrictions. Recommend that before acquisition of a blind, that the item be tested for camouflage effectiveness in both daylight and night-time conditions and that modification of commercial products to enhance IR/thermal suppression be undertaken (as is recommended elsewhere in this book), if necessary.
- Locating a MSS or cache in jungle/rainforest is more difficult than in other environments, as line-of-sight to terrain features or infrastructure is often rare; the terrain and vegetation all looks the same and GPS will often not work effectively; the most likely place to site a MSS/cache is in close proximity to a hilltop, where the vicinity shows no evidence of enemy presence.
- A MSS/cache might be sited at a specific distance and direction from a trail/road junction, if the junction can be clearly differentiated from other junctions in the area, and if the vicinity of the junction is not invested with enemy combatants or local civilians.
- The parent unit should include covert MSS/cache markings in its SOP.
- Possible MSS Items:
  - ○ Weapons, Ammunition, Explosives and Ammunition Magazine Speed Loaders
  - ○ Heavy Duty (e.g. construction) Plastic Trash Bags, subdued in color, various sizes
  - ○ Shelter Items
  - ○ Personal Hygiene Items
  - ○ Rope, String, Various Fasteners, Lightweight Pulley(s)
  - ○ Fire Starter Items

- ○ Stove/Heater (Cold/Wet Climates)
- ○ Water, Water Treatment and Water Containers
- ○ Food and Cooking Items
- ○ Snares/Traps
- ○ Tools: Entrenching Tools, Saws, Machetes, Knives, Chain Saw (with fuel and oil)
- ○ Vehicular Items and Tools (if appropriate): Fuel, Oil, Fuel Stabilizer/Treatment, Transmission Fluid, Tires and Tire Repair Kit, Filters, PLL/Common Repair Parts (e.g. Spark Plugs); Appropriate Repair/Maintenance Tools; Towing Straps
- ○ Assorted Batteries
- ○ Flashlights
- ○ Signal Items: Signal Panels, Mirrors, Strobe Lights, Pen Flares, Smoke Grenades
- ○ Communication items: A Primary and an Emergency Radio
- ○ Maps, GPS, Compasses
- ○ Rucksacks
- ○ Camouflage Nets
- ○ Other Survival Items
- ○ Special Items:
    - ▪ Heavy Weapons; Specialized and Sabotage Munitions and Explosive Devices
    - ▪ Mission Peculiar Items
    - ▪ Clothing Items: Cold Weather or Change of Season and/or Replacement Clothing for Long Duration Missions
    - ▪ Mountaineering equipment
    - ▪ If there is the potential for CBRN operations in the AO/target area, Personal Protection Equipment (PPE) items and decontaminant should be stored in the MSS/cache(s)
- • MSSs/caches must be located where enemy forces or local civilians will not discover them, and where it is generally sheltered from the elements – and yet it must be situated such that the location is identifiable and accessible to the Team. Consider:
- ○ Rough/sloping, and/or heavily vegetated terrain, located away from trails, and adjacent to a singular terrain feature (rock outcropping) to help locate the cache.
- ○ MSSs/caches established in swamp or in areas with permafrost will often be above surface.
    - ▪ In swamps, construct platforms in areas with plenty of canopy and heavy brush entanglements – supplemented with other camouflage.
    - ▪ If the area is infested with crocodiles, hippopotami, etc. the MSS/cache may have to be aloft in trees limbs.
    - ▪ If primates are prevalent in the area, the MSS/cache containers must be made 'monkey-proof'.
    - ▪ Remember that swamps/bogs are present in temperate, tropical/sub-tropical, littoral and cold regions, and that the availability of leaf cover may be limited during certain seasons.
    - ▪ In permafrost areas, cache excavation will be difficult, but the difficulty could be mitigated somewhat if the cache is prepared in advance (e.g. pre-hostility periods, during warmer weather) using appropriate tools.
- ○ MSSs/caches can be buried in a steppe (grassland), but marking and locating the site in an area lacking in notable terrain features becomes a real problem. GPS, optics, range-finders, metal detectors and/or Radio-Frequency Identification (RFID) tags and interrogator instruments may be necessary.
- ○ Mosquito/pest infested areas will often discourage civilian transit and enemy patrolling.
- ○ Lava Beds/Fields:

- Pumice and silica found in lava fields can have an extremely sharp texture that will quickly shred vehicle tires, track pads and footwear. Enemy motorized/mechanized units will not likely cross lava fields. Tracker dogs will not be used in lava fields as their paws would be shredded. Additionally, various wild animals will not enter a lava field; shepherds and their animals will not be found in this environment.
- Team Members can mitigate damage to footwear by wrapping boots in special cladding material, prior to transiting a lava field. Larger sheets can be used as ground cloths for Team personnel.
- Lava beds may form lava tubes, rubble fields and lava domes that are dangerous to cross-country travel (both vehicular and by foot). Some types of lava fields are superb radar reflectors; others are poor radar reflectors – both types could benefit a Team. Ascertain lava field characteristics during mission planning.
- Beware of compass navigation/ location errors in lava beds. Use a GPS to locate a MSS in lava fields (see entry under navigation).

○ Enemy troops and civilian personnel will stay well clear of minefields and other hazardous areas. Consider using false warning signs, using enemy protocols, to deter enemy or local citizen incursions. Ensure these false locations are recorded and subsequently reported to higher headquarters.

○ Shepherds/Farmers (hostile):
- Use areas not suitable for livestock grazing (e.g. near minefields) to mitigate the possibility of accidental discovery by local shepherds or farmers.
- Observe routes used by shepherds/ farmers, then plant booby-traps/ mines (preferably of enemy

Figure 79. Tulelake Lava Field. (*Public Domain*)

Figure 80. A prospective cache site in a Lava Field. (*Public Domain*)

Figure 81. A moderate density thicket of Mountain Laurel will impede enemy movement and maneuver. (*Public Domain*)

manufacture) to kill or maim livestock or personnel. This will cause locals to shift grazing to other areas, and away from the Team target area.

○ Laurel thickets and especially briar patch infested areas will be avoided by enemy troops and civilians.

   ▪ Laurel is a tough shrub with crooked interwoven branches that impede cross-country travel. Laurel is poisonous to animals, so shepherds and farmers will avoid these plants/thickets.

   ▪ Team Members can penetrate briar thickets by using sheets of barrier material.

Figure 82. A moderate density briar patch. Enemy troops will avoid briars. Good for a hide, MSS/cache or Team location. (*Public Domain*)

○ A cache may be buried a specified distance from a permanent/semi-permanent feature or marker such as an abandoned building foundation or grave stone. Use a measuring device (e.g. pre-marked 550-cord) and a compass to measure from the marker to the cache.

○ Successive trips by a Team may be necessary to establish the cache and its contents, or it may require air drop of supplies in the vicinity of the cache location. The Guided Parafoil Air Delivery System-Light or Extra Light (GPADS-L/XL), Container Delivery System (CDS)/ Covert Resupply Dispenser system or similar delivery platforms may be suitable.

Figure 83. An old foundation on abandoned rural property. A good marker for a MSS/cache. (*Public Domain*)

○ If possible, Team Members or higher headquarters should establish caches during pre-hostility periods, perhaps using troops/agents, etc. in the guise of tourists, hunters or under some other cover. Assistance of an intelligence agency may be necessary to this end.

○ In an UW environment, a UW team that is deployed with a Guerilla force, may receive and stockpile SR Team materials in a transient cache, set aside until the SR Team arrives; the Guerilla force might assist the SR Team in the subsequent relocation of the cache, or some of its contents to the vicinity of a planned or actual MSS. It would be unwise to have guerillas know the exact location of the MSS due to OPSEC and/or the threat of theft.

○ Caches should have anti-disturbance features (by SOP) providing a red flag that is detectable from a distance.

○ MSSs/caches should never be located in arroyos/wadis, which are subject to flash floods. At a minimum, desert supplies should include water, rations, fuel, batteries and any other items that would support mission activity tailored to the environment.

- ○ The FOB should possess materials and equipment to prepare water/weather-proof containers to be used at cache sites. 155mm propellant containers, metal ammunition containers, metal drums of varying dimensions, and even wood containers, etc. are suitable for burial of cached items, if they are properly prepared.
- ○ In operations/environments where MSSs/caches are to be employed, the higher-HQ/ FOB might establish a wax dipping tank and plastic bag vacuum heat-sealing equipment. Additionally, the FOB should stock expendables such as commercially-available clothing storage (vacuum) plastic bags, wax, plastic weapons bags, gum/wax-impregnated bags and Cosmoline. Preparation should include placing a thick coat of Cosmoline or equivalent preservative on all external container surfaces for waterproofing and long term storage.
- ○ For longer term storage, some of the container contents (e.g. weapons) might also be coated with Cosmoline. This is particularly relevant to underwater storage or in high humidity/wet environments or in damp soil conditions. Plug the weapon muzzle before immersing in Cosmoline.
- ○ Removal of Cosmoline from equipment requires the application of a solvent (e.g. mineral spirits, kerosene, degreaser), heat (e.g. hot water; steam; hot sun) or a combination of heat and solvent. Solvents are not recommended in field environments as they are a logistics burden. Wrapping the object in a plastic bag (black preferred), using heat from sun exposure, will speed preservative removal in warmer environments.
- An MSS/cache may contain large munition items or items typically used in DA missions. As previously stated, Teams should be multi-functional and capable of executing combat/ DA missions when opportunities present. Where such large items are cached, movement of these items to the vicinity of a target then becomes a concern; the Team may need to deploy with light utility vehicle(s), ATVs/UTVs, pack animals, etc. or may require the assistance of a guerilla force to move the items to staging and firing positions. Examples of operations using large munition items might include:
- ○ Use of MANPADs)/shoulder-launched Surface-to-Air Missiles (SAMs) to interdict enemy air assets operating from enemy base areas, airfields, flying convoy escort, etc.
- ○ Use of aerial/ground rockets to be fired singly or in barrage on enemy base areas/facilities using improvised launch tubes (e.g. PVC pipes) and initiated with electrical delay timers or remote firing devices.
  - ▪ Use in such operations requires that the Team conduct testing and experimentation of various warhead types and fusing options (e.g. Fleshette, High Explosive, WP, High Explosive Anti-Tank (HEAT)) to create firing/range tables and firing direction procedures.
  - ▪ Use to support raids, attack by fire, harassment and interdiction operations.

Weapons TTPs:

- See Primary and Secondary Individual Weapons Maintenance TTPs in this book.
- Team weapons should be selected or tailored to the operation in accordance with a METT-TC analysis and the T/L's CONOPS. As SR Teams are often outgunned and numerically overmatched by enemy forces, it makes sense for the Team to be as heavily armed as possible, while balancing Team Member burden against the need for stealth. Some of the burden can be alleviated by using MSSs/caches, mobility equipment or pack animals.
- In long-duration, deep penetration operations, it may be wise to equip Team Members with enemy weapons. The Team may then re-arm with captured enemy ammunition, which may minimize some resupply risks.
- Silence all sling swivels and accessories.

- Partially tape or cap the weapon muzzle to keep water and dirt from fouling the bore; ensure the tape/cap has air gaps to provide ventilation/drainage. Without ventilation/drainage, condensation can collect in the weapon bore over the duration of a mission, leading to a malfunction.

Figure 84. M26 MASS (stand-alone) with detachable box magazine. (*Public Domain*)

- After a water crossing, and at least daily during a monsoon, point the muzzle downward and slightly ease open the bolt, just enough to break the seal at the chamber, to allow water to drain from the bore.
- Friendly and enemy weapons fires have their own distinctive sounds on the battlefield. Train with these weapons on the firing range and fire them past Team Members so that they are able to distinguish specific sounds. Alternatively, make a recording of various weapons firing and of ordnance exploding. This should include sounds of individual and crew-served weapons, fired from various distances and the impacts of explosive projectiles.
- All Team Members should carry a second or even a third firearm.

Figure 85. A Russian Saiga 12GA Shotgun with a 15-Round Drum Magazine. 30-Round Drums are also available. (*Photo: Pinterest*)

  - Team Members equipped with an under-slung grenade launcher (e.g. M-203) or with an under-slung shotgun (e.g. M26 Modular Accessory Shotgun System) already have a second firearm mounted on their primary weapon. These Team Members should also consider carrying a dependable double-action, semi-automatic pistol.
  - Team Members may have a mission requirement to carry a silenced pistol or non-lethal weapon in addition to their primary weapon.
  - Team Members not equipped with an under-slung grenade launcher/shotgun, might carry either a dependable semi-automatic pistol or a shortened shotgun (such as a short-barreled pump, or the M26 MASS in the standalone configuration).
  - If a shotgun is carried as the primary weapon (e.g. by the point man), the barrel length should perhaps be 20in (standard combat shotgun) with appropriate choke; the weapon should be complemented with dependable box or drum magazines.
  - The recess/hollow under the front and rear pistol grips on some assault weapons can accommodate small items such as cleaning items, survival items, morphine syrettes, etc. If used for such purposes, ensure items are packed so that there is no rattle.
  - In circumstances where a stoppage/malfunction occurs in the Team Member's primary weapon, a pistol or shotgun might be drawn faster than clearing the stoppage/malfunction and reloading the primary weapon in a firefight. Practice this!

- ◦ Remember that in the heat of battle, adrenaline will reduce Team Member motor skills; so rapidly putting a secondary weapon into action must become reflexive. This can only be achieved through practice.
- ◦ Seconds do count; the '21-foot rule' suggests that an enemy could close that distance on a Team Member in 6–7 running paces/slightly over a second. The secondary weapon, if not an under-slung firearm, should be carried where it can be reached and drawn rapidly.
- ◦ Holsters that strap to the upper thigh will leave space on the LBE/vest to carry other pouches/items and offers a rapid draw, but will limit the use of the trousers' bellows pocket. A horizontal-draw holster (mounted on the LBE/vest) is at least as easy to get to (all positions) with a faster draw.
- ◦ A secondary retention/safety strap, or even taping the weapon inside the holster is recommended for rappelling, parachute insertion, or similar activities where friction or sudden shock might inadvertently separate the weapon from the user.
- ◦ Avoid using a plastic clam-shell type holster as it will make a loud noise if it comes in contact with a tree, rock, bamboo, etc. Use a softer holster.

A Joint Service Combat Shotgun Program report on the lethality of shotguns in war states, 'British examination of its Malaya experience determined that, to a range of thirty yards (27.4 meters), the probability of hitting a man-sized target with a shotgun was superior to that of all other weapons.'

The delivery of the large number of projectiles simultaneously may make the shotgun the most effective short range weapon commonly used, with a hit probability 45 per cent greater than a sub-machine gun (5-round burst), and some assault rifles (3-round burst). Multiple shots from a semi-automatic shotguns or bursts from a fully-automatic shotgun will presumably have greater results.

Buckshot, with each pellet notionally as effective as a small caliber handgun round, can be effective at ranges as far as 70meters depending on the barrel choke. [This may be equal to or better than effective handgun range.][2]

Primary Weapon TTPs:

- Check all magazines before going on an operation, to ensure they are clean, properly loaded and functioning. And clean out any debris from the interior of magazine pouches.
- High-capacity drum magazines for various assault rifles/carbines weapon are available. The Team Member may opt to carry only one drum (on the weapon) or drums in lieu of the standard 30-round magazines (the latter requiring special LBE pouches). Only the most reliable of drum magazines should be considered for use by SR personnel, and ammunition within the magazine should not rattle. The advantages include: more ammunition per Team Member; delivery of sustained fire to attain initial fire superiority. Disadvantages: more weight; unwieldy weapon balance. If Team Members opt to each carry a single drum plus standard magazines, the Team should carry a speed loader to later transfer rounds to the drum.
- Mark magazines containing special ammunition (e.g. Subsonic, etc.), with luminescent paint and a tactile mark.
- In snowbound or mountain navigation situations, both hands are occupied with ski poles, rope management and other vital tasks. Therefore, weapon carry and transition to engagement techniques must be modified to accommodate these circumstances.

---

2. Derivative/extract from W. Hays Parks (October 1997). 'Joint Service Combat Shotgun Program'. The Army Lawyer.

- Magazines are best carried in pouches with the butt plate upwards.
- Use 'ranger plates' or field expedient pull tabs of 'Hundred-Mile-An-Hour' tape, or tape that can endure high heat and humidity (or frigid temperatures in winter operation), to facilitate extraction of magazines from LBE magazine pouches. Cover the mouths of spare magazines placed in a haversack/backpack to exclude debris.
- Team Members should frequently practice changing magazines in varying scenarios and conditions (night vs day, static vs moving, defensive vs offensive, Battle Drills, etc.). The ability to rapidly replace magazines supports the necessity to maintain rates of fire/fire superiority, especially when engaged with a numerically superior foe.
- Assault rifles (or submachine guns) that possess a magazine-well generally facilitate rapid swapping of magazines as opposed to the AK series of assault rifles with 'rock and lock' magazine systems.
- The 20-round magazine for the M-16/M-4 series of assault weapons are easier to manipulate than the now standard 30-round magazine, especially for indigenous troops with smaller hand sizes.
- The assault rifle/carbine will typically be carried with the stock retracted/folded, if that feature is provided. If time and situation permits, the Team Member can quietly move the stock to the extended position for aimed or precision shooting; otherwise the Team Member will be firing his weapon with the stock folded/retracted. Practice this.

Shooting Techniques TTPs:

- Aimed shots are preferable when precision is necessary:
  - To avoid collateral casualties
  - To disable key equipment (e.g. radios) or kill key enemy personnel
  - Where vulnerable points on a target must be hit to disable
- SpecOps personnel must develop instinctive firing techniques by developing hand-eye coordination and muscle memory. This requires frequent and realistic weapons training. The goals are to increase <u>speed</u>, <u>combat accuracy</u> (rather than precision) and <u>efficiency</u>.
  - Increasing <u>speed</u> during combat engagements includes:
    - Speed in acquiring the target(s). Includes spotting the enemy/identifying targets and aiming the weapon
    - Speed in getting off the first shots at an enemy
    - Speed in movement
    - Speed in reloading
    - Speed in correcting malfunctions
  - The purpose of <u>combat accuracy</u> is to attain hits on an enemy that will result in an outright kill or make the target unable to pose a theat.
    - Usually, this means that the target box for an enemy combatant is the upper torso, especially during a chance encounter/meeting engagement. If the enemy wears body armor, the target box might be re-designated.
    - Precision aimed shots seldom apply in close quarters combat situations, which are routine in heavily vegetated and rough terrain environments. The collective acts of extending the stock, bringing the weapon to the shoulder, acquiring a stock/cheek weld, bringing the weapon to bear on the target, acquiring a sight picture, etc. that are typical for precision firing, are appropriate when the Team Member has the luxury of time and is at least temporarily stationary or has a stable firing position as would be common when the Team Member is in a defensive position, a deliberate ambush position, when the enemy offers a clear shot (e.g. when the Team Member has the drop on the enemy combatant) or when engaging the enemy at extended ranges. However,

during chance/meeting engagements when the Team Member is not stationary or lacks a stable firing position, when the enemy has the drop on the Team/Team Member, when both the Team Member and the enemy are moving, when the Team Member is taking fire from the enemy at close range and/or from various directions, the shooting techniques associated with precision or marksmanship shooting go out the window.

○ <u>Efficiency</u> in shooting is being able to <u>rapidly</u> attain <u>combat accuracy</u> in a fire fight under the realistic conditions and pressures of combat. The conditions include:

■ Automatically reacting to surprise while enduring the noise and confusion of close combat.

■ Engaging targets at various angles at an instant from the weapon carry position, while the Team Member is moving. During movement, the Team Member should become accustomed to moving his weapon muzzle to track, as much as possible, with his gaze – minimizing time taken to acquire the target and shoot.

■ Rapidly engaging moving and fleeting targets that appear and disappear from view in an instant; and engaging these targets in poor light and/or weather conditions.

■ Maintaining a high rate of accurate fire while moving and promptly bringing the weapon back into action after reloading or malfunction.

■ Acquiring, and hitting targets while under pressure and while adrenaline is coursing through the body. The effects of adrenaline include tunnel vision and reduced motor skills. To correct for tunnel vision (diminution of peripheral vision), the Team Member must 'keep his head on a swivel', rotating his head to scan the area around him. The only remedies for reduced motor skills are repetitive training and to train in situations that induce stress. In the Author's view, after a Team Member experiences several close combat engagements, the effects of adrenaline are lessened; the veteran Team Member can eventually reach a calm and composed state during a firefight.

• Field Manuals dealing with weapon marksmanship may not be authoritative guidance on combat shooting. Although FMs represent prevailing training doctrine for US military personnel, they are periodically revised when other methodologies come into vogue; any FM should be used as a 'guide'. Some FMs largely pertain to marksmanship/precision shooting under controlled firing conditions, rather than in realistic combat simulations/ settings. Further, the infrequency of individual qualification and familiarization firing doesn't prepare the serviceman for close combat shooting.

• Ironically, some FMs devalue instinctive fire. To wit, FM 3-22.9, Rifle Marksmanship states:

'Instinctive Fire. This is the least accurate technique and should only be used in emergencies. It relies on <u>instinct, experience, and muscle memory</u>.'

• Here is the irony:

○ How does the term 'emergencies' apply? When caught in an ambush; when executing a breakout; when conducting an assault in the face of enemy fire; during a meeting engagement with an enemy force; when engaged with a superior enemy force; during an enemy assault on friendly troops; when an enemy is firing from concealment or when the enemy is engaging with fire superiority?

○ Training and drills are the very instruments whereby <u>instinct</u>, <u>experience</u> and <u>muscle memory</u> are to be taught and learned.

'Another of Applegate's training innovations was the use of particularly intense combat firing ranges, which he called the 'House of horrors'. A cross between an obstacle course, a haunted house, and a shooting range, it used a three dimensional layout with stairs and tunnels, pop-up targets, deliberately poor lighting, psychologically disturbing sounds, simulated cobwebs and bodies, and blank cartridges being fired towards the shooter. The range was designed to have the greatest possible psychological impact on the shooter, to simulate the stress of combat as much as possible, and no targets were presented at distances of greater than 10ft (3.0 m) from the shooter.

Applegate also used his house of horrors as a test of the point shooting training. Five hundred men were run through the house of horrors after standard target pistol training, and then again (with modifications in the layout) after training in point shooting. The average number of hits in the first group was four out of twelve targets hit (with two shots per target). After point shooting, the average jumped to ten out of twelve targets hit. Further shooters trained only in point shooting, including those who had never fired a handgun before receiving point shooting training, maintained the high average established by the first group. Similar methods were in use as early as the 1920s and continue to this day, for example the FBI facility called Hogan's Alley.'[3]

Colonel Charlie Beckwith 'procured .45 caliber M2 grease guns which he had the sights sawed off of. We were to learn to shoot instinctively…. Beckwith wanted us to shoot 3x5 cards….'[4]

- FM 3-22.9 further states: 'Soldiers must practice moving with their weapons up until they no longer look at the ground but concentrate on their sectors of responsibility. Soldiers must avoid stumbling over their own feet.' This is another unrealistic statement, with potentially mortal consequences.
  - The High-ready and Low-ready carry positions inherently impair soldiers vision, both laterally and downward. High-ready and Low-ready carry may be perfectly acceptable in the following circumstances:
    - The soldier is in a static position (e.g. ambush, defensive posture) or providing supporting fire during a raid.
    - Difficult terrain and heavy vegetation are not a factor in movement.
    - Stealth is not required.
    - Urban situations or long engagement ranges are present.
    - Whenever magnification/optics are needed to acquire and engage targets.
    - Enemy use of mines/booby-traps is not anticipated.
  - How is a soldier to move with stealth over treacherous ground unless he can determine foot placement?
  - How is a soldier to stealthily negotiate dense vegetation (while bending over or crawling) or difficult terrain (e.g. a steep slope, etc.) while moving with his weapon at High-ready or Low-ready carry positions – and yet avoid stumbling over roots, branches, uneven ground, etc.? Similarly, how might a soldier spot mines, booby-traps and other hazards?

  'Early in World War II, it was found that target shooting skill with the hand gun was not enough for the soldier in combat. It was proved that a man trained only in the target phase of the hand gun was proficient up to the point where he could kill an enemy only when he had time to aim and fire, and providing he could see the sights…. In reality, after the target, aimed-shot phase of training has been completed and the shooter becomes familiar with his weapon, he is only about 50

---

3. *'Point Shooting'*, *Wikipedia, the free encyclopedia*, https://en.wikipedia.org/wiki/Point_shooting#cite_note-33
4. Murphy, 'Blue Light', p.45.

per cent combat efficient, because the conditions under which most combat shooting occurs are entirely different from those presented in the bulls-eye type of training. In a gun battle, the utmost speed, confidence, and ability to use the hand gun from any position—usually without the aid of sights—are paramount. The man who can instinctively handle his weapon quickly and accurately, in varying degrees of light, under all terrain conditions and while under the physical and mental stress and strain of actual combat, stands a good chance of avoiding becoming an object of interest to the stretcher bearer.'[5]

'A 2008 RAND Corporation study that assessed the New York Police Department firearm effectiveness between 1998 and 2006, found the average hit rate during shootouts with suspects was only 18 per cent; when suspects were not shooting back, police officers hit rate rose to only 30 per cent.'[6]

'The actual combat life of the soldier … who may carry a shoulder weapon is often measured in seconds—split seconds. In close-quarter combat, or in-fighting, he must be able to use his weapon quickly, accurately and instinctively. Close-quarter firing, in the case of shoulder weapons, is presumed to be any combat situation where the enemy is not over 30 yards distant and the elements of time, surprise, poor light and individual nervous and physical tension are present…. The aimed shot always should be made when the time and light permit. However, in close-quarter fighting there is not always sufficient time to raise the weapon to the shoulder, line up the sights and squeeze off the shot. Consequently, training only in the aimed type of rifle fire does not completely equip the man who carries a shoulder weapon for all the exigencies of combat. As in combat shooting with the hand gun, he should be trained in a method in which he can use a shoulder weapon quickly and instinctively and without sights.'[7]

• Spotting and engaging a concealed enemy combatant is generally difficult, even in close combat situations, and this task is made even more difficult when both friendly and enemy combatants are maneuvering or moving during the engagement. Further, if the enemy is able to mass automatic weapons fire and achieve fire superiority before the Team can do so, then friendly casualties will quickly mount and a cascading effect of declining Team firepower will quickly occur, leading to further casualties, and so on. Hitting a concealed or fleeting target, in variegated terrain and vegetation, with aimed/precision single or burst fire will logically be less effective than fully automatic fire in close combat situations.

Instinctive Shooting TTPs:

• The shortcomings of instinctive fire noted above can be remedied using techniques applied in realistic, supplemental unit-level training. For a <u>partial</u> solution to instinctive fire marksmanship training, the Author recommends the Rifle Quick Kill technique:[8]
  ○ The technique has the weapon stock seated at the shoulder; both eyes observe over the barrel and are focused downrange – but the sights are not used.

5. Rex Applegate, Fleet Marine Force Reference Publication FM 12-80: 'Kill or Get Killed'. Boulder, Colorado: Paladin Press. p. 98-99. 1976.
6. Bernard D. Rostker, Lawrence M. Hanser, William M. Hix, Carl Jensen, Andrew R. Morral, Greg Ridgeway, Terry L. Schell, 'Evaluation of the NYPD Firearm Training and Firearm-Discharge Review Process' www.nyc. gov/html/nypd/downloads/pdf/public_information/RAND_FirearmEvaluation.pdf
7. Applegate, (1976). 'Kill or Get Killed', p. 98-99.
8. *Principles of Quick Kill*, US Army Infantry School (1996). IT 23-71-1. Paladin Press. ISBN 9780873640657.

- ○ Use of BB guns to fire on metal disks/cans tossed in front of the trainee. This is a cheap and effective method of developing hand-eye coordination. As the BB strikes the metal disk/can, the trainee will be able to instantly (audibly) assess his accuracy performance.
- ○ Once aerial targets are mastered, the soldier can move on to an array of ground targets (disks/cans) using the BB gun and ultimately move on to training with the individual weapon.
- ○ This is fun training that does not require scarce/costly resources. Safety goggles, BB guns, BBs, tin cans, a partner, and an outdoor space are all that is necessary.
- ○ Note that weapon design and characteristics will vary and therefore affect how the Quick Kill technique is adapted to the weapon; some individual weapons will be less adaptable than others.
- Instinctive Pointing Technique.
  - ○ This technique establishes a relationship between the barrel and the eyes when the soldier is in his firing stance. This eyes-to-barrel relationship is maintained during movement and observation; the positioning of the weapon (e.g. stock and forearm) relative to the shooter is also maintained – so that 'the shooter will hit where he looks and where his body points'.[9] Lateral aiming and shooting is done by twisting the upper torso and/or by pivoting of the feet.
  - ○ These relationships are similar in function to the Apache AH-64 Attack Helicopter M230, 30mm Chain Gun when it is 'slaved' to the Integrated Helmet and Display Sighting System, such that the Chain Gun will track with the gunner (or pilot) head movements to point the M230 where he looks.

<u>True Account:</u> **A Master Sergeant assigned to MACV-SOG FOB-2 as the Recon Company 1st Sergeant had previously been assigned to a Provincial Reconnaissance Unit (PRU). It had been his task to train and employ his PRU against Viet Cong infrastructure targets in the I Corps AO. He and his Vietnamese PRU members were stationed on the coast near Da Nang. All PRU team members were equipped with 9mm silenced STEN submachine guns. The Master Sergeant bought several hundreds of 3in glass balls that were commonly used in Vietnam to float fishing nets. The balls were tossed out onto the coastal waters to serve as moving targets as they moved and bobbed with the current and waves. The PRU members swiftly became adept at hitting the balls with two-round bursts using instinctive/ quick/reflexive shooting techniques.**

- ○ Frequent use of Paint-Ball systems and ranges. Initially, fire on stationary targets from stationary positions on a groomed range, again to cultivate hand-eye coordination. Graduate to firing at stationary targets while moving in tactical formation toward the target or laterally. Culminate with force-on-force firing while moving in tactical formation toward or laterally against moving adversaries (e.g. meeting engagement scenarios).
- ○ Initially, live fire, with ball ammunition, on stationary targets from stationary positions on a groomed range; graduating to firing while moving in tactical formation toward or laterally to stationary targets.
- For Battle Drill, the Author recommends:
  - ○ A Crawl, Walk, Run phased approach to range firing, and Battle Drills.
  - ○ Each Phase should first be executed on a groomed range beginning with dry-fire and concluding with live fire. Then, each phase should be executed in terrain and vegetation representative of the AO/Target Area; again beginning with dry-fire and concluding with live fire.

---

9. Applegate, 'Kill or Get Killed', p. 181

- ○ The closer the enemy/target, the less reaction time available. During each phase, engagement ranges should begin with more distant targets, progressing to successively closer targets. This will train Team Members to react more slowly to distant targets and progressively faster to near targets.
- ○ Range firing, especially in the Run Phase, should culminate with live-fire maneuver and movement over terrain, obstacles and vegetation that are representative of the AO.

Because Blue Light used a 'range that ran into an impact area, they could get away with things that simply were not done at Army ranges when Blue Light was activated in November 1977 such as mixing mortars and small arms fire, or fragmentation grenades and smoke grenades.' The range 'was also unique in that they could conduct 180 degree live-fire exercises, like the … technique used when breaking contact with the enemy.'[10]

See the Range Diagram Appendix B.

Close Combat Movement and Engagement TTPs

- The Team will typically move through dissected, heavily vegetated terrain in most operational settings. Team Members should carry their primary individual weapons suspended above the waist from a sling, magazine oriented downward. When moving, especially through vegetation, Team Members will generally be moving at a crouch.
- Individual Weapon Sling: The Author recommends a flexible sling that is convertible from/ to 1-point to 2-point attachment to the weapon, to meet different tactical requirements (e.g. mounted vs dismounted). If the weapon does not have proper attachment points, the Team Member should request weapon sling-point modification. The sling attachment at the weapon stock must not impair weapon handling, aiming or functioning.
- The compact assault rifle, would normally be carried with the telescoping (or folding) stock in the collapsed configuration, especially when negotiating heavily vegetated areas or tight quarters. The stock would typically be extended only for aimed/precision shooting. Instinctive firing techniques must necessarily be used in most engagements where the weapon is in collapsed configuration.
- The Team Member will typically have his dominant/firing hand on the pistol-grip nearly continuously with fingers/thumbs always poised for immediate action, with the thumb or index finger (depending on the weapon) resting on/near the weapon safety. For the M-4, in right hand carry, the thumb will rest on the safety switch and the right index finger will rest alongside the trigger guard. In a close combat meeting/chance engagement, a fraction of a second may decide who survives.
- The non-firing hand will grip the weapon fore-stock as often as possible; however, this hand will also be used to navigate vegetation or obstacles, maintain balance

Figure 86. Kurdish Fighters. Right index finger poised on the receiver, above the trigger guard. Photo suggests training and weapon discipline. (*Public Domain*)

---

10. Murphy, 'Blue Light' p. 42

in difficult terrain and perform a variety of other manual tasks (e.g. compass and/or map orientation).

- Team Member heads will be 'on a swivel', each observing his area of responsibility during movement; observing other Team Members in the tactical formation; observing stealthy and secure hand and foot placement; observing for hazards (e.g. mines/booby-traps, venomous snakes, etc.). The weapon muzzle should generally track with the movement of the head, from its slung carry position.
- The Point Man should not only have sharp senses (especially visual), but he should also possess:
  ○ An abundance of tactical/common sense
  ○ Field craft expertise
  ○ Stealth
  ○ Smallness in stature
    ▪ Smaller point men make a smaller target; if the point man is hit, another Team Member will be better able to drag him away.
    ▪ They are better able to peer under ground level vegetation and perhaps spot an enemy ambush.
  ○ Point Man Armament: In rainforest or jungle environments, consider arming the Team's point man with an enemy weapon (e.g. AK-type) or a shotgun.
    ▪ An enemy weapon in the hands of the Team Point Man, may deter or cause hesitation of an enemy combatant during a chance encounter (e.g. for fear of fratricide).
    ▪ The shotgun was the most effective close combat weapon used by British and Commonwealth patrols (especially SAS's specialized reconnaissance, raiding and counter-insurgency units) during the Malaysian Emergency. Several dependable, detachable, magazine-fed US and foreign semi-/full automatic shotguns are available. Shotguns are also valuable in survival situations.
- Meeting/Chance Engagement Firefight:
  ○ Close combat shooting is dependent on speed, accuracy and efficiency. These shooting essentials are of even greater significance in meeting/chance engagements; moreover the enemy may have the advantage on the Team in shooting from ambush, which the Team must overcome.
  ○ Regardless of which force initiates the firefight, Team Members should not just drop, but lunge, to the ground. Enemy combatants may only be able to see a portion of the Team or Team Member; only able to glimpse the upper torso, for an instant. An inexperienced shooter will fire at what was briefly visible through the vegetation. Lunging for the ground will remove the Team Member from the enemy shot box.
  ○ During a chance encounter/meeting engagement (or an enemy ambush), the enemy will attempt to mass fires on the Team from cover (e.g. behind standing/fallen trees, etc.) and/or from concealment (e.g. from behind a hedge or a stand of bamboo).
  ○ While descending to the ground, Team Members should be thumbing their safeties to full automatic fire. Highly-experienced SOG Team Members were able to lunge for the ground, thumb the safety and shoot – all before hitting the ground. Team Members should commence fully automatic suppressive fire as they bring their weapons to bear on the proximate location of the enemy.
  ○ If a Team Member is spotted by an enemy combatant, enemy initial shots will be directed at the point where the enemy shooter last saw the Team Member; so it is good practice, after hitting the ground that the Team Member should do a roll to get outside the enemy's shot box. Protect weapon optics, if used, when executing the roll.

- ○ In the absence of other instructions from the T/L, as soon as the Team Member is on the ground, he will seek hasty firing positions that offer some degree of cover.
- ○ As soon as possible, Team Members will shift their priority of fire to the chief threat, such as enemy crew-served weapons (e.g. RPGs, machine guns); an even higher priority target might be the enemy radio. Team Members with crew-served weapons must shoot and scoot (e.g. roll a few feet) to new positions as often as possible (especially RPG/Rocket weapons), as the enemy will mass his fires where firing signatures reveal the locations of these key weapons.
- ○ Once the enemy is on the ground and has sought cover, he may only be detectable from muzzle debris (kicked up from the ground) or muzzle flash. When the Team Member spots the muzzle debris/flash, he should place his shots, not at the muzzle of the enemy combatant, but where the shooter actually is, firing behind enemy muzzle flashes or debris clouds.
- ○ Enemy crew-served weapons will be relatively easy to detect in a firefight. These must be suppressed/eliminated as a priority. An RPG's signature during a fire fight consists of: (1) flash and debris cloud to the rear and front of the launcher, (2) a puff of smoke approximately 30m to the front of the launcher as the rocket motor initiates (RPG-7), and (3) a double 'bang' associated with the launcher firing and then the detonation of the round as it impacts.
- ○ When the Team Member's initial magazine is exhausted, the Team Member should, if possible, quickly roll or move to another firing position; the enemy may fire on the abandoned position (located by friendly muzzle flashes and debris clouds). The Team Member begins replacing the empty magazine while moving.
- ○ There is very little grace period during a firefight; by the time initial magazine change occurs, the T/L must analyze the situation and make a decision (within approximately 12-15 seconds); typically, either to have the Team execute a break-contact Battle Drill; or to maneuver on the enemy. Key information that the T/L must instantly evaluate includes enemy's mass of fire and its effectiveness, positional dominance, and enemy immediate reaction (Battle Drill) to contact.
  - If the T/L decides to maneuver on the enemy, one element will provide a base of fire, prioritizing its fires on crew-served weapons and guarding its exposed flank, while the other element maneuvers (using cover and concealment; employing smoke, if necessary) to a position that offers tactical advantage (e.g. superior terrain; enemy flank) and then concentrates fire. Aside from the intra-Team radios, the T/L should have a clearly audible and identifiable signal device (e.g. whistle, star-cluster, etc.), and a Team signal protocol, to control Team/element maneuvers.
  - If the T/L decides to break contact and withdraw, the T/L will use the appropriate signal (by SOP) and the Team will execute its break contact Battle Drill. Use of smoke or CS grenades may be used to facilitate breaking contact, per SOP or at T/L discretion.
  - If the Team is immobilized by casualties, the Team may have to assume a defensive posture. In a COIN mission, the Team may be able to fend off the enemy until CAS or a RF can respond.
- ○ If the enemy is positioned on elevated terrain, he may shoot slightly high. If the enemy is shooting from below the Team, his rounds may strike the ground. Inexperienced SR personnel may have the same tendencies, which must be corrected in training. If the enemy is shooting a weapon that is not very stable in full-automatic firing (e.g. the AK-47), many of his rounds may be off target, unless the shooter is experienced and takes compensatory measures. Fired on full automatic setting, the muzzle of the AK-47 will rise and drift to the left. AK-armed SOG Team Members compensated for this by rolling

the weapon to the left (ejection port oriented upwards), magazine oriented to the right, and then firing from right to left on full automatic.

○ The Team must not allow the shock of a surprise encounter to impair reaction. Hesitation is fatal in close combat. Realistic training is necessary to this end.

○ A 30-round magazine extending from the magazine well of the weapon may be awkward to fire when the shooter is in a fully prone position. It requires the shooter to rise up perhaps 6in, to acquire his target, making him more vulnerable. Firing from cover, behind a fold in the earth or behind a tree, and shifting position, can reduce this vulnerability.

○ Remember that during a chance encounter firefight, the enemy element will also drop to the ground. If the enemy element is not veteran/elite, they will likely lag in massing fires and may not fire effectively. Enemy accuracy and effectiveness of fires may be impaired in the initial moments of a fire fight.

○ Team Members shooting 40mm grenades/RPGs/Rockets must ensure that projectile trajectory through vegetation will not cause fratricide from premature detonation. Team Members so armed must move to acquire the best shooting angle.

○ In any close combat engagement/firefight, Team Members must mitigate tunnel vision. Each Team Member must attend to his role and responsibilities during the engagement. He must keep his head on a swivel, observe his area of responsibility and stay on the alert for enemy attempts to maneuver or for the approach of other enemy elements.

○ Maneuver may be the only way to dislodge an enemy that is occupying a superior position or that has a numerical advantage.

○ If the Team is moving at night, special measures must be employed in close combat firefights. Team Members may not be able to clearly discern terrain and vegetation using NVGs. Muzzle flashes will impair night-vision equipment. Accuracy, without night-vision optics, will diminish substantially. Perhaps the only way to determine if the enemy is equipped with night-vision optics is how accurate and effective their fires have been.

  ▪ It may not be wise to return fire after the initial exchange, as muzzle flashes will reveal Team Member locations and their weapons signatures.

  ▪ If the enemy is downhill, and if the Team Members can see (with NVGs) channels through vegetation, fragmentation grenades should be used.

  ▪ Before withdrawing/moving, the T/L must assess casualties. If the Team has taken more than one serious casualty, the T/L's options may be reduced to maneuvering on the enemy or assuming a defensive posture.

  ▪ If the Team cannot withdraw, the T/L should consider dispatching 2–3 Team Members to flank the enemy element to compel its withdrawal. If these Team Members can get close enough to the enemy flank unobserved, they should use a Claymore mine (angular shot) to achieve instantaneous fire superiority and to maximize enemy casualties. Or detached Team Members can create a disturbance that will either draw enemy fire or deceive the enemy that the Team has withdrawn.

• Train with Multiple Weapons. Simulate a malfunction with the primary weapon and rapidly draw and fire the secondary weapon. Practice this in dry fire and live-fire drills.

• Note that 'authoritative' voices may recommend flawed combat shooting techniques. Some of these practitioners have used flawed combat shooting techniques for so long and with such frequency that they are able to achieve remarkable results notwithstanding the flaws. But most SpecOps personnel are not on the range every day, shooting thousands of rounds at known target arrays and at known distances. The Author recommends that the SR personnel/Team first develop a true perspective of his combat environment and the threat he is facing – and then use an abundance of common sense to maximize his speed, accuracy and efficiency tailored to that environment. If the Team will be operating in terrain that offers extended shooting vistas, as well as close quarters situations, Team Members should train in both instinctive firing and precision firing. If the terrain is heavily dissected and vegetated, training should predominantly focus on instinctive firing.

Foreign Weapons TTPs:

- Foreign (enemy) small arms weapons may sometimes be more effective, lethal and/or dependable than US weapons. For example: RPGs are effective in Anti-Tank (AT), Anti-Materiel and Anti-Personnel (AP) roles. The use or appearance of enemy weapons in the hands of Team Members may deceive an enemy as to whether they are facing friendly or enemy forces – causing them to hesitate. Consider using at least some enemy weapons on SR operations.
- Among combatants trained by the Former Soviet Union (FSU) most are equipped with FSU (Russian) weapons. In many of these circumstances, the RPG serves as the squad 'crew-served' weapon and is often used to initiate contact in ambushes or raids. Because such engagements were initiated with RPGs, the RPG was often responsible for proportionally more US casualties as compared to the AK-47, during the Vietnam conflict.
- RPGs are effective against troops in cover (urban structures or field fortifications) and against hovering, or slow-moving helicopters.
  - A variety of warheads are available for the RPG-7, including AT, AP (with fragment band) and Thermobaric.
  - RPG-7b is the collapsible airborne unit version.
  - The RPG-7 has a bell-shaped metal rear end. It will sound like a bell if it strikes other objects. Consider cutting the bell off to eliminate this problem; it won't affect weapon launches in close combat.

Other Weapons, Attachments and Accessories TTPs:

- Muzzle caps, already discussed, should always be used on rifles/machineguns.
- At least two T/Ls at SOG's FOB-2/CCC (including the Author) found that the slam of the bolt on a suppressed M2A1 .45 caliber grease-gun was too loud to be used in a silent ambush. The T/Ls experimented with gluing either a plastic disk or a felt pad to the face of the bolt. While this remedy mitigated the slamming bolt appreciably, failures-to-fire increased unacceptably. Other submachine guns, which also fired from the open-bolt position, had a similar problem with bolt noise.
- Team Members should keep abreast of equipment and ordnance developments. New, highly effective ordnance may be made available to a Team while it is in its developmental/ prototype stage. Other ordnance may be available from industry, which the Military Services have decided not to pursue or acquire; a number of these items may be in inventory as test articles.

<u>True Account:</u> **A highly qualified SOG T/L took CONUS leave and managed to visit a Honeywell munitions development facility. There, he discovered that Honeywell had developed several unique 40mm grenade cartridges that were not in the military inventory for various reasons. Some of these munitions were added to the SOG 'shopping list'.**

Silencers/Suppressors TTPs:

> '[S]ilencers don't live up to their name, or as they are portrayed in movies in reducing the noise of a gunshot to a deadly whisper. An AR-15… fitted with a top-of-the-line silencer still registers 129 decibels when it is fired with regular ammunition … about as loud as a jackhammer. Attached to some smaller-bore weapons, however, a silencer comes

closer to its stereotype. A silenced .22-caliber gun loaded with special ammunition makes a noise that is 'quieter than an air gun....'[11]

- In the Author's experience, the most silent of pistols was the SOE/OSS 'Welrod' in .32 caliber; but also produced in .22 and 9mm (note that these pistol cartridges are all subsonic). The only sound that could be heard in the .32 version was a much muted fall of the firing pin. The Welrod may still be available to SpecOps, but the weapon has significant disadvantages (including weight, range, rate of fire, and lethality) that normally preclude its use in SR operations.

Figure 87. The Bolt Action OSS Welrod in .32 Cal. The Removable Grip is the Magazine. (*Public Domain*)

Hand Grenade TTPs:

- The spoon on a grenade will make a distinctive 'ping' when the grenade is thrown, revealing user location and telegraphing to the enemy that a grenade is on the way.

Author's Solution:
Learn how to 'roll' the spoon/safety lever off of a grenade, to suppress the sound of spoon release.

- Beware of potential fratricide when using fragmentation (aka defensive) hand grenades on light structures (bamboo/wood). The fragmentation of the grenade may penetrate the walls of the structure and wound or kill Team Members.
- Even in daylight, fragmentation grenades are difficult to employ. The throwing motion will be encumbered by the individual weapon, LBE and rucksack – more so if the Team Member is throwing from a prone or kneeling position. Throwing a grenade so encumbered will substantially reduce throwing accuracy and distance. 'The M67 frag grenade has an advertised effective kill zone of five meter radius, while the casualty-inducing radius is approximately fifteen meters.'[12] Throw <u>downhill</u> and from behind cover – and shed the rucksack, if possible, before attempting to throw a fragmentation grenade.
- The Team Member will make noise (rustling of clothing, shifting of LBE/rucksack) when throwing the grenade. If the enemy alerts to the noise or movement associated with the throwing motion, be prepared to receive weapons fire prior to grenade detonation.
- Throwing a fragmentation grenade uphill is almost always a very bad idea; it is likely to roll back downhill and kill Team Mates.
- Remember that US fragmentation grenades have two safeties: a safety pin for the fuse and a safety clip to retain the spoon. In the heat of battle, the second safety is sometimes forgotten – a wasted grenade that provisions the enemy. Retain at least one safety pin and carry it where it can easily be retrieved; this is useful in the event that the grenade is ultimately not armed/thrown. This safety pin may also be used to 'render-safe' a booby-trap or mine.

---

11. Joe Palazzolo, 'Silencers Loophole Targeted for Closure, ATF Seeks Background Checks for All Members of Weapon-Buying Trusts', Wall Street Journal, October 3, 2013, New York.
12. Field Manual 3-23.30, 'Grenades and Pyrotechnic Signals', Headquarters, Department of the Army, Washington, DC, (2005 revision)

- Use of hand grenades is limited by the throwing range of the Team Member (note that indigenous Team Members may have a weaker throw). There are means to mitigate the range limitation. When in a defensive position, or even in an ambush position, Team Members can use field expedient aids to achieve greater throwing distance. <u>All field expedient aids require training</u>.

Figure 88. A Field Expedient Stick Grenade (with Close-up).

- Stick Grenades:
  - Stick Grenades can be thrown farther than hand grenades.
  - Pick a resilient stick that will not snap in the process of throwing. Carve a groove near one end.
  - Lash the neck or body of the grenade (cylindrical grenades) to the stick, using 550-cord and using the groove to ensure the 550-cord (and the grenade), does not slip from the stick. Double check all knots.
- Grenade Sling. This field expedient can send a hand grenade much further than it can be cast by arm strength. [Similar to TM 31-210 (Obsolete) Improvised Munitions Handbook, HQ, Department of the Army, 1969, Frankford Arsenal, PA. Page 148.]

Figure 89. A field expedient grenade sling.

  - Tie a 12–18in length of 550-cord firmly to the neck (fuse well) of the grenade, ensuring the line does not prevent release of the grenade safety lever/spoon, then tie a large knot at the free end of the 550-cord to assure a firm grip during the launch process. Double check all knots. The line should be prepared in advance of a mission, and should be carried for rapid retrieval and attachment to the grenade.
  - Modify the grenade safety pin to permit easy removal, e.g. only single leg inserted into the safety pin hole of the grenade.
  - The grenade will be whirled around 2–3 times at the end of the cord like a sling and released at the appropriate moment toward the intended target.
    - When a fragmentation/defensive grenade is used, first ensure that the spoon retaining clip is intact and in place, then remove the grenade safety pin. Release the retaining clip prior to launching the grenade; this is best done by a second Team Member.
    - Ensure that there are no obstructions to snag the line or the grenade when twirling it overhead prior to launch; and ensure that there are no obstructions along the estimated

Figure 90. Traditional Sling with Close-up.

flight path of the grenade (and cord) which might entangle the item as it flies toward the target.

- ◦ Remember to release the sling cord within approximately 3 seconds. The grenade must be slung from a vulnerable standing position. If possible, the slinger team should position behind a fold in the earth.
- ◦ This technique should be practiced using training/practice grenades prior to deployment.
- Traditional Sling: This field expedient can be used repeatedly to launch a grenade much farther than tossing it by hand.
  - ◦ Use a 12in x 5in piece of canvas/heavy-duty nylon. Accordion-fold (multiple small folds) the long ends of the rectangle, and firmly attach a 3ft length of 550-cord to each end. The cloth at the each end of the rectangle can be wrapped around a pebble and then secured with 550-cord. This will form the cloth into an elongated cup. Formation of the cup and firm attachment of the cords is <u>essential</u> to retaining the grenade in the sling during the launching process.
  - ◦ At the loose end of one of the cords, tie a large knot. At the loose end of the other cord, tie a non-tightening thumb-loop. The thumb loop should be slightly larger than the circumference of a thumb. Cord lengths should not allow the sling to touch the ground as it dangles from the throwing arm in the launch position.
  - ◦ A Two-man process.
    - ▪ The slinger inserts his throwing hand thumb through the thumb-loop; he then grasps the knotted end of the other cord with the same hand. Launching the grenade requires the slinger assume a standing position; therefore it is best if he launches from a terrain fold that affords protection from direct fire.
    - ▪ He then assumes a balanced position, extends his throwing arm to the rear with the sling perpendicular to the ground, allowing the cup of the sling to dangle from the end of his throwing arm.
    - ▪ When the slinger is ready for loading, the loader removes the grenade pin and places the grenade securely in the cup of the sling. He should grasp the cupped grenade until the slinger is ready to launch – to ensure that the grenade is secured during the heat of combat. When the slinger is ready to launch, the loader removes the safety clip and releases his grasp. The slinger whirls the sling at least two times to gain momentum and then releases the knotted cord from his grasp such that the grenade is released at the proper moment to attain the best trajectory toward the enemy. Centrifugal force retains the grenade in the sling cup until the knotted cord is released. The thumb loop retains the sling in the grasp of the slinger. Be careful that the loaded sling cup does not strike the ground or vegetation while it's being whirled.
  - ◦ The slinger must have a clear launching area and throwing lane so that the sling/grenade is not obstructed during the launching process. Once the grenade is released, it will be less likely to deviate from its trajectory than the Grenade Sling above, as it does not have a cord traveling downrange with the grenade.
  - ◦ This process is easily mastered with practice using training grenades or even rocks.
- Bamboo (or Stick) Thrower (Variants):
  - ◦ <u>Single Pole</u>: Select a 4–6ft <u>green</u> cane/pole with a section interior diameter that is approximately the exterior diameter (2.5in) of the M67 grenade (or the grenade being used). This section will be at the throwing end. The gripping end should taper down (higher on the stem) from the throwing end. Cut the stem cavity at the throwing end below the intra-node in a scalloped configuration; remove the debris, creating an open-ended recess in which the grenade will rest prior to throwing. Tape or a plastic retainer can be used at the bottom of the node, to reinforce the stem. The grenade safety pin

should be modified to permit easy removal, e.g. single leg inserted into the safety pin hole of the grenade. The throwing process is a two man effort: a thrower and a loader. When ready to launch, the loader will remove the safety pin and spoon safety clip (if present), and place the grenade in the recess. If available, a small bungee cord, large rubber band or elastic band may be used to retain the grenade in the recess and hold the spoon in place. The thrower must ensure that the recess is tilted upward at an angle at all times to ensure that the grenade does not drop out of the recess into friendly positions. The thrower then will use the cane/pole as an extension of his throwing arm and launch the grenade toward the enemy, with a motion similar to casting a fishing line. Note that the area around where the Team Member is casting the grenade must be clear of any obstacles that would interfere with the task and the area in front of the Team Member must also be sufficiently clear to allow flight of the grenade toward the enemy, without risking rebounding of the grenade toward friendly troops. This technique can be used from a deliberate ambush position or from a defensive position.

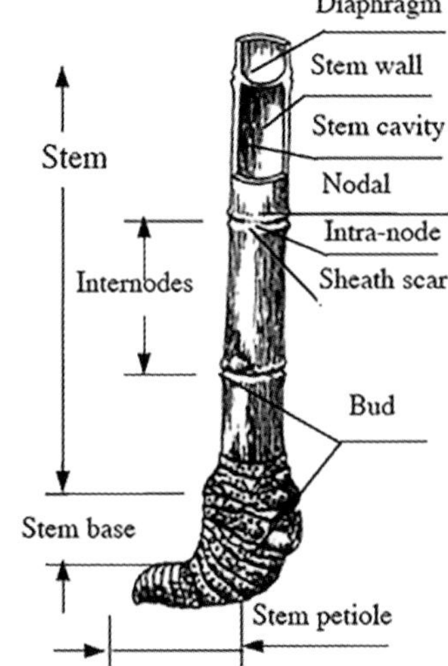

Figure 91. Bamboo Diagram, showing components. (*Public Domain*)

- A variant (pictured) uses a slender, more flexible bamboo pole with an improvised cup, which can be fashioned from a bamboo internode, from a larger diameter bamboo segment, or from any rigid article which can form a cup for the grenade.
- <u>Throwing Pole with Sling</u>: This is a technique adapted from medieval times, where a sling-staff was used to hurl rocks at besieging enemy troops much further than could be achieved in casting by arm. Pole (length may vary) will have one leg of a sling attached to the end of the pole, the other leg of the sling will have a loop or knot. The loop is secured to a peg, notch or nail (head removed) near the end of the pole, or it may be looped directly over the end of the pole as in the nearby illustration.

Figure 92. A variant of the Bamboo Thrower.

This item can be assembled fairly rapidly using a flexible <u>green</u> bamboo (or sapling) pole of approximately 1½ – 2in diameter; cord; and cloth for the sling. Alternatively, a forked branch ('Y'-stick) can be used rather than a pole with peg/notch (see illustration below).

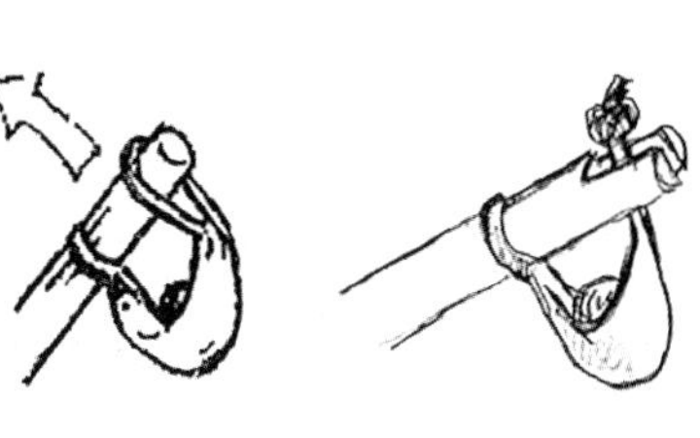

Figure 93. Three variants of a Throwing Pole with sling. (*Public Domain*)

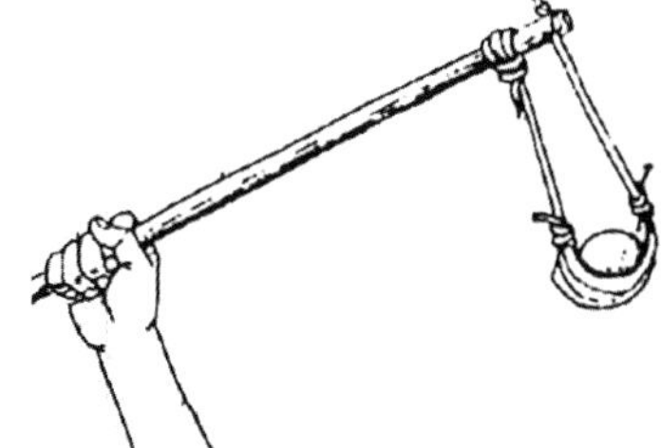

The concept of operations is similar to a shepherd's sling or like that of a medieval hand-trebuchet, where the looped leg will come free of the peg/notch/nail/pole-end as the pole passes approximately 45 degrees from vertical during the throwing motion (using a snapping motion as in casting a fishing lure). Note: The sling must be of such design/material that the grenade fuse assembly, safety clip, or other physical feature will not snag (and be retained by) the sling during throwing. This technique is a two-man operation and is used from a deliberate ambush or defensive position.

Author's Comment: You must seek throwing lanes before you throw any grenade; some throwing lanes may be 'up and over' intervening vegetation.

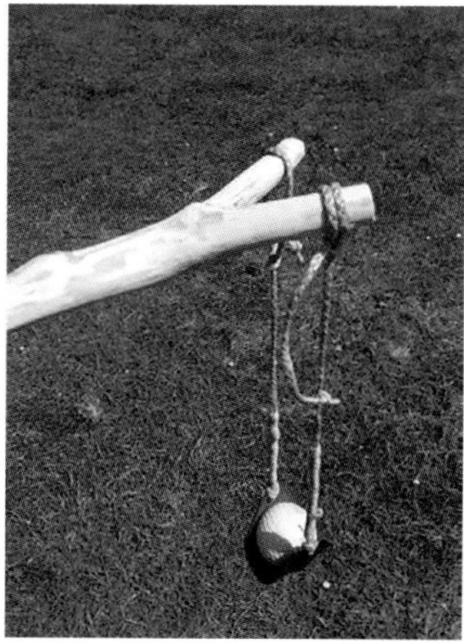

Figure 94. "Y" stick with a sling.

- Team Members should carry a mixture of fragmentation (defensive), CS (Tear Gas), colored smoke, thermite and/or WP (and in urban combat, offensive) grenades on their LBE and on/in the rucksack/haversack:
  - Fragmentation grenades are excellent for inflicting casualties; offensive (concussion) grenades are effective only in enclosed areas; but remember, fragmentation grenades may produce fratricidal casualties. Note: A selective offensive/defensive grenade is in development as of this writing.
  - CS grenades are good for breaking contact, breakout from encirclement and for stopping/ slowing down enemy troops in pursuit of the Team. Further, they create a plume, not unlike that of a smoke grenade, which is useful in concealing Team movement. They will stop dogs from pursuing you, even in wet weather – whereas sprayed CS powder may be ineffective in wet conditions.
  - Remember that CS and smoke grenades have a fuse delay of several seconds as the pyrotechnic mix ignites and the generated plume suspends and spreads. This delay may allow an enemy to evade the cloud or don protective masks. Some foreign and/ or commercially available CS/smoke grenades may act faster, with a more robust and immediate pyrotechnic plume effect or even a burst rather a burn. Additionally, modified US smoke/CS grenades may be specially fabricated, with added incendiary mix, accelerating the burn rate to produce a greater cloud with a more immediate effect – this is not optimal for location marking, but is much preferred when used in Battle Drills.
  - Most, but not all, smoke grenades (especially WP) should be carried in or on the rucksack. If the WP grenade detonates while it is attached to the pack, the rucksack may shield the Team Member from some of the burning phosphorus.
  - Partially camouflage any bright markings on grenades, using subdued spray paint. Leave enough smoke markings to reveal the color.
  - Fold tape through the safety pin rings of grenades and tape the rings to the bodies of the grenades for noise suppression and to help prevent rings from snagging on vegetation.
  - Both safety (cotter) pin legs are bent back along the fuse body during the manufacturing process; when attempting to use the grenade, the safety pin may prove difficult to pull, especially by indigenous troops.

Author's Solution:
Ensure one of the grenade safety pin 'legs' is straightened in advance before mission insertion and the other leg remains bent back along the fuse body.

- Smoke grenades are useful for marking; but they are also useful to screen a withdrawal, maneuver or breakout; this is often neglected by troops in contact.
- Make daily checks on all externally carried grenades, to ensure that the detonator/fuse is not coming unscrewed from the body of the grenade and that safeties are intact.
- Use thermite grenades to destroy Team equipment or destroy enemy facilities, equipment or supplies. In several circumstances, thermite grenades are more effective than demolitions or fragment producing

Figure 95. Failed design. Silenced M-1 Carbine, converted to 9mm.

devices. Remember that the fuse is nearly instantaneous. Also remember that WP, CS, colored smoke and thermite grenades can cause underbrush to catch fire. Take dry conditions into account before using these items.
- The pillar of smoke from WP grenades is capable of penetrating heavy canopy even in wet conditions and is easily observed by aviation assets; however, the plume from regular colored smoke grenades often cannot penetrate heavy canopy. In low cloud cover, fog or heavy mist, WP smoke may not be detectable to aircraft.
- WP grenades are also casualty producing and useful against materiel. Warning: If a WP grenade has even a minor crack in its casing (from being dropped or sharply impacted), the grenade can cook-off – an agonizing way to die.

True Account: SOG Team Members were preparing their LBE and rucksacks for an upcoming mission. The T/L accidently dropped a new, US-manufacture, WP grenade, from less than chest high, as he was withdrawing the grenade from its packaging. As he picked it up, the grenade detonated at crotch level. He managed to walk to the dispensary in absolute agony, trailing a large plume of white smoke from WP particles embedded in his flesh. He died 2–3 days later at the Pleiku field hospital.

- Be aware that some foreign-made munitions are manufactured to a high standard and others are not – PRC pyrotechnic grenades, for instance, are notoriously dangerous. If you are considering the use of foreign munitions for US or indigenous Team Members, ensure that they are safe, that training and operational use instructions are followed and that they are used within the operational envelope for which they were designed. Additionally, any new munitions or weapons introduced into the FOB supply chain, regardless of manufacturing origin, must be thoroughly checked/tested for function and safety before taking them on operations.

True Account: SOG acquired and distributed to its subordinate FOBs, what was commonly referred to as a single-shot, WP grenade launcher; the munition was constructed of a light aluminum tube, with a cylindrical time-delayed WP grenade of approximately 6in in length contained therein; it had a fold-out pistol grip, and fold-out trigger at the end of the munition. A SOG T/L carried one in his rucksack on an operation, but did not use the item. When conducting live Battle Drill training in preparation for his next mission, the T/L noted that the launcher tube had acquired a slight indentation from the previous mission.

The T/L attempted to fire the item on the range during training; the munition initiated with a muffled report, but failed to eject the grenade. He tossed it away and scant seconds later, it detonated. Fortunately, nobody was injured. Apparently the light aluminum tube was too easily impinged under modest pressure and was therefore too hazardous for field use.

<u>True Account:</u> On another SOG acquisition, a number of M1 Carbines were converted from .30 cal to 9mm, and further modified with a cut-off barrel, wire stock, a silencer and optical scope. Two highly experienced T/Ls took one of the weapons to a secluded location to zero-in the scope on a standard target. One T/L fired the weapon, the other spotted. Initial shots fired did not register on the target and the spotter could not spot the strike of a round on the surrounding ground. Bold changes in elevation and deflection were made, followed by more shots, with the same results. The T/Ls checked the weapon and found that all (approximately a dozen) of the 9mm rounds fired had imbedded in the silencer. Miraculously, the weapon did not explode.

Rifle Grenade TTPs:

- Consider rifle-grenades as an add-on capability to Team rifles/carbines (except for those weapons already equipped with 40mm grenade launchers). Several international munitions manufacturers produce rifle-grenades that use a bullet-trap (compatible with a standard ball round vice a crimped cartridge) to propel the grenade from the rifle and they offer a broad selection of warhead types (High-Explosive, High-Explosive Anti-Tank, WP, Smoke, Tear Gas, Illumination, etc.). There are several bullet-trap designs.

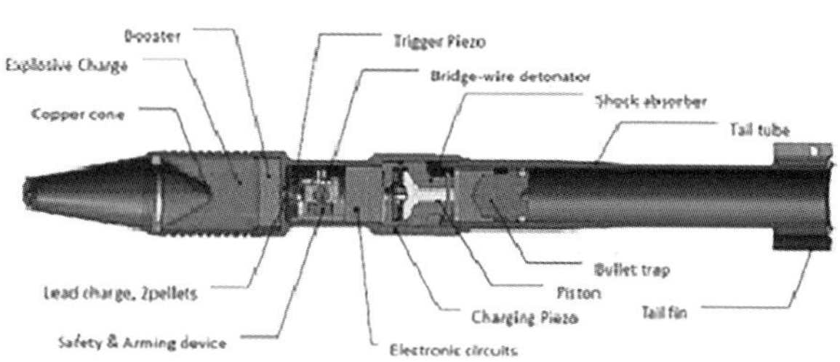

Figure 96. MECAR M200 HEDP bullet-trap rifle grenade. (*Public Domain*)

- Although light in weight and slow in firing rate, modern rifle-grenades can increase the lethality of Team fire and can help the Team attain initial fire superiority, striking enemy crew served weapons and other key targets.
- Rifle-grenades obviously have a greater range than hand grenades. Whereas throwing a hand grenade uphill and through vegetation can be problematic, if not downright suicidal, rifle-grenades can substantially mitigate the fratricide problem.
- Problems/Issues:
  ○ The rifle-grenade should be designed to fit snugly onto the flash suppressor/muzzle of the weapon so that it does not drop off when the weapon is oriented downward. This can be achieved with a muzzle adaptor.
  ○ Some rifle/carbine flash suppressors may have to be replaced in order to accommodate rifle-grenades; or some carbine accessories (flashlight, laser, mounts, etc.) may have to be moved to accommodate the rifle-grenade tail fin/tube assembly.
  ○ The rifle-grenade will need sufficient barrel length (beyond the forward hand guard) to accommodate the grenade. Additionally, the extended length of the weapon, combined with the mounted rifle grenade, will make the weapon more unwieldy while navigating undergrowth.

Missile, Rocket Propelled Grenade (RPG) and Rocket TTPs:

- In respect to RPGs, shoulder fired rockets (and recoilless rifles), beware of the back-blast from the launcher. These systems are generally not designed to be fired from within enclosed spaces. If fired from an enclosed space, the overpressure can cause auditory damage to Team Members.
- If your Team intends to use an anti-tank rocket/RPG, the shooter should displace to another position immediately after the shot; smoke and sound will draw concentrated enemy fire.
- Use of these systems can provide a substantial boost in Team capability and firepower. The broad variety of warheads available in some of these systems would enable the Team to address a broader spectrum of targets.
- Be aware that ultra-hot gasses formed by the burst of AT munitions/warheads can start a fire in dry conditions.
- Until recently, RPGs generally had reusable launchers and the munition often extended from the barrel of the launcher. However, the distinction between modern RPG systems and shoulder fired rockets (e.g. M72 LAW) has diminished; RPG launcher design has been migrating to a one-shot, disposable launcher configuration. A broad variety of shoulder-fired rockets use one-shot, throwaway launchers and are in abundant supply worldwide. Some of these disposable items have objectionable characteristics from an SR perspective.
  - They are cumbersome and often noisy to carry and easily snag vegetation during tactical movement.
  - During Battle Drill, they may take valuable seconds to put into action, allowing the enemy to seek cover and harming Team ability to attain the fire superiority objective early in an engagement. If an RPG/rocket is to be effective in Battle Drill, it must be employed rapidly.
- While the ubiquitous RPG-7 may have been surpassed by technological advances in some shoulder-fired rockets, the system is rugged, economical, soundly-designed and effective; it is in use worldwide and therefore may be a good option for SR Teams. Some comments regarding the RPG-7:
  - The rocket loses accuracy in a moderately brisk crosswind. Not an issue at close combat range.
  - The launcher has a metal, bell-like flare at the rear of the launcher tube. This flare will chime like a bell when it comes in contact with other objects; consider cutting the bell flare off. Note that optical sights available for the RPG-7 are not useful in close range engagements.
  - A paratrooper variant provides a launch tube that breaks down into two sections.
  - A variety of PG warheads are available, including fragmentation, thermobaric, anti-bunker and enhanced effect anti-armor.
  - Night-vision sights are available for the RPG-7.
  - PG-7 ammunition is setback armed and requires an arming distance of 5 meters. The ammunition has a safety cap covering the PG nose; this cap must be removed prior to launch or the fuse will not detonate.
  - A modern-version, US system manufactured for SOCOM, offers substantially greater range and accuracy at reduced launcher weight.

<u>True Account:</u> An experienced SOG T/L armed one of his indigenous Team Members with the RPG-2, largely for anti-personnel application, even though the RPG-7 was available. He reasoned that the additional range of the RPG-7 was not needed in the close environments of the Southern Laos rain-forest, that the RPG-2 weighed substantially less than the RPG-7 and that the ammunition for the RPG-2 was much less cumbersome to carry. The RPG-2, often the 'crew-served weapon' for enemy infantry squads, was formidable in the anti-personnel role

and was responsible for inflicting many casualties on friendly forces in South Vietnam, the T/L decided to 'up-grade' the weapon by wrapping a double-band of flechettes around the PG-2 body to enhance lateral fragmentation.

- Expect advanced munitions/ordnance in the future that will enhance SpecOps capability and lethality. As of this writing, a major US Defense contractor is developing a miniaturized, rifle-mounted, laser-guided missile system with impressive mid-range (estimated 1.5 mi) accuracy, and lethality equivalent to a 40mm grenade. While not yet in the inventory, this R&D effort foreshadows similar systems that will provide SpecOps units lethal standoff capability.
- Man-Portable Air-Defense Systems (MANPADs):
  - These surface-to-air missiles are manufactured by several countries and are proliferating across the world through military (and black market) sales. Simple to employ, they can be used in all levels of conflict to counter aviation superiority or to down adversary aircraft. Rotary wing aircraft are especially vulnerable when not flying nap-of-the-earth.
  - Use of MANPADs will elicit a strong enemy response. The best uses of MANPADs by an SR unit are against lone, unescorted aircraft.

Mines, Booby-Traps and Explosive Devices TTPs:
Policy Caveat: As of this writing, the United States abides with, but is not a signatory to the 1997 Convention on the Prohibition of the Use, Stockpiling, Production and Transfer of Antipersonnel Mines and on Their Destruction. Although the United States is a non-signatory to the Convention the US largely abides by the convention nevertheless.

General Mine/booby-trap/Explosive Device TTPs:

- Huge stockpiles of antipersonnel mines exist worldwide, produced by nations that are either US adversaries or those which may someday become adversaries. Much of these stocks were produced for international sale or as military aid to nations across the political and economic spectrum – and substantial quantities have fallen into the hands of non-state actors. Among the leading providers were/are Russia/Soviet Union, Italy and the People's Republic of China.
- Many anti-personnel mines use few metal components and are difficult to detect.
- Understand/determine why an enemy might emplace mines/booby-traps in a given area. This may help Team Members to notice possible locations of these devices.
  - Harassment and Interdiction (e.g. of friendly UW forces).
  - Protect vital assets/facilities/units.
  - To block enemy (e.g. Team) movements.
  - To channelize an enemy as part of a defensive/barrier plan.
  - To delay an enemy advance or to cover a withdrawal.
  - To create a barrier in conjunction with other operations (e.g. to guard the flanks during offensive operations).
- Upon encountering an enemy minefield, if possible, identify what mines are deployed in the minefield.
  - Anti-tank and/or anti-vehicular mines are normally 'protected' by anti-personnel mines. Blast type mines (e.g. US M-14, Russian PMN or PFM series, BPD-SB-33, etc.) will typically detonate with 5-10 kg of pressure.
  - An enemy will not waste anti-tank mines to protect a rear area facility, as an armor threat is considered unlikely.
  - If a minefield contains tripwire activated booby-traps/mines, a grapnel on a line may be cast forward and retrieved to detonate the device(s).

- ○ Italian mines (e.g. TS-50/VS-50 mines) have been widely sold and are of excellent quality.
  - ○ Former Soviet-bloc client states have large inventories of Soviet mines.
- The Team may emplace mines/booby-traps (with SD features) and then deliberately entice an enemy to deploy and move through the hazardous area.
- Electric and non-electric blasting caps are essential to explosively actuated booby-traps, IEDs and other explosive devices, regardless of origin. The Team cannot exploit these devices, to include captured munitions, on the battlefield without blasting caps. Caps are very lightweight, so there is little reason not to carry a good number of them, especially on deep penetration, long-duration operations.
- The US military has been moving toward Modernized Demolition Initiators (MDI) consisting of pre-capped lengths of time fuse or shock tube – as a replacement for M6 electric and M7 Non-electric initiating systems. While the MDI system is generally safer, more reliable and, in some applications faster to employ, some firing circuits still require electrical initiation. One advantage of the M6 electric system is that electric firing wire can be used repeatedly.
- If the Team encounters an enemy spider-hole, tunnel, cave, etc. that shows recent or continuing occupancy the Team should plant one well-concealed mine/booby-trap at or near the access point. Once the device initiates, the enemy will use another (emergency) exit and will search the immediate area of the detonation to discover more devices. This scenario produces several beneficial opportunities for the Team.
  - ○ The Team can deploy luminescent/IR dye (see use of dyes the section on tracking TTPs), invisible to the naked eye, near the mined entrance. Enemy combatants who search for booby-traps/mines will leave tracks/traces that are undetectable without IR illumination. An enemy trail will reveal the alternate access point; then this access point may also be mined.
  - ○ Use of dye at an access point, may lead the Team to other important finds (e.g. locations of enemy caches, locations of enemy mines/booby-traps, OPs/LPs), etc. Then the Team may mine/booby-trap the area comprehensively.
- The Team may deliberately disturb a road surface to deceive the enemy as to the presumed presence of a mine. This may cause a vehicle or convoy to drive onto the verge, to avoid the presumed mine, to where the actual mine is emplaced. It may also cause the vehicle/convoy to halt, especially if it is at a choke point; this halts the vehicles (and occupants) within an ambush kill zone.
- A Team should generally not use more than one mine/booby-trap (e.g. not in a cluster) at a time, on a road, trail or choke point. If one mine/booby-trap detonates the enemy may conduct a search of the immediate area to detect the location of others; so emplacing more than one device at the same location, is simply supplying the enemy (insurgents). There are several qualifiers to this rule:
  - ○ An insurgent may try to recover the device to add to his inventory; but enemy conventional forces are more likely to Blow-In-Place (BIP) the mine to eliminate the hazard and restore the vicinity/route to usability.
  - ○ If the Team has observation of the seat of the explosion, a command-detonated secondary device may be very effective against other targets converging on the site.
  - ○ If a mine disables a cargo truck, the enemy may attempt to recover both the cargo and the vehicle – placing a number of personnel within a secondary device, far-ambush kill zone.
  - ○ If multiple paths/choke-points must be interdicted to be effective, vary placement and fuse type, or the enemy may 'get wise'.
- Know how to arm/disarm US mines/explosive devices during periods of limited visibility; or any foreign mines/explosive devices the Team may use. Solicit training from EOD technicians.

Use of Mine/booby-trap at Chokepoints (TTPs):

- Select choke points for mine/booby-trap emplacement to increase the probability that the enemy will initiate the device. Use natural resources as field expedient obstacles to create a choke point and/or channelize the enemy into a booby-trap/mine; for instance, many varieties of rattan, a common rainforest liana, are covered in long spines that an enemy combatant will strive to avoid. Rattan lianas can be pulled/cut down and innocuously laced through underbrush to channelize the enemy combatant toward the mine/booby-trap.

- If the Team establishes an ambush site in an area interlaced with trails, the enemy may use a trail to rapidly pursue parallel to the Team's route of withdrawal. Consider using a booby-trap/mine (with a short SD feature) on the most likely route of pursuit or at a trail junction to halt enemy pursuit.

Figure 97. Italian VAR-40 AP Mine. (*Public Domain*)

- Dog Bait: Food, Urine, Feces, Animal Scent/Attractant. Place the bait in a hole or against a confining backstop (e.g. between large tree roots, fallen log) so that the dog (and handler) are channelized in their approach to the bait. Booby-trap accordingly, taking into consideration the differences in dog weight and behavior. Place a mine in anticipation of the handler's position, rather than the dog's.

- If an enemy tracker team leader suspects that he is being channelized, he will attempt to pass around the suspect area. If the SR T/L can predict the path that the tracker team will take, he can place a second device (with SD) to counter this maneuver.

True Account: An experienced SOG Recon T/L wanted to use bait to induce an enemy soldier into stepping on an M-14 anti-personnel mine. The T/L selected bait that he believed would ensure an enemy approach; he used insulting images encaptioned in native text about Ho Chi Minh or 'centerfold' pictures of beautiful models. The T/L mentally placed himself in the role of a wary enemy soldier who would only approach an obvious lure in a circumspect manner. This virtual 'what-if' chess game approach helped the T/L select optimum positioning for both the bait and the mines.

Author Recommendation: Pictures/posters to be used as bait may be designed to resemble enemy advisory/directional signs, propaganda posters, etc. These should use a compelling image and a large font size to attract the enemy's attention; subtext fonts should be in smaller print and/or faintly printed to induce the enemy combatant to dismount from a vehicle and/or to draw near enough to read the document, to bring him into contact with the IED or mine.

Tripwire TTPs:

- Use Teflon® dental floss as trip-wire in snowbound terrain to better shed ice/snow. Other types of floss can be camouflaged to match the prevailing background by drawing the line through a camouflage stick/appliqué or by immersing the floss (un-waxed) in dye during mission preparation. Alternatively, use clear, monofilament fishing line which has a broader application.

- Preferably, place the trip-wire where sunlight will not cast a shadow revealing the wire.
- Tripwire is virtually undetectable to Light Intensification (Starlight NVG) devices. Visual detection is improved when the IR diode light element is illuminated; this will cast a faint light shadow from the tripwire.

Rapid Breaching TTPs:

- Teams may wander into or encounter minefields (e.g. protective minefields around sensitive facilities or high-speed routes of approach) in the course of their operations. A Team may also be pursued into a minefield by an enemy force.
- An SR Team may have to negotiate enemy barrier areas to conduct surveillance, to execute other missions (e.g. a raid), to throw off pursuit, to conduct a break-out or to exfiltrate an area.

<u>True Account:</u> **On a SOG Hatchet Force 'operation in November '68, the unit encountered 6–8 skeletons in a circular perimeter. All that remained was web gear and boots, everything else was gone. Fifty or so feet up the hill from the skeletons was a pile of expended ammo that appeared to be about 300 rounds or more…. The site was about 100ft from a highway. In moving to the high ground to RON [Rest Over Night] our [indigenous/Montagnard] point man stepped on a toe popper [M14 mine] … ; another Montagnard unit member did the same within 10 minutes. [Thereafter,] Montagnard unit members refused to walk point. The next AM, Lt [Name Redacted] volunteered and promptly stepped on a mine as well. [The mines were believed to have been planted by a Vietnamese SOG Team that did not recover, record or report the minefield.]' – Correspondence from Lloyd O'Daniel, formerly of SOG CCC.**

- In long-duration operations, the Team might 'prep' a target area/AO by:
  - Using an MSS to store relevant munitions, equipment and other supplies for anticipated operations.
  - Creating obstacles to channelize enemy, restrict pursuit and deny ground to the enemy.
  - Planting mines/booby-traps to harass the enemy.
  - Pre-position munitions, weapons, etc. at an objective staging area prior to ambush/raid operations.
- In long-duration operations, the Team could penetrate an unobserved enemy minefield with access lanes. Some methods of doing this are covered below. These access lanes can be used to escape enemy entrapment operations at a later date; to establish a hide/surveillance post or defensive position in the midst of a minefield to provide the Team with added security.
- Assuming that the Team finds itself within a minefield or must cross a minefield, the following thoughts and techniques may be useful.
  - If the team is undiscovered and is not being pursued, consider the following approaches.
    - In COIN operations, mines/booby-traps will frequently be planted along routes of approach to a base camp. If the obstacle can be penetrated, the Team may be able to come up on the base camp undetected.
    - When moving along a trail/path, the point man should dangle a long stalk of grass or slender twig in front of him to detect tripwires. The T/L must decide whether deactivation of the mine/booby-trap is worthwhile.
    - When moving along a suspect trail/path, the point man should exercise due caution when crossing over a fallen tree, stepping between tree roots or moving through any choke point.
    - If the point man detects antipersonnel mines/booby-traps, he should mark the mine location, examine for other mines/booby-traps nearby, and select an alternative route past the hazard – to include back-tracking. When back-tracking, retrace Team Member

foot placement, as the Team, on its path in, may have partly penetrated the minefield without initiating any devices.

- Some current AP mines incorporate an anti-disturbance feature, so probing for mines (e.g. with a knife) should be attempted with considerable caution and for determining location only – not for removal. Disarming antipersonnel mines is time-consuming and hazardous.

- If time and opportunity allows, steal one or more large animals from a local farmer or herder and 'stampede' it to cross the minefield. Water buffalo, hogs, cattle, <u>moving at a full run</u>, may make good headway before being crippled or killed in the crossing. The Team may then cross by stepping where the animal had placed its hooves.

- If the situation demands haste, and stealth is no longer appropriate, consider the following approach.

    - Determine the type of antipersonnel mines employed in the minefield, if possible.
    - If any of the mines are tripwire activated, cast a line that is weighted at the end (preferably with a hook), and then, from a covered position, drag the weighted end back across any tripwires. Briefly wait for any fusing time delay to elapse. Detonations will attract enemy attention.
    - Blast type antipersonnel mines require a certain amount of pressure (typically around 10lbs) to initiate. In dire circumstances, drop sufficient weight (attached to a pole) repeatedly on the path forward. Casualties may be taken from the secondary missile hazard, but this danger can be somewhat reduced if Team personnel kneel, while using this technique.

- The enemy may have paths through minefields and barrier wire for ingress/egress for their patrols, or in the case of small outposts (e.g. radio relay), to change shifts and/or to receive supplies (e.g. from a helipad). Observe patrols as they negotiate these barriers. Examination of snow, mud or morning dew may disclose enemy path(s) through wire barriers. Or a commonly used path may be worn through the barrier. Consider using a remote digital camera (e.g. a game camera) to observe the enemy as he negotiates access points/paths through minefields. The camera can be elevated at the end of a bamboo pole/slender tree to obtain a better angle of view and to avoid being detected by the enemy.

- If the Team is using a small minefield/obstacle (e.g. punji stakes) to secure a flank (e.g. during a long term surveillance mission), the Team should consider developing its own path through the obstacle, when time and conditions provide an opportunity. There is obvious risk here, but if the Team is using an obstacle to secure its hide, the Team is already accepting substantial risk (e.g. that the Team may be trapped against the obstacle). In developing the escape route, the Team must:

  - Use vegetation and terrain form to conceal its activity and to shield the Team from enemy fires.
  - Have a reliable means to identify the lane through the obstacle day or night and through various weather conditions.

- If the Team must rapidly penetrate a minefield, to evade pursuit or to escape a trap, consider these options and risks.

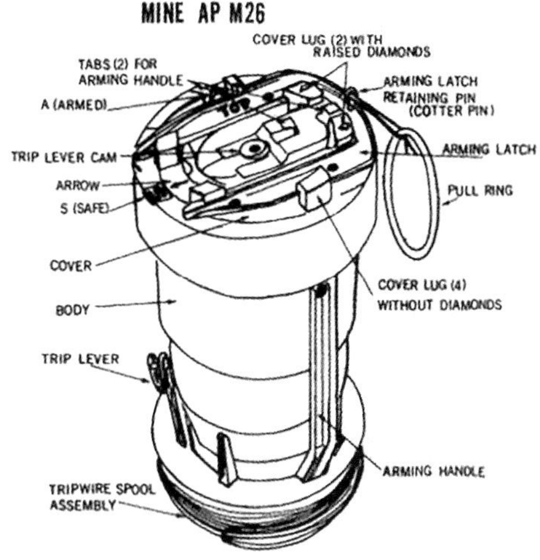

Figure 98. US M-26 AP Mine. Bounding-Type. (*Public Domain*)

- The mine to be most feared during minefield penetration is the 'bounding-betty' type mine. When the mine initiates (fusing options include pressure, trip-wire, tension release, etc.), it causes a fragmentation grenade/can to bound to shoulder height before detonation. It is small comfort that few of these will be in any minefield mix (if at all), as these mines are more expensive than pressure mines. If some of these mines are believed to be interspersed within the minefield, penetrating the minefield may be much more risky an enterprise. The Team will have to resort to more time-consuming techniques to penetrate the minefield. The Team can use a grapnel hook (or a field expedient) and line to clear booby-trap/mine tripwires. This device can be fabricated in the field
- Mine/booby-trap detonation will alert the enemy to the Team presence. A bad situation, as the minefield may be under enemy observation.
- A heavy blanket of snow could be a Team's best friend in penetrating a minefield.
  - Traverse the area on skis or snowshoes to spread the ground pressure of the Team Members. But be wary of possible tripwires attached to bounding mines. Be cautious in using trails in snow broken by enemy insurgents.
  - If possible, transport individual gear across the mined area on sleds/ahkios or a field expedient equivalent, also to spread ground pressure; use several trips if necessary.
  - UTVs/ATVs equipped with treads (or snowmobiles) instead of wheels are preferred in passage of minefields in deep snow, as treads impart less ground pressure than wheels.
- If the minefield is narrow in depth, consider dropping a tree across the area.
- If the Team is using pack animals, consider, in an emergency, driving one across the mined area to rapidly create a lane. To the extent that the animal survives, the Team Members can walk in the animal's hoof prints to cross the minefield.
- Use a stolen vehicle to penetrate the minefield, tying down the steering wheel and weighting down the accelerator.
- Field expedient 'snowshoes' can be used to cross a minefield that is not snow covered. The 'snowshoes' spread the Team Member's bodyweight less than the PSI required to set off a pressure fuse. The Team Member should consider removing his rucksack and placing it on field expedient skids, attaching a line to this bundle, and then pulling it forward every few steps – this will reduce the Team Member's overall bodyweight to further reduce ground pressure.
- During the Chechen resistance against Russia, Chechen fighters would:
  - Sometimes use planks to cross minefields. The planks would disperse the ground pressure to allow fighters to cross a blast-type antipersonnel minefield without triggering a detonation. Alternatively, the Chechens would tie a rock to the end of a pole and then drop the rock on the chosen path to detonate blast-type AP mines. This approach would be used in desperate situations, as the noise of a detonating mine would reveal the presence of the Chechens. Of course, this approach could be suicidal if bounding mines were in the minefield mix.
  - Cast/drop moderately-sized flat rocks, one-by-one in front of a minefield-crossing party to create a lane of stepping stones to evade Russian army Hammer and Anvil operations. One man would take the lead, while others in the party would feed him the rocks. If a rock did not create a detonation, it could be retrieved from the rear of the party to be fed forward to the lead man. The depressions left by the retrieved rocks were considered safe for foot placement for follow-on troops. If the rock initiated a pressure mine, the lead man might endure rock fragment injuries. The lead man must maintain his composure (and balance) when a detonation occurs or he may fall off his stepping stone onto another mine. A safe distance would be maintained between the lead man and those who follow him in the minefield-crossing party in the event that

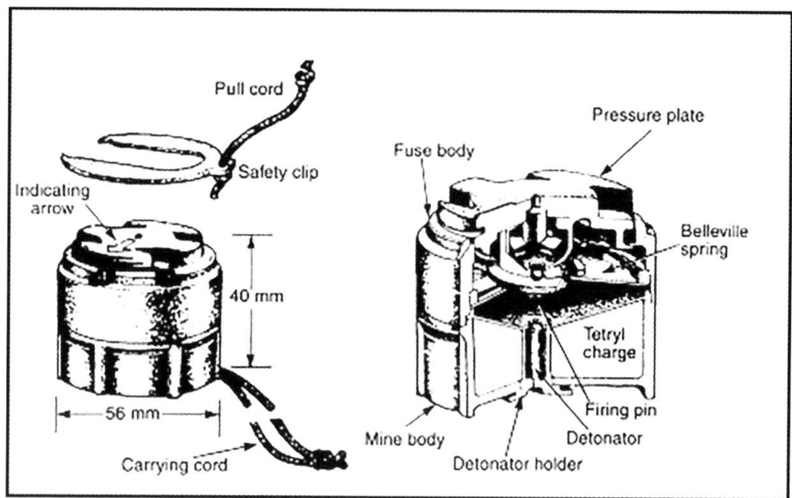

Figure 99. US M-14 (Toe Popper) Blast-Type AP Mine. (*Public Domain*)

a bounding mine was encountered. If the lead man became wounded or killed during the breaching operation, another would take his place.

  ▪ The final option would be for one fighter to walk/run across the minefield to create a path of foot imprints. This was obviously a risky and desperate endeavor, only undertaken as the only course available during a break-out (for instance).

• The enemy's prepared path through a barrier is a choke-point. As the enemy patrol debouches from the foot of the path, the Team may ambush and kill some of the enemy patrol and capture others, the enemy element would not have maneuver room. The same approach may be used on a patrol that is about to return back through the barrier at the conclusion of a patrol. Alternatively (and perhaps better), the Team could plant a mine/booby-trap on the path or at the entrance at the foot of the path; the enemy may ascribe the detonation to an accident.

<u>True Account:</u> **An indigenous soldier, detailed to security duty for the mountain-top Leghorn radio relay site in southeastern Laos, was performing maintenance on the trail leading from the site to the nearby helipad, when he stumbled and stepped off the trail. He stepped on an M-14 'toe-popper' mine propelling him forward, hands extended. His hands set off two more M-14's, flipping him onto his back onto more 'toe-poppers' and killing him.**

• These pathways may be guarded/over-watched by an enemy OP/LP for patrol passage-of-lines purposes. Beware: if a pathway is not under line-of-sight, it may be zeroed in for indirect fire (e.g. mortars) or directional mines (e.g. Claymore-type).

• If the Team intends to use the path, Team Members should take care that the signs of their passage though the barrier area are not readily apparent to subsequent enemy patrols or barrier maintenance crews. Wearing boots with a tread design used by the enemy may not arouse enemy suspicions. If Team Members are not fitted with the enemy tread design, or if they must crawl along the path, the signs of Team Member passage should be eradicated by the tail-gunner.

• Team Members should also take care that enemy patrols do not place/replace mines, trip-flares or booby-traps as they withdraw through the barrier upon their return from patrolling.

• If the Team intends to covertly breach a wire barrier, the Team may first want to determine if AP mines are integrated into the wire defense and how these mines may be circumvented. This can be determined in several ways.

- ○ Animals (e.g. goats) graze freely inside the wire. Sometimes grazing animals are used to keep vegetation from growing up inside the wire and obscuring enemy observation. If animals are not used for this purpose, local civilians or troop labor may be used to clear vegetation; and in areas deemed by the enemy to be secure, fire may be used to clear wire obstacles of vegetation. If the wire obstacle has been recently cleared of vegetation, it is not necessary to observe the actual act to conclude that the wire is free of mines. <u>Warning</u>: Enemy command detonated mines could be used in this setting.

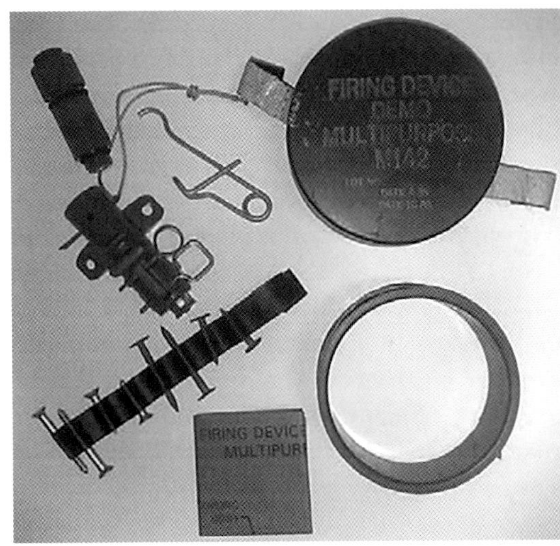

Figure 100. US M142 Multipurpose Demolition Firing Device Components. (*Public Domain*)

- ○ Evidence of enemy troops performing wire maintenance.
- ○ Heavy snow may allow Team Members to traverse an AP minefield.
- ○ If it can be determined that no AP mines are present, the Team may prepare the wire obstacle in advance by covertly cutting then reattaching wire over the course of several days prior to final execution of the breach.
- If the Team intends to explosively breach a wire barrier (e.g. in concert with a guerilla/ partisan unit, or in the execution of a raid), bear in mind that a Bangalore Torpedo will rarely defeat multi-tier concertina wire, and especially razor-wire. A more advanced device must be used. An explosive breach of an obstacle during a raid is generally beyond SR purview, unless the SR Team is in a position to attack a small, isolated, lightly-guarded outpost (e.g. radio relay site; radar antenna site).[13]
- The US M142 Multipurpose Demolition Firing Device has replaced the family of booby-trap firing devices that existed in the US inventory for several decades. This device is very flexible, but it requires training to be effective in the hands of Team Members.
- Do not confuse a fuse-well shipping plug for the actual mine fuse (sometimes packaged separately). Familiarize Team Members on mine features during mission preparation.

<u>True Account:</u> An experienced T/L was assigned a road watch mission in an extremely hot Target Area. The SR Team had been in this same Target Area several times previously, so the T/L knew to expect a tracker team to immediately pursue the Team once it had landed on its LZ and that the tracker team would be working in concert with a much larger unit to execute a 'Hammer and Anvil' operation on the SR Team. The T/L, in preparation for the operation, selected an M14 Mine, an M1 pull-type Firing Device and two Limpet Mine Detonators to use against the tracker team. The Limpet Mine Detonators, with their huge blasting caps, were to be used for SD of the mine/booby-trap, to ensure that the devices would not become a hazard for the next SR Team to patrol the Target Area. The T/L assembled the M14 Mine by replacing the shipping plug with a live detonator; assembled the M1 Firing Device by crimping and sealing a non-electric blasting cap to the device fitting; selected 24-hour chemical delay ampoules for the Limpet detonators, inserted the ampoules into the detonators, and fitted and waterproofed the massive blasting caps to the detonators. A day after insertion, the T/L

---

13. Note: Author possesses intellectual property on breaching munitions

found a suitable location to implant a grenade booby-trap, but discovered that the M1 firing device would not screw into the fragmentation grenade because the cap was too long for the grenade body cap well. On the morning of the second day, the T/L found a perfect site to plant a booby-trap. The Team had descended from a ridge, down a very steep slope, which required Team Members to slide down the hillside on their backsides for perhaps 20 meters. The T/L realized that if the enemy tracker team followed suit by sliding down the same path, they would be unable to stop their descent and would be channelized into a booby-trap. The T/L discovered that the M1 Firing

Figure 101. Limpet Mine Detonator w/ chemical delay ampoule (minus blasting cap and other kit components). (*Public Domain*)

Device would fit into the body of a WP hand grenade. The T/L reasoned that the grenade burst would scatter its burning flakes upward and onto other tracker team members who were following behind their point-man; he further reasoned that the tracker team would be forced to stop their pursuit to care for their wounded. The T/L placed his Team in a perimeter a short distance from the slope; accompanied by the Team tail-gunner, the T/L ascended the slope, assembled the firing device to the WP grenade body, taped the Limpet detonator to the grenade, attached this bundle to a small tree, attached a trip-wire to the firing device and stretched it across the skid trail attaching it to another small tree opposite and then finally camouflaged the grenade. The T/L then crushed the Limpet detonator ampoule and removed the M1 Firing Device safeties. When the T/L returned to the Team perimeter, he transmitted a message to a FAC, advising him that he should bring an air strike on any WP plume that he detected in the vicinity through the remainder of the day. Later that day, the Team heard the distant but gratifying sound of a muffled detonation characteristic of a WP grenade. The Team continued the operation without further interference from trackers.

Claymore Mine Specific TTPs:
The M18A1 Claymore mine is equipped with two blasting cap wells. This allows dual-priming of the mine using both electric and non-electric blasting systems. Each Claymore mine is packed in a two-pocket, canvas sack that also contains: an M4 electric blasting assembly (incorporating a 100ft spool of firing wire) with one end connected to an M6 electric blasting cap and the other fitted with a shunted connecting plug that can be attached to the furnished M57 electric firing device. Packed in every box of six Claymores is an M40 test set (continuity tester).

- Claymores can be 'Daisy-chained' using sections of detonating cord (furnished with the mine kit) that are capped at both ends with non-electric blasting caps, to connect several mines for simultaneous detonation.
- The entire length of the firing wire is often unnecessary, except when the Claymore is to be used in a 'far ambush' or as part of a defensive minefield. In most circumstances (especially in jungle or rainforest environments) the Team Member might consider removing the firing wire from its spool and cutting the firing wire down to reduce its length (and related weight and cube). If the wire is cut down, the wire ends must be reattached (cap at one end, firing

connector at the other), ensuring that bare wire is thoroughly wrapped with electrical tape and that a continuity test is performed on the circuit with the M40 test set. Use elastic tape, rubber bands or bungee cord to 'rigger-roll' (accordion fold) the wire for fast deployment, or wrap the wire around the long axis of the mine using the mine itself as a spool. If a far ambush is intended, at least two Team Members should be designated to deploy the mine and wire. Note that the Team could recover firing wire for reuse and should therefore carry additional electric blasting caps.

Author's Solution:

Carry the Claymore, the 'rigger-rolled' firing wire and the firing device in a cloth bag for rapid deployment, perhaps with the Claymore legs poking through the side of the bag. This configuration will allow the user to rapidly plant the Claymore legs in the earth saving valuable minutes when the Team is fleeing pursuers. Employing the Claymore to deter pursuit, would require the mine to be primed with a non-electric cap (with time fuse and fuse lighter attached) already installed in the cap well. This implies an additional hazard in carrying a primed device that could be detonated by the strike of a small arms round; an acceptable risk. Otherwise, pausing to insert the non-electric cap assembly and securing it in the Claymore with a mine priming adapter, while being closely pursued, would likely be more hazardous than carrying the primed mine.

- User instructions for the Claymore mine indicate that the minimum safe distance for firing the mine is 16 meters. However, Claymores can be fired in much closer proximity to the user when the device is backstopped by a tree or a low earth berm (e.g. spoil from a foxhole) or when the user is firing from adequate cover.

<u>True Account:</u> An eight-man SOG Team was inserted into a target area located in southeastern Laos after midday. The Team moved throughout the remainder of the day and then paused for its evening meal/commo break, prior to final movement to NDP. The Team had deployed four Claymore mines, before breaking out the rations. As Team Members ate their rations, an enemy tracker unit was spotted crawling forward in a line/assault formation. The Team opened fire on the enemy and detonated three of its four deployed Claymores. The T/L, in the process of directing his troops, was unaware that he was standing less than a foot directly behind the remaining Claymore, just as one of the indigenous commandos detonated the device. The detonation tossed the T/L into a full somersault some 4ft in the air, but he immediately sprang to his feet with no ill effects whatsoever.

- To increase the flexibility of the device, and acquire the ability to rapidly shift the aim of a Claymore towards an approaching foe, attach the Claymore to the end of a pole, perhaps using a crossbar. Additionally, a plastic container of CS powder can be positioned in front of the Claymore to create a secondary effect.

Figure 102. Repositioning an aimable claymore mounted on a bamboo stick w/ crossbar. Note: the Team Member will deploy this field expedient from a defensive position or will deploy it and then move back to his protected position (out of frame) to initiate the device at his discretion.

- A Claymore mine can be used in snow cover less than 10cm in depth. In deeper snow, strap the Claymore to a tree above the snow line.
- A Claymore mine can be purposely adapted to contain an internal device (e.g. anti-disturbance, etc.).

Claymore Aim Point TTPs:

- Remember that the Claymore aim point for dismounted troops should be modified for mounted troops. Elevate the mine or raise the aim point for mounted enemy troops/vehicle occupants.
- A solitary vehicle is not governed by convoy tactical speed protocol and will likely move at higher speeds along roads. If the Claymore is located at the verge of the road, the mines destructive arc within the kill zone may not be optimal and the vehicle may pass through the kill zone without severe damage, especially if the Team Member's vision of the kill zone is impaired by vegetation/terrain form. Either move the Claymore back off the road a few meters, to maximize the kill zone, or place it at an angle to the linear feature.
- A similar aiming technique also applies to NDP defense. If a deployed (line formation) enemy is able to approach too close to the perimeter, the Claymore may only take out 2-3 enemy soldiers. But if Claymores are offset at a modest angle, the kill zone will encompass more enemy troops within its kill zone. Some options:
  - Place the Claymore very close to the Team Member's defensive position (e.g. on the opposite side of a tree) to maximize the kill zone.
  - Enhance ability to reorient the Claymore by attaching it to a pole.
  - Place the Claymore at a slight angle to maximize the kill zone. If all deployed angled Claymores are detonated nearly simultaneously, then an adjacent Claymore may have a complimentary effect on the enemy formation within its destructive arc.
  - Place the Claymore very close and at an angle. The T/L must ensure that the destructive arcs of the Claymores do not threaten adjacent Team Members.
- Safety Notes:
  - An old EOD axiom: 'If you can see the explosion, the explosion can 'see' you' – meaning unless you are in protective cover, you can be struck with primary and/or secondary fragments from the explosion.

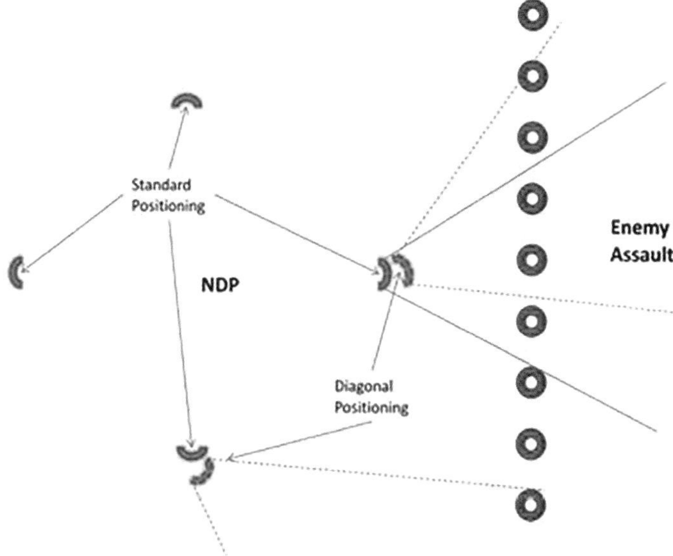

Figure 103. Angular vs Forward-facing Claymore placement (perimeter defense).

○ Explosives effects can often be unpredictable. Always take as many precautions as time and situation allows.

○ Placing Claymores close to a Team defensive position is something of a double-edged sword. The positive aspects are: (1) better kill zone, (2) device more easily deployed/repositioned/recovered, (3) less vulnerable to enemy countermeasures. The negative effects are: (1) the detonation may result in instinctive enemy fires directed at the seat of the explosion, (2) the blast/concussive wave effect of 1¼lbs of explosive so close to Team Members may damage hearing. But the same can be said of incoming RPG warheads (approx. 5.7lbs explosive weight), grenade detonations in confined spaces, etc. On balance, the pros outweigh the cons (Author's opinion).

Claymore/Directional Mine Alternatives TTPs:

• The MM-1 'Minimore' is a miniaturized (⅓-sized) Claymore designed for Special Forces. Note that the explosive is 'field-loaded' with C-4 by the user, lacks the accessories that accompany the M18A1, and has a reduced kill zone. The chief advantages of the MM-1 are smaller size and weight (reducing the load carried by the Special Forces soldier); or allowing the soldier to carry more devices (providing additional coverage/more shots); and increased flexibility (allowing the soldier to direct the blast of multiple devices in varying directions). A field expedient device, using a large soap dish or plasticware and some hardware, can serve the same purpose.

• Another device new to the inventory, as of this reading, is the Mini-Multipurpose Infantry Munition System (M-MPIMS). It is also smaller and lighter than the standard Claymore, but is more powerful than the MM-1.

Comparison of US Claymore-Type Directional Mine			
Device	**Effective Range**	**Fragmentation Zone/Arc**	**Comments**
M18A1 Claymore	50m	At 50M range: 6.5ft High; 50m Wide/60 degree arc	Wt: 3.5lbs Max range: 250m
M-MPIMS	30m optimum	At 30m range: 6.6ft High; 23m Wide	Wt: 2lbs
MM-1 Mini-More	15m	At 15m: 2ft High; 16ft Wide	Wt: 1.3lbs (approx)

• Note that other countries produce munitions similar in function to the US Claymore mine.

• <u>M86 Pursuit Deterrent Munition (PDM):</u> The M-86 'is a small, US anti-personnel mine intended to be used by Special Forces to deter pursuing enemy forces. It functions like a … hand grenade, featuring a [safety] pin and fly-off lever [spoon]…. Once the pin is pulled and the lever has been ejected, a timer is started. After 25 seconds, seven tripwires are launched from the mine [on bobbins] to a maximum distance of 6 meters. The mine then performs an electronic self-check, and is fully armed 65 seconds after the battery is activated. When any of the tripwires [are disturbed], the mine activates, a liquid propellant [which settles to the lowest point of the mine] charge launches [a fragmentation grenade] 1 to 2

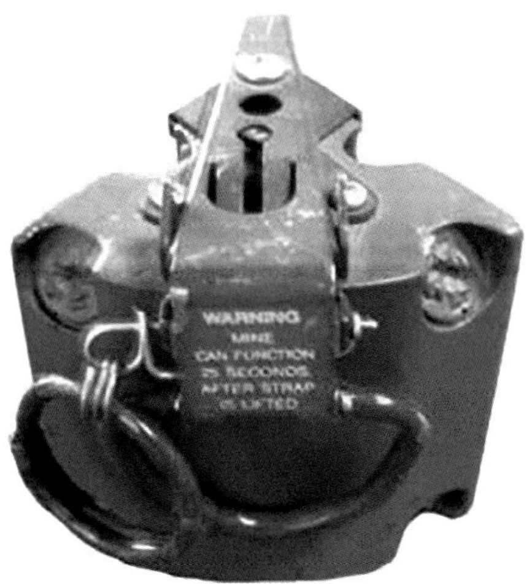

Figure 104. The M-86 PDM; Note imbedded "bobbins". (*Public Domain*)

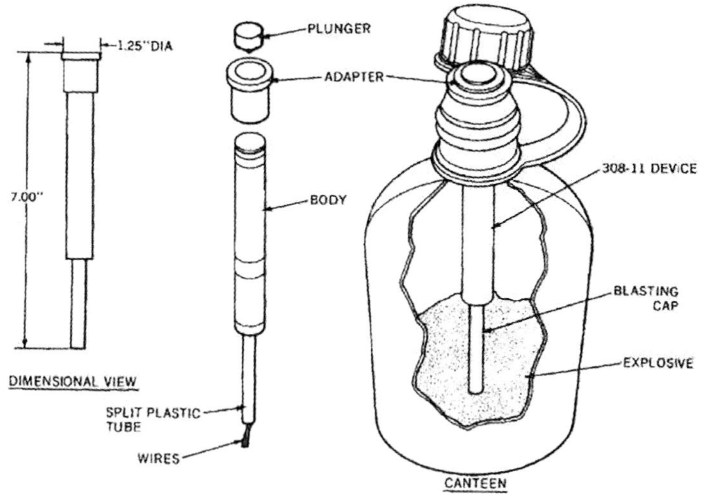

Figure 105. Canteen Booby-trap used by SOG. Note the void within the device to deceptively contain water. (*Public Domain*)

meters into the air. The M43 fragmentation warhead detonates [- projecting 600 fragments into the casualty area]. The mine self-destructs after four hours, or after the battery reaches a certain level [of decay]. If it fails to detonate, the battery will discharge over approximately 14 days, rendering the mine inactive.'[14]

- Enhance your speed of employment, reaction time and ToT with pre-configured mines/booby-traps and demo charges (e.g. EFPs, Shaped Charges), rather than rigging them during a mission.
- Planting of booby-trapped devices (e.g. ammunition, canteen, flashlight, etc.) or even propaganda items is something of an art. Placement of these items must not look too obvious, or as though they were deliberately left by the Team. This means that (1) ideally, the T/L must be relatively certain that the Team is not being tracked – which may take a few days to establish; (2) that the Team ensures that signs of Team passage across danger areas are eradicated; and (3) that the articles are left in the vicinity of enemy troop units or enemy installations or along enemy thoroughfares.
- An enemy soldier will almost always avoid mud or a puddle of water by walking around it; this is where the Team should plant a mine.
- During administrative or tactical movement along a road or trail, the enemy's attention is primarily focused to the front; secondarily to the flanks; rear security is generally allocated to one man (the tail-gunner), if at all; and almost no observation above shoulder level (when the enemy possesses air superiority). These tendencies should be considered in device placement. Examples:
  ○ Position a directional mine/booby-trap behind a tree, a few meters off the linear target area, facing at an angle toward the long axis of the target.
  ○ A device can be placed (along a trail/road, back-trail) up in a tree or attached to a pole to burst overhead.
    ▪ Any Device:
      * Place an elevated device in a position where enemy combatants are unlikely to cast their gaze, and where there is adequate vegetation to conceal the device or unless the enemy's view is otherwise impaired. Vision can be impaired if the enemy is

---

14. 'M86 Pursuit Deterrent Munition', https://en.wikipedia.org/wiki/M86_Pursuit_Deterrent_Munition

entering an area under shadow/heavy canopy from an area of daylight, if the enemy is entering bright sunlight from under heavy canopy, if the enemy is facing directly into the sun, during conditions of fog or mist or if CS or obscurants are being used.

* Use special care to avoid leaving telltale disturbed earth/vegetation or leaving a trail through morning dew, mud or snow when placing the device.

* In snowbound conditions place the tripwire on a down-slope. Enemy trackers, especially ski-troops, will have momentum working against them and will not be able to avoid initiating the device.

- Device with Delayed Function.
    * Place device where ambient noise will obscure any telltale sounds of device functioning (spoon release) and where the enemy is channelized so that he cannot flee the kill zone.
    * Tripwires must not be routed where they may bind on intervening vegetation, etc. If tripwires must be routed through vegetation using a dog-leg between the device and the kill zone, consider using a <u>subdued</u> item (e.g. a safety pin) acting as a 'pulley'.

- Device with Instantaneous Function.
    * Connect an elevated device to the triggering function using subdued and well concealed detonating cord. Both ends of the detonating cord should be capped or a triple-rolled knot may be used as an alternative. A direct tripwire from an elevated device to a trigger may be too hazardous to emplace.
    * Use components of the M142 booby-trap kit for triggering or consider command detonation. Also consider a command detonated secondary device to kill other enemy personnel/elements converging on the site.
    * A Claymore can be attached to the trunk of a tree with subdued bungee cord or local materials (e.g. vines) and aimed downward at an angle using shims. It is best to place the device on the opposite side of the tree to the enemy's expected line of approach to avoid detection. Or the Claymore can be attached to a sapling/young tree or bamboo that can be made to lean and therefore aim the device toward the enemy line of approach. Leafy cover is advised.
    * The device can be placed high on a tree trunk by climbing on the backs of other Team Members. The tree should be climbed from the reverse to minimize telltale sign/debris.
    ◦ Devices should be prepared in advance for rapid deployment.
- Simple Grenade Device: An armed grenade, pulled from a can by a trip-wire, will allow the release of the spoon safety (once the grenade safety pin and clip are removed); the spherical M-67 grenade body does not readily lend itself to the can expedient, but foreign egg-shaped grenades will work well. Alternatively, a tripwire can be attached to a loosened grenade pin (where the grenade is anchored) or to tape that secures the grenade spoon, to similar effect. These are simple, effective devices that are easily and rapidly emplaced. They do have both limitations and compensatory techniques of employment:
    ◦ The telltale 'ping' of the spoon as it releases warns an alert enemy to take cover. The grenade fuse delay allows an alert, experienced enemy to run out of the kill zone. The noise can be suppressed by shortening/cutting down the spoon.
    ◦ Techniques of employment/mitigation:
        - Remove one leg of the safety pin (cotter pin) and straighten the other leg to ease the extraction from the grenade by tripwire. The grenade should be positioned and anchored to allow for a straight pull of the pin.
        - Conceal and/or position the tripwire where it is especially difficult to avoid by an enemy combatant.

- ▪ Employ the grenade where enemy movement out of the kill zone is channelized, impaired or limited.
- ▪ Employ the grenade where other noise (rushing water/cascade; firefight, vehicular movement, etc.) will obscure the sound of activation.
- ▪ A grenade can be concealed in a recess that is cut into a stem cell of bamboo.
- ▪ The grenade can be elevated by bending a sapling/small tree or bamboo and securing it, preferably hidden within leafy foliage.
- ▪ Always use a SD capability.

Tree Mine TTPs:

- Select a tall supple tree or bamboo stalk that is located at the verge of a road. Select the tree/stalk location for optimum camouflage and so enemy vehicle headlights will not be cast on the item.
- Bend this tree or stalk backwards to lash a demolition device or captured munition (e.g. 120mm HE mortar round; 122mm HE artillery projectile, etc.) to the top of the tree/stalk or overhanging limb. Attach a long line (e.g. 550-cord) near the top of the tree/stalk to raise or bend the tree/stalk when necessary.
- If the munition is fused, remove the fuse and stuff the fuse well with C-4 and prepare it for blasting cap initiation. If the fuse cannot be removed, use a securely fastened external C-4 countercharge for initiation. Command/remote initiation can be accomplished in various ways.
- At night, position the tree/stalk; using the line; ensure that the munition is suspended over the road.
- Remotely detonate the mine/device above a target to achieve an air burst.

Enemy Stocks TTPs

- Teams may encounter unguarded or abandoned stocks in the target area. For instance:
  - Unguarded:
    - ▪ Ammunition consumption for some weapons systems may be substantial; for instance, Russian military doctrine/tendencies (and that of client states), favors saturation concentrations by artillery. This suggests that their Ammunition Storage Points (ASPs) may occupy considerable acreage that would be difficult to guard effectively.
    - ▪ Stocks may be established at pre-position sites and uploaded onto cargo vehicles/trailers in anticipation of flexible operations or to tactically disperse stocks. These stocks may be containerized or in break-bulk configuration and may be lightly guarded.
  - Abandoned:
    - ▪ Ammunition stocks that have been attacked by friendly forces may leave large quantities of ammunition scattered about by explosions. Be careful to take only munitions that are unscorched and evidently undamaged. Internally fused projected munitions (e.g. rockets, missiles, AT ammunition) should be avoided, as fusing may have been armed or sensitized during an explosion/fire.
    - ▪ Stocks may have been abandoned by friendly forces when enemy offensive operations or an invasion captures territory where the stocks were located.
    - ▪ Stocks in pre-position sites may have been temporarily abandoned due to OPTEMPO.
  - The Team should consider the following uses for these stocks.
    - ▪ Destruction to prevent use by the enemy or marking the location for destruction by friendly air support.
    - ▪ Contamination (vice destruction) of POL to incapacitate any enemy vehicles that may use it. Would require the Team to use a field expedient contaminant.

- ▪ Ignite the fuel and ambush response.
- ▪ Booby-trap stocks, especially so that the stocks (e.g. ammunition, POL) detonate during or subsequent to movement.
- ▪ Imbed sensors to track the stocks as they are moved to other locations or using units.
- ▪ Use ammunition as IEDs and demolitions. Use captured enemy munitions/projectiles for use as cratering charges or as booby-trap main charges.
- ○ Teams in the field, guerrilla allies (or their auxiliaries/support personnel), can fabricate covert explosive devices that resemble common, benign objects found in the AO. Cast explosives can be melted/steamed out of captured enemy ordnance and poured into molds associated with the physical form of the devices; and dyes can be mixed in or paint applied externally to provide coloration/camouflage. This was done extensively by the British SOE and US OSS during WWII. Detonators can be imbedded into the device and concealed from detection. Example: cast explosive with an imbedded pressure detonator, made to resemble animal dung and used in rural areas where such deposits on rural roads were commonplace.

Author's Note: If this is to be done during a deployment, recommend that Team Engineer (MOS 18C) personnel work intimately with an EOD specialist in the melting/casting process. Melting or steaming out cast explosive is an old EOD technique (WWII) that is no longer in general use by the US EOD community, so while the EOD specialist is not trained to perform this task, he is well trained in associated techniques (e.g. including relevant hazards and precautions associated with explosives).

- ○ Ensure that the Team carries an adequate supply of time delays. Use of delay fusing on operations will minimize ToT and personnel exposure, delay/deter pursuit, provide separation time, allow attack of a broad variety of targets without undue exposure, enhance harassment and interdiction operations and provide a back-up/alternative means (e.g. to electrical or remote) to detonate devices.

EFP TTPs:

- If your mission is, or may include (as a supplementary/opportunistic task), road interdiction/ambush, consider taking/employing an IED (e.g. Explosively Formed Penetrator). Even a small EFP (e.g. 2-inch diameter), if properly constructed, aimed and initiated, can defeat armor vehicles and can destroy locomotives, ammunition, petroleum, electrical generation transformers, etc.
- Note that an enemy may place buffer cars in front of train locomotives to trip pressure activated demolitions, so a command-detonated off-route device, such as an EFP, would counter this ploy.

Mortar TTPs:

- Soft ground or snow can reduce the effectiveness of point detonating (PD) HE rounds by as much as 80 per cent.
- Vintage Soviet or CHICOM HE mortar projectiles, which have proliferated worldwide, will use dependable but terminally inferior PD fuses. In soft ground, many of the casing fragments may be directed upward in a 'V' dispersion pattern because a portion of the round casing becomes buried in the soil before detonation occurs. US mortar fuses use much more lethal PD 'Super-Quick' fuses, promoting a much flatter fragmentation pattern (e.g. lateral fragmentation dispersion). Dropping to the ground when receiving mortar rounds of older manufacture can save Team Member lives, especially in soft ground.

Use of a 60mm Mortar in SR Operations TTPs:

- Large SR Teams (twelve to fifteen personnel) can effectively employ the 60mm mortar in close engagements and Battle Drills.

Advantages and Disadvantages of the 60mm Mortar in Reconnaissance Operations	
Advantages	Disadvantages
<u>Characteristics:</u> The 60mm mortar is an area weapon; this characteristic is a plus when engaging an enemy that is spread out or in locations that cannot be precisely determined. Additionally, the mortar is a high-angle attack weapon that can fire on enemy elements in defilade or behind cover, areas that can't be addressed with direct-fire Team weapons.	<u>Characteristics:</u> The 60mm mortar is an area weapon, not a precision weapon. It has less utility in striking point targets, especially at longer ranges.
The standard US M224 60mm mortar has a range of 3,490 meters. Handheld 60mm mortars, such as the Austrian Hirtenberger M6C-210 60mm 'Commando' mortar, will fire at a reduced range (1,600m) largely due to safety/stability limitations or when firing with a reduced number of propellant increments. But these ranges exceed that of any <u>standard</u> Team weapon.	The maximum effective range is further limited to the target observation capabilities of the Team. Where terrain and vegetation inhibit target observation, the utility of the mortar is diminished accordingly.
The 60mm mortar fires an array of ammunition types, including: High Explosive (HE), Illumination, Smoke (WP), and Infrared Illumination (visible with night-vision goggles).	60mm HE mortar rounds each weigh approximately 3.65lbs. To carry an adequate combat load of mortar rounds requires the dedicated load-carrying capability of two Team Members; other Team Members must often carry the mortar team's rations, additional water, sleeping gear, etc.
Fusing for current US 60mm mortar HE rounds is the M734 Multi-option fuse (featuring proximity burst, near-surface burst, impact burst, or delay burst options) or the M525 Point Detonating (PD) Fuse.	
Rate of Fire:[15] up to 20 Rounds Per Minute (RPM) sustained; 30 RPM in exceptional circumstances and for short periods.	Because the mortarman must take extraordinary care in firing through multiple canopies, he might not attain even a standard sustained rate of fire.
Average Lethal Area[16] for HE Rounds: - PD: 16m (standing enemy in open terrain); 10m (prone). - Proximity: 25m (standing); 14m (prone)	Once the mortar rounds are expended, the mortarman will only be armed thereafter with a sidearm or shotgun; his combat capability therefore diminished.

---

15. Firing rate will vary based on the weapon used and other factors.
16. Probability of Kill (Pk) is estimated on the likelihood of 50 per cent of enemy troops within the fragment/burst area will be killed by the projectile cartridge; number of wounded will exceed Pk..

Advantages	Disadvantages
<u>Close Combat Utility (HE rounds):</u> While doctrinal safety requirements mandate a minimum range of 70m; HE PD rounds can be brought to burst much closer to the Team, especially if terrain form offers some degree of defilade for the Team.	<u>Close Combat Utility (HE rounds):</u> The proximity or near-surface burst fusing option is not recommended in heavy canopy. The rounds will be ineffective, as they will detonate above tree top level. But in the absence of heavy forest cover, this fuse option can be effective in the presence of snow or soft soil, where the PD round has degraded terminal effects.
If the mortarman carefully orients the tube to fire through 'thin' points in multi-layer canopy, avoiding large branches, the rounds will penetrate through small branches and leafy cover and will proceed through its flight trajectory in a generally normal fashion. The M734 Multi-Option Fuse on the M720 series 60mm HE projectile is designed to travel 100m in flight before arming and does not use a bore-riding safety pin – a safety device used in older fusing.	Heavy canopy, may cause some (not all) PD-fused HE rounds to detonate in the branches of trees. If the enemy is closely engaged with the Team, a low airburst detonation might wound friendly troops. Delay fusing will allow penetration of the canopy prior to detonation. But here again, if the enemy is closely engaged with the Team, a low detonation could inflict friendly casualties.
<u>Close Combat Utility (Smoke Rounds):</u> Smoke rounds (WP) are very effective for marking enemy locations for friendly combat aviation assets. Smoke rounds can be brought to burst much closer to the Team than HE rounds and can inflict serious wounds on enemy personnel. Smoke rounds can be used to screen Team movements from enemy observation and fires.	<u>Close Combat Utility (Smoke Rounds):</u> The casualty producing burst of a smoke round is inferior to that of HE.

Mortar Employment TTPs:

- The addition of a mortar to the Team's firepower is advantageous for various engagements.
  - In chance encounters, Team survival is often dependent on attaining and maintaining fire superiority over an enemy unit during firefights. Mortar fire, in conjunction with the array of other Team weapons, may temporarily overmatch the fires of a larger enemy unit and may convince the enemy that they are in contact with a much larger force. Such massed fires may substantially suppress enemy fires and may offer better circumstances for the Team to break contact or maneuver.
  - A mortar may be the only way to bring effective fire on an enemy element located on higher terrain or defilade.
  - In deliberate fires, such as in ambushes, raids or defensive fires, the mortar can inflict substantial casualties on enemy personnel and can damage or destroy materiel and fixed facilities.
  - The support of combat aviation assets is not always timely or even available at all in deep penetrations; therefore, the Team must rely on its organic weapons for survival until these assets arrive.
- If mortar rounds are preset for airburst or PD, ensure that the containers and the rounds themselves are marked so that they can be identified by touch. If the rounds are not preset, fuse markings must be made identifiable in the dark.
- In jungle/rainforest or other limited range environments, mortar rounds are generally pre-configured and fired at 'Zero' charge, to provide direct fire support for chance

engagements at close range; as there is little time to 'cut' propellant charges from the rounds in close combat. Spare propellant charges should be carried separately for longer range engagements. In more open terrain, rounds can be configured with propellant charges attached as more suitable for longer ranges.

- Teams should prefer the use of a lightweight mortar tube with a lightweight base plate designed for hand-held fire. The bipod, sight and standard base plate are generally not carried or used (except in raids). The use of the 60mm hand-held mortar is appropriate when short-range 'line-of-sight' to the target is the norm or when the target is near enough for rapid range estimation. The 60mm mortar may be trigger or drop-fired.

Figure 106. Austrian Hirtenberger M6C-310 60mm "Commando" mortar. (*Public Domain*)

- If the mortarman is an indigenous Team Member, his fires should be closely directed by a US Team Member.

- <u>Extensive</u> training is essential. The mortarman must learn to rapidly identify 'holes' (the absence of large limbs) in overhead canopy, to position the tube under the holes and to angle fire through light to moderate canopy with confidence. The mortarman must learn to estimate tube elevation to bring rounds on target with reasonable accuracy, to adjust fire and to strike close targets with confidence. When the Team enters its NDP (or long halt) location, the mortarman <u>must</u> select his position with an eye to minimizing overhead canopy cover. Note that risk is substantially increased if the mortar is to be used at night under multiple canopies. The T/L should consider the positioning of the mortar element in relation to canopy cover when the Team moves into the NDP location and should personally check mortar positioning. Mortar training is also necessary to build Team confidence in the mortarman's abilities and to integrate mortar fire with Team Battle Drills.

- If Team personnel are behind/within cover, HE rounds can be used in close engagements; otherwise, WP rounds should be used. WP rounds not only are casualty producing, but they provide screening smoke to conceal Team movement.

- The mortar can be effectively used during an assault, flanking movements and other Battle Drills/maneuvers. HE mortar rounds, at zero charge, have even been fired from the hip

Figure 107. J. Walker, Team Leader of RT California. Firing 60mm mortar from standing position; before and after (*Walker*)

during break-contact Battle Drill although this is not recommended. Note the before and after photos above.

- <u>Speed of employment is essential</u>. If the mortarman cannot bring his mortar to bear in time to support the Team during a firefight, the utility of the mortar is substantially reduced. The mortarman might carry four HE rounds, or a mix of HE and WP, with safety pins removed, in canteen covers/bags on the front of his LBE. Older series HE PD fuses are set-back armed with a bore-riding safety pin as an additional safety measure; newer mortar rounds, with advanced fusing, have multiple Safe & Arm features. Rounds must be padded to ensure that they do not knock together during Team Member movement. Immediate action rounds are set to 'zero charge', using only the ignition cartridge for propellant.
- The mortarman should wear a tailor-made rucksack with VELCO®-strip openings and with approximately 10–12 mortar rounds (a mix of WP and HE) carried nose-down in padded interior pouches. External rucksack pouches, that are long enough to encompass mortar round length, might also be used. Some fuse safety pins may/may not be removed at the T/L's discretion; rounds might be set to 'zero charge' in close vegetation/terrain environments. The mortarman carries extra propellant charges separately; these can be reaffixed to the rounds for longer range engagements. The mortar tube is carried on a sling; a bore cover should be attached. Due to the weight of his load, the mortarman will typically carry only a pistol or cut-down shotgun as an individual weapon.
- The 'assistant' mortarman might also carry a tailor-made rucksack with VELCO® openings but he will carry fewer mortar rounds (e.g. 8 x HE). The assistant mortarman may have an additional burden carrying the mortarman's rations, additional water, etc.; he may also have other Team duties. Safety pins may remain inserted in the fuses carried by the 'assistant'. Again, all rounds are set to 'zero charge' if the operations are to be conducted in heavy vegetation and/or close terrain. In an engagement, the assistant mortarman's primary duties are to support and protect the mortarman; he will position himself behind the mortarman, 'rip' open the VELCRO® closures on the mortarman's rucksack, and feed rounds to the mortarman. Other Team Members might carry one mortar round apiece in their rucksacks.
- PVC pipes or munition containers/packaging can be used as field expedient pipe-mortars that can be useful in a variety of scenarios. Use a pipe-mortar to:
  - Cast an improvised line and grapnel (e.g. in mountainous terrain, for stream crossing, etc.).
  - Cast grenades or mortar rounds.
  - Create a barrage effect, with grenades, mortar rounds or rockets launched from improvised mortars, to attack by fire enemy installations, unit encampments, dismounted combatants, etc.
  - Pipe-mortars should be for one-time-use only. A PVC pipe-mortar can be reinforced with wraps of filament tape where propellant combustion occurs.
  - Pipe and munition items should be stored in an MSS cache and brought forward to an objective staging area as required.

Personnel TTPs:

US Personnel Selection:

General Personnel TTPs:

- Remember that friendly casualties may have to be evacuated and replacements delivered in long duration/deep penetration operations. Additionally, specialized attachments/augmentations may also have to be delivered and exfiltrated. Plan for this.

Mental and Character Attributes of SR Personnel:

- US SR personnel should be:
  ○ Volunteers only.
  ○ Of the highest quality possessing the following traits: aggressiveness, risk acceptance, common sense, decisiveness, fitness, mental stability and intelligence.
  ○ Personnel who have successfully completed SpecOps screening and qualification training. Other preferred training and/or experience might include: Reconnaissance, Combat Tracker, Ranger, combat veteran, intelligence coursework.
- SR personnel will often be placed in extremely high risk, high stress situations throughout the duration of each mission. The stresses of being hunted like an animal makes SR missions a particularly severe psychological test. SR Teams will often be pitted against concentrations of enemy combatants who possess significant advantages in fire superiority, mass, greater mobility, a mobilized population and other advantages, while the Team has very limited responsive support (if any) and a diminished possibility of rescue or timely exfiltration. Potential Team Members must be made fully aware of the high risks associated with such an assignment and the challenges of membership in such an elite group. Team Members should also be accorded the opportunity to withdraw from such an assignment, with no career taint or recriminations.
- SpecOps leadership should be made aware of the unique risks and psychological demands of SR operations and should minimize administrative frustrations especially during Team down-time. Staff and support functions should go above-and-beyond in easing the burden on SR personnel. Team Members should also be given special dispensation on minor infractions of discipline.
- SR leadership should be specially attuned to aberrant behavior exhibited by Team Members brought on by operational pressures. When this behavior portends a threat to the Team or to operational success, the leadership must act – but not overreact – to solve the problem. This may be as benign as sending the Team on less hazardous missions, to providing additional down-time; or something more, such as a medical referral and/or reassignment of the Team Members. These latter interventions ideally must be timely so that the individual may be returned to an SR assignment in the future – as the individual's training and experience would remain highly valued.

Physical Attributes of SR Personnel:

- Ideally, US SR candidates should be of medium stature. A tall or heavily built person may be a significant liability to the Team should he become a casualty.
- SR personnel must be especially fit and healthy.
- Ideally, SR personnel will not have any impairment of the senses, especially those of sight, hearing and smell. Of these senses, eyesight is the most important; the Team Member must have vision acuity and very good depth perception. If the Team Member lacks adequate good vision, he may have difficulty in estimating range/distance, and more prone to stumble during movement. If possible, the T/L should have prospective Team Members receive vision testing for acuity and depth perception during a selection process.

Indigenous Team Members TTPs:

- If indigenous personnel are integrated into the Team, consider maintaining one or two overhead positions to replace indigenous Team Members who cannot be available for a mission for such reasons as leave, illness/injury/recovery from wounds, etc.

- Avoid recruiting members of the criminal class for SR Teams. This was done, to some degree, in SOG, with bad results. These personnel typically have no love for authority, no moral compass and their orientation is often for personal gain alone.
- The weeding-out process of weak indigenous SR Team recruits must be resolute during selection (with emphasis on security background checks) and initial qualification training processes. Host nation security services should be informed of any candidates who fail the selection process for cause (especially background checks). Trainees who fail SR coursework or physical criteria, should not be alienated or stigmatized, but might be reallocated for EF/RF recruitment or be offered to the host nation for paramilitary force recruitment.
- A T/L may initially accept a graduate of the training process, and then ultimately decide to 'fire' the new Team Member for any reason during the evaluation period. Dismissal action may recommend elimination from the organization entirely, or may recommend elimination from just the Team. The T/L should state the reason for the firing action for any indigenous Team Member to the S-1 (indigenous personnel) so that another Team or FOB unit does not inherit a serious personnel problem. Serious offenses may be referred to host nation authorities; security protocols should be established for these circumstances.
- The indigenous training process, as in all elite force training programs, should reinforce confidence, power, aggressiveness, and fearlessness.
- Consider the use of 3rd country indigenous commandos as Team Members.
  - They may be already trained, and may be more warlike than other indigenous options. They may have superior skills (e.g. tracking, fieldcraft).
  - A separate language and separate origin may make them more reliable/less vulnerable to enemy attempts to recruit them as agents.
  - Interpreters must be competent.
  - They may be more costly than local indigenous options.
  - They may require accommodations separate from other indigenous commandos.

Presumptive Indigenous Training Cycle:

- Basic: 6 weeks
- Reconnaissance: 5 weeks
- Additional:
  - Position Training: Crew-served weapons, Tail-Gunner, etc.
  - NVG-weapons integration and night movement and maneuvers.
  - Team Level training.
- What to do with indigenous recruits who have failed training or who are no longer field deployable (e.g. due to disabling wounds).
  - Retrain and retain (if possible), for the RF, for FOB security functions or administrative positions.
  - If the persons cannot be placed in support of the SR or FOB, attempt to find a position for them in sister units or in civilian positions with governmental agencies. For instance: governmental security forces or intelligence agencies. Personnel so placed can be valuable sources of information.

Carrying Wounded While Being Pursued/Tracked TTPs:

- During August of 1944, the Germans were preparing to evacuate northern Finland. Operations in this area were arduous even in summer, due to terrain and very limited road networks. 'It took hours to evacuate casualties to the nearest first aid station, and as many as 12 men were needed to escort 1 casualty – 4 litter bearers, 4 relief bearers acting as guards, and four others to carry rations, equipment, and other baggage.'[17] By this example, it is easy to conclude that a Team cannot carry a casualty very far.

- If the Team is in a permissive environment, with friendly air superiority, and where air or water evacuation is possible, some options are:
  - ○ Move to the nearest secure LZ.
  - ○ In the absence of a LZ, conduct extraction through light canopy using 'strings' or ladders.
  - ○ Move to a defensive position and await rescue (by RF, Bright Light Team, SAR, etc.)
  - ○ Move to a MSS where medical (and other) resources are available.
  - ○ Hide/abandon the wounded Team Member's gear to reduce the Team burden.
  - ○ Destroy the gear if recovery will not be feasible; destruction can be effected by a time-delayed explosive/incendiary device, by depositing the gear in a body of water, tossing it down a mountainside or other field expedient means.
- If the Team is in a deep penetration environment, where immediate exfiltration of the WIA is not possible, some points to consider are:
  - ○ The T/L must quickly make difficult decisions: (1) to carry the WIA Team Member, while risking the survival of the rest of the Team, (2) to cache the WIA Team Member for later rescue/retrieval, and (3) whether to leave a Team Member (medic) behind to provide treatment to the WIA.
    - ■ If the T/L decides to carry the WIA, the Team may have to hide/abandon the WIA Team Member(s) gear, rather than bear the additional burden. Remember that some gear has intelligence and operational value.
    - ■ The T/L may have to cache the wounded Team Member(s), leaving behind a Team Medic, depending on severity of the wound(s) and the particulars of the pursuit. Caching of wounded allows Team freedom of action. If this option is to be taken, the decision should be made as early as possible while action is timely and before exhaustion of remaining Team Members sets in. Careful camouflage of the turn-off and trail to the cache site is essential. Note: The Team may only have one medic. The Team cannot afford to leave its sole medic behind, especially with the possibility of additional casualties, so another Team Member might serve. Take GPS coordinates before Team departure.

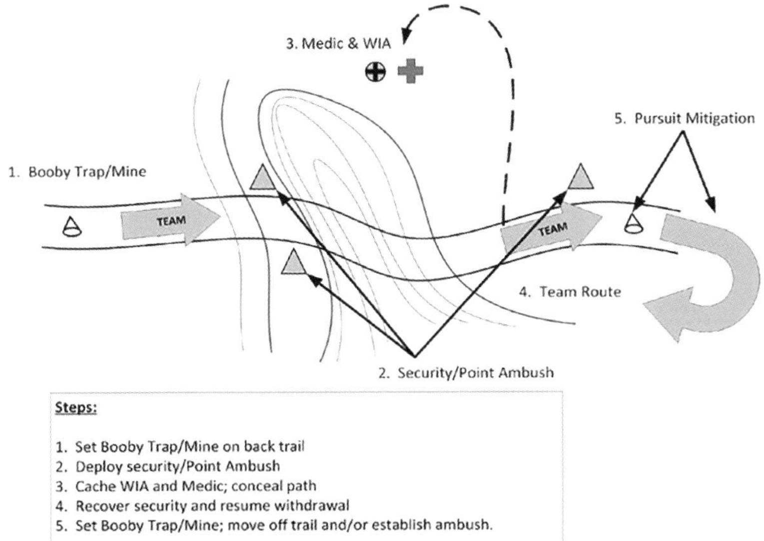

Figure 108. Caching of Team WIA.

---

17. Halder, *Small Unit Actions*, p. 199

- Enemy combatants, especially paramilitaries, insurgents and combatants of nations that are not signatory to international conventions on the treatment of POWs, enemy wounded, etc. may torture and kill SpecOps captives. This is even more likely if the Team Members are not themselves compliant with international conventions (Geneva Conventions; Rules of Land Warfare, etc). It would be commonly known which enemy militaries/paramilitaries allow their combatants to commit torture or other atrocities.

On 14 June 1967, 'the 24th NVA Regiment annihilated a CIDG patrol led by two US Special Forces advisors near Dak To, Vietnam. The bodies of the two Americans were discovered, mutilated. They were cut from groin to head with internal organs exposed. One person who saw the bodies said that the bodies looked like a page from an anatomy book.'[18]

- The gear of the cached WIA Team Member should be left with the individual as it contains food, sleeping items, additional clothing, survival items and individual weapons. The Team Member WIA/KIA cache coordinates must be recorded and transmitted to higher headquarters as soon as possible, for subsequent rescue/recovery action by a Bright Light/RF element.
- If the Team Member is not seriously wounded, but is a mobility burden that risks the survival of Team, he may be left with sterilized land navigation items (map, compass, etc.). If the Team has not returned within a reasonable time (by SOP) to recover the WIA Team Member, he may be able to make his way independently to a RP, MSS, LZ or SERE rendezvous.
- A severely wounded Team Member may be expected to vocalize his pain. The medic must be ready to administer timely medication or other measures to avoid being detected by passing enemy combatants. While ensuring that the WIA Team Member is allowed to breathe freely, muffle his moaning under layers of clothing. If the WIA Team Member cannot be silenced, the medic should assume an overwatch position and then treat the Team Member intermittently.
- The remainder of the Team will lead the pursuers/trackers, away from the cached Team Member and inflict casualties on them if possible. The Team will then return to recover the wounded (and the caregiver), if possible.
  - If a wounded Team Member dies while under care, the medic may move to a designated rendezvous with the Team, after burying/concealing the remains. [See TTPs on Graves Registration.]
- The T/L must try to avoid navigating steep terrain to mitigate the burden on litter-bearers – unless the Team is to move into defensive positions on advantageous terrain or to an LZ.
- When transporting or caching a wounded Team Member, the Team will leave more trace than usual. Other Team Members (especially the tail-gunner) must be diligent in obliterating/obscuring the Team signs.
  - Pad the sling/stretcher with a poncho liner/blanket or place the wounded Team Member within a sleeping/body bag to prevent leaving a blood trail. Suppress a blood trail during movement.
  - The T/L should consider briefly using the existing trail network to get separation from imminent pursuit.
  - The attention of pursuers or trackers is normally to the front, along the trail of the Team, with less attention to the flanks. Do not plant booby-traps/mines in the vicinity of a cached Team Member, unless the Team needs to create time/space from close pursuit.

---

18. Stephen Sherman, ed., *Indochina In The Year Of The Goat – 1967*, Radix Press, Houston, TX, 2016, p. 125

Detonating a booby-trap/mine will cause the enemy to deploy and to observe to the flanks – increasing the possibility of discovery.

- ◦ If the enemy detects a severely wounded Team Member and the Medic in the cache, the Medic should engage the enemy with a Claymore and/or individual weapons and then evade to rejoin the Team – with the expectation that the enemy will provide medical care per international protocols (if this even applies).
- In deep penetration operations, and where a UW environment exists, the Team may be able to solicit the assistance of guerillas/partisan in recovering WIA or even KIA Team personnel.

Graves Registration TTPs:

During operations at strategic depths, where air exfiltration is delayed or not possible, the Team may have to 'cache' or bury a friendly KIA.

- Cache or bury the remains in a location where the enemy is unlikely to find them and where local civilians will normally avoid.
- Time permitting, bury the remains deep enough to suppress decomposition odor that may attract animals. Treat the grave with CS or capsicum powder to deter animals from disinterring the corpse.
- Ensure something metallic is buried with the KIA so that a recovery team can use a metal detector to more easily find the grave site.
- Use GPS, if possible, to record the grave location. If GPS function is not possible, record 10-digit coordinates.
- Leave tripwire strands and other false signs of booby-trapping, near the grave to dissuade the enemy from disinterring the remains. Ensure these false signs are highlighted in the AAR and graves registration reports.
- Leave tell-tales (by SOP) to reveal if the enemy has tampered with the grave or the body.
- Leave markings (by SOP) that a recovery team can use to trace the location of the grave. Record path taken from the grave to a stable land navigation feature and record coordinates (preferably with a GPS) of waypoints where doglegs or course deviations were taken, so that a recovery team may find the remains.

Enemy POW TTPs:

- Do not maltreat POWs. If more than one POW is captured, ensure that they are segregated and not permitted to communicate with each other, from capture through exfiltration through evacuation to higher headquarters. Security and exfiltration TTPs are found elsewhere in this book.
- Intelligence from POWs: POWs will be questioned (post-mission) by professional interrogators who will be seeking answers to PIRs established by higher headquarters; this information will not be timely or of immediate utility to the SR Team such as: trail and road signs, mined/booby-trapped locations, frequency of patrols, presence and location of counter-recon or SOF units, etc. The SR Team should seek relevant information of immediate utility once the POW has been moved to an RP or more secure site. Team O&I (18F) personnel should carry a questionnaire (notebook or IT device) to solicit this information in an effective manner. The FOB S2 should aggressively seek feedback from higher level prisoner interrogations.
- If possible, the T/L/Members should be present during interrogations or should review interrogation reports, to provide context and operational perspective/analysis.
- See POW medical intelligence below.

Medical TTPs:

- Special Forces Medics (18D) are always a shortage MOS. Therefore, an SFOD SR Team should not expect to have two medics assigned.
- The Team medic(s) should tailor medical kit components to the AO and the mission. Remember that ointments and liquids are subject to expansion and contraction based on temperature and altitude (e.g. airborne operations), so add a layer of packaging (e.g. plastic baggies); also add silica gel to pill bottles or condensation will ruin the meds. Ensure medical kit item labels are indelible, easily read and outward facing.
- Medical kit items may feel identical in the dark. In the chaos of combat, it is not enough to memorize the location of medical kit items; tag these items with tactile and luminous markings so that they can be found in the dark. Tagging items are available commercially.
- SR Teams rarely had a medic assigned; when a Team did have a medic, it was on a temporary/mission strap-hanger basis, which was not often done either. Absent a medic, the T/L was issued a small 'One-Zero Kit', consisting of a bare minimum of medical supplies. If a current-day SR Team will not have a medic assigned, then the T/L should consult with a senior FOB medic on the Team's own version of the SOG 'One-Zero Kit' and what contents should be included in it. Further, the T/L, and designated others on the Team, should receive instruction from an FOB medic on the use of the Kit.
- A medical check should be conducted on all Team personnel, to include indigenous Team Members, after return from a long duration mission, to assure they are fit for the next mission. If parasites are endemic to the area, and Team Member infestation is real, confine all Team Members to the FOB for several days prior to launch and ensure they are issued Class 'A' or field rations to ensure they do not acquire intestinal parasites subsequent to a pre-mission medical exam or prior to the mission. The T/L should not hesitate to involve the Team/FOB Medic if there is any question about a Team Member's medical fitness to undertake a mission.
- Battle dressings should be carried on the LBE at a standardized position (by SOP) so that Team Members can access the dressing even in the dark.
- It is recommended that some Team Members carry a one-liter container of IV fluid stowed in their rucksacks along with a starter set, line and catheter/needle. In freezing conditions, this fluid bag may have to be carried in a clothing pocket. Training on IV use, especially pertaining to personnel in shock, with lowered blood pressure, collapsing veins, etc. should be mandatory, especially for Teams deploying without a medic.
- Coughing is difficult to suppress. In the Author's experience, the best cough expectorant/suppressant was Turpin Hydrate (aka GI Gin). This medicine is no longer in the inventory; FOB medical support should request its acquisition or provide a medication of equal efficacy.
- Dextroamphetamines were standard issue in the 'One-Zero Kit'. This controlled substance allowed Team Members to stay awake for long periods of time. Most T/Ls shunned their use, as they sometimes produced hallucinations. The Author recommends the 'Rule of Three' instead (described elsewhere in this book).
- Chances are high that a Team Member will come down with diarrhea during an operation; a bad situation in an OP/LP or Team surveillance/hide. An anti-diarrhea medication must be included in the 'One-Zero Kit'.
- A rush of adrenaline will constrict blood vessels. Subsequently, a Team Member may not initially realize that he has been wounded. Once the adrenaline rush is over, and blood vessels are no longer constricted, more rapid blood loss may ensue. Team Medic(s)/Leaders must understand these effects and verify the condition of Team Members after a firefight.
- Team Members, especially the Team Medic, should include a MedEvac message format in their notebooks/'Quarterback's Wristband'. This document should also contain vital Team

Member information, to include blood type, significant allergies, etc. See message format at Appendix D.

- A 'chase medic' should accompany infiltration and exfiltration operations, e.g. in a chase helicopter, his duties may include:
  - Providing treatment for Team Members (and aircraft crew) if an aircraft is shot down.
  - Providing treatment for WIA Team Members and/or prisoners during extractions.
  - Accompany the Bright Light Team during rescue/recovery operations.
  - Assisting the Team in securing WIA prisoners during extractions.
  - Carrying and providing a spare radio to a Team on the ground.
- When evacuating a casualty or a WIA POW during a string extraction, consider the following advice:
  - Location and severity of the wound may govern whether the casualty may be extracted by string at all and whether other extraction means/methods must be sought. Little, if any, medical aid can be rendered to a casualty extracted by string/ladder, even by a SF medic suspended next to the wounded party.
  - Can the casualty withstand a short duration flight via string? If so, will the extraction helicopter land on an opportunistic LZ in enemy territory to permit loading the casualty into the aircraft cargo bay? Planning for such a MedEvac alternative should be planned, with prospective (lower threat) LZs identified.
  - Medical Intelligence is important. The POW should receive medical treatment for any wounds/injuries before transferring the prisoner to higher authority; treatment should include a reasonably thorough exam. Screening for parasites, infectious disease, nutritional status, scars/tattoos/identifying marks may reveal much about both the prisoner and the condition of enemy combatants in general. This information must be passed on up the chain with the POW.
  - Author recommends the Hanson Rig for string extraction of a WIA Team Member or POW. This substantially enhances security of the patient. The Hanson Rig is described elsewhere in this book.
- Medical Operations at strategic depths suggest that timely MedEvac/exfiltration is often not possible. That is why SpecOps (SFOD) Teams have highly skilled medics who are capable of battlefield surgery.
  - If a Team Member is wounded or severely injured, then the Team Member must be evacuated to a nearby location (a RP) where more effective treatment by the Team medics may begin. The tactical situation will often not permit more than a few minutes in this location, as the enemy may attempt pursuit, maneuver to cut off the Team, summon air assets or artillery support, etc.
  - It is important for remaining Team Members to buy time for the medic and the patient, by deploying booby-traps/mines, by sniping and by conducting point ambushes to slow/deter pursuit.
  - If the Team is planning a deliberate ambush or raid, the possibility of friendly casualties increases, so the Team should prepare a litter or travois in advance, to evacuate a prospective wounded comrade/captive. Litters are discussed elsewhere in this book. The Author recommends that Team Members carry both a Hanson Rig and a 12ft length of rope. These items can be used for travois harnesses/litter slings and various other field expedient purposes. If these items are to be used for load bearing of a casualty, a shoulder pad or field expedient should be used by porters.
  - The likelihood of pursuit requires that the Team medics deliver treatment in increments <u>while moving</u>, until the Team can reach a hide or defensive location. If a cache or MSS

is within a reasonable distance and it contains necessary medical supplies the Team may have to endure a longer march carrying the wounded/injured Team Member.

- ◦ If the Team is operating in an UW environment, friendly guerillas/partisans and/or their auxiliaries may be able to assist with transportation and medical care. US Team Members must have the means to contact these assets in deep operations.

- ◦ The Team and Team medics must train in the field for these realities. During field exercises, lane graders/instructors must put pressure on the Team and medics to treat casualties on-the-move in brief increments. Team medics should arrange their kits and LBE for rapid access for on-the-move incremental treatments.

- ◦ <u>Train for treatment and evacuation of wounded Team Members in all Battle Drills and other tactical engagement scenarios.</u> Medics must train the other (non-medical specialties) Team Members for on-the-move incremental treatments, especially as a casualty may potentially be the Team Medic.

- A standard location of morphine syrettes should be designated by SOP and proper use of morphine should be known by all US Team Members.

Medical Cross-Training TTPs:

Teams may not deploy with the full complement of Military Occupational Specialties (MOSs). Since medics are often in short supply, due to the lengthy training pipeline, needs of higher headquarters, and home station requirements – and they are called upon to perform other duties in the field, such as staffing the FOB clinic, chase-medic rotations, etc., SR Teams will be fortunate to have but one medic available. The depths at which Teams operate, the vicissitudes of weather and the perils of the operational situation mean that timely MedEvac or exfiltration may often not be possible; Teams must therefore have medics and/or medically cross-trained personnel available when they deploy on missions. The FOB commander must confront these realities by insisting that higher headquarters provide a tailored program of medical cross-training provided to non-medic Team Members – or ensuring that such training is provided from existing FOB medical assets.

Once medical cross-training is completed, the Author recommends that these Team personnel, when not committed to a mission, observe wound treatment in the FOB dispensary whenever possible.

Training of the Malayan Scouts (22nd SAS) during the Malayan Emergency included medical training. 'First aid training was important and helped everyone in deep jungle operations, including the aborigines. It was essential that every man in operations understand not only the basics of first aid to the injured, but also general health.'[19]

<u>True Account:</u> A SOG T/L received multiple, severe wounds accompanied by substantial blood loss during an operation in Southeastern Laos. The T/L was equipped with a standard 'One-Zero Kit' with medical supplies including several morphine syrettes. The Assistant T/L applied a tourniquet to staunch the bleeding from a leg wound, then retrieved the kit, selected a syrette at random, plunged the Syrette needle into the T/L's thigh and attempted to squeeze the morphine into the casualty; it didn't seem to extrude. He plucked a second syrette from the kit and attempted to administer morphine a second time, with a similar result. He did not attempt a third iteration, not knowing if any morphine had entered the T/L's system, and fearing an overdose. The Team was eventually exfiltrated, with the T/L in extreme pain

---

19. Shamsul Afkar bin Abd Rahman, *History of Special Operations Forces in Malaysia*, Calhoun: The Naval Postgraduate School Institutional Archive, Monterey, CA, June 2013, p. 43 [quoting [114] Director of Operations-Malaya, *The Conduct of Anti-Terrorist Operations*, Chap. XXIII, 3rd Ed. 1958, 1–7.]

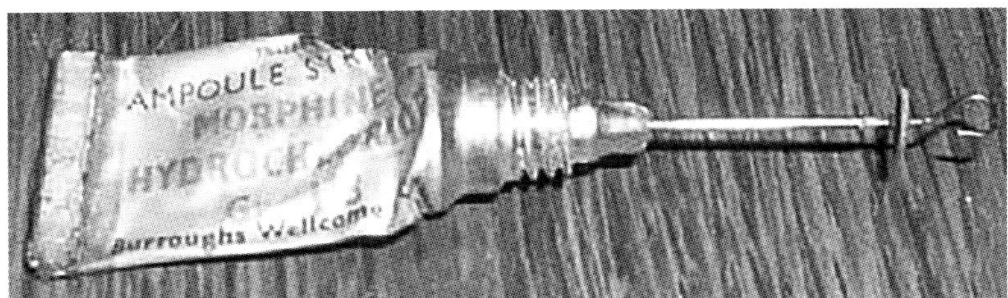

Figure 109. Used morphine syrette. Note rod used to perforate the syrette inserted into the needle channel. (*Public Domain*)

and with considerable blood loss. Upon his return to the FOB, the Assistant T/L turned in the One-Zero-Kit to the dispensary and complained of the ineffectiveness of the syrettes to one of the staff medics. The Assistant T/L then learned several things:

- The syrettes he had selected at random were lacking a looped rod (see illustration), to be used to penetrate the syrette seal; the medic pointed out that other syrettes in the kit were equipped with the rod. None of the US Team Members had been trained in the use and contraindications of morphine.
- He was advised that, had he injected the morphine, the T/L would probably have died due to vasodilation – increasing the casualty's blood loss.
- The Team was not equipped with a proper transfusion kit and most Team Members on other Teams had never been trained in how to use one. Only Teams with a medical Team Member bothered to carry such a kit.

Medical Preventative TTPs:

- Team medics should be familiar with endemic diseases of the AO and with preventative and treatment regimes.
- Some preventative medications (except personal use meds) may be dispensed by the Team medic(s) rather than distributed among individual Team Members. These meds may include anti-malarial tablets. The T/L or Team medic(s) must ensure that each man takes his daily malaria tablet(s) and/or other preventative medications at mealtimes.
- Ensure that Team Members are trained in natural hazards and precautions to be taken in the operational environment. These hazards and precautions would encompass endemic diseases; dangerous animals, invertebrates, and plants, etc. Hazards present in SOG's Southeast Asian Area of Operations included lethal snakes (e.g. cobra, krait, and bamboo viper), venomous millipedes, centipedes, scorpions and spiders.
- Venomous snakes will often be found lying on exposed rocks and hardtop roads to absorb retained heat from solar radiation. Be careful of hand placement during an ascent. Alternatively, in high temperature desert conditions, snakes and other creatures will take shelter from the sun and burrow into cracks, crevices, sand, etc. Be careful when stepping over deadfall during movement.
- In some jungle/rainforest environments and during certain seasons or weather conditions, the ground may become so infested with nocturnal animals and insects that Team Members may have to sleep elevated on platforms, hammocks or among upon tree branches.
- These creatures may invite themselves into the Team perimeter overnight; worse, snakes and scorpions, which become torpid in cooler temperatures, may be inclined to cozy up to a warm human – rolling over in your sleep could be painful … or fatal. When the Team Member awakes he should always remain still for several minutes while only moving his head as he takes stock of his surroundings. The good news is that cool morning temperatures will induce sufficient torpor so that a venomous snake or insect can more easily be killed with a

knife or a stick or it can be driven away by throwing twigs at it; in cold conditions, they go into hibernation. Always shake out sleeping gear (with caution) before stowing.
- Periodically check indigenous Team Members for lice or other infestations.
- In selecting ground for NDP or surveillance, consider choosing sunlit locations that are relatively free of pests (mosquitoes, leeches, etc.). Alternatively, enemy forces are likely to shun infested areas and concentrate on better ground, so pest infested locations should not be automatically discounted – particularly if the Team is adequately prepared.
- Layers of protection, which may be especially necessary in infested areas, include:
  - Skin insect repellant with a high percentage of DEET.
  - Clothing with permeable insect repellant (all clothing layers, from hat to socks).
  - Ground cloth and sleeping gear (e.g. blankets/poncho-liners).
  - Netting (personal and camouflage).
  - Insecticides and insect bait.
  - Avoidance/mitigation of attractants.
- Skin repellant, with a high percentage of DEET (N,N-Diethyl-3-methylbenzamide), applied to the skin and/or clothing works very well to deter insects.
  - Repellant must be reapplied periodically, especially in monsoon conditions, where heavy and constant rainfall will rinse the repellant off. Carry a sufficient quantity.
  - Skin repellant is typically applied only to exposed skin. But it may also be applied to cuffs, beltline, collar and trouser fly to help prevent insects from infiltrating clothing. This should be done prior to launch; regular application of skin repellant on clothing in the field will rapidly deplete the supply.
  - Insects may take refuge from cold weather within piles of fallen leaves and evergreen needles; warming temperatures or the warmth of a human body may arouse them from dormancy. Apply repellant even in cooler conditions prior to taking up a position in dead vegetation.
  - Chiggers are nearly microscopic (.4mm), enabling them to infiltrate through mosquito netting mesh, and certainly able to gain access through breaches in tactical clothing.
- Clothing permeated with permethrin will provide an additional barrier.
  - Tactical uniform shirts worn outside the trousers, offer a large gateway for insect access.
  - Undershirts will often come out from beltlines. Team Members should wear larger sizes or longer length impregnated undershirts to block insects.
  - Consider permeating clothing, ground cloth, etc. with permethrin acquired commercially. Permethrin is effective against mosquitoes, ticks, biting flies, mites/chiggers, leeches and many other insect pests. The application of this repellent is typically sufficient for six weeks or six washings; however, use during monsoon conditions may require routine pre-mission applications. Some clothing items may be purchased with permethrin impregnated for the lifetime of the article.

'Camp in high places, facing the sun.' Sun Tzu

In SOG missions in Laos, a Team could set up on the side of a ridge to perform a road-watch mission where conditions were moist and where the ground would be swarming with hundreds of leeches, and the air swarming with mosquitoes, yet on the opposite side of the same ridge, conditions would be relatively dry and few leeches would be found.

- Precautions in snake-infested territory might include limiting cross-county night movement when some snakes are most active, as it is difficult enough to spot them in daylight; use of NVGs during night movement will generally not be sufficient to illuminate them for detection.

Figure 110. From Left to Right: Cobra, Krait, Bamboo Viper. (*Public Domain*)

- Consider wearing 'snake gaiters'. These full-calf gaiters are made of robust material, including material similar to Kevlar® and are suitable for snake deterrence and for a variety of climates. Properly treated, they may also deter insect access.
- Snake repellants and 'snake-bite' kits are not considered effective. FOB medical support must determine if polyvalent antivenom is available and effective for field use (some antivenom requires refrigeration) in the AO; note that antivenom can have serious side effects.
- Not all snakes are to be found on the ground. The bright green Bamboo Viper has a geographic range from India to China and throughout Southeast Asia, and is most often found poised among green leaves on low lying plants or in bamboo thickets and are therefore able to strike a Team Member on the face, neck or hands. Other arboreal venomous snakes (including the Mamba and certain types of cobras, and a variety of other adders/vipers) are found throughout Southeast Asia, Africa and Central/South America.
- When venomous creatures are spotted during Team movement, all Team Members must be alerted; the Team would be well advised to make a small detour as some snakes are aggressive. Indigenous Team Members are invaluable in spotting these hazards.

> <u>True Account:</u> On several occasions during SOG operations, a mating swarm of wasps or hornets passed across a Team position. During their passage, the wasps/hornets would often briefly land and crawl on Team Members. The only solution to avoid being stung was to immediately freeze in place. Failure to freeze would invite hundreds of wasps/hornets to attack, with potentially lethal consequences. Once a

Figure 111. SOG's Leghorn Relay Site in Southern Laos.

swarm attacked, the only recourse then was for the Team to try to outrun the swarm (very difficult) or to take refuge in water (if a deep stream was nearby). If a Team Member was stung multiple times, it would often result in the Team's extraction and termination of the mission.

Immersion Foot (Trench Foot) TTPs:

- Immersion Foot 'is a medical condition caused by prolonged exposure of the feet to damp and cold. It was a particular problem for soldiers in trench warfare during the winters of World Wars I and II and in the Vietnam conflict. Immersion foot typically occurs when feet are cold and damp while wearing constricting footwear. Unlike frostbite, immersion foot does not require freezing temperatures…. Immersion foot can occur with as little as twelve hours' exposure. Affected feet become numb and then turn red or blue. As the condition worsens, they may swell. Advanced immersion foot often involves blisters and open sores, which lead to fungal infections…. If left untreated, immersion foot usually results in gangrene, which can require amputation.'[20]
- Wet season/monsoon conditions can involve continuous downpours lasting many days in duration; changing socks in these conditions may not be effective in preventing immersion foot. In these conditions, the T/L must:
  ○ Ensure increased surveillance of Team Member's feet.
  ○ Take an operational pause if Immersion Foot is evident.

<u>True Account:</u> The period of 1969 to 1971 produced two seasons of record-breaking monsoon precipitation in Southeast Asia. Notwithstanding these conditions, SOG operations continued. Due to substantial pressure from MACV for intelligence inside Laos, an SR Team would often be inserted through a 'hole in the clouds', sometimes well outside its Target Area and with the actual LZ grid coordinates completely unknown. And often, the Team could not be retrieved at the end of their prescribed mission period. Even though FACs would fly in high-risk, mountainous and socked-in conditions, they often could not locate these Teams even during occasional transient periods of visibility. Subsequently, Teams inserted for a week-long mission might not be exfiltrated for two weeks or more. As the symptoms of immersion foot manifested in the Team personnel, a sophisticated T/L ceased execution of the mission and placed his Team in a 'hide' location near a prospective exfiltration LZ. To facilitate drying of feet and foot gear, the T/L directed the construction of sheltered platforms for all Team Members; Team Members remained on these platforms with their boots and socks removed for several days, until an exfiltration operation could be mounted. At least three Teams from FOB-2, upon their return, required exceptional medical care for the treatment of immersion foot. Several of the Americans from these Teams were MedEvac'd to a military hospital in Okinawa for treatment. Weeks of recuperation time was necessary before American and indigenous Team Members were fit again for duty. Several other Teams were also affected; although their condition was not so serious, these less affected Team Members still required recuperation time.

Miscellaneous TTPs:

Survival TTPs:

General Survival TTPs:

- Look for fresh rabbit scat early in the morning and again at dusk to establish the best location for a snare.

---

20. 'Immersion Foot', http://www.wikidoc.org/index.php/Immersion_foot

- Carry a packet of peanut butter among personal survival supplies. Peanut butter is almost irresistible bait for snares/traps.
- Use a plastic bag as a mitten to minimize scent transferral while handling snare/trap components.
- Some ceramic items (e.g. firebrick, flower pots) can provide superior thermal retention and radiant properties as compared to metal objects.
  - If the Team is located in a cold weather environment near inhabited areas or abandoned buildings, such items may be found and used to provide heat in operational/survival situations. For example, if a surveillance operation is being conducted from an abandoned building in an urban area; a basement room, or a room where an IR signature cannot be detected, can be used to warm up Team personnel when they are off shift, while other Team Members occupy the surveillance station.
  - In snowbound wilderness areas, a snow cave can also be heated with a small improvised ceramic stove.
  - Safety Cautions:
    - Whenever Team Members are clustered in an enclosed space, always ensure occasional venting to evacuate CO_2. Alternatively, create a vent that will suppress heat signature. In a snow cave, use a diagonal tunnel through surrounding snow to cool the vented air. The enemy will not likely use thermal optics to detect venting during daylight hours.
    - Burn candles in a container, to retain the fuel as it converts to liquid and then to vapor as the candle burns. Remember that candles require oxygen for combustion; if the candle sputters, this may signify that it's time to vent. Designate Team Member rotation for fire watch, or risk suffocation.
    - Glazed terra-cotta may give off fumes under heating. Do not use.
- In desert or arid/dry environments ensure that several, or even all, Team Members pack at least one dark-colored, heavy-gauge trash bag.
  - The trash bag can provide multiple functions, to include that of desert water still.
  - The plastic bag can also be used to collect rainwater or condensation. This would be especially important in areas where Team Members risk exposure/compromise when they access water sources; or if they are occupying an isolated hide position.
- Some survival tips:
  - Carry a small, compact survival kit comprised of the following items, beyond items normally carried:
    - Fire starter material for cold/wet regions. In wet regions, dry wood may be impossible to obtain. This may be as simple as cotton balls soaked in petroleum jelly.
    - A flame producing item. Note that a flint will not produce a spark if it is wet; butane lighters will not function below 32°F or if the striker is wet.
    - Snare/animal trap items: string/wire; snare trigger(s); stakes; medium treble-hooks; fish-hooks; sinkers; fishing lures; game attractant (e.g. peanut butter, animal scent/musk).
  - Rather than a pistol, consider a short-barrel shotgun. If carrying a pistol, a silenced pistol would be preferred.
  - A lead fishing weight can be used to cast a line (e.g. for a dead-fall or a snare) over a tree limb; the lead weight (vice other options) is less likely to be caught in tree branches.
  - It may be very difficult to gather water into a canteen or bladder from a shallow water source. A foil or plastic ration bag may be helpful as a funnel/scoop.
  - 550-cord is comprised of seven interior nylon lines encased in a woven nylon casing. The interior lines can be separated from the sheath/casing to be used for snares, fishing lines, etc. Soak 550-cord in an appropriately colored dye to subdue/camouflage interior lines, as required.

- ○ Survival items should also be stored in MSS/caches.
- ○ Note that dead-falls should be five times heavier than the prey for which they are intended.
- ○ If a net hammock is carried by a Team Member, it can be converted to a fish trap/net.
- ○ A baited multi-hook line can collect multiple fish, even at night, but Team Members must periodically retrieve the line before turtles and other fish strip the hook of the captured fish.
- ○ In tidal areas, or where fish are migrating, a weir is an effective method to capture fish when the Team is in a survival mode or when the Team will be stationary in a relatively secure area for several days. Indigenous Team Members may have the know-how to do this.
- ○ Segregate trash at the MSS. Some of it can be used for various purposes if the Team is in a survival mode. For instance:
  - ▪ Food scraps can be used to bait traps or fishing hooks.
  - ▪ Shiny objects might be useful as fishing lures.
  - ▪ Some trash might be useful to lay false trails or to set up enemy combatants for an ambush.
- ○ In extremely low temperatures or blizzard conditions, the Team may have to resort to the 'Human Coil' technique to endure.

> 'A company of men would line up along a rope which was then wound up like a skein of yarn. Thus pressed closely together, and warmed by the heat of each other's bodies, the men would sleep while standing on their feet. Every hour the skein would be unwound and then rewound in such a manner that those on the inside would form the outer ring and those who had been exposed to the cold wind would benefit from the relative comfort of an inside position.'[21]

> This same procedure was used by partisans in the Balkans.

Emergency Stealth Fire TTPs:

- Dakota Fire Pit:
  - ○ Dig two side-by-side holes in the ground to equal depths and connect them at the bottom with a lateral hole. One hole is the vent, the other is the fire pit.
  - ○ Place tinder at the bottom of the fire pit hole. Fuel size ranges from twigs to larger sticks no longer than the depth of the fire pit hole; some smaller tinder should also be stuffed between larger sticks.
  - ○ The Dakota Fire Pit largely shields flames from view, but the heat plume signature is detectable by thermal/IR sensors. The Team must take appropriate measures to conceal the plume signature including: using this fire only during day time, especially dawn and/or during inversion/overcast conditions; avoid green/smoke producing wood; use a fly above the fire to dissipate the plume (also required to shed rain); locating the pit under canopy; screen the pit using terrain form.
  - ○ Not especially good for heat, but superb for cooking. The pit can also be used to heat stones for shelter radiant heating.
- Hoang Cam Fire Pit:
  - ○ See further description elsewhere in this book.
  - ○ Consider prevailing wind when orienting the pit and trench for optimum air flow.
  - ○ The exhaust trench should be deeper than 6in. When excavating, retain sod and earth to re-cover the trench; use green bamboo/sticks, broad leaf vegetation and/or similar materials and then cover with mud, earth and sod.

---

21. Halder, *Small Unit Actions*, p. 248

- ○ The 'stove' element (at or nearest to the fire) can be enhanced by using dry, flat stones (e.g. gathered from a streambed).
- ○ Heat plume suppression measures previously noted.
- Swedish Fire Log:
  - ○ Split a 4ft-long (approx) log section of about 6–12" in diameter, into 4–8 sections lengthwise. Stand one end of the split log in a shallow hole; this will ensure that the sections are kept upright and together.
  - ○ Separate the sections with smaller split pieces, sticks, etc. The log is lit from the top; coals will drop down between the cracks (cracks allow ventilation and oxygen to flow). Make one or two of the split pieces smaller (more slender) than the others; the fire log will burn faster at this point, providing reflected heat.
  - ○ Good for rain soaked areas; will burn for a long time untended. The flat top makes a good cooking surface as long as it remains level.
  - ○ Heat plume suppression measures as above.
- Ensure cooking fires are kept a distance from personnel shelters in the event the heat signature is detected.
- Fire-Making in Wet Conditions:
  - ○ Use dead wood from trees (especially conifers with sap still in evidence) <u>that are still standing or are elevated above the ground</u> (e.g. propped up by other trees). Standing dead wood will be dry on the inside; wood lying on the ground normally will be saturated.
  - ○ Prepare the site by:
    - ▪ Erecting a rain-fly over the fire bed and wood pile locations.
    - ▪ Lay dunnage at the fire bed and wood pile sites to keep firewood off wet ground.
  - ○ Cut firewood should be without bark and with a <u>circumference</u> equal to or larger than 6in (e.g. measured by bringing thumb and middle finger together). Split the dead wood to get to the dry interior. Delay bark removal until the site is prepared, as the exposed dry wood will begin to absorb moisture.
  - ○ If the soil is soaked or snow-covered, use dunnage and a cross-layered bed of sticks for the fire bed. Then lay in tinder with larger sticks/tinder on the bottom (feather these sticks) and smaller tinder on top forming a wedge/teepee.
  - ○ Use several Team Members to speed the process, as cut wood will absorb moisture.
  - ○ Fire Starter for wet conditions: use cotton balls impregnated with petroleum jelly.
  - ○ Teepee additional sticks after the fire starts.
- Consider locating the fire where fog is most likely to linger (stream nearby) to suppress the plume and heat signature.
- Only resort to fire as a necessity. This may include:
  - ○ In deep penetration operations, resupply of food and water may be spotty or not available at all. Team Members may have to subsist on local sources/game, etc. which will require cooking.
  - ○ In severe cold, and in cold-wet conditions, Team Members may require heat to survive the environment.
  - ○ Wounded, injured or ill Team Members may require heat. Battle dressings may require sterilization.
  - ○ Water/snow may require boiling prior to drinking.

Chapter 5

Command and Signal

Command and Control (C2) TTPs:

'The most essential element of combat power is competent and confident leadership.'[1]

- If a T/L is not competent, he will likely not be confident either. Team Members will immediately discover these failings and will not trust a T/L who is incompetent, indecisive or risk averse. The T/L (and subordinate leaders) must constantly work on developing his (their) individual skills and knowledge and mastering TTPs. Team leadership must mentor junior Team Members in developing skills, knowledge and their experience base so they can operate without intensive control/micro-management.
- Differences between LRRP/LRS and SR include:
  - LRRP/LRS procedures exercise tight operational control by higher headquarters; SR Teams are largely on their own; C^2 is exercised by the T/L and in general by control measures contained in plans, geographic boundaries/features, SOPs and ROEs specific to the AO. The T/L should otherwise be free to use his initiative and skills to accomplish missions and functions.
  - LRRP/LRS Teams possess a limited offensive capability (weapons are used for self-defense or to break contact); SR Teams should be better equipped for opportunistic combat and secondary missions.
  - SR Team leadership should be based on merit and experience, rather than rank. This is more easily done outside the typical ODA structure, using a mix of US and indigenous Team Members.
- SOPs are an essential means of FOB and Team C^2. SOPs must streamline, not hinder execution of TTPs, therefore simplicity and discretion is essential. The FOB SOP should define specialized organizational structures, C^2 and support functions.
- FOB command and staff should issue mission type orders; must not micro-manage Team operations in the field or attempt to exercise command through 'remote control', but rather should be highly responsive in supporting Team needs. Teams cannot be tethered to the headquarters for decision-making. Centralized decision-making is highly dependent on communications reliability; it is often not timely in the context of rapidly unfolding, complicated events on the ground; it is also time-consuming and conveys a lack of confidence in the abilities of the SR Team.
- Team Chain of Command is straightforward (organizational structure and rank) and facilitates Span of Control. If the Team must operate where elements are separated by METT-TC considerations (e.g. terrain form, vegetation and functional segregation), then the T/L must ensure Team Members are made aware of revised delegation of responsibilities. The T/L must become comfortable in delegating responsibilities, especially where other Team Members may have superior skills, knowledge and experience (wisdom) within their specialty areas.

---

1. Field Manual 100-5, *Operations*, Department of the Army, Washington, DC 5 May 1986, p. 13.

- The T/L must be able to lead, manage and coordinate many elements and functions, often simultaneously and during periods of high stress and chaotic conditions. These tasks can overwhelm a T/L if he has not developed his subordinates, delegated responsibilities and clearly simplified/defined Team functions (via SOPs, battle drills, etc.). These elements/ functions may include:

• Use of an interpreter to lead, manage, coordinate a Team with indigenous troops.	• Supporting infantry or paramilitary assets.
• Tracker Specialist(s) attachment	• Supporting UW assets.
• Dog Team(s) attachment	• Supporting Intel Community attachment
• Sniper Team attachment	• Combat Controller attachment
• EOD/CBRN Specialist(s) attachment	• Supporting CAS, air delivery and/or FAC assets.
• Supporting Artillery assets and/or a Forward Observer attachment.	• Supporting aerial observer and/or UAV assets.
• Split Team operations	• Heliborne support assets.
• Adjacent, supporting and supported unit status.	• Routine and urgent communications.

The FOB should allocate liaison officers to supporting elements/headquarters whenever possible, especially if the supporting unit is not in a habitual relationship with the SR mission. This liaison function may include:

- Specialized communications equipment enabling the liaison to communicate directly with the FOB, Launch Site and other important communications nodes.
- Survey the supporting unit's decision-making, operational status and operational execution to identify problems that threaten the mission or Team effectiveness/ survivability and to communicate these issues to the supporting unit's staff or commander for resolution or to elevate the issues through the FOB headquarters.

Signal TTPs:

General Signal TTPs:

- The FOB/higher headquarters should refrain from unnecessary communications with the Team; communicating on a planned contact schedule – while continually standing by for high priority messages originating from the Team. The FOB/higher headquarters can establish its real-time information needs through Commander's Critical Information Requirements (CCIR) defined in planning documents and SOPs – and especially in message formats. CCIRs must be distilled to essential command needs and must be designed for brevity and conciseness.
- Team _internal_ communications capabilities, modes and devices include: passwords, signs and countersigns; use of interpreters; hand and arm signals; imagery (photo/video), audible voice and signal devices (to include: noisemakers, firearms or explosives discharge); visual signaling devices; covert trail signs and squad (Team) communicators, tactical tablets (IT) and batteries/power generation. The inventory of internal communications may also include Frequency Modulated (FM) radios, when the Team is acting in cooperation with other units or in separated elements.
- In general, _external_ communications modes and devices typically include: FM/AM/satellite radios, visual signaling devices, imagery (photo/video), communications systems designed to operating in other areas of the frequency spectrum and complimentary equipment/devices

(e.g. specialized antennas, encryption devices, burst devices, GPS/PLS, wire-tapping equipment, target designation equipment, batteries and power supplies, etc.).

- Radios, beacons, sensors and several other Communication and Electronic (C&E) devices require power to operate. Some devices can be turned off to conserve power; other devices must remain on consistently throughout the duration of the mission. Some considerations:
  - A mounted Team's mobility equipment can carry a supply of spare batteries, but the weight of spare batteries will be a substantial burden to a dismounted Team especially, on extended operations.
  - Team mobility equipment power can be used to recharge batteries of Team C&E devices (while the motor is running). Remember to bring sufficient charging cables capable of connecting multiple devices to the mobility equipment power supply.
  - A solar panel has the capability to supply some power to recharge batteries, but this device has notable shortcomings.
    - A solar panel can be worn on the exterior of the rucksack to convert the rays of the sun to provide power while the Team moves cross country. However, the panel will snag on vegetation, may create additional noise during movement, may reflect the glare of the sun and will impede access to the interior and exterior pockets of the rucksack. A solar panel will add to the weight carried by a dismounted Team Member.
    - A solar panel can generally recharge only one device/battery at a time.
    - Solar panel power generated is very limited and may require substantial time to completely recharge the battery/device; if attached to a device that is in operation, may often not keep pace with the power being consumed. Solar panels require sunlight to generate power; its capability is impaired during overcast periods and while direct sunlight is obscured by vegetation/terrain and is nonexistent during hours of darkness.
  - Thermoelectric (Seebeck) generators can convert heat (via fire) to electricity to recharge batteries/devices. This may be feasible during periods when enemy thermal sights are not in use (e.g. daytime, fog, snow, rain).
- Signal/C&E equipment may also include: sensors, camera equipment, and covert marking devices.
- Enemy may have standard signal definitions (e.g. bugle calls, star clusters, etc.). If this information is available, know these.

Cryptography TTPs:

- Signal Operating Instructions (SOI) are classified cryptologic information. As the SOI historically employed by SpecOps has been the one-time-pad (random character encryption), the compromise of a captured Team SOI code was always isolated to the individual one-time pad issued to a specific Team and to the limited number of (sheets) days contained in the pad. Public Key Cryptography (PKC), now in use, makes encryption much more facile, flexible, and perhaps even more secure than the one-time-pad. Further, modern military radios (e.g. Single Channel Ground and Airborne Radio System – SINCGARS) include Integrated Communications Security (ICOM) Models, which provide integrated voice and data encryption – obviating SOI one-time-pads. However, there may be some isolated instances and special circumstances (e.g. UW operations) where SpecOps Forces may operate foreign signal equipment requiring the use of one-time-pads. Regardless of the encryption system employed, if US Team Members are captured (along with an intact communications set) without friendly headquarters awareness of that fact, a captured American could be compelled to reveal the encryption protocol, while it is still in effect, and therefore could be used by the enemy to transmit false encrypted messages that could have a devastating effect on friendly forces.

- Team members must retrieve and/or destroy all cryptologic devices and information from fallen or wounded comrades. This information would include Team Members' notebooks, Quarterback Wristband, etc. which may contain Communications Brevity Codes, Passwords, Signs/Countersigns, special frequencies and any 'dictionaries' of Sensitive and Emergency Visual Signals and Codes. To facilitate the recovery of this information from fallen or wounded comrades, each US or key Team Member should, by SOP, carry maps, notebooks, and SOI, etc. in the same uniform pocket(s).
- If such information is known or suspected to have fallen into enemy hands, higher headquarters must be rapidly notified so that friendly forces may be informed and countermeasures taken.

Team Internal Communications Capabilities, Modes and Devices TTPs:

Interpreter TTPs:
The interpreter is responsible for facilitating verbal communications between the T/L and other indigenous Team Members, for hasty translations of enemy written communications and 'real time' enemy verbal communications.

- The Team indigenous interpreter should be positioned near the T/L in the tactical formation.

> **True Account:** A SOG Team had just crossed a large patch of broken bamboo and upon crossing into the surrounding rainforest, immediately encountered a concealed enemy compound with large structures, signifying a major NVA headquarters. Immediately after this discovery, the Team heard shouting in Vietnamese. The Team's indigenous interpreter remained silent and did not translate the shouts – and the T/L, focused on other matters, did not request a translation. The Team subsequently deployed to assault the compound and its structures in the hope of capturing a senior officer or valuable documents, but before the assault could be launched, the Team was taken under fire and the T/L was severely wounded. After the Team was extracted, it was discovered that the enemy had shouted, 'Americans are in the dead bamboo!' The interpreter had remained silent because he did not understand his responsibilities.

Passwords, Signs/Countersigns TTPs:

- Establish a 'running' password to be used when a Team Member or element is fleeing a pursuing enemy and is approaching another element of the Team. The password should be easy to remember.
- A daily, supplementary or hasty sign-countersign can be based on a number. For instance: the daily number (normally an odd number) might be designated as '9'; the challenge could be any number up to '9'; the password response would be a number that when added to the challenge would sum to '9'. So if the challenge is '2'; the proper response would be '7'.
- Duress codes: Establish duress codes for radio communications to be used in the event of Team Member capture.
  - A radio duress code would be used if the enemy coerces a captured Team Member to transmit a message via a Team radio to lure remaining Team Members or aviation assets into a trap.
  - If a Team Member has been suspiciously missing for a period of time; the Team should perform a simple radio authentication procedure that is innocuous enough to avoid enemy detection.
  - The duress code may consist of a simple word, the absence or presence of which in a message sequence would alert the base station of capture.

- ○ Alternatively, the code protocol could consist of a challenge from the relay/base station operator; and a either a positive response word (indicating a green status) or a negative response word (indicating duress). The relay/base station operator must maintain his composure given a negative response and must be prepared to continue the communications interchange as if nothing is amiss.
- ○ The enemy may coerce a captured Team Member to provide the password response to a challenge in order to close with and assault the Team. The Team should also have an audible/communications internal duress code and a specific procedure for linking up with a missing Team Member.
- ○ The Team, or its parent unit, should have a physical or visual duress code SOP that features a particular hand/arm/body motion that can be used in passage of lines, exfiltration operations or similar circumstances where a Team Member may have been captured and where the enemy is attempting to use them to gain a tactical advantage.
- Armageddon Signal:
  - ○ SR Teams may encounter a situation where indigenous Team Members are actually enemy agents or are otherwise so hostile to US Team Members as they represent a substantial threat (e.g. 'Green-on-Blue' attempts at betrayal/assassination).
  - ○ US SR Team Members should have a special audible phrase and a related, unmistakable hand signal that would alert other US Team Members to the source of the threat, and the requisite action to be taken.

> **True Account:** At SOG's FOB-2, South Vietnamese SR Teams were allocated a specific area of operations, separate from those of US Teams. Typically, these South Vietnamese teams were led by South Vietnamese Ranger officers or NCOs and the remaining Team Members were normally South Vietnamese commandos trained in the same manner as the Montagnard commandos on US Teams. Unlike Montagnard commandos, South Vietnamese commandos were sometimes recruited from the criminal class, many opting for a SOG assignment rather than incarceration. Most of the South Vietnamese Teams assigned to FOB-2 had a very poor mission record. In 1972, a South Vietnamese Ranger lieutenant was newly assigned to lead one of the FOB-2 South Vietnamese SR Teams. Son of a French father and a Vietnamese mother, he held a black belt in karate and had graduated from the prestigious French military academy at Saint-Cyr, returning to South Vietnam to join the military of his native country. Upon his Team assignment and during the term of initial team training that he directed, he became appalled by the quality of the Team Members, their reluctance toward combat, disrespect, indiscipline and many other deficiencies, to the extent that he sometimes resorted to corporal punishment. His first mission ran the entire planned seven-day duration. But his Team Members clearly believed that the lieutenant's mission-focus threatened their lives and they hated him for it. A Vietnamese officer peer at FOB-2 warned him of his jeopardy; he ignored the warning and was assassinated by his Team Members during his second mission – his body never recovered. Had he but one loyal Team Member to alert him during the mission, he might have survived.

Hand and Arm Signal TTPs:

- During operations, and governed by the requirement for noise discipline, Team Members will generally spend the entire mission using hand and arm signals, rarely uttering a word in other than a whisper. Normal volume or loud verbal communications would only be used during such events as fire fights, the execution of ambushes/raids and possibly during helicopter extractions.

- It is the responsibility of the signaler (communications originator) to ensure his message is received.
  - Hand and arm signals between Team Members may be obscured by ambient light, weather conditions and by intervening vegetation or terrain form. Team Members who observe a signal must routinely pass the signal on to other Team Members.
  - If a Team Member is inattentive, he might not see another Team Member trying to pass him a signal. The signaling Team Member should then either use an audible attention-getter (perhaps imitating the sound of small native wildlife) or toss a small twig or pebble toward the inattentive Team Member.
  - If an unwary Team Member is in imminent peril of discovery, it may be necessary for the alert Team Member to initiate a firefight, rather than have Team Members become casualties.
- Army Field Manual 21-60 (Visual Signals) contains some hand and arm signals pertaining to combat formations, Battle Drills and Patrolling. However, not all relevant hand and arm signals are described in the FM. If the Team's higher unit does not have a standard system of hand and arm signals, the T/L should develop a comprehensive and intuitive Team 'dictionary' of hand and arm signals; include the 'dictionary' in the unit/Team SOP; and use the signals during training and field exercises. Ensure that all Team Members have thorough knowledge of the signals and rehearse them (during mission training) prior to every mission, especially with indigenous Team Members. Borrow some gestures from Sign Language dictionaries, especially those that use one-handed signs and those that are most visible. The Team or its higher headquarters should record hand and arm signals on video clips, as many signals involve motion, to train SR personnel to a common standard dictionary.
- By SOP, the Team should have a list of hand and arm signals to recognize separated Team Members/elements. A hand and arm duress and 'Armageddon' signals should be included.
- Hand and arm signals are only useful if they can be seen by Team Members. If the Team has indigenous Team Members (for instance) who are not equipped with functioning night-vision goggles/equipment, then hand and arm signals will not be seen at night. In this situation, the Team might use directional luminescent items and the T/L must contrive universal hand and arm signals that can be used in both day and night.

Audible Voice and Signal Device TTPs:

- A unit/Team SOP may specify the T/L's signal to initiate an ambush, raid or a firefight, such as the discharge of his firearm. In other situations, this type of signal may be inappropriate; for instance, an enemy action may cause Team Members to prematurely initiate contact, where the enemy routinely discharges its firearms from defensive positions, during the conduct of sweeps or during Hammer and Anvil operations.
- The T/L should wear a whistle or other noise-making device (e.g. a small CO2 foghorn) around his neck. These devices are used to signal and direct Team Members in Battle Drills or in other tactical maneuvers (e.g. ambush, raids, etc.), where verbal commands would be muted or confused with battle sounds.

> **True Account:** An experienced SOG T/L was able to have his family mail to him a small CO_2 foghorn (purchased from a marine supply business). After testing the item during training, he decided to use it as a signal device on his next operation. During the operation, his Team engaged in an extended firefight with an enemy platoon. As the enemy platoon was attempting to outflank the Team, the T/L used the foghorn as the signal to break contact with the enemy. The sound caused the enemy unit to immediately cease fire and stop its flanking maneuver. The sound of the foghorn

achieved tactical surprise, as the blast was nearly identical to the sound of a 7.62mm, M134 Mini-gun.

- The Team might use demolition simulators, improvised noise-makers, booby-traps/mines placed along routes of approach to signal an enemy's presence. But remember that the booby-traps/mines, if used, must be equipped with an SD function.

Visual Signaling Device TTPs:

- A T/L may be separated from other elements of his Team by intervening terrain form or vegetation, (e.g. while the Team is deployed in an ambush/surveillance position). If the Team lacks intra-team radios and the T/L cannot otherwise contact the other elements of his Team, he might consider connecting camouflaged paracord from his position to a small, low-growing plant or branch situated with the other element(s); and by using a series of tugs he can silently signal prearranged instructions. Note that the 'dictionary' of instructions signaled by this method will be very limited.
- By SOP, the Team should have a list of light signals to recognize separated Team Members/elements. A duress signal should be included.
- Visual signals appropriate for daytime might include a signal mirror, signal panel, pyrotechnics, tracers, canopy-penetrating pen gun, etc. The use of a signal mirror; strobe light; soldier hand, arm or body signals; and emergency codes depicted by signal panels (or improvised materials) in ground-to-air communication are summarized in FM 21-60. Further information and signaling tradecraft is contained in the paragraph entitled 'Forward Air Controller (FAC)/Air Support Tips' featured in this book.
    - Some of the signals described in FM 21-60 may be modified due to ammunition availability, tactical relevance or other reasons, so the T/L must define the meaning of such signals to Team Members prior to each mission.
    - Note that projected pyrotechnics, tracers, canopy-penetrating pen gun, etc. in ground-to-air signaling should never be aimed directly at friendly aircraft, as it may be taken for ground fire.
    - Also note that a standard pen-gun is unsuitable for operations conducted in areas of continuous or multi-layer canopy as the flare will likely be blocked/deflected by canopy; a canopy-penetrating pen gun is essential.

Covert Signs TTPs:

- Covert signs can be used as trail markings; to warn of mine/booby-trap locations; to leave a pre-arranged signal at an RP; to mark the nearby location of a MSS/cache, to convey a message, to signal danger, etc.
    - The SR HQ or its higher headquarters (e.g. FOB) might establish a common lexicon of covert signs and include this dictionary as an appendix to the unit SOP.
    - Team Members must be trained to know where to look for these signs.
    - It is important to know the enemy's covert trail signs and where to look for them.

Intra-Squad (Team) Communicators/Radios TPPs:

- <u>Intra-Team</u> radios should be used by Teams only under defined circumstances, especially where the devices might impair Team Member hearing/sensitivity to sounds within the physical environment. During normal movement, during rest breaks, communication breaks and NDP, these radios should not be used and earpieces should not be worn. They may be used if elements of the Team are separated, deployed in ambush positions, during fire and movement/maneuver, and similar circumstances, but only specific US Team Members should be using the devices to control Team elements during these events.

- The most current, and future generations of Team communicators will have two (or more) channels, Electronic Countermeasures (ECM), Threat Warning capabilities (alerting the Team to nearby enemy transmissions) and an Intelligence, Surveillance, Reconnaissance (ISR) video interface to UAVs. While this capability is impressive, the demand on battery power will increase, adding to Team burden. Further, <u>over time, 1st Tier threat enemy units may acquire similar capabilities</u>. The Author recommends emphasis on hand and arm signals; Teams should resort to intra-Team communication systems only occasionally and only for essential tasks.
- If the Team is deployed in a covert operation where deniability is essential, older series Team communicators or foreign/commercial hand-held radios may be used. Some short-range, low-power Team Communicators/Radios may emit a tone or a rushing sound whenever a push-to-talk button is pressed, which could reveal the presence of a Team Member or surveillance post. If this option is present on Team communicators, ensure that the tone capability is turned 'off' prior to mission launch and that it stays 'off' for the duration of the operation. Include this on the pre-mission inspection checklist.
- Also ensure that Intra-Squad communicators, when used in proximity to the Team FM radio, do not create a 'feedback' squeal when one of the devices is transmitting nearby. A feedback squeal, occurring when the Team is in close proximity to an enemy, would not only reveal the location of the Team, but it would pin-point vital communications. Consider testing feedback during Team training and during Mission Preparation; note that the potential for an emitted squeal may be specific to radio frequencies in use – which change daily. Therefore, during Mission Prep, run this test against all frequencies (primary and alternate) that are allocated for the mission duration according to the SOI.

Note: See more about signal munitions and signaling a FAC elsewhere in this book.

Photographic and Video Device TTPs:

- Ideally, cameras should have the capability to automatically time and date stamp the image file; stamp the image file with GPS coordinates; and stamp the images and video shots with azimuth/direction orientation. If these features are not provided or operable within the Team's imagery devices, the photographer must record this data separately in a photo log.
- If the Team is equipped with a tactical tablet, some of this logged information may be maintained in the devices memory. Any content maintained on a digital device should be password protected and encrypted. If possible, have an application on the device that will delete/destroy content upon entry of a duress code.

Team External Communications Capabilities, Modes and Devices TTPs:

- SR Teams may have access to and use of an array of HF, VHF, UHF external communication devices of both US and foreign manufacture. Perhaps the most flexible and capable communication system is the FM Joint Tactical Radio System (JTRS), which provides software-defined, low probability-of-intercept encryption of ground-to-ground/ground-to-air/ground-to-satellite communications, multi-network operation, integral GPS and integration with target designators for an array of precision-guided munitions. This radio can operate in single-channel or frequency-hopping modes, has a two-channel simultaneous operating capability, and it receives an automated SOI via an integral Automated Net Control Device (ANCD). As a line-of-sight FM radio, its range is generally limited to

5-10km (depending on weather, atmospherics, location/elevation, antenna type and other factors). The JTRS has variants for CP, manpack/mounted and aviation applications.

- If the SR Team is to operate in the split Team concept, two external Team radio sets will be required.
- Occasionally, the Team may miss a scheduled contact.
  - One missed contact should not be of special concern, other than to ensure an airborne relay capability is positioned in the vicinity for the next scheduled contact.
  - Primary and alternate frequencies and prior day frequencies should be monitored by the relay/base station operators in the event that missed communications is a consequence of technical/administrative problems, a combat encounter or Team emergency.
  - If the Team establishes contact during the second contact window, the station should be verified in accordance with SOI or communications SOP.
  - If a second contact window is missed, this becomes a matter of genuine concern, an airborne relay should orbit in the vicinity of the Target Area and a Bright Light Team or Reaction Force (RF) should be alerted for a rescue/recovery mission.
  - If the Team intends to 'go dark' for some reason (e.g. proximity to enemy forces, etc.), the Team should notify relay/base station operators/higher headquarters of the plan and when the Team communications will be back 'up'.
- If enemy elite units possess radios with transmission detection/warning features, such as are available to US SpecOps, all Team communication devices must 'go dark', except for scheduled contacts using burst mode.
- Establish duress codes for external radio communications to be used in the event of the capture of a Team/Team Member.
  - A radio duress code would be used if the enemy coerces a Team Member to transmit an enemy message via FM or emergency radio, e.g. to lure friendly assets into a trap. If a Team/Team Member has missed a scheduled contact or has been suspiciously missing for a period of time; the relay site or higher-level unit should perform an authentication procedure. The initiation of an authentication procedure may present the opportunity for the Team/Team Member to present the duress code. For example: A relay/base station would ask for an authentication response from the Team/Team Member for which only two code-word answers are possible, (1) indicating all is well, or (2) duress. The authentication query might pose an innocuous question, enough to quell enemy suspicions, again soliciting one of two possible answers.
  - Further, the enemy may coerce a captured Team Member to provide the password response to a challenge in order to close with and assault a friendly element or outpost; a specific duress code for this scenario must also be developed and committed to the unit SOP or OPLAN/OPORD Signal paragraph. Each Team should be assigned a different authentication code.

> A SOE agent team was captured in Belgium upon insertion with radios and codes intact. The Morse code operator was coerced to transmit false messages to SOE Headquarters in Britain; the operator's use of the duress code was not detected at SOE. Subsequent false messages were used to coordinate other team insertions, all of which were captured with equipment and codes intact – and where the Germans established communication stations. In the continuing exchanged of messages, the entire Belgium network was compromised and rolled up by the Germans, with several hundred agents and Resistance personnel captured, without detection by the SOE.[2]

---

2. M.R.D. Foote, *SOE in France*, (digital rendering without pagination).

Virtually the same sequence of events transpired with CIA agent teams, consisting of Vietnamese nationals with northern dialects, which were inserted into North Vietnam. All teams were compromised and 'turned' without CIA detection for years. The compromise of the agent program was not detected until the program was turned over to MACV-SOG, and then not until several months had passed and after SOG had inserted other teams that were similarly compromised. Consequently, SOG converted the agent program to part of a disinformation program, leading the North Vietnamese to believe that a Vietnamese partisan organization and shadow government was active in the North.[3]

IT Devices TTPs:
Team Members may carry tactical laptop computers or tablets that can be extremely useful to the Team. Such devices can also be an impediment. Pros and Cons:

- Pros:
  - Messages and attachments (e.g. photos, video clips, etc.) can be composed, encrypted and compressed in a broad variety of formats, then stored and transmitted from the device via the Team radio. The device must be able to link to the appropriate Team communications systems for secure, rapid (burst) transmission.
  - Substantial intelligence data can be composed and stored on the device, to be later produced during post-mission debriefings.
- Cons:
  - These devices, and the batteries required, represent additional 'weight and cube' to be borne by the Team. In some environments, power can be replenished by a portable, lightweight solar panel, which may already be included in Team mission equipment (especially for deep-penetration, long-duration missions). When these devices are not in use, the device should be shut down as a precaution to eliminate passive signature and to conserve power.
  - Information contained on the device could be of high value, if captured by the enemy. The information within the device should be secure (encrypted) in storage. Ideally, the device should have password protection, to include a duress password that will wipe the memory and/or destroy the memory component.
- A tactical tablet should contain database files, with illustrations, on weapons, vehicles, armored systems, aircraft and other tactical systems (both friend and foe).

Pre-Launch Communications TTPs:

- The T/L should attend the pre-launch pilots briefing. Valuable information issued in the pilots briefing includes: estimated locations and types of anti-aircraft threats; flight frequencies to be used for insertion operations; point-of-no-return coordinates; emergency 'Letter of the Month' to be used in crew/Team evasion and recovery operations (this is a ground-to-air visual signal to be displayed in clearings or LZs using signal panels or other methods).
- By SOP and checklist, ALWAYS conduct a communications equipment and COMSEC materials inspection and conduct communications checks prior to departure for the Launch Site.
- The Launch Site operations officer must ensure that a fully operational back-up radio (complete) is available at the Launch Site at all times. While the operations officer is responsible for the radio and its condition, the SR T/L and the Bright Light T/L, should

---

3. Derivative/extract: Shultz, *The Secret War Against* Hanoi.

perform a daily commo-check shortly after arrival at the Launch Site. The Launch Site spare radio should travel with the chase medic during a launch; if the Team organic radio is damaged or fails after insertion, the chase medic would exchange the radio with the Team. If the chase medic must provide emergency medical assistance (to wounded Team Members or downed aircraft crew, etc.) the medic will need the radio. As the chase medic will be burdened with his medical kit, the LBE of the chase medic must be rigged to accommodate the additional burden of the radio. Prospectively, this may be a chest mount or a shoulder bag.

- By SOP, the Team radio operator(s) should pre-set alternate frequencies on the radio every day at a specified, designated time, so that even in the heat of combat, a quick turn of the dials will switch the radio to the alternate frequency setting.
- C&E device cross-training to Team Members is essential, as the Team communications NCO(s) may become WIA/KIA during an operation. A limited amount of training should also be delivered to the indigenous Team interpreter. All US Team Members should be trained to issue a call for fire and to direct air strikes <u>on the move</u>.

Communications Support TTPs:

- It is of utmost importance that the Team should have excellent communications support and capabilities. Higher headquarters should provide:
  - Top-notch communications equipment in adequate numbers to support the mission.
  - Excellent communications equipment maintenance.
  - Communications redundancy, including emergency radio equipment.
  - Radio Relay capabilities (via relay site(s), Forward Air Controllers, Airborne Command and Control, and/or UAVs) that are continuous and dependable.
  - Radio Relay networks/repeaters that transmit directly to higher headquarters and/or direct support operations staffs. Immediacy of response is essential.

Mission Execution Communication TTPs:

- The FM radio is a line-of-sight communications system. Signal attenuation in dense jungle foliage is commonplace. This problem is compounded if the Team radio is located in low areas in dissected terrain. The Team may have to ascend to elevated terrain to acquire line-of-sight, or have a well positioned relay radio (with a reliable power source) to successfully communicate with a base station/relay site; and the Team may need a special antenna for extended range or directional transmission/reception. A 'noise-suppressed' antenna launcher (a sling-shot might serve), capable of projecting a weighted line above canopy height, at a MSS, could be very useful in deep penetration operations.
- By SOP: Only after the message is composed and encrypted, should the radio operator remove his pack, erect a specialized/long antenna and transmit the message. Upon successful communications exchange (unless the Team is instructed to stand-by for further communications), the Team communicator should dismount and stow the antenna and immediately re-don his pack.
- The radio operator should not erect a long antenna until the message is ready, or almost ready to send. Note that in older and foreign radios (which the Team may find it necessary to use) the long antenna typically has a long, threaded brass end that screws into the radio antenna base. If the Team is attacked while a long antenna is up, the radio operator will have very limited time to unscrew the antenna from the base; if the radio operator moves while the long antenna is still attached, he will almost certainly damage the long antenna and break off the antenna from its base; leaving the broken antenna threaded male end still

screwed into the base. The radio operator will then no longer be able to mount the short antenna to the base – therefore forfeiting the normal transmission range of the radio.

- If the Team operates with the Team radio on, the volume must be turned way down and the radio handset must be positioned close to the radio operator's ear (or an earphone worn). Voice messages are typically whispered by Team Members into the radio handset/microphone, as enemy combatants/tracker teams could be close by.

> 'Recon team leaders would cover their mouth and handset with their hat, put their face to the ground and whisper to minimize their voice.'[4]

- Consider sewing a long slim pocket on the side of the rucksack to store a long/alternate antenna for swift retrieval.
- If an individual emergency radio is lost while the beacon is on (transmitting), it may block all other signals on that frequency until its battery is exhausted.
- Maintain COMSEC and Signal Security. Beware of enemy electronic/signal warfare capabilities. Keep radio traffic to a minimum; use SOI brevity codes for voice transmissions whenever possible, and use burst transmissions for long/data messages. This is especially important if the Team radio lacks frequency-hopping technology; but bear in mind that the energy coming from a transmitting antenna is still detectable, even if the radio has a frequency-hopping capability.

> 'Toward the end of 1941, for instance, the whole SOE organization in the unoccupied zone of France fell into a Vichy police-Gestapo net. In the same year, the Gestapo managed to arrest some SOE operatives in Holland and induced them to transmit fake radio messages to Britain. For over a year, carelessly convinced that the messages were genuine, the SOE continued to parachute agents and supplies into Holland to be promptly taken by the Germans.'[5]

- In a future conflict, an enemy may employ Electromagnetic Pulse (EMP) weapons and/or anti-satellite operations to interdict US communications and intelligence-gathering technologies. Additionally, an enemy may launch combat operations against communications base stations/relay sites. If these interdiction methods are successful, effects on Team operations and mission capabilities will be significant. Higher headquarters should have a plan to deal with these contingencies and to reestablish signal capabilities – to include what steps Teams must take to reestablish communications. The T/L and Team communications personnel must know these plans and must be prepared to operate in a more austere manner. Some points to consider:
  - Designation of planned emergency resupply point(s) or extraction LZs.
  - Locations of cache sites/MSSs.
  - Blind drops to Team last known/specified locations.
  - Suspension of the mission, reduction in mission activities or activation of alternative/contingency missions plans.
  - Evasion routes to covert sites, friendly or neutral states, etc.
  - Procedures to contact friendly guerilla/partisan forces, auxiliaries or intelligence agents.
  - Note: Emergency LZs/DZs, relay sites, caches/MSSs may not be on any map in the Team's possession. During the mission planning process, the T/L should plot azimuths from a planned/known point within the target area through a series of easily identifiable terrain features outside the target area to the contingency destination(s).

<hr>

4. Greenup, 'No Glamour No Glory', p. 107
5. Rothenberg, *Isolating the Guerrilla*, 'The German Experience in World War II', p. 177

'Transmission time is held to a minimum by use of the equipment used for burst transmission; use of prearranged message, brevity, and map coordinate codes; and by transmitting only necessary information. The transmission site of the patrol is changed frequently and, if possible, for each contact…. Each LRRP should incorporate within the encrypted portion of each report an identifying mark (memorized by the patrol) to preclude the enemy from transmitting false reports should a patrol and its cryptographic key be captured.'

FM 31-18, Long Range Reconnaissance Patrol Company, p.10

- Scheduled/routine communications might occur at designated times, for instance, midday and in the late-afternoon or early-evening (prior to moving into the NDP); this may be particularly appropriate if an airborne relay asset must be scheduled to be on station.
- Rather than communicate according to specified daily times, which the enemy may discover, consider communicating at different times that may be specified in the SOI. The chief problems with this arrangement are that of coordination, of perturbations to the conduct of the mission and the commitment of airborne relay assets.
- At the midday break, Team Members should eat their meal in a defensive perimeter, the T/L should compose the midday status report and the radioman should transmit his message; when the meal and transmission are completed, the Team moves to continue the operation.
- At the late-afternoon/early-evening break, the Team Members eat their evening meal and conduct the final scheduled transmission of the day. The minimum content for these transmissions should include: current grid coordinates and Team intentions. When the meal and transmission are completed, the Team moves once again to occupy its NDP location. Do NOT transmit from the NDP except in emergency/urgent conditions.
- If the enemy is extremely close and contact appears imminent, the radioman can covertly transmit to alert supporting communication sites by depressing the push-to-talk button on the FM radio handset to send a covert 'SOS'. Relay site communications personnel, who are monitoring the Team's frequency, should be procedurally trained (by SOP) to respond to the SOS by asking authentication and situational questions that require only 'yes or no' answers by 'breaking squelch' a set number of times. Once relay site personnel have confirmed that the Team is in imminent peril, they would typically notify appropriate air support.
- All US SF personnel should master artillery call-for-fire and CAS airstrike procedures. Mastery includes using these procedures while moving/maneuvering and while under fire. The radio operator should know proximate Team location and/or the relative locations of RPs, in order to better request support and to minimize fratricide risk. These actions should be practiced frequently and training should be conducted under stressful conditions.
- Consider using a 'Quarterback Wristband' to record the daily call sign, to maintain key report formats, to jot down coordinates and to facilitate fire/air support call procedures. Otherwise, record these formats in a notebook. If the notebook contains graph paper, the Team Member will be better able to draw terrain features/profiles and create tables to record information. Remember that the wristband and notebook must be recovered from WIA/KIA personnel along with maps and cryptologic information.
- The radio operator must ensure that the radio is positioned in the rucksack allowing rapid access to controls, connections and antennas.

Countering Enemy Signal Counter-Measures TTPs:

- A sophisticated, technically advanced enemy will employ all possible means to pinpoint the Team location by RDF or will employ other Electronic/Communications Countermeasures (ECCM) to impair Team communications. Team communications personnel must train

on SOI, communication procedures and SOPs to counteract enemy ECCM. Steps to avoid enemy RDF locating operations may include:
  ○ Transmissions should be clear, concise, brief and with minimal frequency.
  ○ Transmit at irregular times. Note: this may not be feasible if the Team is dependent on intermittent or airborne communications relay.
  ○ Transmit using varying frequencies.
  ○ Transmit from different locations.
  ○ Employment of a directional antenna.
  ○ Compression of data before transmission; employing burst transmission, where feasible.
  ○ Consider 'going dark' when in proximity to the target or enemy forces or when in raid/ambush/surveillance positions.

<u>True Account:</u> A SOG SR Team had been assigned an area reconnaissance mission in southeastern Laos and had been moving towards the area of interest for a day-and-a-half. The Team ascended a ridge and stopped at the military crest for its midday meal and scheduled communication. The Team established a perimeter and deployed Claymore mines. The T/L calculated the Team location, drafted a message and handed the draft to the radio operator for encryption and transmission. The radio operator attempted to reach the radio relay site using AN/PRC-25 radio with the standard whip antenna, but was unsuccessful, so he removed his rucksack and double-checked to ensure the frequency was properly set in accordance with the SOI. He then erected the long antenna and was beginning the transmission when first one Claymore and then three others were detonated. The Team had unknowingly established its perimeter next to a high-speed trail running along the ridge-top and an enemy platoon passed by in transit toward Cambodia. The radio operator first retrieved the SOI one-time-pad and draft encrypted message that had been left lying on the ground while the long antenna was being erected; as he then picked up his rucksack and attempted to move down-slope, the erect antenna became entangled with vegetation and the radio operator fell backwards down to the base of the hill, in the process dropping the rucksack which then disgorged the radio downhill until antenna segments and the handset wrapped around vegetation and was caught up-slope. The T/L ran up the steep hill, firing, and kicked the radio downhill causing the antenna to break off at the base and ripping the handset off its cord – rendering the radio useless. Fortunately, US Team Members were equipped with URC-10 emergency radios and were able to call for an extraction.

<u>What Went Wrong:</u>
• The T/L did not check outside the perimeter for the presence of enemy or high-speed routes of approach.
• The radio operator (his first mission), should have pocketed the SOI one-time-pad immediately after encrypting the message.
• Two indigenous Team Members, who were nearest the hill crest and who had initiated the first Claymore, fled downhill and off the ridge, leaving the radio operator alone at the military crest of the ridge. These men were fired upon return to the FOB.
• The radio operator could have separated the antenna segments (connected via antenna shock-cord) before attempting to move downhill. If still impeded, the radio operator should have deliberately broken off the antenna at the base to salvage the radio and handset. The radio operator was insufficiently trained.

Communication Equipment TTPs:

• Have the FOB Communications Section prepare an expedient wire antenna to supplement or replace the standard-issue extended range antenna. Expedient antenna design is depicted in the Ranger Handbook and in Special Operations manuals. The expedient antenna should be easily attached and detached to the external antenna fitting on the radio. As the antenna length should be optimized for operating frequencies, the expedient should be fabricated in sections (with couplings) that can be assembled to provide the proper antenna length or furnished with an adjustable antenna coil.

- Always carry spare radio batteries, but do not remove a spare from its plastic packaging prior to installation. If batteries go dead or become weak, do not discard them while on patrol. Instead, dispose of them by physically destroying the battery and burying the cells; or, throw them into deep water. The enemy can be expected to connect recovered weakened batteries in series to power their own equipment or the enemy can put them to other uses (e.g. to power electrical firing circuits/booby-traps).
- A small battery can recover some of its charge, particularly in cold weather, by placing it in an arm pit or between the legs of a Team Member. A larger battery can gain added life by sleeping with the battery next to the body. Additional life can also be gained by placing batteries in direct sunlight.
- Additionally, carry an extra radio handset and ensure all handsets are waterproofed with plastic wrap.
- All C&E equipment should be inspected for serviceability and function by the FOB communications section prior to issue to the Team. The Team must conduct its own component and system inspection and functional checks upon receipt of the equipment and prior to movement to the Launch Site.
- Handheld or squad radios furnished to US Team members for internal Team communication might not be issued to indigenous troops. While these are low-power, relatively short-range devices, the rules regarding Communications Security/Signal Security (COMSEC/ SIGSEC) still apply, especially when operating in proximity to enemy concentrations or secure facilities.

Wiretapping TTPs:

- When using a wiretap device, do not place the batteries in the set until the device is ready to be placed. If the batteries are carried in the device during the operational movement, they may lose some power even though switches are in the off position.
- Tapping Fiber Optic Cable:
  - Foreign militaries are increasingly employing fiber optic cable for landline communications, particularly to link C⁴ISR units, AA/radar systems and major logistics organizations/ installations, all of which have high data demand. SR Teams might follow the cable to enemy locations, but should exercise care where such lines approach enemy perimeters.
  - Military fiber optic cable has fewer strands (thus a thinner cable) than is common with commercial equivalents. Tactical cable will not be found on or buried in the ground, but will be suspended from tree limbs or existing telephone poles. Line laying teams will typically use a forked pole to suspend or recover the cable. SR Team Members may use the same method to lower the cable to install the tap. If the line is suspended from telephone poles, it may have been installed in advance of military operations to connect major communication nodes in enemy territory; enemy line teams will possess climbing spikes to string additional lines from the main line junctions. Team Members, especially commo personnel, should know what cabling equipment and material looks like.
  - Equipment and methods are available to the SpecOps and Intel communities to tap such cable, which will not register a detectable signal/power loss. The Team will not be able to read encrypted signals or the substantial data transmitted over these lines. The raw data/signal must be collected onto storage media, compressed and transmitted to higher

headquarters via directional antenna to a relay capability (e.g. UAV, satellite) or exfiltrated in some other manner.

- ◦ The Team may have to linger in the vicinity of the tap, to swap media, change batteries, operate the transmitter, etc. The Team must select a hide location and/or employ sensors to detect an enemy approach. And the Team must transmit the data from different locations.

Orders, Reports and Communications Formats TTPs:

- Higher Headquarters staff elements and the SR chain (including the Teams) should rely heavily on SOPs and checklists to reduce staff work, to simplify and streamline procedures and to ensure completeness of planning, preparation, coordination and execution. Orders, reports and communications formats are standardized for these reasons.
- In circumstances where an SR Team must deploy from a home station (or interim location) to an AO or Target Area, requirements for coordination would demand more thorough and elaborate staff work, to include OPLANS/OPORDs, FRAGOs, etc.
- See Formats at Appendix D.

Other Team and FOB Administration TTPs:

- National origin of FOB/Headquarters staff members is an important security consideration. At a minimum, security practices should limit the population of potential agents within the organization.

   'The general rule about staff appointments was only to employ British subjects by birth; or, after the amalgamation with OSS, only British or United States subjects.'[6]

- US FOB/Headquarters and Training staff and Team personnel behavior should be monitored and urinalysis should be considered in suspicious circumstances.
- The FOB should have a Training NCO/officer whose responsibilities should include: range management, Lessons-Learned management, oversight/management of training assets/ equipment, management of common subject lesson plans, training management for special-purpose equipment and coordination with other staff (especially, but not limited to S-2, S-3 and S-4) and higher headquarters.

---

6. M.R.D. Foote. *SOE in France*, no pagination.

Chapter 6

Post Mission Activities TTPs

Debriefing/After Action Report (AAR) TTPs:

'In his battle studies, Ardant du Picq stated … :
The smallest detail, taken from an actual incident in war, is more constructive for me,
a soldier, than all the Thiers and Jominis in the world. They speak, no doubt, for the
heads of states and armies but they never show me what I wish to know – a battalion, a
company, a squad, in action.'[1]

- Perhaps the oddest sound an SR Team Member may hear is that of his own voice after
  a week or more of communicating in hand and arm signals and perhaps the occasional
  whisper; this may last a day or so.
- If the Team is equipped with a tactical tablet, the T/L may find the time (e.g. while the
  Team is occupying a hide position, MSS, etc.) to record mission intelligence/information
  into a standard debriefing format while the information is still fresh in the minds of Team
  Members. During rest and communications breaks Team Members should also mark up
  their maps with observations and with remarks recorded on the back of the map or in a
  notebook. This will pay off during the debriefing/AAR. The best way to do this is:
  - Marking of enemy positions, field fortifications, trails, logistics stores, vehicle parks,
    etc., onto the map using military map/logical symbols, may consume too much space on
    the map, and may obscure terrain features and other essential map data. Instead, mark
    numbers or icons (by SOP) onto the map or overlay/acetate and cross-reference these to
    an information table that is on the back of the map or in the Team Member's notebook.
    Ensure that the Team is secure before pausing to make these postings.
  - As time and opportunity allows, Team Members should compare the observations each
    has collected.
  - This information can later be recorded into a digital debriefing format via a tactical tablet.

Joint FOB/Base Occupancy

- Perhaps the best location for a FOB is at an airfield that is jointly occupied by SR
  Teams, Exploitation Forces, USAF CAS, FACs, UAV components, insertion/extraction
  (helicopter, VTOL, fixed wing) and gunship aircraft. Co-location allows for personnel
  to develop personal and professional relationships; a better understanding of mission
  capabilities and limitations; a degree of joint training; inter-Service/component teamwork
  and common defense.
- During the Vietnam War, Project Delta (B-52) and FACs from the 21st Tactical Air
  Support Squadron (TASS) lived on the same compound and developed strong relationships
  with Delta's SR Teams. This co-location vastly improved operational execution of both SR
  Teams and their FACs and led to early adoption of SR TTP's across Service boundaries.

---

1. Halder, *Small Unit Actions*, p. v

Post-Mission Sustainment/Equipment Maintenance TTPs:

- Immediately following mission debriefing, or no later than the following day, perform the following sustainment activities.
  - Clean and restore to serviceability all weapons and weapon accessories.
  - Clean and/or restore to serviceability all communications gear, communications accessories and propaganda materials. Turn in all C&E materials to include SOI.
  - Turn in and/or replace medical supplies and equipment.
  - Replace all unserviceable personal clothing and equipment to include: uniforms, LBE, rucksacks, etc.
  - Turn in, dispose of, or retain for training any munitions that are less than 'Condition Code A'. Restore normal mission load.
  - Perform maintenance on all mobility equipment and accessories.
  - Clean and restore to serviceability all other Team and special equipment/items.
  - If the Team routinely uses enemy clothing, LBE, equipment on operations, they should be locked up in the US Team room and only be drawn for training, maintenance and operations. Team Members should only use US gear for FOB defense or risk fratricide.
- During Team stand-down, training or FOB duty rotation, train in (or at least discuss) Team response to an attack on the FOB, including actions taken when the attack is launched on the FOB from different directions and with different attack scenarios. Discuss your findings/coordinate with the FOB/Recon headquarters before establishing a Team SOP on this matter.

FOB Security and Defense TTPs:

General Security TTPs:
- If the FOB's mission activities are successful and pose a serious threat to enemy plans and operations, the <u>FOB will be targeted for attack</u>. Since the FOB is a concentration of SpecOps capability, it is a very lucrative target to the enemy. Attacks by fire, to include long range rocketry and mortar fire, pose low risk to the enemy and if the fires are massed and accurate could have a significant result. If the enemy stages a raid on the FOB, they will use <u>elite troops</u> to do so.
- The FOB may have a security force and/or a workforce comprised of local nationals. Be assured that these personnel will have enemy agents among them. If the enemy infiltrates sappers or assaults into the FOB, they will likely target communications, headquarters and billeting structures, and/or any aviation assets within the FOB as their chief priorities. So plan and prepare to:
  - Defend in place (e.g. from the Team room).
  - Use steel wire mesh on the windows, with a gauge and gap sufficient to prevent passage of grenades.
  - Move in the most secure manner and path to the Team's designated defensive positions.
  - Set aside additional arms and ammunition (e.g. Claymores, grenades, etc.) for FOB defense, and carry these items from the Team room to defensive positions.
  - Drill Team Members on techniques for exiting the Team room and in fire and maneuver from Team billets to designated defensive positions.
  - Work to improve Team's designated defensive positions, ensuring that the positions provide overhead cover and can support all-around defensive fires. Consider using CONEX shipping containers, modified to provide firing ports, as the internal structure for the defensive positions/bunkers; this will provide a rigid framework to support sandbagged walls and overhead cover. Recommend that the CONEX be placed atop

dunnage/supports to ensure the floor is above a prospective water line in areas where heavy rainfall (monsoons) may occur.

- A cunning enemy will attack the most lucrative FOB targets. If SR or FOB personnel are routinely assembled at a given place and time, the enemy has been provided an almost irresistible gift. Routine assemblies include: daily administrative formations, PT formations and mess hall lines during mealtimes. Other vulnerable assemblies include occasional entertainment performances. SpecOps personnel attending some assemblies may not be bearing weapons; even if weapons are carried, LBE (with ammunition) will not be worn. A wise T/L will conceive of Team actions under such circumstances and at least discuss responses with the Team.

> 'At 3 A.M. on 23 August [1968], more than 100 NVA sappers penetrated the CCN [SOG Command and Control North] compound, tossing satchel charges and blasting away with AKs. For the next three hours, Green Berets and SEALS from the nearby SOG Naval Advisory Detachment compound combated the NVA raiders. By dawn it was over and the terrible toll was visible: 15 Special Forces officers and NCOs died that night, the greatest single-day loss of Green Berets in Vietnam.'[2]

Sixteen indigenous commandos were also dead.

- When the Team is in a FOB support cycle/rotation, the Team may be assigned night ambush duties outside the perimeter. This is a great opportunity to train in night operations (movement, ambush, raid, battle drill, breakout, etc.). Design this to supplement/reinforce the Team pre-mission training cycle. The T/L should ensure that the training is realistic (without weapon firing). The ambush Team <u>must</u> accurately report its position to the TOC during its night operation to avoid a fratricidal engagement. If the FOB is raided by the enemy, air and fire support assets will normally not be used if the Team location is unknown.
- If resources allow, a Team and/or a RF platoon should be designated for standby, on a rotating basis, to rapidly mount a pursuit of an enemy raid unit. Perhaps the best approach to this is to:
  - Perform an IPB analysis of terrain surrounding the FOB upon establishing the installation.
  - Use the area surrounding the FOB for occasional Team/RF training so that FOB combat assets 'know the ground'.
  - If possible, conduct joint Team and RF training in Hammer and Anvil operations on a periodic basis. If this is not feasible, conduct joint leadership terrain walks.
  - When an attack occurs, the designated pursuit Team should determine the enemy route of withdrawal as quickly as possible. A SR Team on routine perimeter night patrol/ambush duty may be close to the enemy raid position(s) and may be able to engage the enemy. If not close enough to engage the enemy, the Team may be first on the scene and best able to provide information on the enemy route of withdrawal in a timely manner. WARNING: the enemy may be expected to establish security positions and/or ambushes along its route of withdrawal. It may be prudent to initially pursue the enemy along a parallel path, using superior knowledge of the terrain.

Weapons and Munitions for FOB Perimeter Defense:

- If Claymore mines or other anti-personnel command detonation devices are emplaced on the perimeter with lines running to Team defensive positions, consider encasing the firing wire in conduit below the ground surface, or otherwise laying the wires to be protected from

---

2. Plaster, *SOG: The Secret Wars*, p. 193

PD ordnance or sabotage by enemy agents employed as FOB support personnel. Periodically inspect Claymores and/or other lethal devices for sabotage and for serviceability.

- Improvised FOB perimeter defensive weapons:
  - Pole-Mounted Fragmentation Device: This improvised munition can be fabricated as a directional or an omni-directional device. As a directional device, it would function much like a Claymore mine; however, it would contain significantly more explosive and fragmentation pellets. As an omni-directional device, it would be used in defilade areas or for FPF if the enemy has breached FOB defenses. There should be sufficient explosive to propel two layers of fragmentation pellets. This device can use cast explosive. An alternative omni-directional device is to mount an artillery projectile onto a pole.
  - Multiple Rocket Launcher (MRL):
    - Excess, but serviceable aerial (e.g. helicopter) rocket pods can be suspended from a frame and used as a ground-to-ground MRL. It would be fired electrically (battery) and can be rigged for single or ripple fire using a simple control box.
    - The MRL would be able to fire a variety of 2.75in rockets. Ranges will vary based on angle of fire and type of rocket warhead. A rudimentary ranging scale/device, verified by test firing on a firing range, can be fabricated.
    - The impact area and angle of fire must account for the rocket Safe and Arm distance.

More FOB Tips:

- US T/Ls should keep a list of TTPs and Lessons-Learned; these should be discussed post-mission. This is an excellent way to increase the professionalism of Team Members. Collate these Lessons-Learned into a list or add some of them to the Team SOP. These Lessons-Learned should also be shared by the T/L with other T/Ls and/or with the parent SR unit.

'Histories make men wise' – Francis Bacon

- Fabricate a robust 'clothes tree/valet' to mount LBE, weapons and other essential equipment so that they are instantly ready to don in response to an enemy attack.
- Ensure that boots, or other sturdy footwear, are placed in the same location <u>every time</u> so that they can always be located in the dark. Preferably, the footwear should be of a type that can easily be slipped on (without laces) and that will not fly off while running.
- Wear 'pajamas' to bed that use the same camouflage pattern as the standard fatigue/utility uniform. This is to avoid being misidentified as an enemy combatant during an attack. The 'pajamas' may be modified from a spare field uniform.

Appendices:
- A – Glossary/Abbreviations
- B – Notional Tactical Training and Range Complex (w/ Illustration).
- C – Local Weather Indicators
- D – SR Team Orders, Communication and Report Formats
- E – Bibliography, Sources and Further Reading

Appendix A

Glossary and Abbreviations

Glossary

550-Cord – Parachute Cord; now used for many general-purpose tasks.

Abatis – A field fortification consisting of an obstacle formed by fallen trees with the tops and branches of trees facing towards anticipated enemy route of approach.

Advanced Operations Base (AOB) – A small temporary Special Forces operations base established near or within an Area of Operations to command, control, and/or support training or tactical operations. Facilities are normally austere, may include an airfield/airstrip, a pier/anchorage. An AOB is normally controlled and/or supported by a main operations base or a Forward Operations Base. Sometimes referred to as a Launch Site.

Ahkio – A type of toboggan or sledge without runners suitable for carrying goods for people in deep snow.

Air Guard – Surveillance of air avenues of approach toward military units, facilities etc., normally performed by a guard assigned for this function.

Ammunition Supply Point (ASP) – A supply point for munitions/ammunition stores.

Arc Light – A codename/general term for the use of B-52 aircraft sometimes used to support ground tactical operations.

Battle Drill – See Immediate Action Drill

Blue Light – United States military's first counterterrorism unit; precursor to Delta Force.

Bomb Damage Assessment (BDA) – A post-strike ground reconnaissance of an aerial bombardment/major airstrike.

Bright Light Team – SOG recon team serving a weeklong deployment at a launch site as an on-call rescue/recovery force for downed or KIA aircrews, POWs and recon team personnel.

Cache – A hidden or inaccessible storage point used to support covert military operations.

CAR-15 – a carbine version of the M-16 with a telescoping stock and shortened barrel. Precursor to the M4 carbine.

Close Air Support (CAS) – Air action by fixed and/or rotary-wing aircraft against hostile targets that are in close proximity to friendly forces and that require detailed integration of each air mission with the fire and movement of those forces.

Command-And-Control Central (CCC) – also known as SOG Forward Operating Base 2 (FOB 2) located at Kontum, South Vietnam. Responsible for SOG operating areas of southern Laos and northeastern Cambodia.

Command-And-Control North (CCN) – SOG Operating Base near Danang, South Vietnam. Responsible for SOG operating areas in Laos and the demilitarized zone (DMZ).

Command-And-Control South (CCS) – SOG Forward Operating Base located at Ban Me Thuot, South Vietnam. Responsible for SOG operating areas in Cambodia.

Defilade – Protection from hostile observation and fire provided by an obstacle such as a hill, ridge, or bank.

Explosively Formed Penetrator (EFP) – A self-forging warhead application of the Misznay-Schardin effect (aka: Platter Charge).

Forward Air Controller (FAC) – US Air Force pilot flying a small aircraft tasked to control air assets for insertions, extractions, Close Air Support and to identify ground targets for air attack.

Forward Operating Base (FOB) – semi-permanent SOG camp used to station command and staff; Special Forces personnel and indiginous troops; and support operations.

Ghillie Suit – Camouflage clothing designed to resemble the background environment. Typically, a net or cloth garment adorned with loose strips of burlap, cloth, or twine.

Green-on-Blue – a phrase used to describe attacks on US/allied forces by purported friendly or neutral military/paramilitary personnel.

Guerilla – A member of a small, irregular (civilian paramilitary) independent group taking part in irregular fighting, typically against larger regular forces.

High Altitude, Low Opening (HALO) – aka Military Free Fall parachuting.

High Speed Trail – A widened, well-groomed trail designed for rapid foot movement through rugged terrain.

Immediate Action Drill – also known as Battle Drill; series of standardized rapid tactical maneuvers.

Improvised Explosive Device (IED) – A bomb/device constructed and deployed for terrorist or unconventional military/paramilitary use. A booby-trap that uses explosives.

Intelligence Preparation of the Battlefield/Battlespace (IPB) – The systematic process of analyzing the mission variables of enemy, terrain, weather, and civil considerations in an area of interest to determine their effect on both friendly and enemy operations.

Mission Support Site – A clandestine, pre-selected area used as a temporary base or stop over point. The MSS is used to increase the operational range of a Special Operations unit.

Montagnards – Members of highland/mountain tribes of central or southern Vietnam who were extensively employed as indigenous mercenary troops by the US Special Forces.

Motti (Finnish) – A surrounded/encircled military unit or a tactic employed by the Finns designed to encircle an enemy force.

Multiple Rocket Launcher (MRL) – A rocket artillery system capable of rapid/simultaneous rocket launching.

Nightingale Device – a SOG/CIA-developed decoy device using bundles of firecrackers designed to simulate a firefight.

Office of Strategic Services (OSS) – American version of the British clandestine Special Operations Executive (SOE), responsible for espionage, sabotage and other covert operations during World War II; precursor organization to the CIA. MACV-SOG was modeled on the OSS.

One-One (1-1) – A SOG Recon Team positional code used to designate the Assistant Team Leader (US).

One-Two (1-2) – A SOG Recon Team positional code used to designate the Team radio operator (US).

One-Zero (1-0) – A SOG Recon Team positional code used to designate the Team Leader (US).

Partisan – A member of an irregular <u>military</u> force formed to oppose a foreign power/invader through paramilitary/insurgent activity. Often a member of a military force operating on detachment from main force units.

Phoenix Program – US Program, created in 1967, to coordinate intelligence and operations against the Viet Cong Infrastructure (VCI).

Project Delta – Vietnam-era special operations organization (under the 5th Special Forces Group) similar in function to SOGs' C&C activities, but with operations confined within the borders of South Vietnam, aka B-52.

Provincial Reconnaissance Unit (PRU) – An elite paramilitary unit under CIA control, during the Vietnam War.

Red Team – A process conducted by a trained, experienced opposing force, providing commanders/ Blue Team (friendly force) participants an independent capability to continuously challenge plans, operations, concepts, organizations and capabilities in the context of the operational environment and from the adversaries' perspectives.

Rough Terrain Forklift (RTFL) – A forklift designed with large tires for military/construction use over rough, unpaved ground.

Savannah – A grassy plain in tropical and subtropical regions, with few trees.

Shaped Charge – an explosive charge that uses the Monroe effect to focus the effects of the explosive energy used to cut metal, penetrate armor and bore holes through materials.

Special Operations (SpecOps) – Operations conducted in hostile, denied, or politically sensitive environments to achieve military, diplomatic, informational, and/or economic objectives employing military capabilities for which there is no broad conventional requirement. These operations often require covert, clandestine, or low visibility capabilities. Special operations are applicable across the range of military operations. They can be conducted independently or in conjunction with operations of conventional forces or other government agencies and may include operations through, with, or by indigenous or surrogate forces. Special operations differ from conventional operations in degree of physical and political risk, operational techniques, mode of employment, independence from friendly support, and dependence on detailed operational intelligence and indigenous assets. (*Joint Publication 1–02)*

Special Reconnaissance (SR) – aka Strategic Reconnaissance. 'Reconnaissance and surveillance actions conducted as a special operation in hostile, denied, or politically sensitive environments to collect or verify information of strategic or operational significance, employing military capabilities not normally found in conventional forces.' (DoD Dictionary of Military and Associated Terms.)

Steppe – Grasslands/Scrublands generally on treeless grassland plains. Trees may appear near grassland waterways.

Table of Organization and Equipment (TOE) – A document specifying the organization organizational structure, staffing and equipment of US military units.

Tactics, Techniques and Procedures (TTPs) – 'Tactics, techniques, and procedures (TTP) provide the tactician with a set of tools to use in developing the solution to a tactical problem. The solution to any specific problem is a unique combination of these TTP or the creation of new ones based on a critical evaluation of the situation. The tactician determines his solution by a thorough mastery of doctrine and existing TTP, tempered and honed by experience gained through training and operations. He uses his creativity to develop solutions for which the enemy is neither prepared, nor able to cope.' (FM 3-90) A tactic is the highest-level description of unit behavior, while techniques give a more detailed description of behavior in the context of a tactic, and procedures an even lower-level, highly detailed description in the context of a technique.

Tundra – A vast, flat, largely treeless plain found in arctic/sub-arctic regions.

Abbreviations
AA – Anti-Aircraft
AAR – After Action Report/Review
AFV – Armored Fighting Vehicle
aka – also known as
AO – Area of Operations
AOB – Advanced Operations Base (SF)/Aviation Operations Base
AM – Amplitude Modulation
AP/API – Armor Piercing/Armor Piercing Incendiary
APC – Armored Personnel Carrier
APERS – Anti-Personnel
ATACMS – The MGM-140 Army Tactical Missile System
ATV – All Terrain Vehicle
BDA – Bomb Damage Assessment
BIP – Blow-In-Place. Destroy a hazardous item without risking disturbance of the item.
BLT – Bright Light Team. A SR Team with a mission of rescue and/or recovery of friendly personnel.
C^2 – Command and Control
C^3 – Command, Control and Communications.
C^4ISR – Command, Control, Communications, Computers, Intelligence Surveillance and Reconnaissance
C-4 – Plastic High Explosive
CA – Civil Affairs
CAC – US Army Combined Arms Center
CALL – Center for Army Lessons-Learned
CARVER – Criticality, Accessibility, Recuperability, Vulnerability, Effect and Recognizability analyses methodology

CAS – Close Air Support

CBRNE – Chemical, Biological, Radiological, Nuclear, Explosive

CBWs – Chemical Biological Weapons

CCC/CCN/CCS – Respectively, Command and Control Central, Command and Control North, Command and Control South

CCIR – Commander's Critical Information Requirements

CG – Counter-Guerilla

CHICOM – Chinese Communist

CIA – Central Intelligence Agency

COIN – Counterinsurgency is a comprehensive civilian and military effort taken to defeat an insurgency and to address any core grievances

COMSEC – Communications Security

CONEX – A large, steel-reinforced reusable container for shipping military cargo.

CONOPS – Concept of Operations

CONUS – Continental United States

CORDS – Civil Operations and Revolutionary Development Support. US organization created in 1967 to control all American pacification activities, during the Vietnam conflict.

CP – Command Post

CR – Counter Reconnaissance

CS – Tear Gas Riot Agent or Combat Support

CSS – Combat Service Support

CT – Counter-Terrorism/Communist Terrorists

CV – Critical Vulnerability (See CARVER)

CWMD – Combating Weapons of Mass Destruction

DA – Direct Action

DoD – Department of Defense

DoS – Department of State

DS – Direct Support

DZ – Drop Zone

ECCM – Electronic/Communications Countermeasures

ECM – Electronic Counter Measures

EF – Exploitation Force

EFP – Explosively Formed Penetrator

EMP – Electro-Magnetic Pulse

E&E – Escape and Evasion

FAC – Forward Air Controller

FID – Foreign Internal Defense

FM – Field Manual/Frequency Modulation

FOB/FOB2 – Forward Operating Base/ Forward Operating Base 2 (CCC's FOB)

FPV – First Person View

FRAGO – Fragmentary Order

FSU – Former Soviet Union

FTX/STX – Field Training Exercise/Situational Training Exercise

GPS – Geographic/Global Positioning System

GS – General Support

GOTWA – Going, Others, Time, What, Actions. A Team Leader notification to other Team Members to plan, coordinate and communicate T/L excursions from the main body of the unit.

GWOT – Global War On Terrorism

HALO – High Altitude, Low Opening airborne insertion. aka: Military Free Fall

HE – High Explosive

HEAT – High Explosive, Anti-Tank

HEDP – High Explosive, Dual Purpose

HEIAP – High Explosive, Incendiary, Armor-Piercing

HQ – Headquarters
HUMINT – Human Intelligence
IED – Improvised Explosive Device
IFV – Infantry Fighting Vehicle
IO – Information Operations
IPB – Intelligence Preparation of the Battlefield
IR – InfraRed
IT – Information Technology
JSOC – Joint Special Operations Command
KIA – Killed in Action
LBE – Load Bearing Equipment
LED – Light-Emitting Diode
LIDAR (Light Detection and Ranging) – A surveying method that measures distance to a target by illuminating the target with laser light and measuring the reflected light with a sensor.
LLNBL – Lower Left No Bomb Line. The lower left corner of a six kilometer square target area.
LRRP/LRS – Long Range Reconnaissance Patrol/Long Range Surveillance
LSO – Launch Site Officer
LZ – Landing Zone
MAC-V SOG – Military Assistance Command, Vietnam. Studies and Observations Group
MANPAD – Man Portable Air Defense
METT-TC – Mission planning and analysis factors including: the Mission, the Enemy, the Terrain and Weather, Troops and Support Available, Time Available, and Civil Considerations (METT-TC)
MGF – Mobile Guerilla Force
MHE – Material Handling Equipment
MIA – Missing in Action
MILES – Multiple Integrated Laser Engagement System. Generally, a weapon mounted laser-tag system used for simulated force-on-force training engagements.
MOH – Medal Of Honor
MOS – Military Occupational Specialty
MRL – Multiple Rocket Launcher
MSR – Main/Major Supply Route
MSS – Mission Support Site
MILVAN – Military shipping container.
NBCR – Nuclear, Chemical, Biological, Radiological
NCO/NCOIC – Non-Commissioned Officer/ Non-Commissioned Officer In Charge
NDP – Night Defensive Perimeter; sometimes referred to as Rest Over Night (RON). See RAD.
NVA – North Vietnamese Army
NVD/NVG – Night-Vision Device/Night-Vision Goggles
O&I – Operations and Intelligence
OAKOC (aka OKOKA) – Observation and Fields of Fire, Avenues of Approach, Key and Decisive Terrain, Obstacles, Cover and Concealment. OAKOC is used by unit leaders to analyze terrain and the effects of weather on unit operations.
OCONUS – Outside the Continental United States
ODA – Operational Detachment 'A'. aka 'A' Team or Special Forces Operational Detachment (SFOD)
OJT – On-the-Job-Training
OOTW – Operations Other Than War
OP/LP – Observation Post/Listening Post
OPCON/TACON – Operational Control/Tactical Control
OPSEC – Operations Security
OPTEMPO – Operating Tempo (Pace of Operations)
OSS – Office of Strategic Services
PD – Point Detonating
PIR – Priority Intelligence Requirement

PLS – Position Location System

PMCS – Preventative Maintenance Checks and Services.

POL – Petroleum, Oil and Lubricant

POW – Prisoner of War

PRC – People's Republic of China

PRU – Provincial Reconnaissance Unit

PSYOPS – Psychological Operations

QRF – Quick Reaction Force

RAD – Remain All Day defensive/hide position. Daylight equivalent of the NDP.

RDF – Radio Direction Finding

RF – Reaction Force also known as the Quick Reaction Force (QRF). Or Radio Frequency.

ROE – Rules of Engagement

RON – Rest Over Night

RP – Rally Point or Reference Point.

RPD – Soviet Degtyaryov light machine gun, chambered for the 7.62×39mm round (same as the AK-47). Also manufactured by other nations and still in widespread use worldwide.

RPG – Rocket Propelled Grenade

RPV – Remotely Piloted Vehicle or Remote Person View

RRC – Regimental Reconnaissance Company of the 75th Ranger Regiment

RT – Reconnaissance Team

RTFL – Rough Terrain Forklift

S-1 through S-4 – Staff sections in order: Personnel, Intelligence, Operations, Supply

SAM – Surface to Air Missile

SAR – Search & Rescue

SAS – Special Air Service (British SpecOps Regiment)

SATCOM – Satellite Communications

SCG – Security Classification Guidance

SD – Self-Destruct

SEA – South East Asia

SEAL – Sea-Air-Land comprised of Naval Special Operations personnel under the command of US Special Operations Command (US SOCOM).

SERE – Survival, Evasion, Resistance and Escape

SF – US Army Special Forces

SFOD/ODA – Special Forces Operational Detachment/Operational Detachment 'A' ('A' Team)

SFSG – Special Forces Support Group

SIGSEC – Signal Security

SIGINT – Signals Intelligence

SITREP – Situation Report

SLAP/SLAP-T – Saboted, Light Armor Penetrating/Saboted, Light Armor Penetrating – Tracer

SME – Subject Matter Expert

SOE – Special Operations Executive

SOI – Signal Operating Instructions

SOP – Standard Operating Procedures

SOF – Special Operations Forces

SpecOps – Special Operations

SRR – Special Reconnaissance Regiment of the UK Special Forces

SR – Strategic or Special Reconnaissance

SRR – British Special Reconnaissance Regiment

SWOT – Strengths, Weaknesses, Opportunities and Threats analysis methodology

TACAIR – Tactical Air Support

TACON – Tactical Control

TEL – Transporter, Elevator and Launcher

TICs – Toxic Industrial Chemicals

TIMs – Toxic Industrial Materials
T/L – Team Leader
TM – Technical Manual
TOE – Table of Organization and Equipment
ToT – Time on Target
TTPs – Tactics, Techniques and Procedures
UAV – Unmanned Aerial Vehicles
UTV – Utility Terrain Vehicle
UW – Unconventional Warfare
VR – Visual Reconnaissance
WIA – Wounded in Action
WMD – Weapon(s) of Mass Destruction
WP – White Phosphorus

Appendix B

Notional Tactical Training and Range Complex

Remote, isolated areas located in allied and/or host nation territory may be available for fairly unrestricted tactical training use by SpecOps units. Some of these areas may be located in contested or 'free fire' zones; such an area was reserved for SOGs' CCC/FOB2 live-fire training. Similar areas may be available within the US, but these areas will have a spectrum of constraints/restraints (environmental, safety, dry vs live fire and combined arms, etc.) imposed or may have terrain limitations.

The following map* (next page) shows an ideal Tactical Training and Range Complex that has the following characteristics.

- A variety of Features, Terrain and Locations including:
  - Isolated with surrounding protective terrain features.
  - Rugged, dissected/ravines; valleys; hilly to mountain ridges.
  - Variety of vegetation: Forest/Jungle; high grass, bamboo, shrubs.
  - Road (improved; unimproved) and Trail network.
  - River and stream (including intermittent) network.
  - Swampy areas.
  - Abandoned village, buildings and cultivated areas.
  - Abundant areas for Battle Drills over variations in terrain and vegetation.
  - Sufficient Area for CAS and Artillery/Mortar impact zones and Demolition Range
  - LZ/DZ locations.
  - Abundant locations/features for training in small scale raids and all types of ambushes.
  - Sufficient size for Exploitation Force training.

- The Complex could easily be modified with the following Features.
  - Mock minefield.
  - Mock targets: insurgent camp; vehicle-park; communications relay, AA site, unit bivouacs and field fortifications, logistics points, etc.
  - Helicopter landing pad.
  - Range Control and outbuildings.

- The Complex could support a spectrum of training activities to include.
  - SR Team and RF/EF tactical navigation and movement (mounted and dismounted) over varied terrain and types of vegetation.
  - Stream and river crossings.
  - Battle Drills (Dry and Live Fire)
  - Raids and ambushes (all types)
  - LZs/DZs: Insertions/extractions (landing/string), Rough Terrain Personnel and Resupply Drops
  - Blue Team-Red Team exercises (Dry)
  - Demolition training.
  - Employment of integrated RF/EF Force and combined arms live-fire CAS and artillery fire.

* Note: Contour interval is 20m.

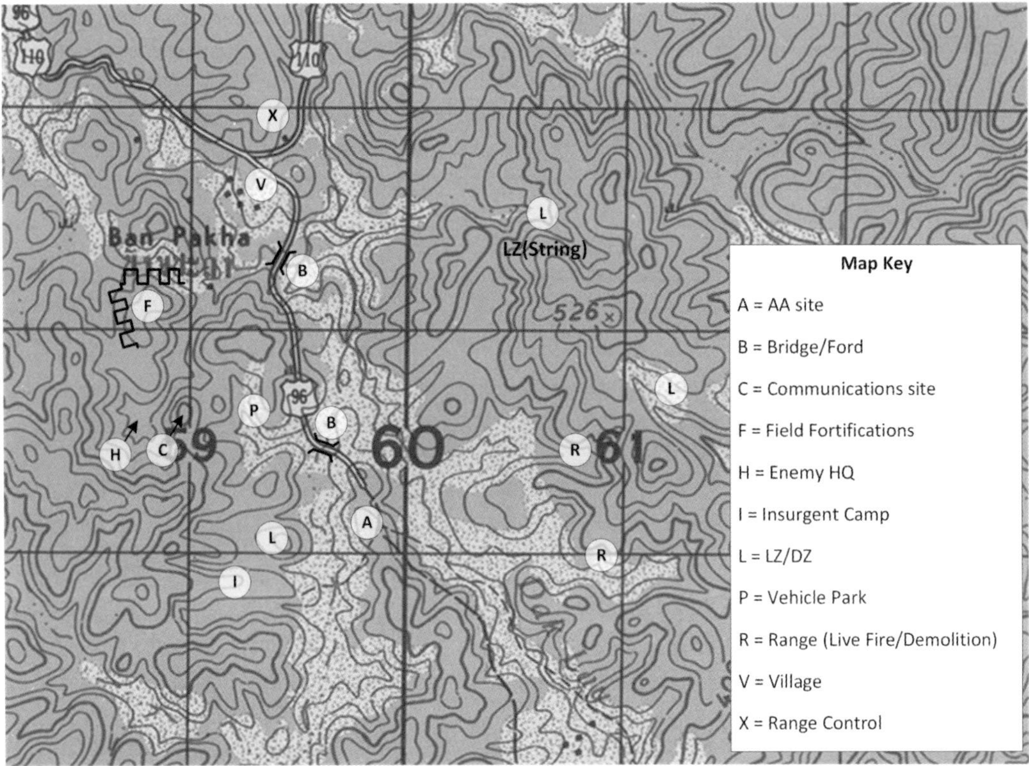

Figure B1. Notional Tactical Training and Range Complex.

Appendix C

Local Weather Indicators

General:

Weather forecasting provided by higher headquarters will certainly be useful in pre-deployment planning and preparation and prior to movement to/from the target area. But regional conditions and conditions within an Area of Operations (AO) may be substantially different to what actually occurs within a specific Target Area, so Team Members must be able to predict local weather for planning of its operations, post-insertion, while the Team is on the ground in the Target Area. No single weather factor should prove conclusive in predicting local conditions; a mutually-reinforcing correlation of factors is required. Team Members should correlate as many of the following weather indicators for reasonable accuracy.

- Typical prevailing seasonal temperature and precipitation conditions vs current actual conditions.
- Typical prevailing wind conditions that apply to the Target Area through all seasons vs current actual conditions.
  - This should be provided by higher headquarters.
  - A shift in wind direction may indicate the approach of a front.
  - Wind speed estimation. An instrument for this purpose is unnecessary (see the Beaufort Scale below).
- Times of dusk and dawn for the duration of the mission.
  - This should be provided by higher headquarters.
- Moon phases/illumination throughout the mission window.
  - This should be provided by higher headquarters (see sample table below).
- Cloud formations.
  - Cloud altitude and height will be affected by region (e.g. Temperate, Tropical) and the Jet Stream (see examples and descriptions below).
- Barometric pressure.
  - Broad utility, but especially important for predicting weather changes in mountainous terrain.
  - Shows current and predictive atmospheric conditions (e.g. High pressure vs Low pressure; inversion, neutral, lapse, etc.). When these conditions are inconsistent with time of day and/or prevailing weather norms this is one predictor of an approaching front.
  - Small, dependable and durable digital personal barometers are available commercially.
- Other phenomena: for instance, horizon coloration, moon/sun halos, animal and even insect behavior, terrain effects, etc. See METT-TC Weather-Based TTPs in core text.

Wind Conditions:

Beaufort Wind Scale Table					
Beaufort Wind Scale	Average Wind Speed (Knots/ MPH)	Wind Description	Probable Wave Height in Feet*	Sea State	Land Condition
0	<1/<1	Calm	0	Glassy	Calm. Smoke rises vertically [or remains static (inversion)].
1	1-3/1-3	Light Air	0.3	Rippled. No foam.	Smoke drift indicates wind direction. Leaves and wind vanes stationary.
2	4-6/4-7	Light Breeze	0.7	Smooth Wavelets that do not break.	Wind felt on exposed skin. Leaves rustle. Wind vanes begin to move.
3	7-10/8-12	Gentle Breeze	2	Large wavelets that start to break. Scattered whitecaps.	Leaves and small twigs constantly moving. Light flags extended.
4	11-16/13-18	Moderate Breeze	3.3	Small waves with breaking crests. Fairly frequent whitecaps.	Dust and loose paper raised. Small branches begin to move.
5	17-21/19-24	Fresh Breeze	6.6	Moderate waves of some length. Many whitecaps. Some spray.	Branches of a moderate size move. Small trees in leaf begin to sway.
6	22-27/25-31	Strong Breeze	9.8	Long waves begin to form. White foam crests very frequent. Some airborne spray.	Large branches in motion. Whistling heard in overhead wires, boughs.
7	28-33/32-38	Near Gale	13.1	Foam blown from breaking waves. Moderate spray.	Whole trees in motion. Effort needed to walk against the wind.
8	34-40/39-46	Gale	18	Moderately high waves with breaking crests and blowing foam. Considerable airborne spray. Somewhat reduced visibility.	Some twigs broken from trees. Progress on foot seriously impeded.

Beaufort Wind Scale	Average Wind Speed (Knots/ MPH)	Wind Description	Probable Wave Height in Feet*	Sea State	Land Condition
9	41-47/47-54	Strong Gale	23	High waves with some crests rolling over. Dense blown foam. Large amounts of airborne spray. Reduced visibility.	Some branches break off trees and some small trees and road signs blow over.
10	48-55/55-63	Storm	30	Very high waves with overhanging crests. Large patches of foam. Considerable tumbling of waves with heavy impact. Large amounts of airborne spray.	Trees are broken off or uprooted. Structural damage likely.
11	56-63/64-72	Violent Storm	37.7	Exceptionally high waves. Very large patches of wind driven foam. Large amounts of airborne spray. Severe reduction in visibility.	Widespread vegetation and structural damage likely.
12	64+/73+	Hurricane	46+	Huge waves. Sea completely white with foam and driving spray. Greatly reduced visibility.	Severe widespread damage to vegetation and structures. Debris and unsecured objects hurled about.

Selected Cloud Types (Rain Clouds)

Mid-Altitude Cloud Formations

Altostratus

- Gray or bluish sheet-like clouds that are thin enough to reveal the sun as if seen through a mist.
- They do not produce a halo effect.
- They do not produce shadows of objects on the ground.
- May cause very light precipitation.
- Be observant, as Altostratus may presage Nimbostratus with more severe weather.

Nimbostratus

- A thickening of an Altostratus with a lowering cloud base. Lower clouds are often seen hovering beneath or merging with this cloud at its base.
- Dark gray cloud layer with falling rain or snow, often heavy.
- Blots out the sun.
- Accompanied by generally continuous rain or snow.

Low Cloud Formations

Stratus

- A low formation similar to mid-altitude Altostratus as its gray cloud is uniform; not normally thin enough to reveal the sun.
- Produces hill fog.
- May gain sufficient thickness to produce drizzle, ice crystals or snow grains.
- As it breaks up, blue sky will be revealed.

Cumulonimbus

- The classic thunderstorm cloud; a heavy, dense and often very dark cloud in the form of a mountain or huge tower. The upper portion is nearly always flattened resembling an anvil or vast plume.
- Beneath the cloud base or in the vicinity, other scattered clouds may often be seen to merge with the base.
- Produces heavy precipitation, and may produce hail and tornadoes.

Source (Public Domain): https://www.weather.gov/jetstream/basicten

Figure C1. Altostratus. (*Public Domain*)

Figure C2. Nimbostratus. (*Public Domain*)

Figure C3. Stratus. (*Public Domain*)

Figure C4. Cumulonimbus. (*Public Domain*)

Phases of the Moon

Principal and Intermediate Phases of the Moon									
Phase	Northern Hemisphere	Southern Hemisphere	Visibility	Mid-phase standard time	Average moonrise time	Average moonset time	Northern Hemisphere	Southern Hemisphere	Photograph
New moon	Disc completely in Sun's shadow (lit by earthshine only)		Invisible (too close to Sun)	Midday	6 am	6 pm			Not visible
Waxing crescent	Right side, 1–49% lit disc	Left side, 1–49% lit disc	Late morning to post-dusk	3 pm	9 am	9 pm			
First quarter	Right side, 50% lit disc	Left side, 50% lit disc	Afternoon and early evening	6 pm	Midday	Midnight			
Waxing gibbous	Right side, 51–99% lit disc	Left side, 51–99% lit disc	Late afternoon and most of night	9 pm	3 pm	3 am			

Phase	Northern Hemisphere	Southern Hemisphere	Visibility	Mid-phase standard time	Average moonrise time	Average moonset time	Northern Hemisphere	Southern Hemisphere	Photograph
Full moon	Completely illuminated disc		Sunset to sunrise (all night)	Midnight	6 pm	6 am			
Waning gibbous	Left side, 99–51% lit disc	Right side, 99–51% lit disc	Most of night and early morning	3 am	9 pm	9 am			
Third quarter (or last quarter)	Left side, 50% lit disc	Right side, 50% lit disc	Late night and morning	6 am	Midnight	Midday			
Waning crescent	Left side, 49–1% lit disc	Right side, 49–1% lit disc	Pre-dawn to early afternoon	9 am	3 am	3 pm			

Appendix D

SR Team Orders, Communications and Reporting Formats (Samples)

Warning Order		
The Team Leader (T/L) will receive a Warning Order from higher HQ, and then issue his own Warning Order to Team Members supplementing the original warning order with details. The Team Leader should first obtain maps, review other information (e.g. target folder/intelligence reports, if available; aerial photos, etc.) and then mark up his own map (master) with relevant information. Revisions to the warning order will occur as a result of updated information, VR (if available) information and other factors that may emerge during the Team preparation period. Additional sub-paragraphs may be necessary based on such factors as environmental and terrain conditions, etc., where these matters are not covered by SOP. The chief briefing aid should be the Team Leader's (master) map. Once fleshed-out, this document will serve as the 'talking-paper' for the Mission Briefing.		
Para #	**Item**	**Explanation**
	Warning Order #	May include Mission Code Name. Otherwise, Self-explanatory.
	Reference	When the T/L receives map sheets, he should ensure that all map sheets are of the same edition and are current. When the T/L issues his Warning Order, he will distribute these map sheets to Team Members, who should mark up the sheets to be identical to the T/L master. The master map should not be 'cropped' prior to the launch date (if then).
	Time Zone	Provided if the mission is to take place in another time zone.
	Task Organization	If the Team is to receive attachments, these personnel should be present during the T/L Warning Order. All other information regarding attachments will be briefed in sections 1.e. and 3.c-d. If the Team is to be attached to a larger force (e.g. in COIN missions), so state, and refer to paragraph 5.
1.	Situation	Subparagraphs should be very brief, citing who, what, where and why (where these facts are known).
1.a.	Commander's Intent	Summarized/briefly stated.
1.b.	Area of Interest/Operations	Refer to 'uncropped' master map.
1.c.	Target Area	Target Area should be designated with a box on the master map.
1.d.	Enemy Forces	Use the master map to brief the enemy situation in the target area. Who, what, where, if known. Weather and light data/forecasts (tabular).
1.e.	Friendly Forces	Mission of next higher or participating units (e.g. UW force support; counter insurgency attachment of the Team). Location of Target Area boxes for Teams operating nearby during mission window; this information will not be posted to the master or Team Member maps (OPSEC). Bright Light, RF units in support. Units providing aviation and fires support (and types of fires).

Para #	Item	Explanation
1.f.	Attachments & Detachments	Responsibilities of attached elements. At the Team level, detachments will rarely be seen.
2.	Mission	Who, what, when, where and why. Will include Priority Intelligence Requirements (PIR) and primary and supplementary missions in general order of priority.
3.	Execution	
3.a.	Concept of Operations (including maneuver)	T/L CONOPS. Use the master map to brief operational phases, general area of insertion, general line of movement, key locations/ danger areas, etc. Subject to change and increasing levels of detail throughout the preparation phase.
3.b.	Tasks to Other/ Subordinate Units	Tasks and plans of friendly COIN, UW, exploitation forces. Tasks for attached elements (if any) may be stated here.
3.c.	Coordinating Instructions	Many details will be covered under SOPs. Coordinating instructions regarding other units (e.g. COIN/UW, exploitation forces) operating in proximity to the Team's Target Area. Environmental considerations.
3.c.(1)	Timeline/Schedule	Best in tabular format, using reverse planning (1/3-2/3 rule). Cites When, What, Where, Who.
3.c.(2)	General Guidance	Cite deviations to SOPs. Cite Team Member assignments for Team planning and preparation phase.
3.c.(3)	Security and Deception Guidance	Based on higher HQ guidance and on T/L supplementary instructions; may include PsyOps guidance.
3.c.(4)	Fire Support Plan	Fire support resources, includes very long range fires. Location of Reference Points.
3.c.(5)	SERE	Not briefed until later in the planning and preparation cycle. Includes a primary and 'alternate' plan.
4.	Sustainment	
4.a.	Logistics	
4.a.(1)	Maintenance	All weapons, equipment, as required. Mobility equipment pre-mission services and equipment upgrades.
4.a.(2)	Transportation	Transportation services required for training and preparation.
4.a.(3)	Supply	All classes of supply that pertain to the Mission, Team composition, CONOPS and individual/special mission equipment. Training as well as mission items (and resupply) to be drawn. Basic Loads and supplementary items by SOP or cite deviations thereto. Special uniform/equipment/items and mission environment-specific items. Use pre-printed lists/forms for requisitioning. Established/planned prepositioned, MSSs/caches locations and stocks.
4.b.	Personal Services Support	Wills and personal effects disposition instructions (updates).
4.c.	Health Support	Pre-mission medical screening, Drawing of Team aid kits, individual and Team medical items.
5.	Command and Signal	
5.a.	Command	Chain of Command by SOP or cite deviations thereto.
5.b.	Control	SOPs in effect. Rules of Engagement that may apply.
5.c.	Signal	Drawing of signal equipment, signal devices and SOI. Cite deviations to SOP, if any. Emergency signals (e.g. SERE letters for the duration of the mission). Signs/countersigns. Special code-words. Special radio procedures.

Notes: Use checklists shown in the Ranger Handbook or similar SpecOps sources. These include intelligence, operations, fires support checklists.

SALUTE+ Report		
Line #	Item	Explanation
1	SizE	Number of Troops or Equipment/Vehicles Observed
2	Activity	Self Explanatory
3	Location	Map Sheet and Coordinates
4	Unit/Uniform	Enemy Unit by Type (e.g. Infantry, Artillery), if this can be determined. Bumper markings may specifically identify the unit(s) observed. Otherwise Report particulars of uniform; this may be sufficient to later determine the type of unit.
5	TIME	DTG (Local)
6	EQUIPMENT	Identify specific equipment if possible.
7	+ TEAM INTENTIONS	Self Explanatory.
Notes: The SALUTE Message Format may be used instead of an NBC-1 Report Format. If used for this purpose, provide local weather conditions (especially wind speed and direction) if possible.		

NBC-4 Format (Contamination Report)		
Line #	Item	Explanation
1	Date and Time:	DTG. Self-explanatory
2	Unit:	Unit Making Report
3	Event:	Type of Incident: Nuclear, Biological, Chemical
4	Alpha:	NBC Strike Serial Number: Assigned by Higher HQ.
5	Hotel1:	Type of Burst; Biological/Chemical Agent and Persistency (**P**=Persistent; **NP**=Non-Persistent)
6	Kilo:	Not Required.
7	Quebec1:	UTM/Six-grid Coordinates at the site of the Reading/Sample Taken. Cite Air/Liquid Sampling.
8	Romeo:	Not Required.
9	Sierra1:	DTG of Reading/Initial Identification or Initial Sample Taken.
10	Time:	DTG of Observation
11	Narrative:	To Clarify Report and to Report Team Intentions
12	Authentication:	Self-Explanatory. Automatically done with Joint Tactical Radio System
Notes: Initial Report is sent FLASH Precedence using the SALUTE Message Format. SR Team may be directed to make an ID (if equipped to do so) or take readings/samples. Line 1 Used for Chemical Hazard Reporting		

MedEvac Request Format		
Para #	Item	Explanation
1	Location of LZ:	Not normally provided in the clear. MedEvac aircraft normally rendezvous with the FAC some distance away from the Target Area.
2	Team Identification:	Call Sign or Code Word
3	No. of Casualties by Precedence:	A=Urgent (w/in 2 hrs); **B**=Urgent Surgical (w/in 2 hrs); **C**=Priority (w/in 4 hrs); **D**=Routine (w/in 24 hrs).
4	Special Equipment Needed:	A=None; **B**=Hoist; **C**=Extraction Equipment; **D**=Ventilator; E=Defibrillator

Para #	Item	Explanation
5	No. of Casualties by Type:	L=Litter +#, A=Ambulatory +#, K=KIA
6	Tactical Situation:	N=No enemy in area; P=Possible enemy in area (use caution); E=Enemy in area (caution); X=Enemy in area; armed escort required.
7	LZ Marking:	A=Panels; B=Pyrotechnic signal; C=Smoke signal; D=None; E=Other (e.g. strobe, landing lights, IR, etc.)
8	Casualty Nationality/ Status:	A=US Military; B=US Non-military; C=Military Non-US; D=Civilian Non-US; E-EPW
9	NBC Contamination/ Communicable Disease:	N=Nuclear/Radiological; B=Biological (Cite agent if known); C=Chemical (Cite agent if known).

Call for Fire Format (Artillery, including Mortars)		
Para #	**Item**	**Explanation**
1	Team I.D./Call:	Establish radio contact with FAC or artillery unit. Cite; 'Fire Mission'
2	Mission Type:	Cite: 'Adjust Fire – Not Observed'
3	Location:	'Shift From RP AB1234, Right 200, Up 300' Note: Direction to target not provided (deviation – See Notes)
4	Target I.D.:	'Enemy platoon under canopy moving along a ridge top NE-SW'
5	Team Intentions:	'Moving to alternate location' Not Required.
6	Authentication:	If required or provided by FAC/relay. Automatically provided by JTRS.
7	Fire Adjustment and/or Results after impact:	Results Not Required (but much appreciated by the supporting unit).

Notes: This is a modification to the Shift-From-A-Known Point Call for Fire procedure to be used when the Team is employing fire support while on the move (e.g. while the Team is being pursued) and/or when the Team cannot see its target. As this procedure deviates from normal procedure, it must be coordinated with the supporting unit. See Supporting Fires TTPs regarding 'Danger Close' procedures.

Close Air Support (CAS) Request Format		
Para #	**Item**	**Explanation**
1	Team Identification:	Call Sign or code word
2	Warning Order:	Example: Code Word or Cite 'Tactical Emergency'
3	Target Location:	Grid Location; Shift From a Known Point (e.g. RP); or Direction and Distance From Marker. Example: 'Enemy 50M South of smoke.'
4	Target Description:	Type and number of targets, activity or movement; point or area targets, include desired results on target and time on target.
5	Team Location (and Activity):	Example: 'From Marker, 100M North; moving North; enemy in pursuit.' Team location is not disclosed 'in the clear', unless the Team is using a smoke marker for ID.
6	**Navigation Recommendations:**	Example: 'Make pass East to West.'
7	Threats:	Example: '12.7mm Machine Gun on Ridge to the South and small arms fire from target location.'
8	Navigation Hazards:	Example: 'Beware High tension wires to North.'
9	Team Intentions:	Example: 'Break Contact and move to LZ for extraction.'

Team Leader's Recon (GOTWA)		
Item #	**Item**	**Explanation**
1	<u>G</u>oing:	Where is the T/L (and his party) <u>Going</u>.
2	<u>O</u>thers:	What <u>Others</u> will accompany the T/L or designated Team Member.
3	<u>T</u>ime:	<u>Time</u> that T/L or designated Team Member will be gone.
4	<u>W</u>hat:	<u>What</u> to do if the T/L or designated Team Member does not return on time. **[by SOP]**
5	<u>A</u>ctions:	What <u>Actions</u> will be taken on enemy contact while the T/L or designated Team Member is gone. **[by SOP]:**
5.a.	If the T/L or designated Team Member becomes engaged with the enemy, the. . .	
5.a.(1)		**T/L or designated Team Member will . . .**
5.a.(2)		**Team will . . .**
5.b.	If the Team becomes engaged with the enemy, the. . .	
5.b.(1)		**T/L or designated Team Member will . . .**
5.b.(2)		**Team will . . .**

OBSTACLE INTELLIGENCE (OBSTINEL)/MINEFIELD/BOOBY-TRAP REPORT		
Item #	**Item**	**Explanation**
A	Map Sheet(s):	Self-Explanatory
B	Date and Time Discovered/ Emplaced:	If enemy minefield, make an estimate of when the enemy emplaced the mines. If Team emplaced, record DTG.
C	Type of Minefield:	AT, AP or Mixed.
D	Grid Location of Mine/Booby- Trap/ Minefield Limits:	If possible, 8-digit coordinates.
E	Depth of Minefield:	If known.
F	Estimated Time to Clear Minefield:	If Team emplaced, note self-destruct time/lifecycle. Otherwise 'unknown'.
G	Estimated Material and Equipment to Clear Minefield:	If Team emplaced with self-destruct, 'None Required'. Otherwise 'unknown'.
H	Routes for Bypassing the Minefield:	Self-Explanatory.
I-Y	Grid References of Lanes and Lane Widths (Meters):	To include markers/tell-tales.
Z	Additional Information:	For Example: specific mines discovered/emplaced; Types of mines discovered/emplaced (e.g. bounding, EFP, directional); types of booby-traps; type(s) of self-destruct used, obstacle under observation/ covered by fields of fire.
<u>Notes</u>: Normally, other information is sought for mine/minefield emplacement by US forces. The Team should only be using mines equipped with serf-destruct features.		

Bibliography

Periodicals

Greenup, Al, COL (Ret), 'No Glamour No Glory: A Ground Pounder's Perspective of Project Delta Forward Air Controllers in the South East Asia War', *The Drop*, Special Forces Association, Fayetteville, NC. Winter 2017.

Grove, Thomas, 'Russian Special Forces Seen as Key to Aleppo Victory: Low-profile ground deployments show importance of battle to Kremlin', *Wall Street Journal*, New York, 16 December, 2016.

Jablonsky, David, 'National Power', Parameters, US Army War College Quarterly – spring 1997.

Jumper, Roy Davis Linville, 'Malaysia's Senoi Praaq Special Forces,' *International Journal of Intelligence and Counterintelligence*, Vol. 13, 2000.

Kesling, Ben and Barnes, Julian E., 'US Army Rethinks Strategy As Threats Shift', *Wall Street Journal*, https:\\www.WSJ.com/articles/u-s—army-outlines-new-approach-to-shifting-battlefields-1507521602. 9 October, 2017.

Kluge, P.F, 'Director Sam Peckinpah: What Price Violence', *Life Magazine*, Vol. 73, No. 6, 11 August, 1972.

Murphy, Jack, 'Blue Light: America's First Counter-Terrorism Unit', *The Drop*, Special Forces Association, Fayetteville, NC, Summer 2016.

Palazzolo, Joe, 'Silencers Loophole Targeted for Closure, ATF Seeks Background Checks for All Members of Weapon-Buying Trusts', *Wall Street Journal*, New York, 3 October, 2013.

Parks, W. Hays. 'Joint Service Combat Shotgun Program'. *The Army Lawyer*. October 1997.

US Army, Combined Intelligence Center-Vietnam, Handbook: 'What A Platoon Leader Should Know about the Enemy's Jungle Tactics', US Army: October, 1967.

Vongsavanh, Soutchay, Brigadier General, *RLG* [Royal Laotian Government] 'Military Operations and Activities in the Laotian Panhandle', *Indochina Monographs*, US Army Center of Military History, Washington, DC, 1981.

Military Field/Technical Manuals

Field Manual 3-05, Army Special Operations Forces, Headquarters, Department of the Army, Washington, DC, 20 September 2006.

Field Manual 3-07.22, Counter Insurgency Operations, Headquarters, Department of the Army, Washington, DC, 1 October 2004.

Field Manual 3-22.9, Rifle Marksmanship, Headquarters, Department of the Army, Washington, DC, (with change 1, dated 10 February 2011).

Field Manual 3-23.30, Grenades and Pyrotechnic Signals, Headquarters, Department of the Army, Washington, DC, (2005 revision).

Field Manual 12-80, Kill or Get Killed, Author: Applegate, Rex., Boulder, Colorado: Paladin Press, 1976.

Field Manual 21-60, Visual Signals, Headquarters, Department of the Army, Washington, DC, 30 September 1987.

Field Manual 31-16, Counterguerilla Operations, HQ Department of the Army, Washington DC 19 February 1963.

Field Manual 31-18, Long Range Reconnaissance Patrol Company, HQDA, Washington, DC, August 1968.

Field Manual 31-20-5, Special Forces Special Reconnaissance Tactics, Techniques and Procedures, HQDA, Washington, DC (E-print copyright Clandestine Press, 2001–2008), 23 March 1993.

Field Manual 31-23 (ID), Special Forces Mounted Operations, Tactics, Techniques and Procedures, Department of the Army, United States Army John F. Kennedy Special Warfare Center and School, Fort Bragg, 3 March 1998.

Field Manual 34-36, Special Operations Forces Intelligence and Electronic Warfare Operations, Department of the Army, Washington, DC, 30 September 1991.

Field Manual 100-5, Operations, Department of the Army, Washington, DC, 5 May 1986.

IT 23-71-1, Principles of Quick Kill, US Army Infantry School, Ft. Benning, GA, 1996.

Joint Publication 1–02, Dictionary of Military and Associated Terms, Headquarters, Department of Defense, Washington, DC, as amended through 15 February 2012.

TC 31-34-4, 'Special Forces Tracking and Countertracking', Headquarters, Department of the Army, Special Operations Press, September 2009.

Technical Manual 31-210 (Obsolete) Improvised Munitions Handbook, HQ, Department of the Army, Frankford Arsenal, PA. 1969.

Books

Ankony, Robert C., *Lurps: A Ranger's Diary of Tet, Khe Sanh, A Shau, and Quang Tri*, revised ed., Rowman & Littlefield Publishing Group, Lanham, MD, 2009.

Beckwith, Charles, *Delta Force: The Army's Elite Counterterrorist Unit*, Avon Books (HarperCollins), New York, 1983.

Beckwith, Charles A. and Knox, Donald, *Delta Force, A Memoir by the Founder of the US Military's Most Secretive Special Operations Unit*, HarperCollins Publishers, 12 February 2013.

Bergen, Peter L., Manhunt, *The Ten Year Search for Bin Laden from 9/11 to Abbottabad*, Crown Publishers, New York, 2012.

Charters, David A. and Tugwell, Maurice, eds., *Armies in Low-Intensity Conflict: A Comparative Analysis*, McLean, VA: Pergamon-Brassey's International Defense Publishers, 1989.

Comber, Leon, *Malaya's Secret Police 1945–60: The Role of the Malayan Special Branch in the Malayan Emergency*, Institute of South East Asian Studies: Singapore, 2008.

de B. Taillon, J. Paul, *The Evolution of Special Forces in Counter-Terrorism: The British and American Experiences*, Westport, CT: Praeger Publishers, 2001.

Director of Operations-Malaya, *The Conduct of Anti-Terrorist Operations*, Chap. XXIII, 3rd Ed. 1958.

Durkin, J.C., *Malayan Scout SAS: A Memoir of the Malayan Emergency, 1951*, Gloucestershire, UK: The History Press, 2011.

Foote, M.R.D., *SOE in France: An Account of the Work of the British Special Operations Executive in France 1940–1944*. Louis Gros, London, Her Majesty's Stationary Office, 1966.

Gooley, Tristan, *How to Read Water, Clues and Patterns from Puddles to the Sea*, The Experiment, LLC, 2016.

Huang, J.H., *Sun Tzu: The New Translation*, William Morrow and Company, Inc., New York, 1993.

Hurth, John D., *Combat Tracking Guide*, Stackpole Books, Mechanicsburg, PA, 2012.

Jowett, Philip and Snodgrass, Brent, *Finland at War 1939–45*, Osprey Publishing, 5 July 2006.

Lucas, James, *KOMMANDO, German Special Forces of World War II*, Cassell Military Classics, London, 1985.

Mace, Garry R., *Dynamic Targeting and the Mobile Missile Threat*, Department of Joint Military Operations, US Naval War College, Newport, RI, 17 May 1999.

Mackenzie, Alastair, *Special Force*, I.B. Tauris & Co Ltd: London, 2011.

Moyer, Mark, *Phoenix and the Birds of Prey*, University of Nebraska Press, Lincoln & London, 1997.

Murray, Williamson, *Air War in the Persian Gulf*, Nautical & Aviation Pub Co of America; 1 July 1995.

Nagl, John A., *Learning to Eat Soup with a Knife: Counterinsurgency Lessons from Malaya and Vietnam*, Chicago: The University of Chicago Press, 2002.

Plaster, John L., *SOG: A Photo History of the Secret Wars*, Boulder, CO, Paladin Press. 2000.

Plaster, John L., *SOG: The Secret Wars of America's Commandos in Vietnam*, Simon & Schuster, New York, 1997.

Prados, John, *The Blood Road: The Ho Chi Minh Trail and the Vietnam War*, John Wiley & Sons, Inc., New York, 1999.

Saal, Harve, *SOG MACV Studies and Observation Group: Behind Enemy Lines*. Edwards Brothers, 1990.

Sherman, Stephen, ed., *Indochina In The Year Of The Goat – 1967*, Radix Press, Houston, TX, 2016.

Shultz, Richard H., Jr., *The Secret War Against Hanoi: The Untold Story of Spies, Saboteurs, and Covert Warriors in North Vietnam*, (Harper Collins, New York, 2000).

Singlaub, John K., and McConnell, Malcolm, *Hazardous Duty: An American Soldier in the Twentieth Century*, Summit Books, New York, 1991.

Švābe, Arveds, ed. *Latvju Enciklopēdija* (in Latvian), Stockholm: Trīszvaigznes, OCLC 11845651 (1950-55).

van Horne, Patrick and Riley, Jason, *Left of Bang: How the Marine Corps' Combat Hunter Program Can Save Your Life*, Digital Version, Black Irish Entertainment, LLC. 2014.

Waltz, Edward, *Knowledge Management in the Intelligence Enterprise*, Altech House, Inc., Norwood, MA, 2003.

Studies and Reports

Andrews, Marshall, Isolating the Guerrilla, 'The American Civil War', Historical Evaluation and Research Organization, 2233 Wisconsin Avenue, N.W., Washington, DC, 20007, 1 February 1966.

Armstrong, Richard N., Lieutenant Colonel, US Army, Soviet Operational Deception: The Red Cloak, US Army Combat Studies Institute, US Army Command and General Staff College, Ft. Leavenworth, KS, 1986.

Banner, Gregory T., The War for the Ho Chi Minh Trail, Master's Thesis, US Army Command and General Staff College, 1993, (DTIC, AD-A272 827).

BDM Corporation, A Study of Strategic Lessons-Learned in Vietnam, Vol. 1, The Enemy, McLean, Va.: (1979).

Cosmas, Graham A., CMH Pub 91-7-*1*, *MACV – The Joint Command in the Years of Withdrawal, 1968–1973*, Center for Military History, US Army, Washington, DC, 2007.

Centre for Defence and International Security Studies (CDISS), Lancaster University, '1990: The Iraqi Scud Threat,' http://www.cdiss.org/scudnt3.htm.

Halder, Franz. General Wehrmacht, ed., Small Unit Actions during the German Campaign in Russia, DA Pam 20 – 269, Department of the Army, Washington, DC, July 1953.

Kelly, Danny M., Major (USA), Master's Thesis: 'The Misuse of the Studies and Observation Group as a National Asset in Vietnam', Command and General Staff College, Ft. Leavenworth, KS. 2005.

Kelly, Danny M., Major (USA) tape recorded interview of John L. Plaster, Command and General Staff College, Ft. Leavenworth, KS, 7 January 2005.

Kelly, Danny M., Major (USA) tape recorded interview of General (Ret) John K. Sinlaub, Command and General Staff College, Ft. Leavenworth, KS, 4 March 2005.

Killblane, Richard E., 'Convoy Ambush Case Studies', Transportation Corps, Fort Eustis, VA, 2006.

Perry, William J., 1996 Annual Defense Report, Chapter 22, Special Operations Forces, Headquarters, Department of Defense, Washington, DC, 1996.

Rahman, Shamsul Afkar bin Abd, History of Special Operations Forces in Malaysia, Calhoun: The Naval Postgraduate School Institutional Archive, Monterey, CA, June 2013.

Raine, Linnea P., Isolating the Guerrilla, 'The Philippine Insurrection, November 1899-July 1902', Historical Evaluation and Research Organization, 2233 Wisconsin Avenue, N.W., Washington, DC, 20007, 1 February 1966.

Rosenau, William, Special Operations Forces and Elusive Enemy Ground Targets: Lessons from Vietnam and the Persian Gulf War, RAND, Santa Monica, CA, 2001.

Rosenau, William. US Air Ground Operations Against the Ho Chi Minh Trail, 1966–1972 (PDF), RAND Corporation. Santa Monica, CA, 2000.

Rostker, Bernard D.; Hanser, Lawrence M.; Hix, William M.; Jensen, Carl; Morral, Andrew R.; Ridgeway, Greg; Schell, Terry L. 'Evaluation of the NYPD Firearm Training and Firearm-Discharge Review Process' www.nyc.gov/html/nypd/downloads/pdf/public_information/RAND_FirearmEvaluation.pdf.

Rothenberg, Gunter E., Isolating the Guerrilla, 'The German Experience in World War II'. Historical Evaluation and Research Organization, 2233 Wisconsin Avenue, N.W., Washington, DC, 20007, 1 February 1966.

Toppe, Alfred, Brigadier General, Wehrmacht, Night Combat' CMH Pub, 104-3 (formerly DA Pam 20-236, June 1953), Center of Military History, United States Army, Washington, DC, Facsimile Edition, 1982, 1986.

Unnamed German Generals and General Staff, Military Improvisations During the Russian Campaign', CMH Pub, 104-1 (formerly DA Pam 20-201, August 1951), Center of Military History, United States Army, Washington, DC

Unnamed former German Panzer Army G4. Department of the Army Pamphlet No. 20 – 240, Historical Study, Rear Area Security in Russia, The Soviet Second Front behind the German Lines, Department of the Army 25, DC, 31 July 1951.

Unnamed authors and contributors, Historical Study, German Anti-Guerrilla Operations in the Balkans (1941–1944)' Department of the Army Pamphlet No. 20-243, Department of the Army, Washington 25, DC, 5 August 1954.

Other Media

'316th Cavalry', http://www.benning.army.mil/armor/316thCav/RSLC/index.html.

'Ancient History Encyclopedia' (website), http://www.ancient.eu.com/image/166/.

'Anonymous French Soldier', HTTPS://InMilitary.com/A French Soldier's View of US Soldiers in Afghanistan (aka A Nos Freres D'Armes Americain), 6 January 17.

'Army Reconnaissance Course', https://www.benning.army.mil/Armor/316thCav/ARC/.

'Battle of Mang Yang Pass', Wikipedia, https://en.wikipedia.org/wiki/Battle_of_Mang_Yang_Pass.

'Direct Action (military)', Wikipedia, https://en.wikipedia.org/wiki/Direct_action_(military). 4 August 2017.

'French Indochina, The First Vietnam War', http://www.gia-vuc.com/frenchindochina.htm, Fall 2016.

'French Indochina', http://www.gia-vuc.com/gcma%20history.htm.

'Immersion Foot', http://www.wikidoc.org/index.php/Immersion_foot.

'Lauri Torni, aka Larry Thorne', Wikipedia, https://en.wikipedia.org/wiki/Lauri_T%C3%B6rni#Vietnam_War_and_death, Fall 2016.

'Majewski raid on Mai-Guba along the Soviet Murmansk rail line during the Continuation War (January 1942)'. http://www.ww2incolor.com/finnish_forces/Major_TimoJohannesPuustinen_Knight%23117.html, Fall 2016.

'Manhunt: the Search for Bin Laden', © 2013, Home Box Office, USA.

'Point Shooting', Wikipedia, the free encyclopedia, https://en.wikipedia.org/wiki/Point_shooting#cite_note-33, Fall 2016.

'Puustinen Raid (March 1943)', http://www.ww2incolor.com/finnish_forces/Major_TimoJohannesPuustinen_Knight%23117.html.

'Special Reconnaissance', Wikipedia, https://en.m.wikipedia.org/wiki/Special_reconnaissance, No date.

Resources and Additional Reading

Tracking:

Hurth, John D., Combat Tracking Guide, Stackpole Books, Mechanicsburg, PA, 17055, 2012.

Foot Care:

'Immersion Foot', http://www.wikidoc.org/index.php/Immersion_foot.

Land Navigation:

Declination: http://www.ngdc.noaa.gov/geomag-web/#declination.

Munitions:

'Pursuit Deterrent Munition', https://en.wikipedia.org/wiki/M86_Pursuit_Deterrent_Munition.

Dear Reader,

We hope you have enjoyed this book, but why not share your views on social media? You can also follow our pages to see more about our other products: facebook.com/penandswordbooks or follow us on X @penswordbooks

You can also view our products at www.pen-and-sword.co.uk (UK and ROW) or www.penandswordbooks.com (North America).

To keep up to date with our latest releases and online catalogues, please sign up to our newsletter at: www.pen-and-sword.co.uk/newsletter

If you would like a printed catalogue with our latest books, then please email: enquiries@pen-and-sword.co.uk or telephone: 01226 734555 (UK and ROW) or email: uspen-and-sword@casematepublishers.com or telephone: (610) 853-9131 (North America).

We respect your privacy and we will only use personal information to send you information about our products.

Thank you!